Extreme Weather

Extreme Weather

Forty Years of the Tornado and Storm Research Organisation (TORRO)

EDITED BY

ROBERT K. DOE

WILEY Blackwell

This edition first published 2016 © 2016 by John Wiley & Sons, Ltd.

Registered Office
John Wiley & Sons, Ltd, The Atrium, Southern Gate, Chichester, West Sussex, PO19 8SQ, UK

Editorial Offices
9600 Garsington Road, Oxford, OX4 2DQ, UK
The Atrium, Southern Gate, Chichester, West Sussex, PO19 8SQ, UK
111 River Street, Hoboken, NJ 07030-5774, USA

For details of our global editorial offices, for customer services and for information about how to apply for permission to reuse the copyright material in this book please see our website at www.wiley.com/wiley-blackwell.

Library of Congress Cataloging-in-Publication Data

Doe, Robert, author.
Extreme weather : forty years of research / by the Tornado and Storm Research Organisation ; Robert K. Doe.
     pages   cm
  Includes bibliographical references and index.
  ISBN 978-1-118-94995-5 (cloth)
1. Climatic extremes–Great Britain.   2. Weather.   I. Tornado and Storm Research Organisation, author.   II. Title.
  GB5008.G69D64 2015
  551.55′4–dc23
                                        2015018780

A catalogue record for this book is available from the British Library.

Wiley also publishes its books in a variety of electronic formats. Some content that appears in print may not be available in electronic books.

Cover image: 17 April 2008, Dawlish, Devon, U.K. © Matt Clark

Set in 9/11pt Times by SPi Global, Pondicherry, India

Printed in Singapore by C.O.S. Printers Pte Ltd

1   2016

Contents

Notes on Contributors

Paul R. Brown
Paul has had a lifelong interest in weather and climate, part of the time in a professional capacity, the rest as an amateur. He has been associated with TORRO since its foundation and a formal member of the organisation since 1998. Since 2006, he has been a staff member assisting the head of the Tornado Division, G. Terence Meaden, in compiling current and historical reports of tornadoes and other whirlwinds.

Matt Clark, B.Sc. Hons. (Reading)
Matt studied meteorology at the University of Reading, obtaining a B.Sc. in 2005. The degree course included 1 year of study at the University of Oklahoma. Since 2005, Matt has worked in Observations Research and Development at the Met Office in Exeter. In recent years, he has published a number of observational studies of damaging thunderstorms and cold fronts in the United Kingdom. Matt has been a member of the Royal Meteorological Society since 2003 and a member of TORRO since 2006.

Robert K. Doe, B.A. Hons. (Exon), Ph.D. (Ports), FRMetS, MRI
Robert graduated with a Ph.D. from the University of Portsmouth, where he specialised in coastal storm climatology. Robert is a director and treasurer of the Tornado and Storm Research Organisation (TORRO). He is a fellow of the Royal Meteorological Society (FRMetS) and a member of the Royal Institution (MRI). He was editor-in-chief of the *International Journal of Meteorology* (2002–2006) and has published research on meteorological phenomena including tornadoes, waterspouts, floods, snowstorms, ball lightning, coastal storms, climate and risk. He is an honorary research fellow in the School of Environmental Sciences at the University of Liverpool.

Derek M. Elsom, B.Sc. (Birm), M.Sc. (Birm), Ph.D (Oxford Brookes), FRMetS
Derek is professor emeritus at Oxford Brookes University following his retirement in 2012 from his position as professor of geography, dean of the Faculty of Humanities and Social Sciences and pro vice-chancellor. His research has focused on weather hazards, especially tornadoes, hailstorms and lightning in the United Kingdom and on evaluating the effectiveness of alternative strategies for air quality management in urban areas, including high-pollution episodes (smogs), in various countries. His national and international research has been published in more than 120 journal papers, chapters, reports and books, and he has presented papers at many national and international conferences. Professor Elsom joined the Tornado and Storm Research Organisation (TORRO) in 1980 and raised international awareness of TORRO's pioneering tornado research when he presented a paper at the 12th International Conference on Severe Local Storms at San Antonio, United States, in 1981. Together with Dr G. Terence Meaden, he organised a succession of TORRO conferences at Oxford Brookes University from 1985 onwards. He was director of TORRO from 1994 to 2004. Professor Elsom's current research is concerned with assessing lightning impacts in the United Kingdom and increasing public awareness and understanding of the lightning risk.

Adrian James, M.A. (Cantab), MCLIP
Adrian joined TORRO in 1975, writing the monthly thunderstorm reports for the *International Journal of Meteorology* between 1988 and 1990, and was archives director of the TORRO Ball Lightning Division from 1991 to 2000. He is currently assistant librarian at the Society of Antiquaries of London.

Peter Kirk, B.Sc. Hons. (Reading)
Peter received a B.Sc. in meteorology from the University of Reading in 2006. His interests include severe weather around the world including tornadoes and tropical cyclones, as well as general meteorology across the United Kingdom. He has been a member of TORRO since 2004 and an associate fellow of the Royal Meteorological Society since 2006. He is a manager of the UKWeatherworld forums and is part of the UKWeatherworld Warnings Team, issuing warnings of severe weather across the United Kingdom.

Paul Knightley, B.Sc. (Reading), M.Sc. (Birm), FRMetS
Paul has been fascinated by all aspects of meteorology from an early age. He joined TORRO in 1992, and his interest with tornadoes and severe thunderstorms grew. In 1998, he visited the central United States to go storm chasing, and has been 15 times to date, where he has seen numerous severe thunderstorms and tornadoes. He undertook a masters' degree in meteorology in 2000 and has since been a weather forecaster at MeteoGroup UK, where he is now forecast manager. Paul took on the role of issuing severe weather forecasts for TORRO in the early 2000s and has been doing it since. He analyses the forecasts each year and publishes the results in the *International Journal of Meteorology*. Paul is the current head of TORRO.

John Mason, B.Sc., M.Phil. (Aber)
John is a geologist with a long-standing interest in weather extremes through his photographic work. He has recorded in detail several notable extreme weather events in his home area of mid-Wales, including flash flooding and tornadoes, and continues to do so in collaboration with TORRO. He is also an active member of the international team that writes for and manages the award-winning climate change website *Skeptical Science*.

G. Terence Meaden, M.A. (Oxon), M.Sc. (Oxon), D.Phil. (Oxon), FRMetS
Terence is a professional physicist, meteorologist and archaeologist with undergraduate and doctoral degrees in physics and a master's degree in archaeology from Oxford University. He has held academic posts at Oxford University; Grenoble University, France; and Dalhousie University–Halifax, Canada, as a professor of

physics. He has researched tornado climatology for over 40 years. In 1972, he established the International Tornado Intensity Scale based on the Beaufort Scale; in 1974, the Tornado and Storm Research Organisation (TORRO); and in 1975, the *Journal of Meteorology*. He has been a FRMetS since 1957 and his publications exceed 300, with tornado track investigations spanning the period 1967–2007. It was because of TORRO's huge database on the incidence of British and Irish tornadoes that he was invited to be a consultant for the new-build nuclear plant industry regarding tornado hazard and damage risk.

Bob Prichard, B.A., FRMetS
Bob has had a lifelong interest in the weather and was a forecaster in the Met Office for most of his career until he took early retirement in 2004; he broadcast forecasts on the BBC for the Met Office for nearly 25 years. When S. Morris Bower retired from running the Thunderstorm Census Organisation, Bob sought to maintain some form of monthly thunderstorm reports, initially through the pages of the Climatological Observers Link (COL) bulletin in 1974 and then through the *Journal of Meteorology*. He has now returned to providing a monthly thunderstorm report to the COL bulletin. He was editor of the Royal Meteorological Society journal *Weather* from 2009 to 2013.

Tim Prosser, M.Eng. (Oxon)
Tim has been a member of TORRO for 10 years and has a longstanding amateur interest in weather phenomena. His academic background is in mechanical engineering, and he is currently an I.T. specialist with an interest in Geographical Information Systems (GIS). He has carried out a number of site investigations for TORRO, including a recent investigation into damage caused by a downburst and tornado near York.

Mike Rowe, B.A. Hons. (Exon)
Mike was a teacher of geography and history until his retirement in 2010. He has been a keen weather enthusiast since childhood, with a particular interest in historical weather events. While at university, he began to compile historical weather data and, finding more tornado cases than he expected, decided to concentrate on them. He joined TORRO in 1975 and for many years wrote the monthly reports of the Tornado Division.

David Smart
David is a research associate of the UCL Hazard Centre, University College London. His interests are studying and understanding all forms of severe and 'high-impact' weather, particularly using numerical modelling and simulation. Some recent publications focus on cold-frontal misocyclones and tornadoes in the United Kingdom and the nature and origin of extreme winds in severe extratropical cyclones ('windstorms').

Mark Stenhoff, B.Sc. Hons., M.Phil. (Lond), CPhys, MInstP, FRMetS, FRAS
Mark wrote his 1988 thesis on ball lightning. He was scientific director of the TORRO Ball Lightning Division from 1985 to 1992. He is the author of *Ball Lightning: An Unsolved Problem in Atmospheric Physics* (Stenhoff, 1999).

John Tyrrell, B.A., Ph.D., FRMetS
John was a senior lecturer at University College Cork until 2009, where he taught and researched in the Department of Geography. Upon retirement, his research activities have continued as a research associate. He was previously at the universities of Aberystwyth, Nairobi, Lusaka and Grahamstown. He has investigated tornadoes in Ireland since 1995 and published results in national and international scientific journals. John was head of TORRO for 2 years and is currently the regional coordinator for Ireland.

Jonathan D.C. Webb, B.A. Hons (Bris)
Jonathan has enjoyed a huge interest in the skies and weather from an early age. He is Thunderstorm Research Director of the Tornado and Storm Research Organisation (TORRO), having joined the TORRO team in 1985. He has also been a member of the Royal Meteorological Society since 1982 and recently completed a 5 year term on the editorial board of *Weather*. His published research includes case studies of thunderstorm episodes and associated severe convective weather, also summaries and analyses of the TORRO research databases of hailstorms and lightning damage. He has also contributed to published research on temperature and precipitation extremes in the U.K.

Richard Wild, B.Sc. Hons., Ph.D., FRMetS, FRGS, MAE, MCFSS
Richard is the weather services commercial manager and forensic/senior meteorologist at WeatherNet Ltd. He is responsible for weather forecasting, weather warnings and legal-related weather consultancy work. Richard has a B.Sc. (Hons.) in geography and a Ph.D. investigating the spatial and temporal analysis of heavy snowfalls across Great Britain. He is a fellow of the Royal Meteorological Society and has produced 40 research articles about snow/snowfalls/blizzards in academic publications including the *International Journal of Meteorology* and *Weather* and also 2 books. He is a staff member of the Tornado and Storm Research Organisation (TORRO) where he is the research leader of the Heavy Snowfalls Division and has held this post since it was established in July 1998.

Foreword

This volume celebrates 40 years of meteorological research and publications by the Tornado and Storm Research Organisation (TORRO) since its launch in 1974. TORRO's members include weather forecasters and other meteorological professionals, researchers and academics as well as amateur weather enthusiasts. All share a passion for the study of severe weather.

TORRO was founded for the purpose of systematising the collection and analysis of information on tornadoes and waterspouts occurring in Britain and Ireland, but this soon extended to thunderstorms and damaging hailstorms. As the number of TORRO members expanded, data collection and research embraced lightning impacts, ball lightning, snowfalls, rainfall deluges and temperature extremes. TORRO has focused on severe weather that poses significant risks to people and their activities in Britain and Ireland. TORRO's pioneering research ensured that Britain became the first European country to have a realistic assessment of the risk that tornadoes posed. In the early 1990s, TORRO developed the first forecasts of tornadoes in these countries and has continued to improve the forecasting of tornadoes and severe weather ever since.

For 40 years, TORRO has enabled many hundreds of amateur weather enthusiasts to contribute to improving the meteorological understanding of severe weather. The incidence and impact of some severe weather phenomena, often very localised, cannot be readily measured using standard meteorological stations but, instead, require individuals to undertake prompt visits to locations of possible tornado and hailstorm damage to document the type of damage, interview eyewitnesses and determine path width, length and intensity. TORRO's regional network of members, supported and advised by TORRO directors, has enabled this to happen. Members promptly report the occurrence of severe weather events to TORRO directors, and a thorough investigation is quickly coordinated. TORRO's weather enthusiasts have also searched historical journals and documents, academic journals, mass media reports and other sources to identify and document past severe weather events to add to TORRO's databases of tornadoes, waterspouts, hailstorms, lightning impacts, ball lightning, heavy snowfalls and rainfall deluges. Cross-checking and confirmation of this archive research have enabled the compiling of extensive databases including a 1000-year database of some 3800 tornado and/or waterspout events and 2500 hailstorms. These extensive databases have enabled TORRO to provide new international intensity scales and revised national risk probability assessments regarding these severe storm threats.

Providing open access to all the meteorological information that TORRO has collected, and the research insights it has gained, has always been a priority. A founding principle was to acquire and disseminate knowledge for the general public good. TORRO databases and research findings have been made readily available to other researchers through their publication, notably, in the *Journal of Meteorology* (which became the *International Journal of Meteorology* in 2005), *Atmospheric Research*, and the Royal Meteorological Society's *Weather*. The *Journal of Meteorology*, launched in 1975, has regularly published monthly and annual listings of severe weather phenomena and annual reviews from the TORRO directors. All this provides key information for current and future researchers. Many university students – the next generation of meteorological professionals – have benefitted from access to this readily available information for their dissertations.

Since 1985, TORRO has organised annual conferences and meetings, notably at Oxford Brookes University, to disseminate and discuss the latest research findings. Involvement of international researchers with the Oxford conferences subsequently resulted in the setting up of the biennial 'European Tornadoes and Severe Storms' conferences, the first convened in Toulouse, France, in 2000. This was co-sponsored by TORRO, and TORRO members presented several papers including one outlining the first compilation of a European tornado climatology. TORRO encouraged and supported the development of other national organisations to focus on severe thunderstorm and tornado research in Europe, along similar lines to TORRO. Sharing its databases with other European organisations, TORRO has contributed significantly to improving risk assessment and understanding of tornadoes and hailstorms at the European scale which, in turn, is helping to improve understanding of these severe weather extremes in other parts of the world too. Much progress has been made but new and challenging research questions continue to arise. For example, the First International Summit on 'Tornadoes and Climate Change' was held in Crete, Greece, in May 2014 with several TORRO researchers participating.

TORRO remains as relevant today as it did when it was founded 40 years ago. It has contributed many improvements and new insights into our understanding of severe weather phenomena in Britain and Ireland and throughout Europe as a whole. Revised risk assessments are now widely available for key severe storm phenomena. However, severe weather continues to pose a threat to our lives and activities. There are many research questions about severe weather which remain, and new ones are emerging. Consequently, TORRO's network of professionals and amateur weather enthusiasts will need to be even more active in future in systematic data collection, analyses and investigations and in ensuring the widest possible dissemination of its databases and research findings for the benefit of current and future generations.

G. Terence Meaden (Founder and Head of TORRO: 1974–1994, 2005–2007)
Derek M. Elsom (1994–2004)
John Tyrrell (2008–2009)
Paul Knightley (2010 to the present).

Past and present heads of the Tornado and Storm Research Organisation (TORRO)

Preface

This book celebrates 40 years of the *Tornado and Storm Research Organisation* (TORRO). It is proudly published in association with the *Royal Meteorological Society*, of which many contributors to this volume are either associate fellows or fellows. The past 40 years has seen a wealth of severe weather research, valuable data collected and publications produced by the research group. The aim of this text is to present selected highlights of the main research areas. It is thematically arranged into three sections – *tornadoes*, *thunderstorms and lightning* and *extremes* – and opens with an introductory chapter by G. Terence Meaden on researching extreme weather in the United Kingdom. In this chapter, Meaden presents personal recollections on the founding of TORRO and positions severe weather research. He explains the history of the research group, the development of the International T Scale and Hailstorm Intensity Scale and the dissemination of research through publication and conferences and considers future directions of severe weather research.

The *tornadoes* section begins by discussing the history of tornadoes and examines a unique database of events from AD 1054 to 2014. Chapter 2, by Paul R. Brown and G. Terence Meaden, discusses etymology and terminology of the word tornado, examines historical sources and presents maps of the geographical distribution of all known tornadoes by strength from the 11th to the 19th centuries. Matt Clark and David Smart author Chapter 3 on supercell and non-supercell tornadoes in the United Kingdom and Ireland. This chapter examines the types of weather system that produce tornadoes in the United Kingdom. It presents insightful synoptic situations associated with tornadoes, with a climatology by synoptic type and illustrated case studies. Chapter 4 presents the frequency and spatial distribution of tornadoes in the United Kingdom, written by Peter Kirk, Tim Prosser and David Smart. It details a new spatial density analysis with informative mapping and highlights and discusses certain anomalies in detail. Chapter 5 examines tornado extremes in the United Kingdom. Written by Mike Rowe, it presents many decades of data collection and analysis into the earliest, longest, widest, severest, and deadliest tornadoes. Chapter 6 by John Tyrrell is dedicated to site investigations of tornado events. This chapter contains a wealth of information on site investigation methods, how to assess tornado intensity and how to interpret results.

The *thunderstorms and lightning* section starts with some epic thunderstorm event analysis. Chapter 7 written by Jonathan D.C. Webb is titled 'Epic Thunderstorms in the United Kingdom and Ireland'. Webb presents detailed discussion on 15 'epic' thunderstorm events and outlines the causes and effects of each storm. Chapter 8 is a unique account of thunderstorm observing. Written by Bob Prichard, it is a lively reflection on more than 55 years of his personal thunderstorm records. He discusses the most interesting and unusual storms and comments on the changing nature of communications for performing such observations. Chapter 9

written by Jonathan D.C. Webb and Derek M. Elsom is an authoritative guide to severe hailstorms and details the establishment of a hail database and the Hailstorm Intensity Scale. It looks at extreme hailstorms events and their impact, frequency and distribution, along with the synoptic and meteorological conditions which generate them. Lightning extremes are introduced by Derek M. Elsom and Jonathan D.C. Webb with a chapter on lightning impacts in the United Kingdom and Ireland. It outlines lightning as a weather hazard, research into lightning and lightning impacts, an analysis of lightning incidents causing injuries and deaths and the frequency with which lightning strikes a person. The chapter concludes with discussion on locations to avoid during thunderstorms and impacts on aircraft, motor vehicles and animals. Chapter 11 summarises ball lightning research in the United Kingdom by Mark Stenhoff and Adrian James. Beginning with definitions and early beliefs about lightning and ball lightning, this chapter highlights that much is known about ball lightning and that there is a wealth of information available. It examines particular case studies in detail and the direction of ball lightning research.

The *extremes* section starts with a chapter on forecasting extremes. Paul Knightley looks at the forecasting of extreme weather, particularly thunderstorms and tornadoes. He presents an overview of modern forecasting and forecasting techniques and examines an ingredient-based methodology. Knightley also examines TORRO forecasts with in reference to particular extreme events. Chapter 13 discusses historical extreme events. The author, Robert K. Doe, examines flood events in the United Kingdom for the years AD 1–AD 1300. Using a wide variety of sources, Doe discusses these with reference to their reliability, locations and descriptions and to the causes and impacts of the events. Chapter 14 is on extreme rainfall and flash flood events in the United Kingdom and Ireland. The authors, John Mason, Paul R. Brown, Jonathan D.C. Webb and Robert K. Doe, discuss these events with reference to synoptic patterns and selected case studies. The text details severe convective and dynamic rainfalls and hybrid rainfalls – dynamic precipitation with embedded convective cells – and is illustrated with case studies. Chapter 15 examines heavy snowfalls, written by Richard Wild. In this chapter, Wild analyses spatial and temporal heavy snowfall over a 139-year period. There is specific focus on heavy snowfall days and heavy snowfall events along with reference to Lamb Weather Types. With selected case studies, this presents a detailed understanding of heavy snowfalls.

The *Appendix* is divided into the following sections – firstly, supplementary data relating to tornadoes, hail, temperatures and snowfall; then photographs from a number of meetings, conferences and symposia over the decades; and, lastly, a new and unique map of tornadoes in the United Kingdom by Tim Prosser. This innovative map shows the distribution of tornadoes using the complete TORRO dataset for the years 1054–2013.

This book presents a flavour of TORRO research interests. The book would not have been possible without the dedication of hundreds of TORRO members, not just in the United Kingdom, but worldwide. The members have been hugely supportive and encouraging over the decades, especially with the supply of observations and data and post-storm rapid response site investigations. It is important to reiterate that the research group has been voluntary, and this book shows that many have been extremely dedicated. Lastly, the research objectives and outcomes highlighted within this book would not have been possible without the vision and conceptualisation of Dr G. Terence Meaden, to whom this book is a testament.

Robert K. Doe
Editor

About the Companion Website

This book is accompanied by a companion website:

www.wiley.com/go/doe/extremeweather

The website includes:

- Pdfs of appendix A from the book for downloading

1

Researching Extreme Weather in the United Kingdom and Ireland: The History of the Tornado and Storm Research Organisation, 1974–2014

G. Terence Meaden

Kellogg College, Oxford University, Oxford, UK

1.1 Introduction: The Early Years

TORRO was launched in Britain in 1974 as the Tornado Research Organisation. The seeds had been sown a long time earlier, in 1950, when I became a 15-year-old amateur meteorologist at grammar school in Trowbridge, Wiltshire. Before then, I had decided to become a scientist or archaeologist for which a university education would be needed, and so I had embarked on preparing the necessary background. Because physics is the quintessential science, I made it my forte, while being attracted none-theless to the problem-solving challenges of archaeology for its discoveries of the prehistoric human past. By 1948 the school-teachers knew that Oxford University was my ambition. In April 1950, Trowbridge Boys High School purchased a Stevenson Screen, thermometers, a rain gauge and a subscription to *Weather* Magazine. New vistas appealed. By January 1951, I had bought a good rain gauge and built my first thermometer screen for a home station. I also subscribed to *Weather* and have done so ever since. The Meteorological Office's *British Rainfall* accepted my rainfall data from 1951 to 1966, and, because I enjoyed thunderstorms and studying tornadoes, I joined Morris Bower's Thunderstorm Census Organisation.

For some years, I ran two weather stations. In February 1951, the editor of the county newspaper *Wiltshire Times* announced that its weather correspondent was leaving the county and asked for a replacement. I immediately supplied a detailed monthly weather report and became the newspaper's weather correspondent for the 3 years until I went to Oxford. This initiative at writing for the press and reporting the weather was appreciated by the schoolteachers and Oxford dons, and undoubtedly helped me, age 17, at the interviews for the entrance examination in January 1953.

At Oxford I read physics where I got to know meteorologist Professor Alan Brewer (1915–2007) who proposed me as a Fellow of the Royal Meteorological Society (FRMetS) in December 1957. Doctoral research followed (1957–1961) (Figure 1.1) and then a post-doctoral fellowship funded by the Atomic Energy Research Establishment at Harwell. In 1963 I left to pursue low temperature physics research at the University of Grenoble in France and afterwards got a tenured position as a Professor of Physics at Dalhousie University in Halifax, Canada. I even took my Stevenson screen and Snowdon rain gauge to Canada, encouraged by the prospect of measuring very low winter temperatures, savage wind chills and high cyclonic rainfalls from which the Maritime Provinces suffered.[1]

Back in Oxford, in 1972 and 1973, I specialised in studying the incidence of tornadoes in Britain. I was impressed by the number of well-documented historical cases scattered through volumes in the Radcliffe Science Library. Britton (1937) in his chronology of British weather to the year 1450 aided research for the medieval centuries, and Ralph Edwin Lacy's paper in *Weather* 1968 appealed for its listing of tornadoes from 1963 to 1966. Among

---

[1] The wettest 24 hours I recorded was a deluge in December 1970 with over 7 in. (180 mm) of rain.

Extreme Weather: Forty Years of the Tornado and Storm Research Organisation (TORRO), First Edition. Edited by Robert K. Doe.
© 2016 John Wiley & Sons, Ltd. Published 2016 by John Wiley & Sons, Ltd.

Figure 1.1 *Terence Meaden in 1961 (left) and in 2013 (right).*

the long runs of magazines, George Symons' *Meteorological Magazine* included hundreds of storm cases and many tornado events in the early issues from 1865 to 1914.

Seeing that tornado and thunderstorm research was not being done by the Meteorological Office, here was a worthy area of study with obvious prospects of long-lasting usefulness. At the time I thought that a series of severe-storm books might develop, but instead a weather magazine (*The Journal of Meteorology*) was launched.

It felt necessary to compile data on severe-storm and tornado events as they were being reported countrywide in order to inspect fresh damage with minimum delay, so I subscribed to a press-clipping service using search words like tornado, whirlwind, waterspout, freak wind and hurricane wind. The database soon held hundreds of tornado cases. As best as possible, details of track length, direction, path width and damage were summarised. Regarding the latter, an expanded version of the Beaufort scale at first proved helpful for estimating wind speeds of the mostly weak tornadoes, but something more specific and relevant to the reality of characteristic tornado damage was needed. At this point, in 1972/1973, it proved easy to devise a practical tornado intensity scale that would maintain the advantages offered by the scientific strengths of Beaufort, and would, moreover, be achievable by introducing only a minor rearrangement of the basic Beaufort velocity formulation. The International T-Scale was born.

1.2 International T-Scale: Theoretical Basis

The Beaufort scale is the recognised scientific international wind-speed scale. It began two centuries ago in 1805, for use at sea. A land-based version followed later. The International T-Scale was designed in Europe for practical use with weak and strong tornadoes, whereas Fujita's system was initially planned to characterize strong North American tornadoes in view of potential hazards to nuclear power stations.[2]

The T-Scale is as universally valid as the Beaufort scale because the T-Scale is simply the universal Beaufort scale in a form rendered practical for tornado-strength winds. It embraces the digit range from T0 to T10 where each T number stands for tornado maximum strength (Table 1.1). The Beaufort scale is supported by the authority of the World Meteorological Organisation, so the T-Scale stands to benefit similarly. The basic velocity equation is the same. The T-Scale avoids the big code numbers that would arise if the ordinary Beaufort scale was used to assess tornado winds.

The Beaufort scale was formalised in the 1920s at international meetings by agreeing that velocity, $v = 0.837 B^{3/2}$ metres per second – the measurements made at 10 m above ground level.

Hence the T-Scale formula:

$$v = 0.837(2T + 8)^{3/2} \text{ or } v = 2.365(T + 4)^{3/2}$$

Thus, the T-Scale is the Beaufort scale modified for tornado wind-speeds in which T0 equals B8 gale force at 18.9 ms⁻¹ (for 3-second gusts).

This automatically makes T2 *exactly equal* to hurricane-speed B12 or 34.8 ms⁻¹.

Note that T-Scale and B-Scale numbers are exactly equivalent in whole digits.

[2]An American scale by Prof. T.T. Fujita was devised (1972/1973) to assign maximum known strengths to US structures including, especially, nuclear power stations and other vulnerable high-risk structures (Fujita, 1973). The T-Scale or International Tornado Intensity Scale, (1972/1973), was intended to apply to all the world's tornadoes (Meaden, 1975–1976). Previously, the whole world had made use of the Beaufort Scale to assess wind speeds at least up to hurricane-speed B12 but beyond as well.

Table 1.1 *The international tornado intensity scale.*

T-scale	Wind speeds	Characteristic Damage Intensity (Descriptive; for General Guidance only)
T0	Light tornado 17–24 ms^{-1}	Loose light litter raised from ground in spirals. Tents, marquees seriously disturbed; the most exposed tiles, slates on roofs dislodged. Twigs snapped; trail visible through crops
T1	Mild tornado 25–32 ms^{-1}	Deckchairs, small plants, and heavy litter become airborne; minor damage to sheds. More serious dislodging of tiles, slates, chimney pots. Wooden fences flattened. Slight damage to hedges and trees
T2	Moderate tornado 33–41 ms^{-1}	Heavy mobile homes displaced, light caravans blown over, garden sheds destroyed, garage roofs torn away, much damage to tiled roofs and chimney stacks. General damage to trees, some big branches twisted or snapped off, small trees uprooted
T3	Strong tornado 42–51 ms^{-1}	Mobile homes overturned/badly damaged; light caravans destroyed; garages and weak outbuildings destroyed; house roof timbers considerably exposed. Some big trees uprooted or snapped
T4	Severe tornado 52–61 ms^{-1}	Vehicles levitated. Mobile homes airborne/wrecked; sheds airborne for considerable distances; entire roofs removed from some houses; roof timbers of stronger brick or stone houses totally exposed. Strong trees can be snapped or uprooted
T5	Intense tornado 62–72 ms^{-1}	Heavy motor vehicles levitated. More serious building damage than for T4, yet most house walls usually remaining. The oldest, weakest buildings may collapse entirely
T6	Moderately-devastating tornado 73–83 ms^{-1}	Strongly built houses lose entire roofs and perhaps a wall or two; more of the less strong buildings collapse, as with wooden frame houses
T7	Strongly-devastating tornado 84–95 ms^{-1}	Wooden frame houses wholly demolished; walls of stone or brick houses may collapse or be beaten down. Steel-framed warehouse-type constructions may buckle slightly. Locomotives can be thrown over. Considerable debarking of trees by flying debris
T8	Severely-devastating tornado 96–107 ms^{-1}	Motor cars hurled great distances. Wooden-framed houses and their contents dispersed over great distances; stone or brick houses irreparably damaged; steel-framed buildings buckled
T9	Intensely-devastating tornado 108–120 ms^{-1}	Many steel-framed buildings badly damaged. Locomotives or trains thrown and tumbled great distances. Complete debarking of any trees left standing
T10	Super tornado 121–134 ms^{-1}	Entire frame houses and similar structures lifted bodily from foundations and carried some distances. Steel-reinforced structures may be badly damaged

T	0	1	2	3	4	5	6	7	8	9
B	8	10	12	14	16	18	20	22	24	26

This makes the T-Scale wholly practical and especially useful for rating the weaker tornadoes from T0 to T6. Most European tornadoes range from T0 up to T6 or T7, while all the world's tornadoes span the range T0 to T9 or T10. The paper by Meaden *et al.* (2007) provides a full explanation.

The F-Scale, and its revised version the EF-Scale, ranges from 0 to 5. Fujita set his force F1 at minimum hurricane speed (not at Beaufort's average hurricane speed). This means F1 = Beaufort 11.5 (instead of B12) and ensures that there is never whole-number equivalence, for example F2 = B15.37; F3 = B19.22; F4 = B23.06; F5 = B26.91.

Even for the United States, 92% of tornadoes are recorded as F0, F1 and F2, while more than 98% are F0, F1, F2 and F3. Only between 1 and 2% have been rated F4, and only one in a thousand reaches F5. Worldwide, much fewer in percentage terms reach F4 and F5. Indeed, most of the world's tornadoes are only F0, F1 and F2 – which is better expressed as T0–T5.

The T-Scale was put to strict and timely use when preparing the Sizewell tornado risk study in 1984–1985 (Meaden, 1985) and recently when tornado risk calculations were needed for Electricité de France Energy Plc at Hinkley Point in 2010–2011 (Meaden, 2011). It is important to note that if TORRO's 40 years had been spent applying the F-Scale to assess British tornado

strengths, the risk calculations for British nuclear power stations would not have been possible through an insufficiency of available intensity-data points. To summarise, a truly universal scale requires a sound theoretical basis to deserve universal acceptance – and it should apply readily to all the world's tornadoes. The decimal T-Scale meets the necessary strict scientific requirements, just as the MKS system does for scientific work everywhere. It is a truly international scale.

1.2.1 Hailstorm Research

The compilation of hailstorm statistics is a major division of TORRO. Initiated by Michael Rowe, the databank has been well maintained by Jonathan Webb for three decades. Important publications in international journals have resulted, including collaborations with international scientists (e.g., Sioutas *et al.*, 2009). The hailstone size scale and the Hailstorm Intensity Scale created by TORRO have proved very helpful in Britain and worldwide (Webb *et al.*, 1986; 2009) (see Chapter 9 for details).

1.2.2 Temperature Extremes for the British Isles

In 1984, TORRO published a list of Britain's highest temperatures for every day of the year (Meaden and Webb, 1984). This led to an invitation to update and republish the study (including the lowest known temperatures) in the Royal Meteorological

Society's Millennium Year's *Weather* (Webb and Meaden, 2000). Since then, the paper has had a demonstrable positive influence on the presentation of weather information by forecasters and press reporters countrywide. Similar research culminated in the publishing of daily extreme values for rainfalls across Britain (Ross, Webb and Meaden, 2009).

1.3 Tornado Research Organisation

The Tornado Research Organisation was the original name and hence the acronym TORRO. The launch of *The Journal of Meteorology* soon followed, starting with the first issue in October 1975. At once, weather-forecaster Bob Prichard began providing monthly thunderstorm reports – which he is still doing 39 years later – and so the name was changed to the Tornado and Storm Research Organisation and the acronym TORRO retained.

The intention was that TORRO would be *a self-funded research body serving the international public interest*. The main objective was to ensure the publication and dissemination of the huge amount of data on severe storms and tornadoes that was being gathered. The quantity of information was so vast that no existing magazines could publish everything in any detail. Soon after launching TORRO, I got to know Mr. Michael W. Rowe and a long fruitful collaboration began. Independently, he had been gathering tornado data in much the way that I had. We merged our databases and shared everything new that came to our attention. The content of the database stretched back to the 11th century. The earliest date for a tornado known for the United Kingdom is 1091 (London) and for Ireland 1054 (Rosdalla, County Westmeath).

Later, in the 1990s, David Reynolds was helpful during the years when he was working for his PhD at the University of South Wales, and Paul Brown in recent years has been brilliant at industriously rewriting and extending the tornado/whirlwind-related database. His archival research of newspapers has greatly augmented the known cases for the 18th and 19th centuries. By 2014 we could say that the totality of archived British and Irish events number more than 3800 tornadoes and waterspouts. Additionally, there are many hundreds of reported funnel clouds and other whirlwinds (land devils, water devils, eddy whirlwinds, fire devils, gustnadoes). Rowe (1999) published a summary article on tornadoes in the British Isles to 1660. The current number of tornadoes being *reported and recorded* annually for the British Isles averages nearly 50; and so one may ask 'how many more are still being missed – unseen in the countryside or taking place at night'?

The general work of the Tornado and Storm Organisation has been reviewed several times, for example by Elsom *et al.* (2001) and Meaden and Rowe (2006). Besides the tornado work, data collections were made of the following:

- Severe hailstorms and details of big and giant hail (*see* Webb, 1993)
- Lightning injuries and deaths (Derek Elsom and Jonathan Webb)
- Thunderstorm occurrences and distribution (Bob Prichard, Jonathan Webb and Keith Mortimore)

- Thunderstorm rains (various, including Jonathan Webb and John Mason)
- Heavy snowfalls (Richard Wild)
- Ball lightning (initially Michael Rowe and Terence Meaden, then Mark Stenhoff, Adrian James, Peter Van Doorn and Chris Chatfield)
- Coastal storm impacts and floods (*see* Doe, 2002)
- Historical cases of other early storms in Britain (e.g. during the invasion by Julius Caesar in 55 BC, *see* Meaden, 1976) and early weather monitoring in Britain (e.g. in the years 1269–1270 and 1337–1343, *see* Meaden, 1973). Also Webb (1987) for Britain's highest daily rainfalls by county and month.

In 1985, TORRO executives included Michael Rowe, Bob Prichard, Derek Elsom, Jonathan Webb, Keith Mortimore, Adrian James, Chris Chatfield and Albert Thomas. By 1990/1995, additional executives and key personnel were Peter Matthews, Alan Rogers, Wendy Rogers, Ellie Gatrill-Smith, Mark Stenhoff, David Reynolds, Ray Peverall, and Richard Muirhead. Figure 1.2 shows the attendance at the June 2002 staff meeting.

By 2005 further personnel were Paul Knightley, Helen Rossington, Nigel Bolton, Robert Doe, Matthew Clark, Tony Gilbert, Stuart Robinson, Stephen Roberts, Richard Wild, John Mason, Sam Jowett, Mark Humpage, Stuart Robinson, Peter Van Doorn, Paul Domaille, Ian Loxley, Ian Miller, John Tyrrell, John Wilson, Chris Warner, Samantha Hall and Martin Collins. Figure 1.3 shows the staff at the August 2005 staff meeting and Figure 1.4 in November 2006. Since 1994, the Heads of TORRO have been either academics (Prof. Derek Elsom, Oxford Brookes University, followed by Dr John Tyrrell, University College Cork) or, as now, a professional meteorologist and forecaster (Paul Knightley).

1.4 The Inaugural Issue of *The Journal of Meteorology*

October 1975 saw the first issue of *The Journal of Meteorology* (Figure 1.5). The issue for July/August 1985 was the hundredth. The intention was to publish for the public benefit details of every tornado event known for Britain from 1960 onwards and to report major site investigations regarding tornadoes and damaging hailstorms, and this has been done ever since.

Upon its launch, the journal had a warm reception from the renowned, much-liked Professor Gordon Manley (1902–1980) (Figure 1.6). He wrote saying how pleased he was at this enterprise and the prospect of such storm research advancing the cause of meteorology and weather forecasting. Professor Hubert Lamb (1913–1997) (Figure 1.6), too, immediately supported the young journal, and at the time of TORRO's first conference (1985) (see Figures 8.1–8.9) and the journal's 100th issue, he wrote a timely piece about the importance of independent meteorological research (Lamb, 1985).

The mounting information about the high frequency of tornado occurrences was impressive. The journal would ensure that TORRO would publish all incoming data about new and past events, and this has been achieved in some detail for the

Figure 1.2 *Staff meeting in Devon, 8 June 2002; (left to right) Wendy Rogers, Terence Meaden, Derek Elsom, Robert Doe, Tony Gilbert, Stephen Roberts, Nigel Bolton, Harry McPhillimy (TORRO Archives).*

Figure 1.3 *Staff meeting, 6 August 2005, in Devon, hosted by Alan and Wendy Rogers. From left to right,* standing: *Chris Chatfield, Mike Rowe, Alan Rogers, David Bowker, Jonathan Webb, Robert Doe, Nigel Bolton, Derek Elsom, Ian Loxley, Richard Wild, Steve Roberts and John Mason;* kneeling: *Wendy Rogers, Terence Meaden, Paul Knightley, Peter Matthews, Tony Gilbert and Ian Miller (TORRO Archives).*

Figure 1.4 *Staff meeting in Oxford, 18 November 2006; (left to right) Richard Pearson, Alan Rogers, Wendy Rogers, Jonathan Webb, Chris Chatfield, Robert Doe; (front row) Mike Rowe, Peter Matthews, Terence Meaden, Derek Elsom (TORRO Archives).*

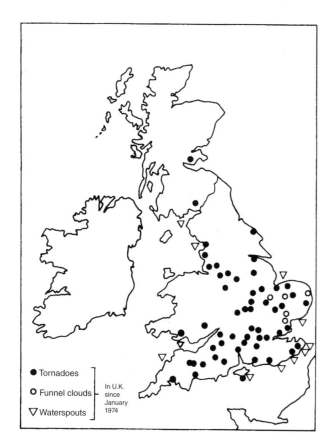

Figure 1.5 *The front cover of the inaugural issue of The Journal of Meteorology (October 1975) depicted the locations of tornadoes, funnel clouds and waterspouts in the United Kingdom between January 1974 and mid-September 1975.*

years 1960–2013, also the principal data from the TORRO Hailstorm Division for the period 1980–2013 by Jonathan Webb. The Blizzard and Snowstorm Division has published summaries from 1870 to 2013 (Richard Wild), while the Lightning Deaths and Injuries Division has published everything from 1985 to 2013 (Derek Elsom and Jonathan Webb). Through the work of Bob Prichard and the TORRO Thunderstorm Division, the journal published British thunderstorm general reports monthly from 1975 to 2013. Moreover, the Thunderstorm Census Organisation (TCO), founded by Mr. Morris Bower of Huddersfield in 1924, was taken over by TORRO after his death.[3] Editors of the Journal since 2002 are as follows: Robert Doe, 2002–2006; Samantha Hall, 2006–2011; Paul Knightley and Helen Rossington, 2012 to present. *The Journal of Meteorology* was renamed *The International Journal of Meteorology* in 2005 (Figure 1.7).

1.5   Storm-Damage Site and Track Investigations

Hundreds of site investigations have been done by TORRO members in the 40 years from 1974 to 2014, studying tornado tracks and damage. The first sites studied were earlier, in 1966 at Headington (north Oxford) and in 1967 (Trowbridge to Melksham, Wiltshire) during brief spells when I was visiting England (Meaden, 1984). In addition to

---

[3] TORRO has in its collection all TCO thunderstorm records from 1924 to 1980.

Figure 1.6 *Prof. Gordon Manley (1902–1980) and Prof. Hubert Lamb (1913–1997), friends of TORRO. Images courtesy of Royal Holloway, University of London and the University of East Anglia.*

Figure 1.7 The International Journal of Meteorology, *celebrating its 30th anniversary in September 2005.*

tornado site investigations, TORRO members have examined occurrences of storm damage by lightning, hail, rainfall deluges and snow. Ball lightning has received much serious attention (see Chapter 11).

To exemplify in the case of tornadoes, consider the day of 3 November 2009 for which Tony Gilbert had forecast the possibility of tornadoes in the south of England. Because numerous sites needed to be investigated, the following TORRO members from the regional network shared the tasks:

Matthew Clark	Coombe Bissett (S. Wiltshire) S.W. of Salisbury
Brian Montgomery	Charlton All Saints (S. Wiltshire) S. of Salisbury
Brian Montgomery	Romsey/Timsbury (West Hampshire)
Daniel Mellor	South of Timsbury (West Hampshire)
Nigel Paice	North of Romsey (West Hampshire)
Tony Gilbert	Ampfield (Hampshire)
Helen Rossington with Paul Knightley	West Meon → Owslebury (Hampshire)
Matthew Clark	Several areas of mid-Hampshire including Monkswood
Sarah Horton	Northchapel (Hampshire)

The paper by Clark (2010) records the study behind the meteorological and site investigations. More generally, countrywide, a network of regional coordinators exists, details of which can be found on the TORRO website together with information as to other aspects of TORRO's work. By the late 1990s, tornado forecasting became a feature of TORRO's activities. At various times, forecasters have been David Reynolds, Tony

Figure 1.8 *Tornado at King's Heath, Birmingham, 28 July 2005. Photograph supplied to TORRO. © Ian Dunsford, Birmingham City Council.*

Gilbert, Nigel Bolton and Paul Knightley. Examples of site investigations are indicated next.

1.6   Birmingham Tornado of 28 July 2005

A thousand properties were damaged and many destroyed during the Birmingham tornado of 28 July 2005 (Figure 1.8). Over 50 homes were condemned for demolition. Figure 1.9 shows the tornado track through the city of Birmingham, with reference to previous events.[4] Indeed, this was not the first time Birmingham experienced such a severe tornado. Figure 1.10 shows Pathé news from an event on 14 June 1931. Sadly this resulted in a fatality in Sparkhill, Birmingham, where a woman, sheltering in the doorway of a shop, was killed instantly by a falling wall. Property damage was extensive, but confined to a narrow swathe. Many roofs were completely lost, with houses and shops being demolished (Figure 1.11).

Seventy-four years later, in 2005, the TORRO forecaster issued the following warning:

TORRO TORNADO WATCH No. 2005-020

TORNADO WATCH issued at 1400 UTC Thurs. 28 July 2005

VALID: 1400–1900 GMT on Thursday 28th July 2005 for the following regions of England: North and East Midlands, Lincolnshire, Yorkshire

SYNOPSIS: Thunderstorms have developed across the Midlands, and will track north-east. CAPE, shear and veer on Larkhill and Brize Norton ascents suggest possible tornado development to strength T3. The S.E. edge of the storm is approaching Birmingham, and is expected to track N.E. to the Humber and be the focal point of any development.

THREATS: Risk of isolated tornadoes to T3 in path from Birmingham to Lincolnshire. © Copyright TORRO 2005

1.7   TORRO Conferences

TORRO has held conferences in Oxford annually since 1985 and biannually since 1995.[5] Figures 1.12, 1.13 and 1.14 show selected events in 2003 and 2006. Staff also meet regularly to

---

[4] Other major tornadoes of recent years include the Selsey tornado of January 1998 investigated by Meaden (1998) and Matthews (1999) and the T5 tornado of 7 December 2006 at Kensal Rise, West Central London, investigated by Kirk (2013).

[5] *Tornadoes and Storms* I, II, III, IV, V (being TORRO conference proceedings 1985–1999) [ed. G. T. Meaden and Derek Elsom]. Special Issues of the Journal of Meteorology, Artetech Publishing Company, Bradford-on-Avon, Wiltshire.

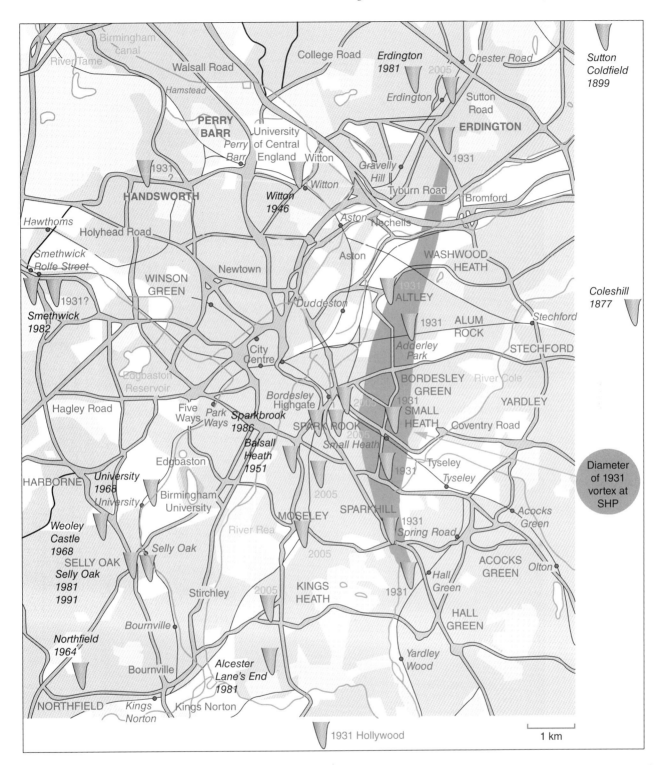

Figure 1.9 *The map shows the tornado track of 28 July 2005 through the city of Birmingham. Maximum intensity was T5/6 on the International T-Scale. Sites of tornado damage in preceding years are indicated too. Chief investigators were Ian Brindle, Matthew Capper and Peter Kirk. Details are given in papers by Knightley (2006), Kirk (2006), Pearson (2006), Smart (2008) and Meaden and Chatfield (2009). (See insert for colour representation of the figure.)*

Figure 1.10　*News footage of the Birmingham tornado of 14 June 1931. © British Pathé.*

Figure 1.11　*Structural property damage as a result of the Birmingham tornado of 14 June 1931. Shops and houses on the Coventry Road received severe damage. © British Pathé.*

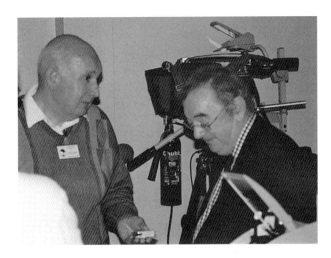

Figure 1.12 *TORRO's Alan Rogers with Weather Forecaster Ian McCaskill at the 2003 annual autumn conference in Oxford, UK (TORRO Archives).*

oversee the effective running of the organisation (Figure 1.15). TORRO members have attended most meetings of the *European Conferences on Severe Storms* that began in 2002. More recently, in May 2014, Drs. Robert Doe and Terence Meaden presented at the *First International Summit on Tornadoes and Climate Change* held in Chania, Crete (see photographs in Appendix B).

1.8 The Future

It is intended that the work of TORRO will continue far into the future. TORRO's mission statement is steadfast and decisive because the organisation is a privately-supported research body *serving the international public interest*. The dangers posed by tornado-strength winds are an eternal problem for architects involved with designing engineered structures including ordinary house buildings at one extreme and nuclear-plant new buildings at the other. Calculating return-period risks is at the heart of these

Figure 1.13 *The TORRO conference of 2006. Top picture, audience. Lower picture, speakers (sitting, left to right), Chris Chatfield, Tony Gilbert, Nigel Bolton, Michael Rowe, Peter Van Doorn, Paul Knightley, John Tyrrell, Jonathan Webb, (at back, standing) Derek Elsom, Terence Meaden, John Mason (TORRO Archives).*

Figure 1.14   *Question time in Birmingham in 2006, on the first anniversary of the T5/6 tornado of 28 July 2005: Paul Knightley and Chris Chatfield at the right, Terence Meaden at left. © Robert Doe.*

Figure 1.15   *TORRO Staff meeting of 29 June 2012; (left to right) Jonathan Webb, Robert Doe, Paul Knightley (and James Glaisher (the sculpture)) at the meetings room of* The Royal Meteorological Society in Reading, *UK. Photograph by Terence Meaden.*

problems for which the maintenance of a sound database is essential. Another principal objective is improved forecasting of tornadoes for which a climatology is needed and a better understanding of atmospheric dynamics. The question of whether tornadoes will become more frequent in response to climatic change has already been raised (*First International Summit of Tornadoes and Climate Change* at Chania, Crete, May 2014), and to answer it, the best possible tornado databases are needed.

As in the past, TORRO's fortunes depend on having a thriving, pro-active membership, and strong scientific leadership. All weather enthusiasts, whether amateur or professional, are welcome to join and partake in the activities.

Acknowledgements

The Directors of TORRO wish to thank the hundreds of members and professional meteorologists who, over the many years, have contributed to TORRO's undoubted success. The encouragement and support of professional meteorologists in universities, research institutes and weather forecasting have been much appreciated and repeatedly acknowledged.

Additional information

In addition to the mentioned publications in *The (International) Journal of Meteorology* and the reference list below, the following are also key TORRO publications:

Meaden, G.T. (1985) *Tornadoes in Britain (with assessments of the general tornado risk potential and the specific risk potential at particular sites including Sizewell).* Nuclear Installations Inspectorate, Great Britain. Health and Safety Executive, Tornado and Storm Research Organisation, 131pp.

Meaden, G.T. (ed.) (1990) *Ball Lightning Studies*. Artetech Publishing Co., Bradford-on-Avon. 84pp.

Stenhoff, M. (ed.) (1992) *Proceedings Ball Lightning Conference, Oxford.* Tornado and Storm Research Organisation (TORRO), Oxford.

Bolton, N., Elsom, D.M. and Meaden, G.T. (2003) Forecasting tornadoes in the United Kingdom. *Atmos. Res*, **67–68**: 53–72.

Meaden, G.T. (2011) *Assessing the Tornado-Risk Potential for Coastal Somerset at Hinkley Point in Southern Britain*, for EDF Energy PLC and Nuclear New Build Generation Company Ltd.

For more detailed information about TORRO, readers should visit the TORRO and the *International Journal of Meteorology* web sites at:

www.torro.org.uk

www.ijmet.org

References

Britton, C. E. (1937) *A meteorological chronology to A.D. 1450.* Geophysical Memoir, no. 70, HMSO, London.

Clark, M. R. (2010) A survey of damage caused by the Monkwood, Hampshire, U.K. tornado of 3 November 2009. *Int. J. Meteorol.*, **35**, 291–297.

Doe, R. (2002) Towards a local coastal storm climatology: Historical perspectives and future prediction. *J. Meteorol.*, **27**:268, 117–124.

Elsom, D. M., Meaden, G. T., Reynolds, D. J., Rowe, M. W. and Webb, J. D. C. (2001) Advances in tornado and storm research in the U.K. and Europe: The role of the Tornado and storm research organisation. *Atmos. Res.*, **56**, 19–29.

Fujita, T. T. (1973) Tornadoes around the world. *Weatherwise*, **26**, 56–62, 78–83.

Kirk, P. (2006) A mammoth task: The site investigation after the Birmingham tornado 28 July 2005. *Int. J. Meteorol.*, **31**, 255–260.

Kirk, P. (2013) Kensal rise, London tornado site investigation 7 December 2006. *Int. J. Meteorol.*, **38**, 244–252.

Knightley, P. (2006) Tornado-genesis across England on 28 July 2005. *Int. J. Meteorol.*, **31**, 243–254.

Lacy, R. E. (1968) Tornadoes in Britain during 1963–1966. *Weather*, **23**, 116–124.

Lamb, H. H. (1985) TORRO and the importance of independent meteorological research. *J. Meteorol.*, **10**, 180–181.

Matthews, P. (1999) The Selsey tornado of 7 January 1998: A comprehensive report on the damage. *J. Meteorol.*, **25**, 197–209.

Meaden, G. T. (1973) Merle's weather diary and its motivation (AD 1337–1343). *Weather*, **28**, 210–211.

Meaden, G. T. (1975–76) Tornadoes in Britain: Their intensities and distribution in time and space. *J. Meteorol.*, **1**, 242–251 (based on a lecture at London University to the Royal Meteorological Society in 1975).

Meaden, G. T. (1976) Late summer weather in Kent, 55 BC. *Weather*, **31**, 264–270.

Meaden, G. T. (1984) The Trowbridge-Melksham tornado and severe local storm of 13 July 1967: An addendum to M.E. Hardman's 'The Wiltshire storm'. *J. Meteorol.*, **9**, 288–290.

Meaden, G. T. (1985) *A study of tornadoes in Britain, with assessments of the general tornado risk potential and the specific risk potential at particular regional sites.* Prepared at the request of HM Nuclear Installations Inspectorate, Health and Safety Executive. Warrington, Lancashire. 131pp. A privately commissioned report.

Meaden, G. T. (1998) Selsey tornado, the night of 7–8 January 1998. *J. Meteorol. U.K.*, **23**, 41–55.

Meaden, G. T. (2011) *Assessing the tornado risk potential for coastal Somerset including Hinkley Point in Southern Britain*. Nuclear Site Licensing. EDF Energy plc. 167pp. A privately commissioned report.

Meaden, G. T. and Chatfield, C. R. (2009) Tornadoes in Birmingham, England, 1931 and 1946 to 2005. *Int. J. Meteorol.*, **34**, 155–162.

Meaden, G. T. and Rowe, M. W. (2006) The work of the Tornado and storm research organisation: Tornadoes as a weather hazard in Britain. *Int. J. Meteorol.*, **31**, 237–242. [Paper given at the Birmingham University Conference of July 2006.]

Meaden, G. T. and Webb, J. D. C. (1984) Britain's highest temperatures for every day of the year, 1 January to 31 December. *J. Meteorol.*, **9**, 169–176.

Meaden, G. T., Kochev, S., Kolendowicz, L., Kosa-Kiss, A., Marcinoniene, I., Sioutas, M., Tooming, H., and Tyrrell, J. (2007) Comparisons between the theoretical versions of the Beaufort Scale, the T-Scale and the Fujita Scale – and a proposed unification. *Atmos. Res.*, **83**, 446–449.

Pearson, R. (2006) Images of a disaster: A selection of Birmingham tornado photographs. *Int. J. Meteorol.*, **31**, 261–265 and inset pages I–IV.

Ross, N. A., Webb, J. D. C. and Meaden, G. T. (2009) Daily rainfall extremes for Great Britain and Northern Ireland. Part 1. *Int. J. Meteorol.*, **34**, 57–69; Part 2, 75–81.

Rowe, M. W. (1999) 'Work of the Devil': Tornadoes in the British Isles to 1660. *J. Meteorol*, **24**, 326–338.

Sioutas, M., Meaden, G. T. and Webb, J. D. C. (2009) Hail frequency and intensity in Northern Greece. *Atmos. Res.*, **93**, 526–533.

Smart, D. (2008) Simulation of a boundary crossing supercell – a note on the 28 July 2005 Birmingham U.K. event. *Int. J. Meteorol.*, **33**, 271–274.

Webb, J. D. C. (1987) Britain's highest daily rainfalls: The county and monthly records. *J. Meteorol.*, **12**, 263–266.

Webb, J. D. C (1993) Britain's severest hailstorms and hailstorm outbreaks 1893 to 1992. *J. Meteorol.*, **18**, 313–327.

Webb, J. D. C., Elsom, D. M., Meaden, G. T. (1986) The TORRO hailstorm intensity scale. *J. Meteorol.*, **11**, 337–339.

Webb, J. D. C., Elsom, D. M. and Meaden, G. T. (2009) Severe hailstorms in Britain and Ireland: A climatological survey and hazard assessment. *Atmos. Res.*, **93**, 587–606.

Webb, J. D. C. and Meaden, G. T. (2000) Daily temperature extremes for Britain. *Weather*, **55**, 298–314.

Part I
Tornadoes

2

Historical tornadoes in the British Isles

Paul R. Brown[1] and G. Terence Meaden[2]

[1] *Tornado Division, Tornado and Storm Research Organisation (TORRO)*, Bristol, UK*
[2] *Kellogg College, Oxford University, Oxford, UK*

2.1 Introduction

From its foundation TORRO has always sought not only to investigate and document contemporary tornadoes and other whirlwinds but to compile a record of as many historical events as possible going back to the earliest identifiable tornadoes. Before the advent of modern science the few surviving records come to us from the medieval chroniclers, and their descriptions sometimes require careful interpretation. From the 17th century onwards accounts begin to appear in the scientific journals, and from the 18th century the most useful sources become the newspapers of the time.

From 1974 for a period of 12 years TORRO usefully subscribed to press-clipping services as a way of obtaining newspaper reports of whirlwinds. Since then the innovation of the internet has greatly facilitated the task of searching old newspapers (and other publications), enabling this to be done electronically in a hundredth of the time it would take to do the same thing manually. There are, however, limits to the thoroughness of this method, relying as it does on the automatic recognition of electronically scanned words, which are not always clear enough to be identified correctly.

Another research method employed some years ago by Michael Rowe on behalf of TORRO was to make appeals through local and regional newspapers for information about whirlwinds of the past that readers might have stored away in their memories. This resulted in a good response (in letters received through the post), but because many of these correspondents, then in their 70s or 80s, were recalling events from 50 years or more earlier, their memories were naturally no longer sharp enough to be precise about dates, and many of the tornadoes discovered this way (from the first half of the 20th century) can only be dated to within a year or two at best.

* http://www.torro.org.uk/

2.2 Etymology of the Word *Tornado*

Tornado was originally a seaman's word, coined by 16th century sailors sailing down to the west coast of Africa thence taking the trade wind route across to the West Indies (on plundering missions following Columbus's accidental discovery of these islands while searching for a route to the Orient). When in tropical waters they occasionally encountered severe thunderstorms of a kind unfamiliar at home, in which there were violent squalls of wind sometimes going right round the compass. They came to call these storms *ternados*; but the change of a single letter to give us the modern *tornado* is not as straightforward as it might seem. Although both words have a Spanish look to them, neither is recorded as such in that language. Spanish does, however, have *tronada* meaning a thunderstorm and *tornar* meaning to turn; and Oxford's conjecture (OED, 1926) is that *ternado* was a clumsy adoption of *tronada*, to which an improvement was later attempted by associating it with *tornar* to give *tornado* (and other variant spellings).

Be that as it may, the word at this time carried no implication of the violent small-scale vortex usually associated with a tornado nowadays, as shown by the earliest references in print:

1. 1556 W. TOWERSON in Hakluyt *Principal Navigations, Voyages* ... (1589). The 4-day we had terrible thunder and lightning with exceeding great gusts of raine, called Ternados.
2. 1599 HAKLUYT *Principal Navigations, Voyages* ... II. ... for we had nothing but Ternados, with such thunder, lightning and raine, that we could not keep our men drie 3 hours together.
3. 1634 SIR T. HERBERT *Travels into Africa and Asia* ... We crost the Æquator, where we had too many Tornathoes; 1638 ed. ... wee were pesterd with continuall Tornathes; a variable

Extreme Weather: Forty Years of the Tornado and Storm Research Organisation (TORRO), First Edition. Edited by Robert K. Doe.
© 2016 John Wiley & Sons, Ltd. Published 2016 by John Wiley & Sons, Ltd.

weather, compos'd of lowd blasts, stinking showers, and terrible thunders.
4. 1697 DAMPIER *New Voyage round the World* (1699). We had fine weather while we lay here, only some Tornadoes or Thunder-showers.

The first suggestions of revolving winds are met with in 17th century references, but there is still no clear evidence of a spinning vortex in these descriptions:

1. 1625 PURCHAS *Pilgrims* II. The sixteenth, we met with winds which the Mariners call, The Turnadoes, so variable and uncertaine, that sometime within the space of one houre, all the two and thirtie severall winds will blow. These winds were accompanied with much thunder and lightning, and with extreme rayne.
2. 1626 CAPT. SMITH *Accidence … for all Young Seamen*. A gust, a storme, a spoute, a loume gaile, an eddy wind, a flake of wind, a Turnado, a mounthsoune, a Herycano.
3. 1656 BLOUNT *Glossographia*. *Tornado*, (from the Spanish *Tornada*, a returne, or turning about) is a sudden, violent and forcible storme of raine and ill weather at sea, so termed by the Mariners; and does most usually happen about the Æquator.
4. 1710 J. HARRIS *Lexicon Technicum* II. *TORNADO* is the Name given by the Seamen for a violent Storm of Wind, and sometimes followed by Rain; it usually swifts or *turns* about to almost all Points of the Compass, whence I suppose its name.

During the 17th and 18th centuries, when the Mississippi region of North America was being explored, the pioneers sometimes experienced the terrifying whirlwinds characteristic of that region; and either through a misunderstanding of the seaman's term or through a deliberate adaptation of it, began to call these storms *tornadoes*. Thus by the 18th century we find:

1. 1755 JOHNSON *Dictionary of the English Language*. TORNADO, … a hurricane; a whirlwind.
2. 1770 GOLDSMITH *Deserted Village*. While oft in whirls the mad tornado flies.
3. 1788 COWPER *Negro's Complaint*. Hark! He answers!——Wild tornadoes Strewing yonder sea with wrecks, Wasting towns, plantations, meadows.

And by the 19th century, definite references to the modern type of tornado:

1. 1849 LYELL *Second Visit to the United States* II. This tornado [17 May 1840] checked the progress of Natchez, as did the removal of the seat of Legislature to Jackson.
2. 1883 *Encyclopaedia Britannica* XVI. … the region of most frequent occurrence of tornadoes [in the United States] is the region where a large number of the cyclones of the United States appear to originate.

So for a long time we have two related but different meanings of the word coexisting, and references in the literature could be to either. Those in British newspapers predominantly relate to the seaman's tornado (at least in connection with storms in this country) until well into the 19th century, although it is clear that tornadoes of the whirlwind type were becoming well understood by the late 1700s, if not earlier.

By the 20th century the old meaning of *tornado* was obsolescent (perhaps because of the passing of the era of the sailing ship), except in West Africa where (we are told by the reference books) it still carries a meaning akin to what modern meteorologists would probably call a mesocyclone or microburst; and nearly all *tornado* references now are (or are intended to be) to the modern whirlwind type of tornado. There is, incidentally, sometimes uncertainty whether the plural of *tornado* should be *-s* or *-es*: either form is in fact acceptable but *-es* is the preferred spelling.[1]

2.3 Terminology

It follows from what has been said that, even apart from its (rather frequent) metaphorical use, the word *tornado* cannot always be taken literally in pre-20th century accounts: tornadoes of the kind TORRO is interested in are more often described in some other way. The commonest word is the generic term *whirlwind*, but that of course is not restricted to tornadoes, and can mean a dust devil or perhaps an eddy whirlwind; or there might be nothing in the description to indicate a spinning vortex of any kind, as distinct from a squall or severe gust. Other terms sometimes found for tornadoes in old newspapers are *hurricane, waterspout, typhoon, tempest, cyclone, storm, land spout*.

Another word that causes more trouble than might be expected is *waterspout*. This started as a non-meteorological term, dating from at least as far back as the 14th century, for a spout, etc. through which water is discharged, such as what we now usually call a drainpipe for removing water from a roof, and continued in that use until quite recently.[2] It is clear, however, that the atmospheric phenomenon we call a waterspout must have been observed by sailors from the earliest times of sea voyages, but the first print references in English to that use of the word date from no earlier than the 18th century:

1. 1738 T. SHAW *Travels, or Observations Relating to Several Parts of Barbary*…. Water Spouts are more frequent near the *Capes* of *Latikea, Greego*, and *Carmel* than in any other Part of the *Mediterranean* Sea.
2. 1788 *Volney's Travels through Syria and Egypt* I. … and hence will result those columns of water, known by the name of *Typhons* and *water-spouts*.
3. 1815 J. SMITH *Panorama of Science and Art* II. When a whirlwind happens at sea, or over the surface of water, it forms the phenomenon called a *water-spout*.

---

[1] There is no fixed rule in English for the plural of words ending in *o*. The most that can be said is that those in general use, and those that are (or appear to be) of English origin, usually take *-es* in the plural; those that are technical or not often used, and those that are (or appear to be) foreign, usually take *-s*. But this does not get us very far with *tornado*, which could belong to either category.
[2] Nineteenth-century newspapers are full of accounts of burglars shinning up and down waterspouts and of householders hauled before the magistrates for not keeping their waterspout in good repair and allowing it to drip on passers-by.

The Bible, from the 17th century (1611), has 'Deepe calleth unto deepe at the noyse of thy water-spouts' (Psalm 42), but this probably does not mean atmospheric waterspouts (OED, 1928). The quotation, however, is thought to have encouraged the adoption of the term in meteorological use, because it was often believed that the column reaching from cloud to sea was actually water pouring out of the cloud (as from a watering can), or water drawn up from the sea to the cloud, rather than merely water droplets suspended in the air; and although as early as the 1750s the remarkably fertile mind of Dr Benjamin Franklin was suggesting that whirlwinds and waterspouts were essentially the same, differing only in the surface over which they passed, the belief in a column of water persisted, at least among non-scientists, for much longer.

And one misconception led to another. When a funnel cloud was seen over land, but not reaching the surface, its visual similarity to a waterspout at sea led people to believe that it, too, was water pouring out of the cloud, but in this case somehow suspended in mid-air and not reaching the ground; and it was watched apprehensively lest it should 'burst' and release its contents, thereby flooding the area beneath it. Reasoning backwards from this misconception, if a sudden unexplained deluge of water rushed down a valley (in what we now call a flash flood) it was assumed that somewhere up in the hills (unseen) a waterspout must have burst and discharged its supposed supply of water on the ground, whence it cascaded down the valley.

Thus we find three meteorological definitions of *waterspout* in use: (i) a tornado over the sea, (ii) a funnel cloud not reaching the ground, (iii) a cloudburst. It is interesting that in some old accounts in which a funnel cloud is reported over land (but described as a waterspout), a whirlwind on the ground is also mentioned, yet the writer rarely seems to make the connection between the two, assuming it was nothing more than coincidence that they occurred together.

It should be added that some present-day meteorologists eschew the term *waterspout*, preferring to call such whirlwinds *tornadoes*, the same as those over land, the rationale being that both have the same physical causes and it is just a matter of chance whether they happen to form over land or over water. While scientifically sound, the objection to this reasoning is that it conceals the differences in their effects: a tornado nearly always causes some degree of damage, whereas a waterspout, unless it passes over a boat or reaches the shore, leaves no trace of its passage. Moreover, statistics of waterspouts are not directly comparable with those of tornadoes because a waterspout can only be recorded if seen (usually from the sea shore), whereas a tornado can often be inferred from its after-effects even if not observed in action, including for instance in the hours of darkness or in daylight when shielded by rain or cloud.

2.4 Accuracy and Completeness of the Records

Tornadoes cannot be recorded instrumentally. Modern Doppler weather radars are capable of indicating where a tornado might have occurred in association with mesocyclones or mesovortices, which can also be monitored for indications of tornadoes as they occur. But visual evidence of either the tornado funnel or the damage track remains the only way in which a tornado can be confirmed. Discovery of historical tornadoes therefore depends on four factors: (i) the tornado or its effects must have been witnessed; (ii) the information must have been committed to permanent record (on paper) in a comprehensible form; (iii) the record must have survived to the present day; (iv) the source must have come to our attention.

Even in modern times not all tornadoes find their way into the TORRO records, the weaker ones and those in sparsely populated areas being the ones most likely to escape notice. Attempts have been made in the past to quantify the shortfall between the number of tornadoes recorded and the number that actually occur, using modern data (e.g. Matthews, 2003; Holden and Wright, 2004), but differing methodologies make it difficult to draw reliable inferences from the results. But whatever the shortfall in the modern records, it is clear from Figure 2.1 that this gets progressively worse before the 20th century (it should, however, be noted

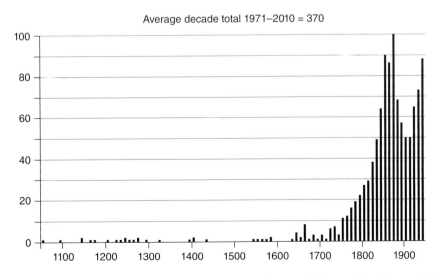

Average decade total 1971–2010 = 370

Figure 2.1 *Decadal totals of known tornadoes from* AD *1051 to 1950 (see the note in the text with reference to the dip* circa *1900).*

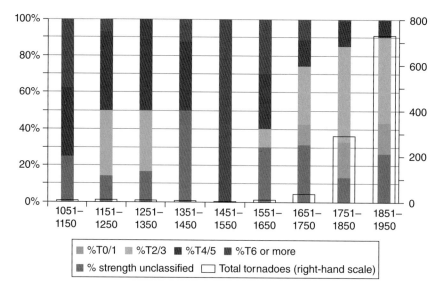

Figure 2.2   *Cumulative percentages of tornadoes by T-strength for each 100-year period from AD 1051 to 1950 (left-hand scale). Also showing totals of all tornadoes in the same 100-year periods (right-hand scale). (See insert for colour representation of the figure.)*

that searches of historical sources are still going on and are likely to yield further additions to the totals in due course). Conversely, the subjective nature of tornado identification means it is likely that a few entries have been falsely classified as tornadoes which would have been interpreted differently had fuller information been available; but these should (we hope) be much fewer than those of true tornadoes unknowingly omitted.

A feature to be expected, and which is borne out by the statistics, is that the tornadoes of greatest severity are those most likely to be documented in extant historical sources. Thus tornadoes of, say, T4 and higher constitute a much greater proportion of the earliest totals than is the case in more recent times, as shown in Figure 2.2. It should also be noted that, even when a modern site investigation has been carried out, the assessment of the T-strength of a tornado is not an exact science but is a best estimate from the information available. Such estimates become much more uncertain when trying to understand what someone wrote hundreds of years ago, and the strengths assigned to historical tornadoes, especially those from medieval times, should always be treated with caution. The principal ways in which the strength of a tornado is assessed are its effects on either buildings or trees. But whereas tree damage (at least for similar age, root structure, and species of tree) should remain constant over the centuries for a given strength of tornado, the same cannot be said for building damage, and it is a moot point how much allowance should be made for changes in building standards over the years, and whether any allowance should be made for the age and condition of a building (if known) at the time of damage. Furthermore, the potential for trees being damaged by a tornado will naturally have been greater in earlier centuries when the country was much more wooded than it is now, and *vice versa* for buildings as they continue to spread over more and more of the land.

A question arises over how to count tornadoes that descend to ground intermittently over a long track: should each descent to ground be counted as a separate tornado, or should the whole track be counted as a single long-lived tornado, even though it

was not on the ground at all points? There is no simple answer to this. Our practice, however, tends to be to count separate tornadoes if the gaps in the track exceed the lengths of ground contact, but to treat the whole track as one tornado if the gaps are shorter (remembering that gaps may be more apparent than real – it might just be that no-one investigated those parts, or that there was nothing there capable of sustaining damage). The way in which these tornadoes are dealt with can have a significant effect on the statistics.

Dates can be a cause of difficulty in old newspaper records. When they are given by the date of the month, e.g. 10th inst. or 28th ult., that is usually clear enough (especially if they also state the name of the day); but more often than not only the day of the week is given, such as 'Monday last' or 'Thursday week' or 'on Saturday'. This ought still to be reasonably clear if we can be sure that the newspaper is the primary source, but very often other newspapers copy the report from the original source (which might be an agency) and repeat it over the coming days or even weeks; and they usually do so verbatim, making no allowance for the fact that what was 'Monday last' in the original has become 'Monday week' or even 'Monday fortnight' by the time they publish it. This is a common cause of dating errors by whole weeks, because it is often difficult to decide whether the report to hand is the original or merely a copy of it from days or weeks before.

2.5   Analysis of Historical Tornadoes

Figure 2.1 shows the number of tornadoes recorded in each decade from AD 1051 to 1950, together with the modern (1971–2010) average decade total. The pronounced dip in numbers about the turn of the 19th/20th centuries is mainly, if not wholly, an artefact of the data: at the time of compilation these decades had not been searched as thoroughly in old newspapers as the earlier ones.

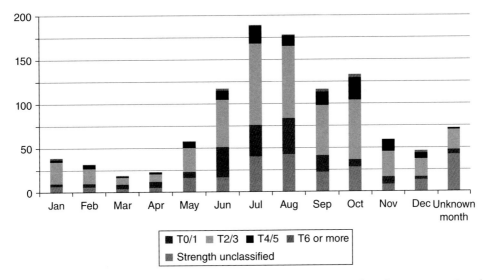

Figure 2.3 *Monthly totals of tornadoes by T-strength from* AD *1054 to 1950. (See insert for colour representation of the figure.)*

Figure 2.2 shows the cumulative percentages of tornadoes within given ranges of T-strength for each 100-year period from AD 1051–1150 to 1851–1950 (left-hand scale), together with the actual numbers of all tornadoes in the same 100-year periods (right-hand scale). This demonstrates how, when totals are very low (before about 1750), the records are dominated by the strongest tornadoes (purple and red shadings), while these make up a far smaller percentage when totals are much higher.

Figure 2.3 shows the grand total of tornadoes by T-strength in each month of the year during the period AD 1054 to 1950. This clearly still shows the same summer and autumn peak, and spring minimum of occurrence, that is apparent in the data for modern times,[3] despite the much smaller numbers of known tornadoes in these early years.

In Figure 2.4a–e we present the geographical distribution of tornadoes by T-strength for each century from the eleventh to the nineteenth. In Figure 2.4a and b combined distributions are shown for the 11th–13th and 14th–16th centuries respectively, the annotations next to the plots indicating the relevant century. When the strength spans two ranges it is plotted as the lower range, for example T3/4 is plotted in orange. Where the beginning and end of the track are known only the starting point is shown. A few early tornadoes are of unknown location (e.g. 'somewhere in England') and cannot therefore be plotted geographically; these are indicated in the upper right insets where applicable, in which the dots represent the number of such tornadoes by strength in the relevant century. The Shetland Isles have been omitted from the maps because no historical tornadoes are known for this area, and the one historical tornado (of unknown strength) in the Channel Islands (19th century) is not shown. The red plot in Lincolnshire in Figure 2.4c (17th century) marks what is probably the severest tornado on record for the British Isles at force T8/9 (Welbourn, 23 October 1666).

2.6   Examples of Historical Tornado Reports

We now proceed to give a few sample tornado reports from past centuries to show how the style of description evolves over time.

[3] Further details are presented in Chapter 4.

1. AD 1054 April 30 (April 24 Old Style), Rosdalla, County Westmeath

    Although there are one or two earlier reports that might have been tornadoes, for example AD 991 Lough Hackett, County Galway (O'Donovan, 1856), the first one deemed certain enough to count in the records occurred (also in Ireland) at Rosdalla, County Westmeath, probably on 30 April 1054. It is described in *Chronicum Scotorum*, translated from the Irish by W. M. Hennessy (1866):

    A tower of fire was seen at Ross-Deala, on the Sunday of the festival of [Saint] George, during the space of five hours; black birds innumerable going into and out of it; and one large bird in the middle of it; and the little birds used to go under its wings when they went into the tower. They came out and lifted up, into the air, the greyhound which was in the middle of the town, and let it fall down again, so that it died immediately; and they lifted up three garments, and let them down again. The wood, moreover, on which the birds perched fell under them, and the oak whereon the birds alighted was shaking, together with its roots in the ground.

    [Original]

    ｋｔ. Cloicceċ ceneꝺ ꝺꝼaicꝼin 1 Roꝼꝼ ꝺeala ꝺia ꝺomnaıǵ ꝼele ᵹiuꝛᵹi ꝼꝛıa ꝛe cuıᵹ núaiꝛ. eoın ꝺuꝼa ꝺıaıꝛmıċċe ınn ocuꝛ aꝼꝼ, ocuꝛ aoın en móꝛ ına meꝺon ; ocuꝛ ceᵹoıꝼ ꝼo a cluımꝼıꝺe na hén beᵹa an can ceᵹoıꝼ ıꝛın cloıcceċ. ᴛancuccuꝛ amac ᵹuꝛ coᵹbaccuꝛ an coın baoí ꝼoꝛ láꝛ an baıle ınaıꝛꝺe ıꝛın aıeꝛ, ec caꝛlaıᵹꝛıc é ꝛıuꝛ conꝺeaꝛbaılc ꝼo ceꝺoıꝛ, ec cuaꝛᵹaꝛaccuꝛ cꝛı bꝛuıc ꝺılenꝺ ınaıꝛꝺe, ocuꝛ ꝛa leᵹꝛıoc ꝼíuꝛ ꝺoꝛıꝺıꝛı. aⁿ caıll ıaꝛum ꝼoꝛꝛan ꝺeꝛeccuꝛ na heóın ꝺa ꝛoċaıꝛ ꝼoıċıꝺ, ocuꝛ an ꝺaıꝛꝺe ꝼoꝛꝛan ꝺeꝛıoccuꝛ na heoın, ꝛo baoí ꝼoꝛ cꝛıoċ cona ꝼꝛemaıꝺ hı calmann.

    At first sight this might not seem much like the description of a tornado, but Rowe (1989) has shown that the references to 'tower of fire' (funnel cloud), 'birds' (swirling debris or scud cloud) and the lifting of objects can confidently be interpreted as evidence of a tornado. He also clarifies the date, which is given variously as 1052, 1054 or 1055 in different sources. The impossible duration of 5 hours (9 hours in another source) is simply an error, either of transcription or for 'at the fifth (or ninth) hour' of the day. A slightly embellished version of the same account is given by Joyce (1911) in which the oak tree is said to have been 'torn by the roots from the earth'. Rosdalla (or Ross-Deala) was in what is now the parish of Durrow, between Killbeggan and Tullamore. (There is insufficient information to assign a T-strength to this case.)

2. 1141 May 19 (May 12 Old Style), Wellesbourne, Warwickshire. T5/6

The language of the Irish example above is fanciful, almost supernatural, and the chronicler has no notion that he is describing a whirlwind. A century later we have a direct reference to a whirlwind (*ventus turbinis*) in *Florentii Wigorniensis Chronicon*, by John of Worcester, edited by B. Thorpe (1849); we are indebted to Rowe (1999) for the following translation from the Latin:

[1141] On the fourth day before the octave of the Ascension of the Lord about the ninth hour of the day, at a village called Wellesbourne, distant one mile from Hampton, a village of the Bishop of Worcester, a violent whirlwind arose, and a most foul darkness reached from the earth to the sky. Striking the house of a priest called Leofred, it razed his outbuildings to the ground and

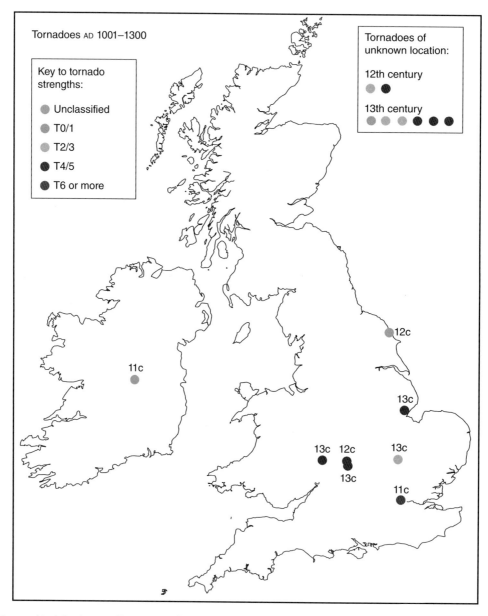

Figure 2.4   *(a) Geographical distribution of known tornadoes by strength for 11th–13th centuries (see text for the meaning of the upper right inset). (See insert for colour representation of the figure.)*

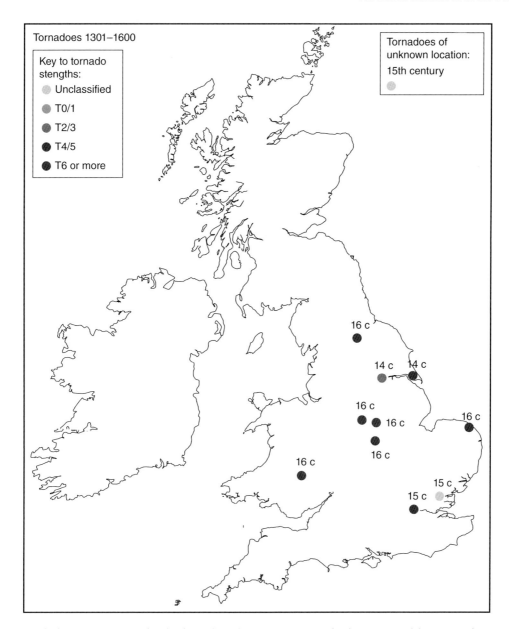

Figure 2.4 *(Continued) (b) as in Figure 2.4a but for the 14th–16th centuries (see text for the meaning of the upper right inset). (See insert for colour representation of the figure.)*

smashed them to pieces. It also removed the church roof and threw it across the River Avon. It also threw down in a similar way almost 50 houses of the peasants and left them uninhabitable. There also fell hail up to the size of pigeons' eggs, by the blows of which a woman was killed. At this spectacle all present were filled with terror and dismay.

[Original]
Siquidem quarta feria ante octavam Ascensionis Dominicæ [11 Maii], circa nonam diei horam, apud villam quæ Walesburna dicitur, distans ab Hamtonia, episcopi Wigornensis villa, miliario uno, ventus turbinis vehemens exortus est, et caligo teterrima, pertingens a terra usque ad cœlum, et concutiens domum presbyteri, cui nomen Leovredus, et officinas ejus, omnes solo tenus prostravit et minutatim confregit, tectum quoque ecclesiæ abstulit, et ultra Avenam flumen projecit. Domus etiam rusticorum fere L. simili modo dejiciens, inutiles reddidit. Grando quoque ad magnitudinem ovi columbini cecidit, cujus ictibus percussa quædam fœmina occubuit. Hoc viso, qui affuerunt admodum exterriti fuerunt et conturbati.

Rowe (1975) explains the erroneous date of 11 May (Old Style) inserted by Thorpe in the Latin version above.

3. 1323 July 2? (June 24? Old Style), Cowick, East Riding of Yorkshire. T3
A not uncommon theme before the Enlightenment was for a whirlwind to be attributed to the Devil or other evil spirit, as

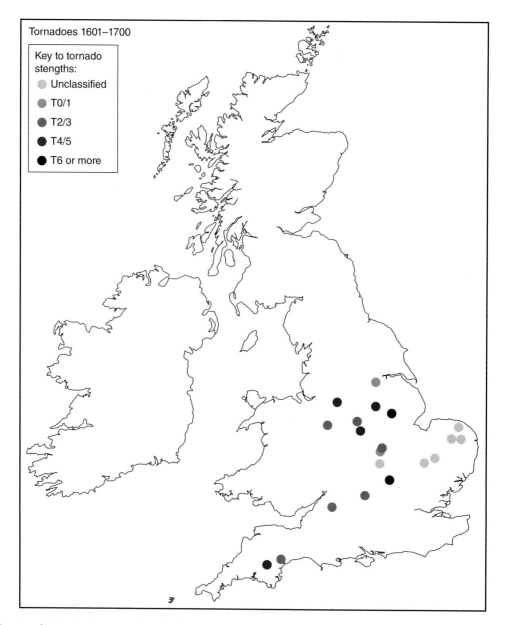

Tornadoes 1601–1700

Key to tornado
stengths:
○ Unclassified
○ T0/1
● T2/3
● T4/5
● T6 or more

Figure 2.4 *(Continued) (c) as in Figure 2.4a but for the 17th century. (See insert for colour representation of the figure.)*

in this account from the *Flores Historiarum* (various authors), edited by H. R. Luard (1890).

[1323] About the festival of St John the Baptist [24 June] an evil spirit ... in a furious storm and dark whirlwind struck Cowick, not far from Selby, while the King [Edward II[4]] was there with his favourite, the younger Hugh Despenser. Firstly, at the ninth hour, while the King and his followers were at the breakfast table, the storm covered the place with a darkness like night, so that in the hall guests could hardly see. Secondly, it tore up by

the roots oaks of wonderful size, and other trees, twisting some in two at the middle, some it split from top to bottom and all the debris was carried aloft. Thirdly, there was a horrible splashing of the encircling waters ... such that the water was lifted from the bottom of the channel and sprinkled beyond the houses of the manor ...

[Original]
Spiritus vero nequam, caput totius iniquitatis, qui sibi obtemperantibus quandoque fallaciter assistit blanditiis demulcendo, quandoque terrores timores incutit formidabiles comminando, hic igitur circa festum sancti Johannis Baptistæ in tempestate furoris et turbine tenebroso manerium de Couwik, non longe distans a cœnobio de Selby, ubi rex cum nequissimo consiliario suo Hugone

---

[4] Edward had by this time attracted opprobrium on account of his male lovers, and the wording of the original Latin text reflects, more so than our translation, the chronicler's disapproval of the king's licentiousness.

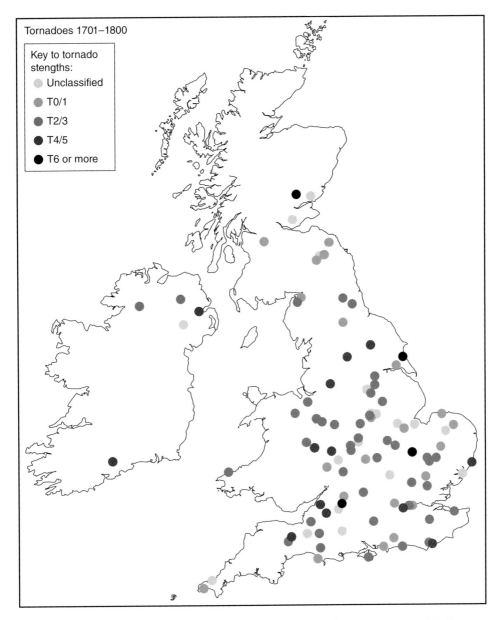

Figure 2.4 *(Continued) (d) as in Figure 2.4a but for the 18th century. (See insert for colour representation of the figure.)*

Dispensario juniore tunc temporis degebat, his modis aggressus est. Primo locum illum et manerium hora nona, rege cum suo parasito in mensa prandente, obtexit subito caligine tenebrosa; ita quod in aula ministrantes præ nimia obscuritate vix discumbentes possent intueri. Irruit etiam secundo in ipsa nigredinis densitate cum tanto impetu, ut quercus miræ magnitudinis cæterasque arbores radicitus avulsit, quasdam vero in duo per medium distorquens et quasdam a summo usque deorsum findebat, ac omnia in altum aera volando sublevans frustratim confugit. Tertio horribili fragore in aquis circumfluentibus impetum suæ nequitiæ tam violenter ab alto præcipitavit, ac ipsas aquas a profundis alveorum ferens in sullime ac ultra domos manerii et circa crepidinem annorum respergens, ut visibus intuentium profunda gurgitum aquis videbantur carere. Talibus ergo signorum indiciis rex cum suis turbatus et conterritus locum continuo deseruit, vita pæne desperatus.

4. 1558 July 17 or 21 (July 7 or 11 Old Style), Sneinton, Nottinghamshire. T7
   By the 16th century we have original accounts in English (with old spellings), as in the following from Holinshed's *Chronicles* (1586); other sources, for example Stow (1615), give the date as either 7 or 11 July Old Style:

   Also this yeare [1558] within a mile of Nottingham, was a marvellous tempest of thunder, which as it came through two townes, beat downe all the houses and churches, the bels were cast to the out side of the churchyards, and some webs of lead foure hundred foot into the field, writhen like a paire of gloves. The river of Trent running betweene the two townes, the water with the mud in the botome was carried a quarter of a mile, and cast against the trees, the trees were pulled up by the roots and cast twelve score [feet?] off.

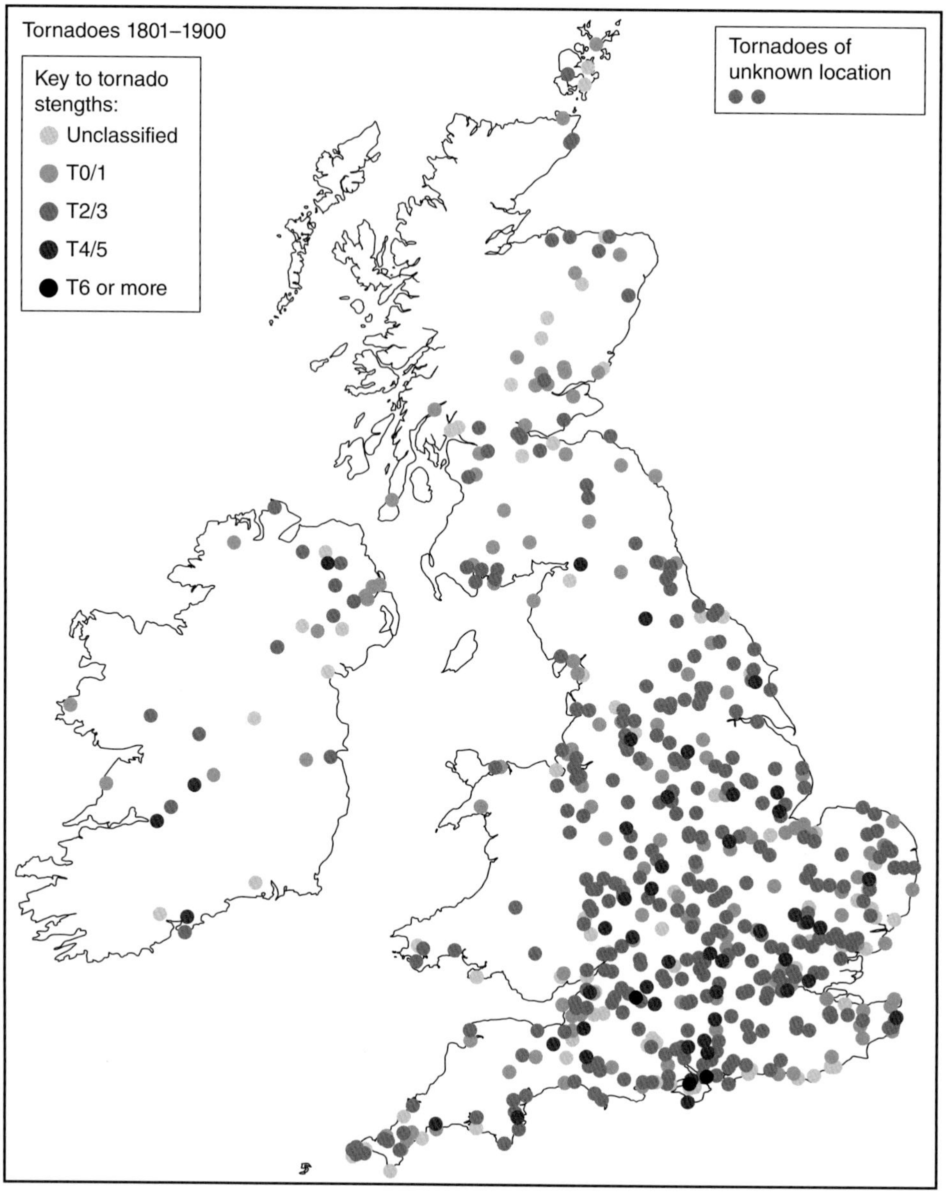

Figure 2.4   *(Continued) (e) as in Figure 2.4a but for the 12th century (see text for the meaning of the upper right inset). Contains Ordnance Survey data © Crown copyright and database right 2015. (See insert for colour representation of the figure.)*

Also a child was taken forth of a mans hands two speares length hie, and carried a hundred foot off, and then let fall, wherewith his arme was broken, and so he died. Five or six men thereabout were slaine, and neither flesh nor skinne perished; there fell some haile-stones that were fifteene inches about, &c.

5.   1669 November 9 (October 30 Old Style), Ashley, Northamptonshire. T3
   After the mid-17th century some quite detailed accounts begin to appear. The following is by J. Templer (1671) in *Philosophical Transactions of the Royal Society* (original spellings and italics retained):

   Octob. 30. 1669. Betwixt five and six of the Clock in the evening, the wind Westerly, at *Ashley* in *Northamptonshire* happen'd a formidable

Hurrican, scarce bearing sixty yards in its breadth, and spending itself in about seven minutes of time. Its first discern'd assault was upon a Milk-maid, taking her pail and hat from off her head, and carrying her pail many scores of yards from her, where it lay undiscover'd some days. Next, it storm'd the Yard of one *Sprigge*, dwelling in *Westthorp* (a name of one part of the Town,) where it blew a Waggon-body off the axle-trees, breaking the wheels and axle-trees in pieces, and blowing three of the wheels so shatter'd over a wall. This waggon stood somewhat cross to the passage of the wind. Another waggon of Mr. *Salisburies* marched with great speed upon its wheels against the side of his house to the astonish-ment of the inhabitants. A branch of an *Ash-tree* of that bigness that two lusty men could scarce lift it, blew over Mr. *Salisburies* house without hurting it, and yet this branch was torn from a Tree, an hundred yards distant from that house. A Slate was forced upon a window of the house of *Samuel Templer*, Esq;, which very much bent an Iron-bar in it; and yet 'tis certain, that the nearest place, the Slate was at first forc'd from, was near two hundred yards. Not to take notice of its stripping of several houses; one thing is remarkable, which is, that at Mr. *Maidwells* Senior it forced open a door, breaking the latch, and thence marching through the entry, and forcing open the Dairy-door, it over-turn'd the milk-vessels, and blew out three panes or lights in the window; next it mounted the Chambers, and blew out nine lights more. From thence it proceeded to the Parsonage, whose roof it more than decimated; thence crosseth the narrow street, and forcibly drives a man headlong into the doors of *Thomas Briggs*. Then it passed with a cursory salute at *Thomas Marstons*, down to Mr. *George Wignils*, at least a fourlongs distance from *Marstons*, and two fourlongs from *Sprigg's*, where it plaid notorious exploits, blowing a large hovel of pease from its supporters, and setting it cleverly upon the ground, without any considerable damage to the thatch. Here it blew a gate post, fixed two foot and an half in the ground, out of the earth, and carried it into the fields many yards from its first abode.

6. 1741 September 19 (September 8 Old Style), Bluntisham, Cambridgeshire. T6
    Another detailed account from *Philosophical Transactions of the Royal Society* is by S. Fuller (1739–1741):

Cambridge, Sept. 9. 1741
Yesterday was the most violent Hurricane of Wind in these Parts, that ever was known since the Memory of Man. *Cambridge* was not in the midst of the Hurricane, so that it has escaped very well. I happened to be paying a Visit to Dr. *Knight*, a Cotemporary of yours, of our College, who lives at *Bluntsham* in *Huntingtonshire*, about 10 Miles North-west of *Cambridge*. We were in the midst of the Hurricane; but, by getting into the strongest Part of the House, we escaped without any great Danger. The Morning, till half an Hour after Eleven, was still, with very hard Showers of Rain: At half an Hour after Eleven it began to clear up in the South, with a brisk Air, so that we expected a fine Afternoon: The South-west cleared up too, and the Sun shining warm drew us out into the Garden. We had not been out above 10 Minutes, before we saw the Storm coming from the South-west: It seemed not to be 30 Yards high from the Ground, bringing along with it a Mist, which rolled along with such incredible Swiftness, that as near as we could guess, it ran a Mile and an half in half a Minute: It began exactly at 12 o'Clock, and lasted about 13 Minutes, Eight Minutes in full Violence: It presently unhealed the House we were in, and some of the Tiles, falling down to Windward, were blown in at the Sashes, and against the Wainscot on the other Side of the Room; the broken Glass was blown all the Room over, the Chimneys all escaped; but the Statues, which [were] on the Top of the House, and the Balustrades from one End to the other, were all blown down. The Stabling was all blown down, except Two little Stalls, where, by the greatest Fortune in the World, stood my Horse and the Doctor's. All the Barns in the Parish, except those that were full of Corn quite up to the Top, were blown flat upon the Ground, to the Number of about 60. The Dwelling-houses escaped to a Miracle; there were not above a Dozen blown down out of near 100. The Alehouse was levelled with the Ground, but by good Luck not a Soul in it. If the Storm had lasted Five Minutes longer, almost every House in the Town must have been down; for they were all, in a manner, rocked quite off from their Underpinnings. The People all left their Houses, and carried their Children out to the Windward Side, and laid them down upon the Ground, and laid themselves down by them; and by that means all escaped, but one poor Miller, who went into his Mill to secure it against the Storm, and was blown over, and

crushed to Death betwixt the Stones and one of the large Beams: I saw him taken out. All the Mills in the Country are blown down: I do not hear of any more bodily Mischief; only one Miller at *Willingham*, so much bruised, that they hardly expect his Life. Hay-stacks and Corn-stacks are some quite blown away, some into the next Corner of the Field. The poor Pigeons, that were catched in it, were blown down upon the Ground, and dashed to Pieces; one of which I found, myself, above half a Mile from either House or Hedge. Where-ever it met with any boarded Houses, it seemed to exert more than ordinary Violence upon them, and scattered the Wrecks of them for above a quarter of a Mile to the North-east, in a Line: I followed one of these Wrecks myself; and, about 150 Yards from the Building, I found a Piece of a Rafter, about [lacuna] Feet long, and about Six Inches by Four, stuck upright Two Feet deep in the Ground; and at the Distance of 400 paces of my Horse, from the same Building, was an Inch Board, Nine Inches broad, 14 Feet long: I am convinced, that these Boards were carried up into the Air; for I saw some, that were carried over a Pond above 30 Yards; and I saw a Row of Pales, as much as Two Men could lift, carried Two Rods from their Places, and set upright against an Apple-tree. Pales, in general, were all blown down, some Posts broke off short by the Ground, others torn up by the Stumps. The whole Air was full of Straw: Gravel-stones, as big as the Top of my little Finger, were blown off the Ground in at the Windows; and the very Grass was blown quite flat upon the Ground. After the Storm was over, we went out into the Town, and such a miserable Sight I never saw: The Havock I have described; the Women and Children crying, the Farmers all dejected; some blessing GOD for the Narrowness of their Escape, others wondering how so much Mischief could be done with one Blast of Wind, which hardly lasted long enough for People to get out of their Houses. I talked to Two People, that were out in it all the Time, who said, that they heard it coming about half a Minute before they saw it; and that it made a Noise something resembling Thunder, more continued, and continually increasing.

I saw a Man in the Afternoon, who came from *St. Ives*, who says, the Spire of the Steple, which is one of the finest in *England*, is blown down, as is the Spire of *Hemmingford*, the Towns having received as much Damage as *Bluntsham*. There was neither Thunder nor Lightning with it, as there was at *Cambridge*, where it lasted above half an Hour, and consequently was not so violent. Some few Booths in *Sturbridge-fair* were blown down. The Course of the Storm was from *Huntington* to *St. Ives*, *Erith*, between *Wisbich* and *Downham* to *Lynn*, and so on to *Suetsham* [Snettisham]: We have heard nothing of it farther to the South-west than *Huntington*, nor farther North-east than *Downham*. Very few Trees escaped: The Barns that stood the Storm, had all their Roofs more damaged to the Leeward Side than to the Windward. We are in great Hopes the Storm was not general; I am apt to think it was much such a Storm as ran through *Sussex* about 10 Years ago.[5] The Storm was succeeded by a profound Calm, which lasted about an Hour; after which the Wind continued pretty high, till 10 o'Clock at Night.

7. 1760 September 23, Curriglass, County Cork. T4
    Also in the 18th century we begin to get useful reports in newspapers. At first these are usually contributed by correspondents but later they are more often written by the journalists themselves. The following is from the *London Evening Post* of 16–18 October 1760:

[5] [Bexhill Down, 31 May 1729.]

An Extract of a Letter from the County of Corke.

On Tuesday the 23d of September last, at Break of Day, there were terrible Claps of Thunder near Curriglass in this County, and soon after a most violent Hurricane arose, which did not continue above two or three Minutes. It came from the South, or rather from the South South West. The Track in which it ran did not exceed four-score Yards in Breadth, in some Places it does not appear to have been so much. But where-ever this impetuous Stream or Current of Air came it bore down all before it; many Houses have been greatly damaged, one House and Barn near us, have been entirely blown down, and the Thatch and Wattles of the Roof carried to a Quarter of a Mile's Distance. Ricks of Faggots have been overturned, and scattered along the Fields. Stacks of Hay and Corn have been carried away, and the greater Part irrecoverably lost. One may trace the Course, the Storm went by Hay, Straw, and Thatch, which lie scattered along the Country, and the Grass, Stubble, Fern, &c. are laid flat on the Ground along the Path which the Tempest took, as if a rapid Stream of Water had run over it. The Church at Mogealy has had a narrow escape, the Storm passed just by the East End of it, and has only shattered one of the Windows, but has carried away the Tops of two Ash-Trees which grew on the Church-Yard Ditch. Where the Tempest began, or where it ended, is more than I can tell. I traced it about six Miles, and find it went over the Black Water towards the Arigan Mountains; in my Way I met with an high Furze Ditch, which run East and West in that Part of it against which the Hurricane came, the Furze are entirely torn up by the Roots: I measured the Chasm or Gap which was made, and found it to be about 33 or 34 Yards wide.

8.   1859 December 30, Calne, Wiltshire. T6

The reader will have noticed that the word *tornado* appears in none of the examples quoted so far, and even *whirlwind* occurs in only a few. The favoured word for a tornado among the men of letters in the 17th and 18th centuries seems to have been *hurricane*. In the following very long press report from the *Devizes and Wiltshire Gazette* of 5 January 1860 (from which some repetition and irrelevancies have been omitted) both *hurricane* and *tornado* are used interchangeably[6]:

THE HURRICANE AT CALNE

On Friday there was a complete hurricane at Calne, preceded by a tremendous hailstorm—the pieces of ice which fell being so large and sharp as to cut the hands of some of those who were so unfortunate as to be exposed to them. Rain, too, afterwards fell in torrents, accompanied with thunder and lightning, the flashes following one another in rapid succession, and rendering the darkness during the intervals still more appalling.

With the rain and hail the violence of the elements increased; and it was not long before the hurricane arose, and in a few minutes effected a destruction which, to be credited, should be witnessed. Its power, however may perhaps be imagined when we state that stately elm trees of a girth of from 6 to 10 feet were snapped asunder; others uprooted, tearing up some 10 tons or more of earth with them; and others, again, completely dismembered—the bare trunks alone standing in gloomy solitariness. 'The monarch oak, the patriarch of the trees', appears alone to have possessed sufficient

strength for this encounter with the elements, and its strength was severely tested. Many of its branches were completely twisted, but still it nobly braved the storm. Buildings, too, that happened to stand within its course, were swept away as chaff before the wind …

This tornado appears to have commenced in a field belonging to the Marquis of Lansdowne, about a mile from Calne on the Devizes turnpike road … Its course was then towards the Rookery Farm, over Quemerford Villa … It then bore on to Blacklands … Its seemingly capricious course can then be traced circularly towards Cherhill … In passing over the open land to Yatesbury, the fury of the storm seemed, if possible, to increase, rolling and roaring onwards, according to an eye-witness, like a tremendous eddying wave, and bearing aloft everything before it … The storm passed on, leaving devastation behind and carrying destruction to everything in its course. Half a dozen men, working in a barn which was blown down, belonging to Mr. Eyles of Monkton, were buried some time beneath the *debris*, but were fortunately extricated uninjured.—Other farm buildings, with two labourer's cottages, and a skilling containing ten oxen, belonging to the same gentleman, were also destroyed, the oxen escaping. The Church suffered so severely, as to prevent the services being performed on Sunday; a quantity of the roofing was carried away, and one of the windows shattered. Of the old Vicarage house, but three rooms remain standing, and the village school has lost its bell turret. Mr. Reed's house and many of his ricks were much damaged. About 50 large elms were torn up, some of which were laid in opposite directions. The blast continued its course over the Temple farm to Ogbourne, where it seems to have expended itself, after having effected a sweep of more than 10 miles, taking in a breadth of 3 to 400 yards.

We have been favoured with the following from our correspondent at Calne, Mr. Graham, the master of the Middle School:

SATURDAY

The neighbourhood of Calne was yesterday visited by one of the most destructive storms ever witnessed. It appears to have been a regular tornado, having a curvilinear motion, and progressing at a rapid rate. Symptoms of an approaching storm were visible at 1 o'clock p.m. The whole atmosphere became thick and heavy. It was so dark that it was scarcely possible to read without artificial light. Presently the vault of heaven was lit up by vivid flashes of lightning accompanied by loud and sudden claps of thunder, which together with the big hail stones falling thick and fast, and the roaring of the mighty wind, produced a scene at once awful and sublime. The whirlwind appeared to commence on the outskirts of Bowood-park, near Quabb's farm, passing on to the Rookery Farm, marking its path by trees blown down and others snapped off in the middle, thence to Quemerford-villa, carrying away the roof of the stable; but when it reached Mr. Slade's mill at Quemerford, it carried away the chimneys, stripped off the spouting and lead work, and blew some of the slate tiling at least 400 yards from the spot. The adjoining skilling, in which cattle were feeding, was literally deprived of its roof, and the stone tiles of which it was composed strewn about the road in pieces as fine as could be produced by a stonebreaker's hammer. Pieces of freestone coping a yard in length were blown away, and a large tree near the millstream is torn completely out of the earth, having at least 30 square feet attached to its roots. At Blackland-park, belonging to Mr. Marshall Hall, the destruction of property was immense. Part of the roof of the lodge was blown away. Hundreds of trees are lying on the ground, and it is a remarkable fact that trees 8 and 10 ft. in circumference were snapped in sunder like

---

[6] Purists might object that *hurricane* is here misused, but such objection is ungrounded. Although the word began as the term for a tropical storm of the Caribbean, its use has long since been broadened to mean any exceptionally strong wind, in addition to the more restrictive definition placed upon it by meteorologists.

matchwood; while others, especially the heavy-topped firs, 30ft. in length, were blown out of a plantation across the turnpike-road, and into an adjoining field. The park and plantations form a complete scene of desolation. Twenty or 30 men were engaged all the afternoon clearing the road, which was quite blocked up. The property of G. H. Walter Heneage esq., at the Hail Farm, which is in the occupation of Mr. Arnold, sustained much damage. The skilling, a very substantial building, was levelled to the ground, and a fine row of elms, the growth of a hundred years, are dragged up by the roots and spread abroad in wild disorder. At Cherhill, about three miles from Calne, the storm was terrific. Whole ricks of wheat were carried away, and some of the sheaves to such a distance as to render it extremely doubtful whether the proper owner will ever see them again. A large tree fell across the shed, and the waggon it contained was cut in two. Several cottages were more or less blown down. We are happy to state that amid all this destruction not a single life, either of man or beast, has been lost. One old man had a narrow escape. He had gone under a small shed for shelter, when a large tree came down with a crash across the roof; but, strange to say, the poor fellow crept out of the *debris* safe and sound. One little outhouse at Yatesbury was borne quite away, not a vestige of it to be seen. The amount of damage done to barns and ricks belonging to Mr John Tanner, cannot be less than £1,000. Such a storm was never before known. It is quite impossible at present to state accurately the full amount of damages sustained.

MONDAY

Since writing the above account, which appeared in this morning's *Times*, I have carefully examined the line of route taken by the storm, and find the devastation greater than any one can imagine.

There is scarcely a single house in Cherhill but has suffered in some measure. Some noble old trees have been blown down near the church. Mr. John Neate has suffered severe losses. There were 8 cows in a small place enclosed by hurdles behind Mr. Neate's house, and across this place several trees fell; but the cattle remained uninjured. The noise of the storm at this place is described as being so great that it was impossible to hear the crashing of the falling trees, though they were within a few yards of the door. From the quantity of hay and straw blown about from the neighbouring ricks and thatched roofs of the cottages, the atmosphere became so thick and hazy as to resemble a heavy fall of snow, thus producing partial darkness. The roof of one cottage was completely and neatly lifted from its four walls … Passing on to the village of Yatesbury we find the cottages here present even a more dilapidated appearance than at Cherhill … The whirlwind left the Rectory undisturbed, and travelled along the fir plantation on the right, fancifully selecting here and there one, which was either snapped in two or dragged up by the root. But it is at Mr. Tanner's where the work of destruction appears to be greatest. Here two large barns were blown down, from the ruins of which a boy was extricated, fortunately more frightened than injured. The windows of Mr. Tanner's house were blown in or otherwise smashed by the flying boughs. The wind entered with a yelling roar, causing the tables to cut fantastic capers around the room. It will scarcely be credited, yet we have it from the best authority, that a cow was raised from the ground and placed down again in the middle of an adjoining pond. One of Mr. Tanner's waggons was lifted out of a field over a hedge 15 feet high into the barn yard, clearing a cart at the same time, which was under the hedge. The trees and hedge-rows have a most picturesque appearance, being clothed with hay and straw, blown from the different ricks. It is almost impossible, by words, to give a correct notion of the effects produced by this storm: they must be seen to be fully realized …

For the following we are indebted to the Rev. A. C. Smith, the highly-respected Vicar of Yatesbury:

It may interest your readers to hear some particulars of a most furious tornado which swept over a long, but narrow district of our Wiltshire downs on Friday last (Dec. 30th), and though of scarcely three minutes duration, and extending but 300 or 400 yards in breadth, swept a clear and most perceptible path in its onward progress—tearing up by the roots, and snapping short off some of the largest elms and other trees, unroofing barns, stacks, and cottages, and hurling men and cattle to the ground, and dashing them furiously to and fro, and rolling them over and over in its rough embrace.

It occurred at about half-past one p.m., and began its devastation about a mile to the south of Calne, coming up from the west. Its course was then for about ten miles in nearly a straight line for E.N.E., passing through Quemerford, Blacklands Park, (where it demolished about 150 large trees), the villages of Cherhill, Yatesbury, and Monckton, when ascending the highest ridge of the Marlborough Downs, it terminated its career at Ogbourne, either having by that time expended its fury, or shaped its course for the open track of country, where it could leave no trace of its visit. Those who saw it coming up over the open down describe it as a thick volume of smoke or a dense cloud rushing through the air; and so appalling was its appearance, and so terrific the roar of its approach, that all the villagers retreated within their houses, apprehending some unwonted catastrophe. In an instant the whirlwind was upon them, ushered in by a most vivid flash of lightning, and an instantaneous clap of thunder; and so sudden and furious was its onset, so loud and deafening its roar, so strange and unearthly the darkness, (not unlike that attending the annular eclipse of the sun last year) so terrific the crash of falling roofs, (tiles and rafters and thatch seeming to fill the air, while the windows were beaten in by hailstones of an unusual magnitude), that many thought the judgment day had arrived, and others believed an earthquake was demolishing their houses.

But in less time than it will take to read the account of it the hurricane had passed by; and then when the frightened villagers emerged from their cottages, what a sight met the eye on all sides! the largest trees torn up by the roots, large branches snapt off and carried on twenty yards from where they fell; large barns in ruins, ricks demolished, their own homes unroofed, and their gardens filled with straw; fallen chimneys and tiles; and all this havoc effected in three minutes of time!

But though the storm passed through three villages in its course, though it occurred in the very middle of the day … not a single life was lost, nor did any serious accident occur to either man or beast; for though several men and boys were buried under the ruins of fallen barns they were promptly extricated from their perilous position, with no worse result than sundry bruises and an exceeding terror, and even the large cart horses which were whirled across the yards and then dashed against the sheds, and the cows which were thrown down into the ponds, appear none the worse for their conflict with the storm.

One remarkable peculiarity attending the progress of this hurricane, and proving it to have been a *tornado* in the true sense of the word, is the strange position in which it left the trees it had uprooted, and the eccentric partiality it shewed in its attacks: thus some of the largest trees which stood close together lie facing one another, proving that they were blown down from diametrically opposite directions; and some of the houses, though in the direct line of the storm, exhibit their western fronts perfectly unscathed, while the opposite sides are beaten in and the roofs carried off; and again in a square garden, the east and west walls were both blown inwards, and the east no less than the west sides of the corn stacks were hurled to the ground,—the wind appearing to have wrapped round them, and carried off the sheaves in all directions, and to an incredible distance.

The scene of desolation presented by Blacklands Park and the three villages on which the hurricane expended its fury, is indeed sad to behold, and the loss of property must be very considerable, in addition to the irreparable loss of a vast number of large trees which no money can replace, and which are especially valuable in the bleak down district, where they are so much needed for shelter from the wind. But though in these high exposed situations we are often assailed by boisterous breezes, yet in the memory of the oldest inhabitant no tradition of anything resembling such a storm as that of Friday exists, with the single exception of a like whirlwind (though perhaps of scarcely so violent a nature) which devastated the adjacent village of Cliffe Pypard, about four years since.[7] These two storms, occurring in almost the same district, and within so short a time … seem to be phenomena which deserve the attention of the scientific.

No description can at all fully depict the amazing force of this hurricane, the effects of which must be seen to be appreciated; though perhaps a few details may in some degree show its fury:— Thus many of the fine elms at Blacklands, the Hale Farm, and Cherhill exhibit their trunks broken off in the middle, their heads and large limbs evidently twirled round by the whirlwind like the twisting of a withe; at Yatesbury, a large broad-wheeled waggon was taken up by the wind and whirled over a high hedge, and deposited on its side a dozen yards or more from where it stood, and Mr. Tanner from his window saw his large cart horses carried off by the storm from one end of the yard to the other; while at Mr. Eyles' farm at Monckton the substantial roof of a long and perfectly new cattle shed was lifted off the walls which supported it in a solid mass.

These few facts may, perhaps, in some slight measure, mark the extraordinary violence of this tempest; but while we marvel at its power, and mourn over the desolation it has caused, and sympathise with the sufferers, by far the foremost of whom is my friend and neighbour Mr. Tanner, we must not omit to mark the wonderful Providence which preserved life amid such great dangers, nor forget that such awful visitations are sent as warnings by Him who 'walketh on the wings of the wind.'

It must be acknowledged that most newspaper accounts are not as admirably detailed as this one, but then most tornadoes do not reach force T6.

2.7   Concluding Remarks

In this chapter we have explained some of the problems of identifying tornadoes from historical sources. We have considered the sources available to the modern researcher, and how best to interpret the language and style of the old records in present-day terminology. In illustration of this we have quoted a few examples of tornado reports from previous centuries. We have also noted that despite the diligent (and continuing) efforts of TORRO in this field, our historical tornado data remain unavoidably incomplete relative to those for recent times. Nonetheless, the TORRO database contains the most complete listing of historical (and current) tornadoes available for the British Isles.

References

Fuller, S. (1739–1741) Concerning a violent hurricane in Huntingtonshire [*sic*], Sept. 8. 1741. *Philos. Trans. R. Soc.*, **41**, 851–855.

Hennessy, W. M. ed. (1866) *Chronicum Scotorum*, 280–281. London. Longmans, Green, Reader, and Dyer.

Holden, J. and Wright, A. (2004) UK tornado climatology and the development of simple prediction tools. *Q.J.R. Meteorol. Soc.*, **130**, 1009–1021.

Holinshed, R. (1587) *The Chronicles of England, Scotlande and Irelande*. 2nd ed. Vol. **3**, 1142. London.

Joyce, P. W. (1911) *The Wonders of Ireland*, 32–33. London. Longmans, Green.

Luard, H. R. ed. (1890) *Flores Historiarum*. Vol. **3**, 216–217. London. HMSO.

Matthews, P. (2003) An investigation into the influence of the Isle of Wight on waterspout and tornado frequency and the effect of population densities on recording events. *J. Meteorol.*, **28**, 213–225.

Murray, J. A. H. and Craigie, W. A.; Bradley, H. and Onions, C. T. *A New English Dictionary on Historical Principles [Oxford English Dictionary]*. Vol. 10, parts 1 and 2 (1926, 1928). Oxford. Clarendon Press.

O'Donovan, J. (1856) *Annals of the Kingdom of Ireland by The Four Masters*. 2nd ed. Vol. **1–2**, 727. Dublin. Hodges, Smith.

Rowe, M. W. (1975) The dates of two early British tornadoes. *J. Meteorol.*, **1**, 103.

Rowe, M. W. (1989) The earliest documented tornado in the British Isles: Rosdalla, County Westmeath, Eire, April 1054. *J. Meteorol.*, **14**, 86–90.

Rowe, M. W. (1999) 'Work of the Devil': tornadoes in the British Isles to 1660. *J. Meteorol.*, **24**, 326–338.

Stow, J. (1615) *The Annales of England*, 634. London.

Templer, J. (1671) A relation of two considerable hurricans [*sic*], happen'd in Northampton-shire. *Philos. Trans. R. Soc.*, **6**, 2156–2157.

Thorpe, B. ed. (1849) *Florentii Wigorniensis Chronicon*. Vol. **2**, 131. London. Sumptibus Societatis.

---

[7] [22 September 1856.]

3

Supercell and Non-supercell Tornadoes in the United Kingdom and Ireland

Matt Clark[1,2] and David Smart[3,4]

[1] Tornado and Storm Research Organisation (TORRO)*, Exeter, UK
[2] Met Office, Exeter, UK
[3] Tornado and Storm Research Organisation (TORRO)*, London, UK
[4] UCL Hazard Centre, University College London, London, UK

3.1 Introduction

The scientific documentation of convective storms in the United Kingdom began in earnest in the 19th century, often in the form of detailed eyewitness accounts of exceptional and destructive thunderstorms and related phenomena, including tornadoes and large hail. In the 20th Century, the introduction of new technologies such as radiosondes, weather radars and satellites allowed increasingly detailed observations to be made of storm environments, structures and life cycles. Since the late 20th Century, high-resolution numerical weather models have provided a wealth of new information, including insights into key aspects of storm dynamical behaviour, that could not have been gleaned from observational data alone.

A useful distinction can be drawn between storms that produce damage and those that do not. Indeed, understanding why some storms produce damage and others do not has been, and continues to be, one of the key aims of thunderstorm research. Where the differences are understood, the occurrence of damaging storms can be better predicted, and their negative impacts may be mitigated. Severe thunderstorms are defined by TORRO as thunderstorms that produce any of the following:

- One or more tornadoes
- Hail of diameter 20 mm or greater
- Wind gusts of 25 ms^{-1} (50 knots) or greater

* http://www.torro.org.uk/

Although these criteria are designed to distinguish between storms that produce damage and those that do not, the occurrence and severity of damage may also depend on whether these phenomena occur in isolation or together (e.g. hail near 20 mm in diameter is more likely to produce damage if accompanied by strong winds) and the duration of their occurrence at any one location. As flooding is a phenomenon that is not uniquely associated with convective storms (e.g. Hand *et al.*, 2004), the meteorological definition of a severe thunderstorm includes no criteria concerning rainfall rates or rainfall totals. However, in the United Kingdom and Ireland, as elsewhere, flooding comprises many more of the instances of thunderstorm damage than do tornadoes or large hail. Localised flooding is possible with any storm that produces heavy precipitation; the likelihood and severity of flooding depends on the intensity and duration of the rainfall (e.g. Doswell *et al.*, 1996), in addition to other non-meteorological factors such as the nature of the topography and the ground hydrology.

In this chapter, we describe the basic life cycle of convective storms and the organisational structure of different types of storm. We briefly discuss the influence that the storm type has on the likelihood of flooding, large hail, severe wind gusts and tornadoes. We then look specifically at storms that produce tornadoes in the United Kingdom and Ireland, as this has been at the core of TORRO's work since its inception. By considering the synoptic situations associated with tornadoes in these countries, in conjunction with radar data and other observations, some inferences are made about the proportion of tornadoes associated with different storm types. Case studies of some recent tornadic storms will be described, using observations and model simulations, in order to further illustrate the various types of tornadic storm and the different synoptic situations in which they occur.

3.2   Basic Structure and Life Cycle of a Storm Cell

Since the early days of their study, it was recognised that thunderstorms have a basic structure comprising an ascending current of relatively warm, buoyant air (the updraught) and a descending current of relatively cold, dense air (the downdraught), in which the heavy rain and hail fall (Byers and Braham, 1948; Byers and Battan, 1949; Ludlam and Scorer, 1953). Together, these form an idealised unit or *cell* of convection. Thunderstorm cells, by definition, produce lightning. However, they are only a subset of all precipitating convective cells. Some convective cells are too shallow, or too weak, to produce the separation of electrical charge required for lightning to occur, though their basic updraught–downdraught structure is the same as in a thunderstorm cell. A good example is the often heavy, but usually non-electrified, shower which commonly occurs over the United Kingdom in the autumn and winter months. Such non-electrified convection is often less vigorous than electrified convection and, with a few exceptions, usually does not produce damage.

Convective cells undergo a life cycle comprising three stages: *developing, mature and dissipating* (Figure 3.1). In most convective cells, this life cycle takes around 30–60 minutes to complete. In the developing stage, the cell consists entirely of an updraught – a bubble of relatively warm, moist air, which is more buoyant than the surrounding air and therefore rises. As the air rises, it expands and cools; water vapour within the air condenses, cloud droplets and eventually precipitation, begin to develop aloft. As the precipitation particles grow in size within the cloud, they can no longer be suspended by the updraught and start to descend towards the surface. Eventually, the loading of the updraught by precipitation and the cooling associated with melting and evaporation of some of that precipitation lead to the development of a downdraught. With the formation of a downdraught, the cell reaches its mature phase, during which the heaviest precipitation reaches the ground. As the cell reaches the dissipation stage, the downdraught becomes dominant, overwhelming the updraught and leading to the demise of the cell.

When the cold downdraught reaches the ground, it spreads out from the cell. Near the leading edge of this turbulent downdraught air (the *gust front*, so called because its passage is often accompanied by a distinct increase in wind speed), lifting of the surrounding air occurs. If conditions are favourable, this can lead to the triggering (*initiation*) of new cells.

3.3   Storm Mode: An Overview of Single-Cell, Multicell and Supercell Convection

The term *storm mode* describes the organisational structure of convective cells or groups of cells. Classification by mode is useful because the potential for severe weather, including flash flooding, large hail, damaging wind gusts and tornadoes, is different for

different storm modes. Therefore, if the storm mode is recognised, the potential for severe weather can be better predicted. Three basic modes exist: single cells, multicells and supercells (Markowski and Richardson, 2010). The storm mode is primarily controlled by the amount of vertical wind shear (i.e. changes in wind speed and/or direction with height) in the environment within which the storm is growing (e.g. Weisman and Klemp, 1984), though it may also be influenced by the nature of interactions between storm cells and the presence or absence of mesoscale features such as convergence zones. In practice, the distinction between modes is not always clear, since the structure of real storms is often complex and may evolve significantly as a storm goes through its life cycle, encounters differing environmental conditions or interacts with neighbouring storms. In general, however, in weakly sheared environments (i.e. relatively little change in wind speed or direction with height), single cells are favoured (Figure 3.2). Single cells consist of only one updraught–downdraught pair. Because of the relatively weak wind shear, the downdraught falls through the updraught as the cell matures, thereby weakening and eventually overwhelming the updraught (as in Figure 3.1c). As a result, single-cell storms tend to be short-lived. Severe weather is unlikely, though it is possible for a short window in the early mature phase of the cell's lifetime if the convection is particularly vigorous. Initiation of new cells may occur along the gust front of single cells, but there is no preferred flank for new development; consequently, where clusters of cells develop, they do not show well-organised structure.

In conditions of stronger vertical shear (greater than about 20 knots difference between winds at the surface and at 6 km above ground level), multicells are favoured. Multicells comprise a group of cells, of varying maturity, in which new cells are repeatedly triggered on a preferred flank of the existing cells, usually in close proximity to them (Figure 3.3). This results in an organised cluster of storms which is able to persist for much longer than any individual component cell. The interaction of the vertical wind shear with the gust front of existing cells is responsible for the initiation of new cells on the preferred flank of the existing cell. The cluster persists for as long as conditions remain favourable for the triggering of new cells along the gust front. Multicell clusters can grow to a much larger size than an individual convective cell. A Mesoscale Convective System (MCS) is a type of multicell storm in which thunderstorms form a contiguous precipitation area with a horizontal scale greater than or equal to 100 km in at least one direction (AMS, 2014). MCSs are a fairly rare occurrence in the United Kingdom. For example, Lewis and Gray (2010) found a mean annual frequency of 2.3 during 1998–2008. However, the inter-annual range was large; for example, no events occurred in 2002, while six occurred in 2001. The frequency of smaller clusters of multicell storms which do not meet MCS size criteria is likely to be considerably higher.

MCSs often exhibit a linear (*squall line*) structure, as the gust fronts of individual cells merge to form a continuous line; lifting and the triggering of new cells then occurs all along the leading edge, maintaining the line of convection. Tornadoes and locally damaging wind gusts occasionally occur along the leading edge of such systems. Figure 3.4 shows an example of an MCS which exhibited predominantly linear structure over southern England on 22 October 2013. The developing and newly mature cells, which are associated with the broken line of highest rainfall rates,

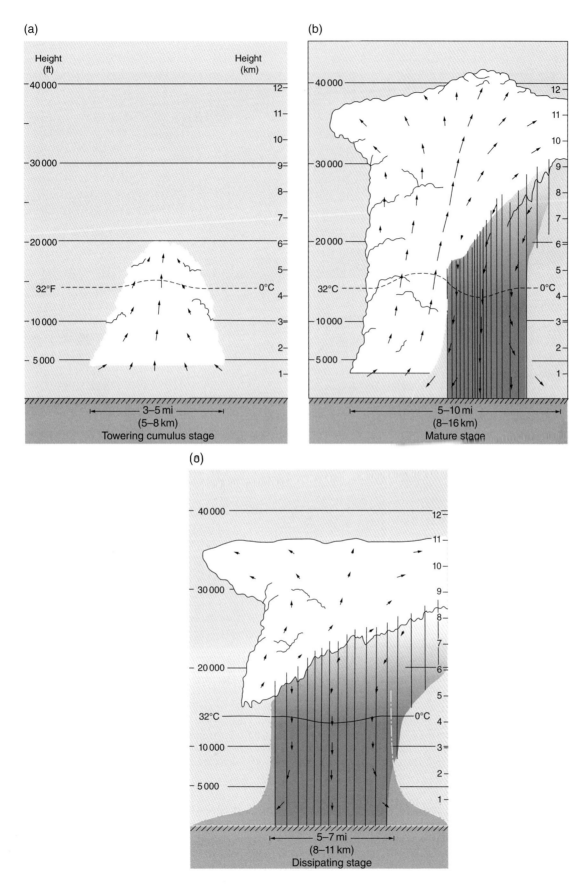

Figure 3.1 *Life cycle of a convective cell. (a) Developing stage (also known as the 'towering cumulus' stage), (b) mature stage and (c) dissipation stage. Vertical and horizontal dimensions are typical of cells growing in midlatitude, continental regions in the warm season. Image by Paul Markowski and Yvette Richardson (2010). © John Wiley & Sons Ltd.*

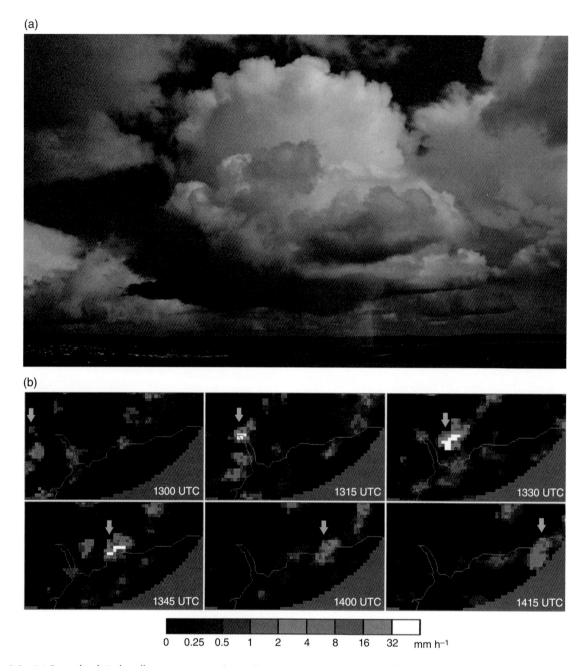

(a)

(b)

1300 UTC   1315 UTC   1330 UTC

1345 UTC   1400 UTC   1415 UTC

0   0.25   0.5   1   2   4   8   16   32   mm h⁻¹

Figure 3.2   (a) Example of single-cell convection near Exeter, Devon, at 1310 UTC on 10 April 2012. The view is towards the east and the cell is about 9 km distant. At the time of the photograph, the cell is just entering the mature phase of development, as evidenced by the newly developed narrow shaft of precipitation visible under the base of the cell. © Matt Clark (b) Sequence of 1-km resolution radar rainfall imagery between 1300 and 1415 UTC on 10 April 2012, showing the development and dissipation of the photographed cell (marked by the grey arrow in each panel). © Crown Copyright Met Office. (See insert for colour representation of the figure.)

are located towards the eastern, leading edge of the system. Rainfall associated with the older, decaying cells merges to form a larger area of lighter precipitation on the north-west flank of the system. This is often referred to as the *stratiform* rainfall region.

The potential for damage is somewhat higher with multicell storms than with single-cell storms. In particular, multicell storms can produce flash flooding, owing to the potential for relatively prolonged periods of very heavy rainfall over affected locations, especially where several cells form and track over the same area

(sometimes referred to as *cell training*). Prominent examples of flash flood-producing multicell storms in the United Kingdom include the Hampstead storm of 14 August 1975 (Keers and Wescott, 1976; Miller, 1978) and the Boscastle storm of 16 August 2004 (Barham, 2004; Doe, 2004; Burt, 2005; Golding *et al.*, 2005).

In conditions of very strong vertical wind shear (>40 knots difference between the surface and 6 km above ground level), and provided that cells are sufficiently separated in space so as not to merge into larger convective systems, supercells may form.

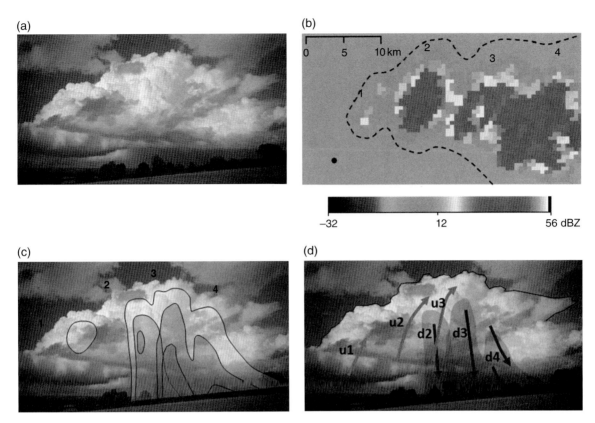

Figure 3.3   *Multicellular convection near Radstock, Somerset, on 29 July 2013. (a) Photograph of the storm at 1805 UTC, as viewed from the south-west. © Matt Clark. (b) Radar reflectivity at 1806 UTC. Black circle shows approximate location from which the storm was photographed. © Crown Copyright Met Office. (c) Idealised structure of the precipitation areas associated with each cell. Yellow, orange and pink shades denote reflectivity greater than 20, 30 and 40 dBZ, respectively. (d) Idealised schematic of updraughts (red arrows), labelled u1-u3, and downdraughts (blue arrows), labelled d2-d4, within each cell comprising the multicell storm. Cell (1) is in the developing stage, cell (2) is in the early mature phase, and cell (3) is in the mature phase (note that it is partially obscured by cell (2) in panels (a), (c) and (d)) and is associated with the heaviest precipitation at ground level. Cell (4) is in its weakening phase. (See insert for colour representation of the figure.)*

Supercells are long-lived, essentially single-cell, storms in which there is *continuous* development along a preferred flank (distinct from multicells, in which new development is episodic). Supercells contain a long-lived updraught which rotates about a vertical axis (Figure 3.5). The rotating updraught and its associated region of low pressure, collectively referred to as the storm's *mesocyclone*, are typically around 3–8 km in diameter. Storm rotation can be detected by Doppler radars, which are able to observe not only precipitation intensity but also information about the direction and strength of the winds within a storm. The rotation within the mesocyclone produces a distinct pattern, or *signature*, of winds (Brown and Wood, 1991). This is sometimes referred to as a *velocity couplet*. For classification as a supercell, the mesocyclone must extend through at least one-third to one-half of the storm's depth (in order to distinguish it from the shallow vortices that may occur in other types of convection) and persist for at least 20 minutes (Moller *et al.*, 1994; Markowski and Richardson, 2010). Although a supercell is essentially a single entity – comprising a long-lived, well-defined updraught – it can persist much longer than an ordinary single cell, owing to the highly organised manner in which the air flows through the storm. Lifetimes of 2–4 hours have been observed on occasion in the United Kingdom (e.g. Browning and Ludlam, 1962; Knightley, 2006; Clark and Webb, 2013; Westbrook and Clark, 2013). Such lifetimes are quite commonly observed in supercells in the United States.

The supercell was first identified as a specific class of convective storm in a study of a severe thunderstorm that occurred over southern England (the 'Wokingham' storm of 9 July 1959; Browning and Ludlam, 1962), though the vast majority of subsequent research has been conducted in the United States. The Wokingham storm was observed by radar from the time of its formation over the English Channel. After crossing the Dorset coast, it intensified substantially and attained a quasi-steady structure, subsequently producing a long, continuous swathe of large hail across parts of central southern and south-eastern England (Figure 3.6).

Supercells are a fairly rare type of storm, because the environments that favour their formation occur only under certain, specific, synoptic conditions. However, when such storms do form, the potential for large hail, tornadoes and damaging non-tornadic wind gusts is considerably higher than with other types of storm. Since supercell storms are very much more likely to produce severe weather than non-supercell storms, and because they are structurally and dynamically distinct from other thunderstorms, it is useful to classify storms as either supercells or non-supercells. Similarly, it can be instructive to classify tornadoes as either supercellular or non-supercellular, where supercellular tornadoes are defined as those occurring in association with a supercell's mesocyclone, specifically.

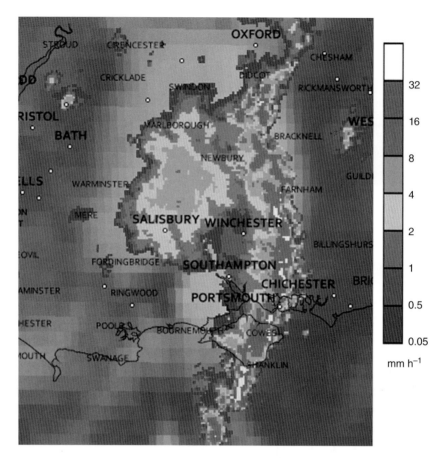

Figure 3.4   *Radar rainfall rate and satellite infrared image of a mesoscale convective system exhibiting linear morphology over southern England at 2000 UTC 22 October 2013. The system was moving towards the north-east. Panel shows data over an area of width 150 km. © Crown Copyright Met Office. (See insert for colour representation of the figure.)*

3.4   Tornadoes in Supercell and Non-supercell Storms

The formation of tornadoes (*tornadogenesis*) requires two basic ingredients: a source of rotation (i.e. spin) in the air, which is termed *vorticity*, and a mechanism by which the vorticity may be intensified. Intensification occurs when the air is stretched in the direction parallel to the axis of rotation. For example, for air rotating about a vertical axis (in the same sense as the rotation of a merry-go-round), stretching in the vertical direction may occur underneath a convective cell, where updraught strength increases with height (i.e. below the level of maximum updraught strength). Stretching of vertical vorticity is also associated with horizontal convergence of air, which acts to narrow and deepen the column of rotation. Smaller vortices, such as dust devils, occur in the absence of updraughts associated with deep convection, but they rarely attain damaging intensities and are generally very short-lived.

The association of severe local storms, including supercells, with strong environmental vertical wind shear has long been recognised (Dessens, 1960; Newton, 1960; Byers and Atlas, 1963; Browning, 1964). The shear is associated with horizontal vorticity (i.e. rotation about a horizontal axis; in the same sense as the rolling motion of a rolling pin). As air flows into the updraught of the storm, it is tilted. As this occurs, the horizontal vorticity in the air is also tilted into the

vertical. This results in a counter-rotating pair of vortices aloft, straddling the storm updraught (Figure 3.7). When winds veer with height, the cyclonic vortex dominates and becomes collocated with the updraught. The resulting deep, rotating updraught – the mesocyclone – is the defining characteristic of a supercell. Since the pressure within the vortex is low, the storm updraught is augmented below the level of maximum vortex strength (which is usually several kilometres above ground level), and air is drawn into the region of relatively low pressure. For this reason, updraughts within supercell storms can be exceptionally strong (sometimes $>50\,\mathrm{ms}^{-1}$; Markowski and Richardson, 2010). This energy helps to explain their association with very large hail and other types of severe weather.

The exact mechanism of tornado formation within supercells is not fully understood, though it is now widely accepted that downdraughts play a crucial role in the development of storm-scale rotation close to the ground and in the final phases of intensification to tornadic strengths (e.g. Klemp and Rotunno, 1983; Davies-Jones and Brooks, 1993; Wicker and Wilhelmson, 1995). This process does not occur in all supercells. In the United States, it is estimated that only around 25% of radar-observed mesocyclones produce tornadoes (Trapp *et al.*, 2005). The fraction of supercells that produce tornadoes in the United Kingdom is not known.

In certain circumstances, tornadoes occur in association with non-supercellular convection. The fraction of non-supercell storms

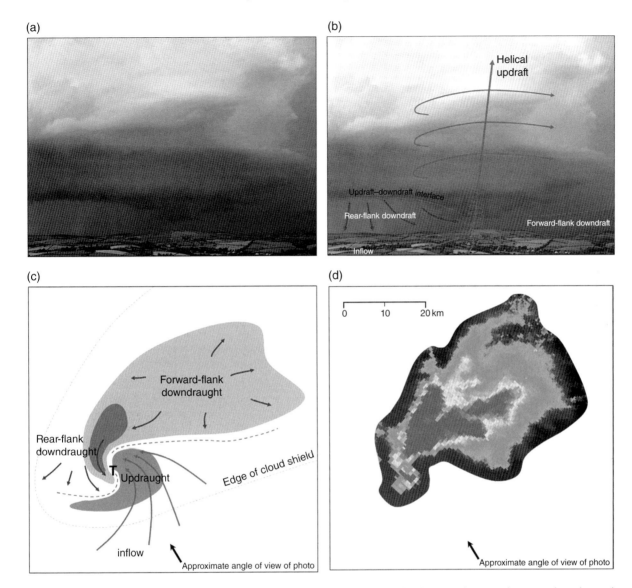

Figure 3.5 *(a) Photograph of a supercell thunderstorm near Newtownards, Northern Ireland, on 31 July 2006. The view is from the south-east. Photo © Martin North. (b) Annotated view of the same storm, with idealised representation of the main airflows within the storm. Red arrows indicate inflow and updraughts. Blue arrows indicate downdraughts. © Martin North (c) Idealised schematic of a supercell, showing the updraught and downdraught regions, and storm-relative low-level winds. Dark grey shading denotes typical location of large hail. Light grey shading denotes location of smaller hail and rain. 'T' indicates location of tornado, where present (after Lemon and Doswell (1979)). (d) Radar reflectivity image of a large hail- and tornado-producing supercell which occurred over the Midlands on 28 June 2012. Reflectivity scale is given in Figure 3.3. © Crown Copyright Met Office. (See insert for colour representation of the figure.)*

that produce tornadoes is thought to be much lower than for supercells. However, because the overwhelming majority of convection is non-supercellular, non-supercell tornadoes can comprise a significant fraction of all tornadoes. Non-supercell tornadogenesis may occur in a variety of ways. One mechanism is stretching by the convective updraught of pre-existing vertical vorticity originating near ground level (Figure 3.8). Common sources of vertical vorticity include sea breeze fronts, synoptic-scale frontal boundaries, the gust fronts of storm cells (particularly where gust fronts from more than one cell collide) and other wind zones where air parcels of differing speeds and directions converge. Generally speaking, tornadoes occurring in this type of set-up are weak and short-lived. However, because they tend to form in newly developing cells,

sometimes before much precipitation has been produced (e.g. Wakimoto and Wilson, 1989), they may be relatively easy to observe visually. The rotation is typically rather shallow, originates near ground level and occurs on horizontal scales well below that of the parent storm cell; this is in stark contrast to supercells, in which the tornado is embedded within a column of rotation that is several kilometres wide within the storm updraught (the mesocyclone).

In multicell storms that have distinctly linear structure (hereafter *linear convective systems*), a different mechanism of non-supercell tornadogenesis may occur (Trapp and Weisman, 2003; Weisman and Trapp, 2003). The density difference between the cold downdraught air behind the leading edge of the line and the warmer environmental air ahead of it is associated with horizontal *baroclinic*

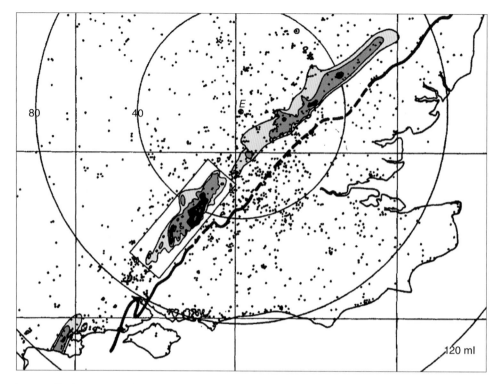

Figure 3.6 *Distribution of hailstone size in the 9 July 1959 'Wokingham' storm. Yellow, orange, black and pink shaded areas denote hail diameters exceeding 0.64, 1.27, 2.54 and 4.13 cm, respectively. Dots indicate individual hail observation points (442 also lie within the rectangle surrounding the most severely affected area). The thick black line shows the radar-observed locus of the right flank of the main storm. Browning and Ludlam (1962) was the first study of a supercell thunderstorm, paving the way for the recognition of supercells as a distinct class of severe thunderstorm. Adapted from Figure 3.2 of Browning and Ludlam (1962). Reproduced with permission of John Wiley & Sons Ltd. (See insert for colour representation of the figure.)*

Figure 3.7 *Schematic showing the development of vertical vorticity by tilting of the horizontal vorticity associated with the environmental vertical wind shear by the updraught within a supercell thunderstorm. The tilting initially results in the development of a pair of counterrotating vortices at mid-levels within the storm. When winds veer with height (as is typically observed within a warm sector in the northern hemisphere), the cyclonic vortex becomes dominant with time and becomes collocated with the storm updraught. Image by Paul Markowski and Yvette Richardson (2010). © John Wiley & Sons Ltd.*

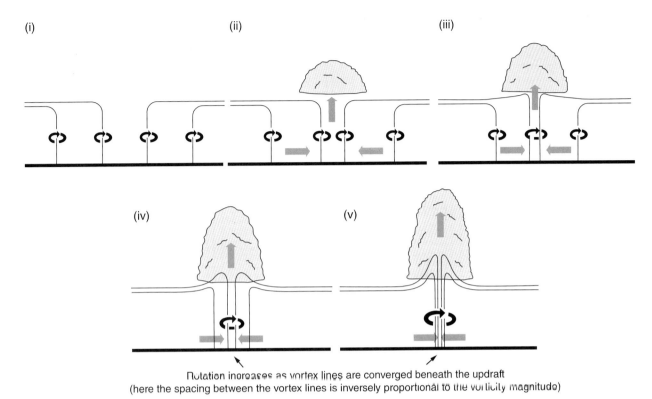

Rotation increases as vortex lines are converged beneath the updraft
(here the spacing between the vortex lines is inversely proportional to the vorticity magnitude)

Figure 3.8   *Tornadogenesis associated with stretching of vertical vorticity by the updraught of a developing convective cell. (i) A zone of vertical vorticity is present before initiation of the cell in which the tornado develops. (ii) A cell initiates above the convergence zone and an updraught develops. The developing updraught is associated with increasing horizontal convergence near ground level. (iii and iv) The updraught continues to develop, and the stretching provided by the updraught and the horizontal convergence sharpen the convergence zone, and the magnitude of vertical vorticity below the storm updraught increases. (v) Given sufficient stretching, a tornado may develop in the zone of intense vertical vorticity underneath the updraught of the cell. Image by Paul Markowski and Yvette Richardson (2010). © John Wiley & Sons Ltd.*

vorticity. Localised regions of intense updraught or downdraught along the line may tilt this horizontal vorticity into the vertical, forming counterrotating pairs of vortices near the leading edge of the system. Alternatively, a similar mechanism may act upon the horizontal vorticity associated with the environmental vertical shear within which the linear convective system is growing, again resulting in counterrotating vortex pairs. Over time, the cyclonic vortex tends to dominate due to the Coriolis effect (Trapp and Weisman, 2003). Where such vortices find themselves under an updraught, vorticity stretching and tornadogenesis sometimes occur. Although the tilting process is rather similar to the tilting that produces a vertical vortex in a supercell, the vortices along linear convective systems differ from supercell mesocyclones in that they are usually relatively shallow, evolve rapidly and do not become collocated with the storm updraught for long periods of time; indeed, simulations have shown that such vortices may become collocated with a local updraught minimum along the leading edge of the line in their latter stages of development (Trapp and Weisman, 2003).

Another type of weather system that sometimes produces tornadoes is the Narrow Cold Frontal Rainband (NCFR). As the name implies, these systems occur along strong cold fronts (and, more rarely, occlusions) associated with Atlantic low pressure systems moving across, or to the north of, the UK. NCFRs consist of a narrow line of strong, relatively shallow updraughts (usually around 2–3 km deep at most in the United Kingdom) and an associated line of intense precipitation that resembles the convective line within linear convective systems (Browning and Pardoe, 1973; Houze *et al.*, 1976; James *et al.*, 1978; James and Browning, 1979). Unlike linear convective systems, the updraughts in NCFRs are forced largely by strong horizontal convergence across the frontal zone. Buoyant instability is usually weak or non-existent. Despite this lack of instability, updraughts at the NCFR leading edge can be strong (up to $20\,\mathrm{ms}^{-1}$ in extreme cases; e.g. Carbone, 1982), owing to the strength of the horizontal convergence at the surface cold front. At the leading edge of the NCFR, the wind veers abruptly, sometimes by 90° or more. Where the wind veer is large, and the pre- and post-frontal winds are strong, the zone of wind veer may become extremely narrow (less than 1 km wide in some cases), forming a sheet of intense vertical vorticity along the leading edge of the rainband (Figure 3.9a). Over time, the vortex sheet may roll-up into more discrete centres of rotation called misocyclones, with typical diameters of 1–4 km (Figure 3.9b and c). It is thought that these vortices occur as a result of horizontal shearing instability along the vortex sheet that constitutes the frontal zone (e.g. Carbone, 1982, 1983; Smart and Browning, 2009). In some cases, misocyclones produce tornadoes and other instances of localised wind damage. As the vortices form, distortions occur in the NCFR as a result of advection of precipitation by the flow around the vortex, sometimes resulting in *line segment-and-gap* or *core-and-gap* morphology (e.g. Hobbs and Biswas,

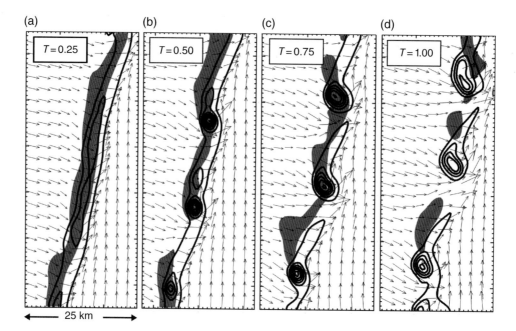

Figure 3.9 *Near-surface vertical vorticity (black contours; positive values only plotted; contour interval = 0.02 s⁻¹) and rainwater mixing ratio (>2 g·kg⁻¹ shaded grey; indicative of areas of heavy rainfall), from a high-resolution, nested, real-data model simulation of the 24 September 2007 NCFR, illustrating the process of vortex-sheet roll-up. T denotes the time (in hours) after initialisation of the fine (500 m) grid. At T = 0.25 h, a nearly uniform sheet of vertical vorticity exists, accompanied by an unbroken line of heavy precipitation just behind (west of) the axis of maximum near-surface convergence. By T = 0.50 h, discrete vorticity centres have begun to develop along the line. As the vorticity centres develop further, the line of heavy rainfall becomes highly segmented. Note the swathes of strong westerly winds on the southern flanks of each of the vortices at T = 0.75 and T = 1 hour. © David Smart.*

1979; James and Browning, 1979) (see Figure 3.9c and d). The situation is complicated, however, by the fact that core-and-gap morphology is not uniquely associated with misocyclones; gaps have been shown to occur as a consequence of other processes, including along-front differences in the angle between the front and the low-level pre-frontal flow (Browning and Roberts, 1996) and the formation of wave-like disturbances in the vertical winds near the leading edge of the cold front (Kawashima, 2007).

In a 2003–2010 climatology of cool-season (September–February) linear convective systems in the UK, mostly comprising NCFRs, an average frequency of 12.8 NCFRs per season was found, though with large variations from year to year (Clark, 2013). Frequencies were highest over central and southern England; 27.8% of NCFRs produced at least one tornado. Owing to the great horizontal extent and longevity of some NCFRs, numerous tornadoes occasionally occur in a single event, though the probability of a tornado at any given point along the line is very small, even in these cases. Outbreaks of 10 or more tornadoes in the United Kingdom (of which there were three in the period 2003–2012) are nearly always associated with such systems. NCFR tornadoes are typically short-lived, but they are occasionally strong (e.g. Buller, 1978; Clark, 2010). An outstanding example is the T5 tornado which hit Bicester, Oxfordshire, on 21 September 1982 (Elsom, 1983). Identification of tornado damage can be difficult in NCFR cases, owing to the possibility of narrow swathes of non-tornadic wind damage occurring in close proximity to tornado damage (Smart and Browning, 2009). Furthermore, minor wind damage, which could mask the tracks of weaker tornadoes, may occur in association with the generally strong winds associated with the sometimes-intense low-pressure systems within which NCFRs are embedded. For this reason, many weak NCFR tornadoes probably

go unreported, though it is also true that non-tornadic wind damage may occasionally be misclassified as tornado damage.

3.4.1 Synoptic Situations Associated with Tornadoes in the United Kingdom

Tornadoes are known to occur in a variety of different synoptic situations in the United Kingdom (Bolton *et al.*, 2003). Since supercell thunderstorms require very specific conditions (i.e. a combination of sufficient buoyant instability and strong vertical wind shear), they tend to occur only in a limited number of synoptic situations. The two situations in which most supercell thunderstorms (and associated tornadoes) occur are:

1. The Spanish Plume
   Between May and October, warm, moist and unstable air masses, often of continental origin, may be advected into the United Kingdom ahead of cold fronts or occlusions moving in from the west or south-west (Figure 3.10a). Where this situation occurs in conjunction with strong flow, as might be associated with a deep mid- to upper-level trough situated to west of the United Kingdom (Figure 3.10b), vertical wind shear may become sufficient for supercell development. It is the combination of warm, moist air near the surface, which often forms a well-defined tongue extending into the United Kingdom from the south or south-east, and an upper-level trough moving from the west or south-west towards the United Kingdom, that constitutes the *Spanish Plume* (Carlson and Ludlam, 1968; Morris, 1986; Gray and Marshall, 1998; Bennett *et al.*, 2006; Lewis and Gray, 2010). Another

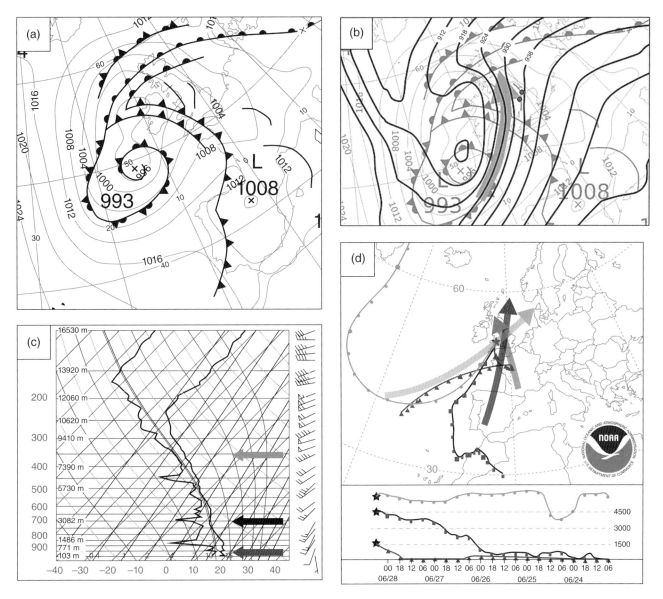

Figure 3.10 *Schematic illustrating the* Spanish Plume *synoptic set-up over the United Kingdom on 28 June 2012. (a) Met Office surface analysis chart for 1200 UTC 28 June 2012. Isobars are drawn at 4 hPa intervals and fronts are drawn using standard symbols. Crown Copyright Met Office. (b) 300 hPa geopotential height at 1200 UTC 28 June 2012 (blue contours; contour interval is 60 m; contour labels indicate the height, in tens of metres, of the 300 hPa pressure surface). The blue arrow indicates the axis of strongest winds at 300 hPa. The red dots indicate locations of the two tornadoes that occurred within ±2.5 hours of analysis time. (c) 0000 UTC Nottingham radiosonde data, illustrating conditions typical of the warm sector and associated* plume. *Bold black lines indicate the environmental temperature (right-hand line) and dew point temperature (left-hand line) profiles. Red, blue and green arrows indicate the approximate heights of the end points of the trajectories marked in (d). Courtesy of University of Wyoming. (d) Trajectories of air at 1500 m (red), 3000 m (blue) and 6000 m (green) above ground level over the 5-day period ending 0900 UTC 28 June 2012. Arrows indicate idealised trajectories of airflows at these heights in a typical Spanish Plume set-up. © Courtesy of NOAA Air Resources Laboratory. (See insert for colour representation of the figure.)*

key feature is a layer of relatively warm air at mid-levels that originates over the elevated terrain of the Iberian Plateau or North Africa (Figure 3.10c and d). During the warmer half of the year, insolation under clear skies heats the ground strongly in these regions, which warms the lowest layer of the atmosphere (the *boundary layer*). This warm air is drawn poleward in south or south-south-westerly flow. As the air mass moves north over France, it becomes detached from the surface, resulting in the development of an *Elevated Mixed Layer*

(EML) – a layer of relatively warm, well-mixed air whose base is located above ground level. The temperature profile associated with the EML is distinctive (Figure 3.10c); the base of the feature is marked by a temperature inversion (the *capping inversion*, or *lid*), with very steep lapse rates (i.e. rapid decreases in temperature with height) through the layer above the inversion. Convection often develops within the unstable layer associated with the EML. Such convection may be intense, producing copious lightning and, occasionally, large hail. However, tornadoes do not occur, because the base of the convective cloud and the source of the updraught air are located well above ground level (i.e. the convection is *elevated*). The presence of the capping inversion acts as a barrier to surface-based convection, allowing heat and moisture to build within the already warm and moist lowest layer of the atmosphere. Where strong insolation is present, or gradual ascent in the middle levels of the atmosphere ahead of the advancing mid- to upper-level trough occurs, the inversion is weakened, and surface-based convection may eventually 'break through' the cap. In the United Kingdom, surface-based convection is often suppressed by the mid-level cloudiness associated with elevated thunderstorms, which can limit insolation and deplete the reservoir of energy available for convection. However, on the relatively rare occasions that surface-based convection does occur, it can be especially intense, and surface-based supercells and associated tornadoes are possible. Many, if not all, of the instances of giant hail in the United Kingdom (>50 mm diameter) are thought to have been associated with supercells occurring in this type of synoptic situation. A classic example of this type occurred on 28 June 2012 (see Section 3.10).

2.   Showery air masses in the cold sectors of low-pressure systems

Tornadic supercells may also occur in the showery air masses that exist behind low-pressure systems moving east or north-east across, or close to, the United Kingdom. Instability in these situations is typically lower than in the *Spanish Plume* scenario described above, and the convection is often correspondingly weaker. However, the instability may become relatively large inland in the summer, given strong insolation, or near windward coasts in autumn and winter, when particularly cold air masses move over relatively warm seas. As in the Spanish Plume scenario, the vertical wind shear in these post-frontal air masses is often too weak for supercell development. However, in some situations, such as where subtle troughs are embedded within the flow (Figure 3.11), or where there is a mesoscale boundary induced by convergence, shear may be locally enhanced to the extent that supercell development becomes possible. The relatively low tropopause height that is typical of cool-season polar maritime (Pm) and arctic maritime (Am) air masses is associated with relatively shallow storms: the tops may be around 5–8 km above ground level, compared to a typical height of 10–12 km above ground level for warm-season storms occurring in pre-frontal environments. Supercell storms occurring in cool-season, post-frontal environments have accordingly sometimes been referred to as *low-topped* or *shallow* supercells. It should be noted, however, that in spite of their relative shallowness, the basic structure and dynamics of these cool-season supercells are apparently no different to those of their larger, warm-season counterparts.

Non-supercell tornadoes may occur wherever there is deep, moist convection and consequently have been observed in a wider variety of synoptic situations than supercell tornadoes. Indeed, non-supercell tornadoes have been observed in those situations that could support supercells; for example, the T5 'London' tornado

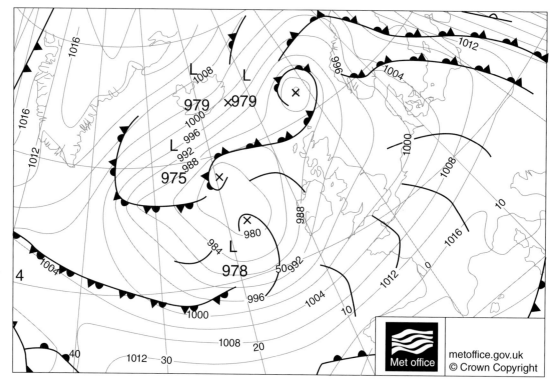

Figure 3.11   *Surface analysis chart for 0600 UTC 16 December 2012, showing an example of a returning polar maritime air mass with embedded troughs. Two tornadoes occurred in west Cornwall on this day, associated with a thunderstorm developing ahead of the trough located over the south-west approaches and Biscay at analysis time, which moved north-east through the day. © Crown Copyright Met Office (2012).*

of 7 December 2006 formed in association with an intense linear convective system in a strongly-sheared returning polar maritime (rPm) air mass (Clark, 2011). Tornadoes have occasionally been observed in Spanish Plume set-ups where the dominant storm mode is multicellular rather than supercellular, for example, in cases where MCSs form. However, the large majority of non-supercell tornadoes in the United Kingdom are associated with two additional synoptic types. The first is within shallow low-pressure areas or *cols* (areas of weak pressure gradients between high- and low-pressure systems) where surface winds are light and variable in direction. Given sufficient instability, slow-moving showers and storms are triggered by outflows from previous showers, or where sea breeze fronts and other zones of converging winds occur. The second situation in which non-supercell tornadoes are frequently observed is along frontal rainbands (usually NCFRs, as discussed previously), associated with low-pressure systems moving eastwards across, or to the north of, the United Kingdom.

3.5 Towards a Climatology of Tornadoes by Synoptic Type

In order to provide an estimate of the relative frequency of tornadoes in the different synoptic types described above, tornadoes listed in the TORRO database for the period 2003–2012 were assigned into synoptic categories, using archived 6 hourly Met Office surface analysis charts and radar data. The 10-year analysis period corresponds to the availability of the radar data; imagery has only been archived continuously since 2003. The synoptic categories are as follows:

- Pre-frontal tornadoes are defined as those occurring ahead of (generally to the east or north-east of) a cold front or occlusion which is moving towards or across the United Kingdom. These were subdivided into tornadoes occurring in association with tropical continental (Tc) and tropical maritime (Tm) air masses, depending on whether the air over the United Kingdom originated from the near continent (e.g. southerly, south-easterly and easterly flows for Tc air masses) or from the ocean (south–south-westerly to south-westerly flows for Tm air masses).
- Post-frontal tornadoes are defined as those occurring within Pm, rPm and Am air masses. Collectively, these comprise the showery air masses behind the frontal systems associated with travelling midlatitude depressions. In rPm flows, a southerly component is evident, for example, where the polar air mass has extended around the western and southern side of the depression and is starting to return poleward on the south-east flank of the depression (Figure 3.12a). In Pm air masses, the flow is generally westerly or north-westerly with a source over the North Atlantic (Figure 3.12b). Am air masses are defined as those having origins at high latitudes (defined here as greater than ~70°N), generally associated with northerly flows over and to the north of the United Kingdom (Figure 3.12c).
- Tornadoes occurring within slack surface pressure patterns are defined as situations in which surface pressure gradients are very weak (generally less than about 4–6 hPa 1000 km^{-1}) and in which there is no predominant flow direction or clearly defined air mass source (both at the surface and aloft). Examples include

slow-moving, weak depressions (Figure 3.13a), cols (the area of slack pressure gradients between two high- and two low-pressure areas; Figure 3.13b) and, occasionally, weakly defined anticyclones or ridges of high pressure (Figure 3.13c).
- Tornadoes directly associated with frontal rainbands (mostly NCFRs) are defined as those that can unambiguously be attributed to a frontal rainband. These were sub-classified by frontal type (cold front, occlusion or warm front). Note that events occurring along pre- or post-frontal troughs located close to, but clearly separate from, the front were placed into the pre-frontal and post-frontal categories, respectively.

Events that did not easily fall into the above types were listed as unclassified. Over the 10-year period, only 3 events (0.7%) were listed as unclassified. These tornadoes all occurred near to occlusions, but their position relative to that of the front could not be confidently determined. The distribution and climatology of tornadoes in the United Kingdom and Ireland is further discussed in Chapter 4 of this volume.

3.6 Monthly and Annual Frequencies of Tornadoes by Synoptic Type

Figure 3.14 shows the monthly tornado frequencies for each synoptic type, and Table 3.1 shows the percentages of tornadoes associated with each type annually and in each season. Tornadoes occurring in slack pressure and pre-frontal situations are responsible for relatively few tornadoes over the analysis period (14.0 and 11.5%, respectively) and are almost exclusively a warm-season phenomenon. Of the pre-frontal tornadoes, 74% were associated with tropical continental (Tc) air masses, and 26% with Tropical maritime (Tm) air masses. The absence of pre-frontal tornadoes outside the warm season is likely a consequence of the fact that pre-frontal air masses tend to be stable in the cool season, having travelled north over increasingly cool waters, or over a cold continent, inhibiting the development of convection. The lack of cool-season, slack pressure pattern tornadoes probably reflects the rarity of such synoptic situations outside of the warm season, though it may also be a consequence of the fact that light winds favour radiative cooling over land in winter, leading to stable conditions which inhibit convection. Post-frontal tornadoes are an almost year-round phenomenon, though in common with pre-frontal and slack pressure pattern tornadoes, frequencies are highest in the summer months. In contrast, frontal tornado frequencies peak strongly in the autumn and early winter, with comparatively few such events in the summer. This reflects the fact that the strong cold fronts exhibiting NCFRs are most common in the autumn and winter.

Taking the year as a whole, post-frontal and frontal tornadoes are dominant (Table 3.1); 73.8% of all tornadoes were associated with these two types over the 10-year analysis period. Taking tornadoes of intensity T2 and stronger only (i.e. those events that are more likely to be associated with significant damage to property), the dominance of frontal and post-frontal events is even more striking (Table 3.2): 86.0% of all such events were associated with these types. Very few tornadoes in slack pressure situations attained T2 or greater intensity. Pre-frontal situations were responsible for a larger but still relatively small fraction (~10%) of all T2+ tornadoes.

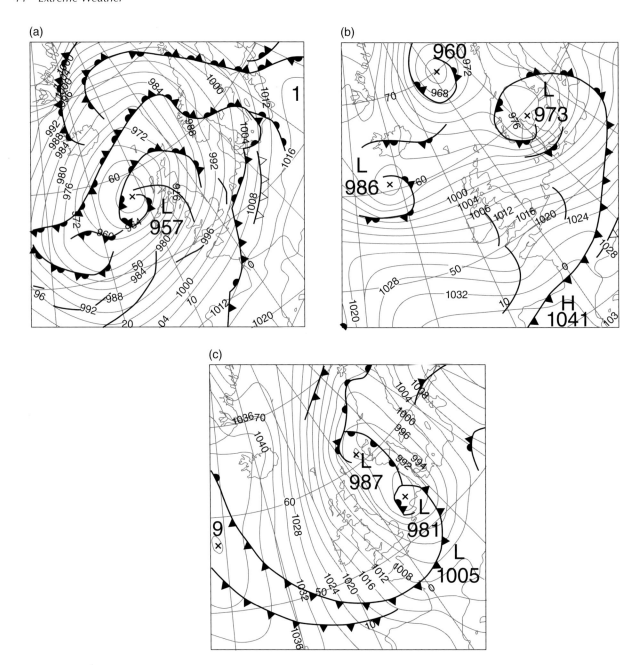

Figure 3.12   *Surface analysis charts showing examples of (a) returning polar maritime, (b) polar maritime and (c) arctic maritime air masses. (a) 1800 UTC 6 January 2014, (b) 1200 UTC 1 January 2007 and (c) 0000 UTC 25 November 2005. © Crown Copyright Met Office (2014).*

3.7   Spatial Distribution of Tornadoes by Synoptic Type

Cool-season, post-frontal tornadoes primarily occur close to coasts, especially the Channel coast of southern England and the southern coasts of East Anglia and South Wales (Figure 3.15a). This reflects the fact that instability tends to be highest over the sea at this time of year, therefore exposing windward coasts to the most intense convection, especially where sea surface temperatures are highest. In contrast, warm-season, post-frontal tornadoes show a more even distribution over inland areas of England,

Wales and Ireland, though frequencies are much lower in Scotland (Figure 3.15b). Since there is no obvious meteorological reason why such tornadoes should be less common over Scotland, it is possible that the lower frequencies in this area reflect the lower population densities here, compared to elsewhere in the United Kingdom; in sparsely populated areas, tornadoes are less likely to be witnessed and reported (e.g. Mulder and Schultz, 2015).

Some interesting differences are apparent in the spatial distributions of rPm tornadoes, specifically, in the cool and warm seasons (red dots in Figure 3.15a and b). For example, warm-season rPm tornadoes occur mostly over the Midlands and northern

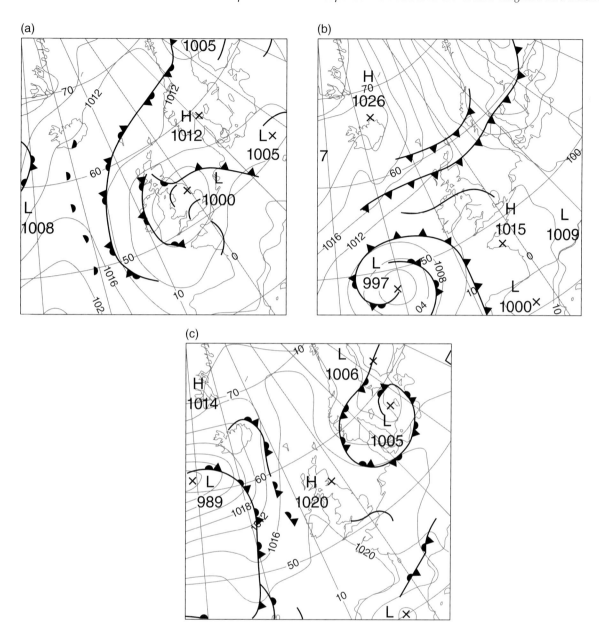

Figure 3.13 *Surface analysis charts showing examples of (a) a weak depression, (b) a col and (c) a weak ridge of high pressure. (a) 1200 UTC 18 August 2006, (b) 1200 UTC 12 June 2007 and (c) 1200 UTC 1 August 2005. © Crown Copyright Met Office (2014).*

England but are rare to the south of a London to Bristol line (perhaps with the exception of locations right along the English Channel coast). This reflects the fact that southern areas have relatively short land tracks in the southerly and south-westerly wind directions typical of rPm flows: cells developing over land therefore tend to mature and produce tornadoes only after they move north or north-east of the area, into the Midlands or northern England. The same reasoning can perhaps explain the distribution of pre-frontal tornadoes, which are largely confined to the Midlands, parts of eastern and northern England, and the central lowlands of Scotland, where land tracks are substantial in flows with a southerly component (Figure 3.15c). In contrast, cool-season rPm tornadoes tend to occur largely over southern England and South Wales (red dots in Figure 3.15a), owing to

the proximity of these areas to the relatively warm waters of the English and Bristol Channels. Relatively few such events occur north of the Midlands, though local maxima are apparent over the central lowlands of Scotland and over the central part of Ireland. The central Ireland maximum is not easy to explain given the general tendency for convection to occur near the warmer seas at this time of year.

Tornadoes associated with frontal rainbands occur with a fairly even distribution over much of England, Wales and Northern Ireland but are much less common over Scotland and western parts of Ireland (Figure 3.15d). Clark (2013) showed that NCFRs were most common over central and southern parts of the United Kingdom; the distribution of frontal tornadoes therefore likely reflects the differences in the frequency

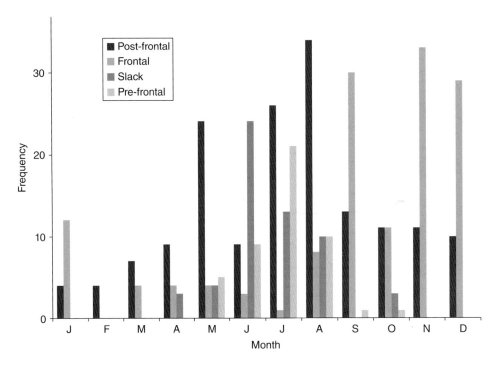

Figure 3.14 *Monthly tornado totals by synoptic type during the period 2003–2012. (See insert for colour representation of the figure.)*

Table 3.1 *Seasonal frequency of tornadoes by synoptic type, expressed as a percentage of all tornadoes in each season.*

Season	Post-frontal	Frontal	Slack	Pre-frontal	Others
Spring (MAM)	61.5	18.5	10.8	7.7	1.5
Summer (JJA)	40.8	7.1	27.8	23.7	0.6
Autumn (SON)	30.7	64.9	2.6	1.8	0.0
Winter (DJF)	25.0	74.0	0.0	0.0	1.0
Annual	39.7	34.1	14.0	11.5	0.7

Table 3.2 *Seasonal frequency of tornadoes rated as ≥ T2 intensity by synoptic type, expressed as a percentage of all ≥ T2 intensity tornadoes in each season.*

Season	Post-frontal	Frontal	Slack	Pre-frontal	Others
Spring (MAM)	66.7	19.0	0.0	14.3	0.0
Summer (JJA)	44.1	14.7	8.8	32.4	0.0
Autumn (SON)	20.5	79.5	0.0	0.0	0.0
Winter (DJF)	25.5	72.5	0.0	0.0	2.0
Annual	37.2	48.8	2.3	10.9	0.8

of the NCFRs with which they are generally associated. However, the frequency differences may be exaggerated by the relatively low population densities in western Ireland and much of Scotland.

Tornadoes in slack pressure situations show a conspicuous cluster over the East Midlands (Figure 3.15e). Heat lows sometimes develop inland in slack pressure situations where strong insolation occurs; weak flow into the centre of the heat low may influence the location and strength of convergence lines. One hypothesis is that the Midlands cluster reflects the position of a frequently-occurring convergence line in slack

pressure situations. It is not clear, however, whether the cluster is unique to the relatively short analysis period chosen or whether it is a more general feature of the spatial distribution of this type of tornado. A longer period of analysis would be required to investigate how robust the feature is. Excluding the Midlands cluster, many of the slack pressure pattern tornadoes are situated within close proximity to the coast (within ~30 km), which may reflect the importance of sea breeze convergence lines for the initiation of convection in favourable synoptic situations.

3.8 Morphology of Tornadic Storms

The morphology of tornado-producing convection was determined by inspecting composite radar rainfall imagery corresponding to the time and location of each tornado within the TORRO tornado database over the 2003–2012 period. Where the storm responsible for the tornado could be identified unambiguously, it was classified as a cell, cluster, broken line or line. Table 3.3 describes the classification criteria, and an example of each type is shown in Figure 3.16. Of the 408 tornadoes recorded in this period, morphologies could be determined for 368 (90.2%). The distribution of morphologies varied strongly with synoptic type (Table 3.4). For example, linear structures (i.e. lines and broken lines) were responsible for 97.8% of frontal tornadoes, but only 13.0% of slack pressure pattern tornadoes. Cellular morphologies (i.e. clusters and cells) dominated in slack pressure pattern, post-frontal and pre-frontal situations, being responsible for 87.0, 70.6 and 81.8% of tornadoes in each type, respectively. Over all synoptic types, cellular morphologies were responsible for 48.4% of tornadoes, and linear morphologies for 51.6%.

Figure 3.15   *Locations of tornadoes by synoptic type. (a) Cool-season (October–March) post-frontal tornadoes, (b) warm-season, post-frontal tornadoes, (c) pre-frontal tornadoes, (d) tornadoes associated with frontal bands and (e) tornadoes in slack pressure patterns. (See insert for colour representation of the figure.)*

Table 3.3   *Criteria for the classification of convection morphology from composite radar rainfall rate imagery.*

Morphology	Classification Criteria
Cells	Discrete maxima of rainfall rate. Individual maxima are connected by rainfall rate $<1.0\,\mathrm{mm\,h^{-1}}$ (usually $0.0\,\mathrm{mm\,h^{-1}}$)
Clusters	Discrete maxima of rainfall rate that are generally connected by lighter regions of rainfall rate ($\geq1.0\,\mathrm{mm\,h^{-1}}$) and in which the individual maxima show little overall spatial organisation
Broken lines	Discrete maxima of rainfall rate that are generally connected by lighter regions of rainfall rate ($\geq1.0\,\mathrm{mm\,h^{-1}}$) and in which the individual maxima are wholly or predominantly arranged along one or more lines
Lines	Unbroken, or nearly unbroken, line of high rainfall rates ($\geq8.0\,\mathrm{mm\,h^{-1}}$) in which embedded rainfall rate maxima are only weakly defined

3.9   Association of Supercells with Giant Hail

Significant falls of hail are another characteristic feature of supercell storms. Incidences of 'giant' hail (here defined as hail $\geq50\,\mathrm{mm}$ in diameter) are thought to be almost always associated with supercells (Markowski and Richardson, 2010). Where large hail falls over continuous, long, narrow swathes, the occurrence of a supercell may be inferred with even more certainty, since it is

highly improbable that any other type of storm could *continuously* produce such large hail over an extended track. Over the period 1952–2012, only 12 cases of hail greater than or equal to 50 mm diameter are known in the United Kingdom (Webb *et al.*, 2009; Clark and Webb, 2013). These cases all occurred between May and September, in association with pre-frontal Tm or Tc air masses. The development of large hail in supercells is a consequence of the exceptionally strong updraughts of these storms,

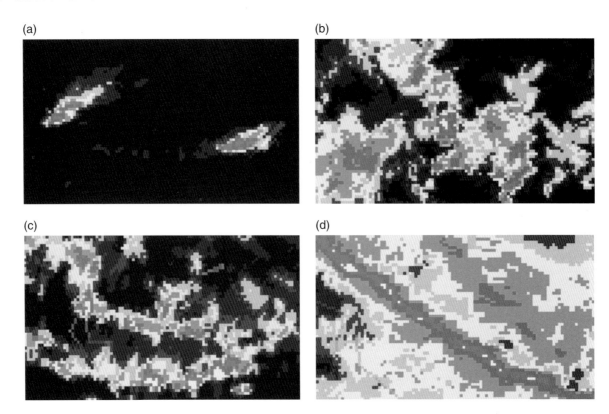

Figure 3.16   *1-km resolution radar rainfall imagery showing examples of four morphologies of convection: (a) cell, (b) cluster, (c) broken line and (d) line. Radar rainfall rate scale is given in Figure 3.2. © Crown Copyright Met Office. (See insert for colour representation of the figure.)*

Table 3.4   *Tornado totals, as a function of storm morphology and synoptic type, over the period 2003–2012.*

Morphology	Post-frontal	Frontal	Slack	Pre-frontal	Others	All Synoptic Types	All Synoptic Types (%)
Line	21	129	3	6	0	159	43.2
Broken line	19	7	3	2	0	31	8.4
Cluster	25	2	12	6	1	46	12.5
Cell	71	1	28	30	2	132	35.9

together with their highly organised structure. The intense updraught is associated with a region of weak radar echo called the *vault*; the vault contains air which contains large concentrations of supercooled cloud droplets but few larger precipitation particles. Hailstones that are located close to the edge of the vault are able to grow rapidly since they have access to the un-depleted cloud water in the vault (Browning and Foote, 1976; Browning, 2014). The combination of very large instability that is characteristic of Spanish Plume synoptic set-ups and the aforementioned dynamic and structural features unique to supercells likely explains the near-exclusive association in the United Kingdom of giant hail with supercells occurring in this type of synoptic situation.

Supercells in rPm and Pm flows tend to be of relatively limited vertical depth, owing to the lower tropopause height in these air masses. CAPE is more modest (often in the range 100–1000 J kg^{-1}), because dew point temperatures are lower and the convective layer is shallower. Hail occurring in supercells developing in this type of environment is therefore not usually of exceptional size. However, it may still be uncharacteristically large for the season and larger than would normally be expected given the relatively modest CAPE; for example, highly localised falls of hail 25–40 mm in diameter were

observed in association with the cool-season tornadic supercells of 26 January 1984 in south Devon (Bailey and Mortimore, 1984) and 16 December 2012 in west Cornwall (Clark and Pask, 2013).

3.10   Case Studies of Supercell and Non-supercell Tornadoes

Three recent cases of tornadic storms in the United Kingdom are now described in more detail. The cases presented include a warm-season tornadic supercell, a cool-season tornadic supercell and an NCFR. The cases have been selected on the basis that they have been well documented by TORRO site investigations and because good observations data exist.

3.10.1   Case 1: The Cold Front of 29 November 2011

During 28 November 2011, frontal systems associated with a deep low-pressure area centred near Iceland moved into northern and western parts of the United Kingdom. By 0000 UTC on the 29th, a small wave had developed on the cold front just west of

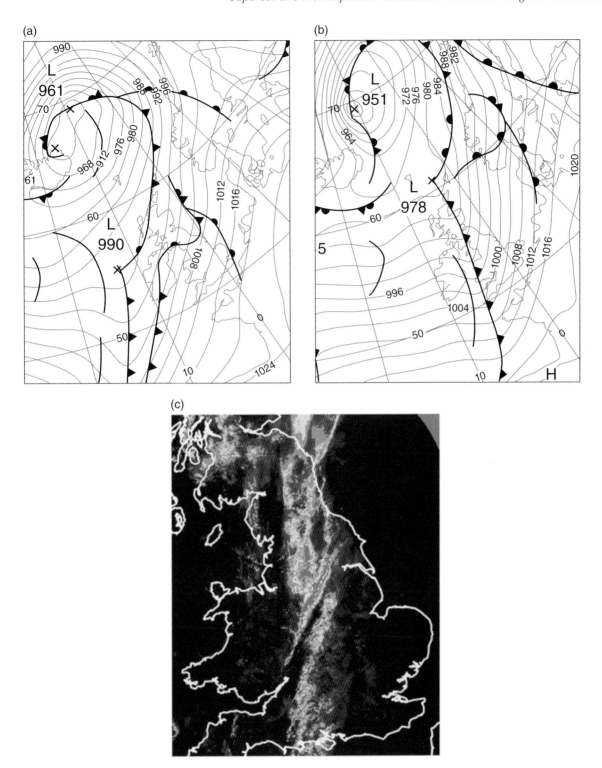

Figure 3.17   *Surface analysis charts for (a) 0000 UTC 29 November 2011 and (b) 1200 UTC 29 November 2011. (c) Composite radar rainfall data at 1500 UTC 29 November 2011, showing the narrow cold frontal rainband. Radar rainfall rate scale is given in Figure 3.2. © Crown Copyright Met Office (2011). (See insert for colour representation of the figure.)*

the United Kingdom and Ireland, as a new area of low pressure began to develop along the front to the north-west of Ireland (Figure 3.17a). This low moved north-east very rapidly over the following 24 hours, deepening from 990 hPa at 0000 UTC on the

29th to 961 hPa by 0000 UTC on the 30th. The cold front to the south of the track of the low became sharply defined as it surged east and south-east across the United Kingdom (Figure 3.17b), and an NCFR developed (Figure 3.17c). The strongest part of the

NCFR moved eastwards across North Wales, northern England and the North Midlands between 1200 and 1800 UTC on 29 November. Several tornadoes occurred, in addition to at least two instances of non-tornadic damaging winds. One of the tornadoes, at Breighton in Yorkshire, produced a damage track at least 12 km long and was captured on security camera as it moved through a builders' yard. Another tornado affected the Heaton Moor and Hyde areas of Greater Manchester; roof damage occurred to properties along the track of the tornado, and the gable end of one house in Hyde was badly damaged. This tornado was rated as T3 intensity.

As the cold front crossed northern England, it was marked by a near-90° wind veer, from southerly ahead of the front to westerly behind (Figure 3.18). In some NCFR cases, winds fall light immediately behind the cold front. However, in this case, post-frontal winds were at least as strong as pre-frontal winds, implying particularly strong horizontal convergence across the front. The frontal passage was marked by a 3 or 4°C drop in temperatures and

a near-instantaneous pressure increase of 1–3 hPa. Most of the tornadoes occurred along the track of a mesoscale bulge in the NCFR (as is evident in Figure 3.18). The bulge moved east–north-east through the far north of Wales and then across the southern half of northern England between Liverpool and Hull. The cross-frontal temperature decrease, pressure surge, horizontal convergence and vertical vorticity were generally largest along the track of this bulge. The bulging configuration of the line occurred due to a local acceleration of the front, which appears to have been associated with the development of a maximum in post-frontal wind speeds.

The tornadic part of the NCFR was not sampled by Doppler radar. However, the probable development of misoscale (1–4 km diameter) vortices (misocyclones) can be inferred from the radar reflectivity data. A number of small distortions developed in the line, some of which subsequently evolved into line gaps. Examples of the gaps and distortions at 1410 UTC are shown in Figure 3.19. The distortions occur where flow around

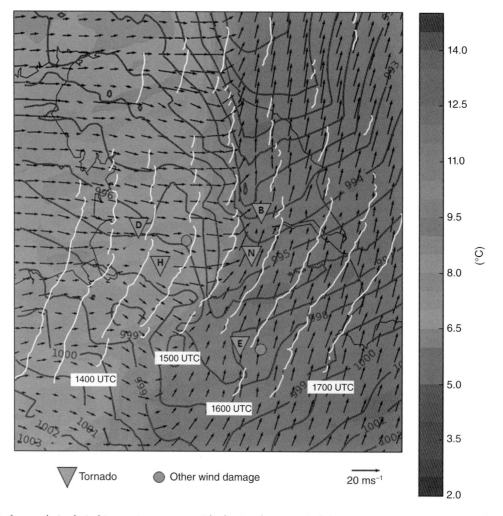

Figure 3.18 Surface analysis of wind (arrows), temperature (shading) and pressure (solid contours) at 1500 UTC 29 November 2011. Analysis has been constructed using 1-minute resolution data from the Met Office's network of automatic weather stations. Locations of NCFR segments at 30-minute intervals between 1300 and 1700 UTC are indicated by the white solid lines. Tornado locations are indicated by inverted triangles. Letters indicate location of tornado report. D = Darwen, Lancashire; H = Hyde and Heaton Moor, Cheshire; N = New Rossington, South Yorkshire; B = Breighton, East Yorkshire; E = East Leake, Leicestershire. Non-tornadic wind damage locations are indicated by the yellow circles. (See insert for colour representation of the figure.)

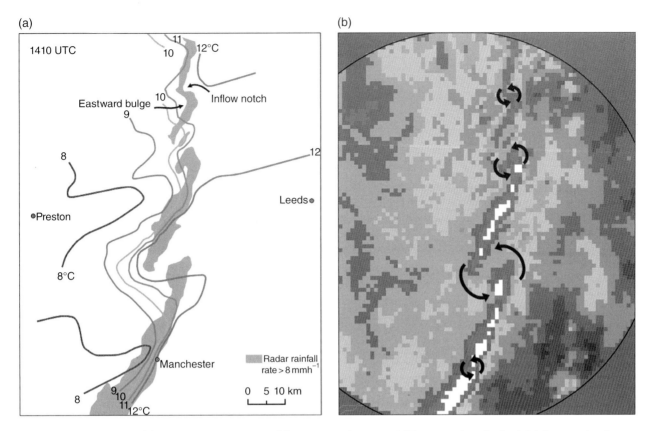

Figure 3.19 *A closer view of the segment–gap structure of the 29 November 2011 NCFR over northern England. (a) Contoured surface temperatures and (b) radar rainfall rate at 1-km resolution. Scale is as in Figure 3.2. Grey shaded areas in (a) denote regions of radar rainfall rate greater than 8 mm h^{-1}. Black arrows in (b) indicate the probable locations of cyclonic vortices along the line. © Crown Copyright Met Office. (See insert for colour representation of the figure.)*

a vortex influences the progress of the leading edge of the line on a localised scale, creating a small eastward bulge to the south and an inflow notch to the north. Line gaps sometimes occur as the vortices mature and decay (as in the case of at least two of the four vortices marked in Figure 3.19b), during which time they may also increase in width. In the absence of Doppler radar data, the association between vortices and tornadoes cannot be confirmed in this case. However, it is apparent that the tornadoes generally occurred along the bulging part of the line where new perturbations and gaps along the line were developing. In some cases, gaps formed in the same line-relative location as a tornado, shortly after the time of tornado occurrence. For example, note the line perturbation that formed east–north-east of the Heaton Moor tornado by 1500 UTC and an evolving gap to the east–north-east of the Breighton tornado by 1600 UTC (Figure 3.18).

The 29 November 2011 NCFR is one of a number of analysed cases which suggest that mesoscale variations in the post-frontal winds may influence where tornadoes and other wind damage occur along an NCFR (Clark and Parker, 2014). Where post-frontal winds are stronger and orientated nearly perpendicular to the NCFR, the convergence across the front may be especially intense, resulting in a narrower, sharper line of wind veer across the front. In the 29 November 2011 case, the break-up of the NCFR into segments separated by fairly regularly-spaced line gaps near the apex of the bulging part of the NCFR over northern England is suggestive of the occurrence of horizontal shearing instability

along the line of sharp wind veer (e.g. as in Figure 3.9), though in the absence of more detailed observations and model simulations, this cannot be proven. Horizontal shearing instability has been invoked to explain the formation of misocyclones, some of which were tornadic, in other studies of NCFRs. For example, Smart and Browning (2009) suggested that horizontal shearing instability was likely responsible for the development of vortices in a real-data, numerical model simulation of a tornadic NCFR that affected the United Kingdom on 24 September 2007, and Carbone (1982, 1983) suggested that this mechanism was responsible for the formation of a tornadic vortex along an intense cold front in California. Further modelling and observational studies are needed to better understand how the formation of vortices occurs in this type of event.

3.10.2 Case 2: The English Midlands Supercells of 28 June 2012

On 28 June 2012, a number of supercell thunderstorms developed over the Midlands, within a pre-frontal air mass characterised by very high surface temperatures and dew point temperatures (Figure 3.10a). Elevated thunderstorms had occurred early in the day over parts of south-west England and Wales. While these storms and their associated cloud cover prevented the subsequent development of surface-based convection over many northern and western parts of the United Kingdom, clearer skies to the south-east allowed temperatures to climb over much of the Midlands

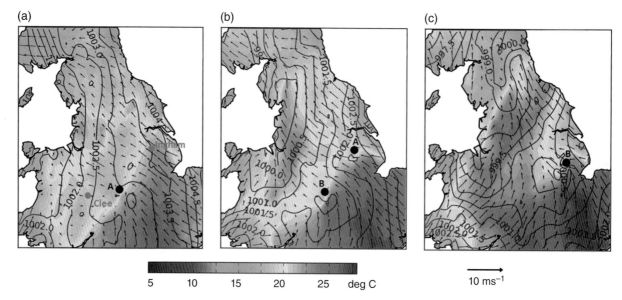

Figure 3.20 *Analysis of surface temperature (shaded), pressure (solid contours) and winds (arrows) for (a) 1000 UTC, (b) 1200 UTC and (c) 1400 UTC 28 June 2012. 'A' denotes the location of the first surface-based supercell of the day, which produced the tornado in Leicestershire. 'B' denotes the location of the second surface-based supercell, which produced the swathe of large hail through Leicestershire and Lincolnshire and the tornado in Lincolnshire. Reproduced with permission from Clark and Webb (2013). © John Wiley & Sons Ltd. (See insert for colour representation of the figure.)*

and southern England through the morning (Figure 3.20). The 1200 UTC sounding from Nottingham sampled the environment within which the strongest storms were developing (Figure 3.21). In addition to the very large CAPE, the sounding reveals strong vertical wind shear. The vector difference between winds at the surface and 6 km above ground level was 42 knots, which is sufficient for supercell development. The tephigram shows a capping inversion at around 900–925 hPa (marked in Figure 3.21). Surface temperatures of around 24°C were required for air parcels originating close to the surface to overcome the inversion, given the observed surface dew point temperatures of around 21°C. These temperatures were realised relatively early in the day over the South and East Midlands, and surface-based thunderstorms began to develop from mid-morning. The initial development occurred on the south-east flank of a cluster of elevated storms over the West Midlands, where unobstructed access to the increasingly warm surface air over the South and East Midlands allowed a particularly intense cell to develop. Doppler radar showed strong rotation within the cell by 0930 UTC, and a T4 intensity tornado occurred between Newbold Verdon and Cropston Reservoir in Leicestershire at about 1030 UTC (Figure 3.22).

As this supercell moved north-eastwards, it encountered cooler surface air and soon weakened. However, new surface-based cells were already beginning to develop over the South-west Midlands. As these cells moved north-east and intensified, several became supercellular in nature, exhibiting deep rotation for a time. In general, the merger of these cells with other convective cells resulted in supercellular structures transitioning into multicellular storms after a relatively short period of time (less than an hour). However, one of the rotating cells remained isolated for around 2 hours. This cell persistently exhibited a distinct hook echo and a velocity couplet (Figure 3.23), the radar signatures of the rotating mesocyclone of the storm, which extended to at least 8–10 km

above ground level. During this period, the cell produced a continuous, 110-km-long swathe of hail greater than 20 mm in diameter, in addition to a T2 intensity tornado that produced a 12-km-long trail of damage over Lincolnshire (Figure 3.24). Hail 40 mm in diameter occurred over two swathes; the first, of length about 75 km, extended from Coventry to just south-west of Grantham. The second, much shorter swathe occurred just south and east of Sleaford, at around the time that the tornado was occurring. At several points along the storm's track, hailstones exceeding 75 mm in diameter fell. The largest reported stones were in the Burbage and Melton Mowbray areas (both in Leicestershire), where hail up to 90 mm along the longest axis fell. As a measure of the rarity of such a storm in the United Kingdom, it may be noted that there are only 12 confirmed occurrences of hail greater than 50 mm in diameter in the 60 years ending in 2012. Only 15 hailstorms are known for which the swathe of large hail extended to over 100 km in length in the 210-year period ending in 2010 (Clark and Webb, 2013; also see Chapter 9 of this book).

As the supercell moved north-east over Lincolnshire, it too began to merge with other storms developing to its south, and by mid-afternoon, an expanding multicell cluster of storms had evolved over Lincolnshire. These moved out into the North Sea by late afternoon. Although these storms were intense, no further instances of large hail or tornadoes were reported.

3.10.3   Case 3: The West Cornwall Supercell of 16 December 2012

December 2012 was a mainly unsettled and mild month across the United Kingdom, with numerous low-pressure systems affecting the country. During the 14th and 15th, a complex system of depressions moved slowly north-east across the North Atlantic and north-west Europe. Behind the associated fronts, an rPm air

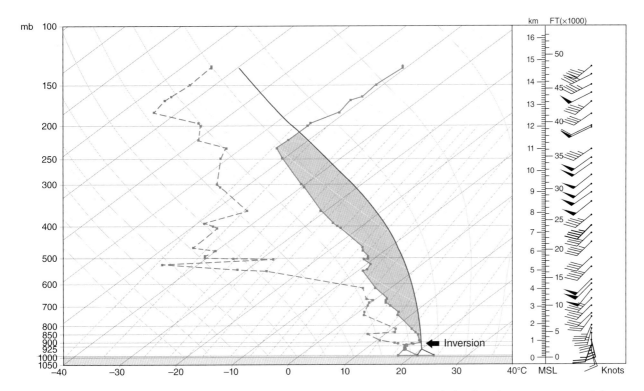

Figure 3.21   *Tephigram showing 1200 UTC 28 June 2012 Nottingham radiosonde data, modified with surface temperature and dew point temperature as observed immediately south-east of the Leicestershire–Lincolnshire supercell. Solid and dashed red lines show the environmental temperature and dew point temperature profiles. Blue line shows the parcel curve. The 'positive area', indicating the CAPE, is shaded orange. Adapted from a plot created using RAOB software. (See insert for colour representation of the figure.)*

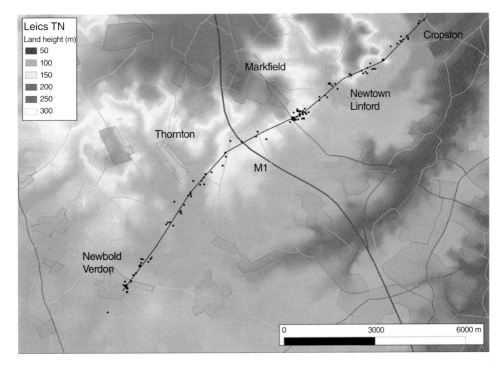

Figure 3.22   *Damage track of the T4 Leicestershire tornado of 28 June 2012, as revealed by a site investigation conducted by TORRO member Tim Prosser. Black dots indicate individual instances of damage, and the black line is the estimated track of the centre of the tornado circulation. The colours indicate the altitude of the terrain (see key, inset). Contains Ordnance Survey data © Crown copyright and database right 2015. © Tim Prosser. (See insert for colour representation of the figure.)*

(a)   (b)

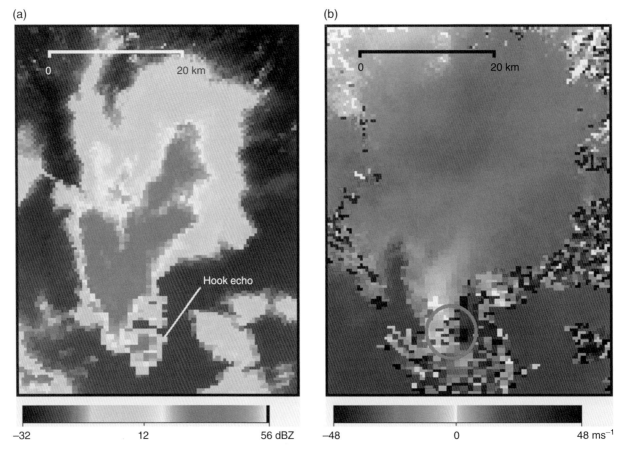

Figure 3.23   *Radar reflectivity (a) and radial velocity (b) at 1321 UTC 28 June 2012, showing the supercell which generated the swathe of large hail in Leicestershire and Lincolnshire and the tornado in Lincolnshire. The velocity couplet, indicative of storm rotation, is marked by the pink circle in panel (b). © Crown Copyright Met Office. (See insert for colour representation of the figure.)*

mass moved in from the Atlantic (Figure 3.11). This air mass, having originated over Canada several days previously, was warmed strongly in its lower layers as it crossed the relatively mild waters of the Atlantic Ocean. By the time it reached the United Kingdom, the air was unstable over a deep layer, as shown by the 0000 UTC Camborne (Cornwall) radiosonde data. Modification of the Camborne sounding using 0900 UTC surface temperature and dew point temperature yields 670 Jkg⁻¹ of CAPE (Figure 3.25). Within the unstable air mass, numerous showers and isolated thunderstorms occurred.

The vertical wind shear within the rPm air mass was generally rather weak. For example, the Camborne 0000 UTC sounding shows 17 knots vector difference between winds at the surface and winds at 6 km above ground level, which is below the range of values typically associated with supercells (see Section 3.3). However, wind profiler observations revealed that the magnitude of the wind shear fluctuated throughout the day, in association with the passage of subtle troughs embedded within the rPm flow. Just ahead of one such trough, surface winds over Cornwall backed and decreased slightly in speed, while upper-level winds strengthened. The magnitude of the 0–5 km vertical wind shear correspondingly increased to over 40 knots just ahead of the trough at about 0900 UTC. Therefore, conditions were favourable for supercell thunderstorms over a limited area, just ahead of the advancing trough.

By 0830 UTC, several of the showers developing close to and ahead of the trough came into range of the Doppler radar network, as they advanced towards south-west England (Figure 3.26). Velocity couplets were apparent in association with several of the convective cells, indicating the presence of updraught rotation (highlighted by the red circles in Figure 3.26b). Most of these cells remained over the sea, but the southernmost cell made landfall at Sennen, on the western tip of Cornwall, around 0900 UTC. A very small hook echo was evident at the south-western edge of the cell between 0840 and 0930 UTC (e.g. as can be seen in Figure 3.26a). A waterspout was sighted from Sennen just before the cell made landfall. As the cell moved through the Sancreed area at about 0915 UTC, it produced a tornado which resulted in a damage track 4.2 km long and 20–30 m wide (Figure 3.27). The damage was mainly to trees, though roof damage also occurred at Treganhoe Farm, just east of Sancreed. The maximum intensity of damage was rated at T2–3. Just north of Treganhoe Farm, the cell produced hail 25–40 mm in diameter. A second tornado occurred just south of St Erth at 0930 UTC, after which the cell rapidly weakened.

It is interesting to compare the size and structure of the tornadic supercells of 28 June 2012 and 16 December 2012, together with another example which occurred over Oxfordshire on 7 May 2012 (Westbrook and Clark, 2013). Radar reflectivity

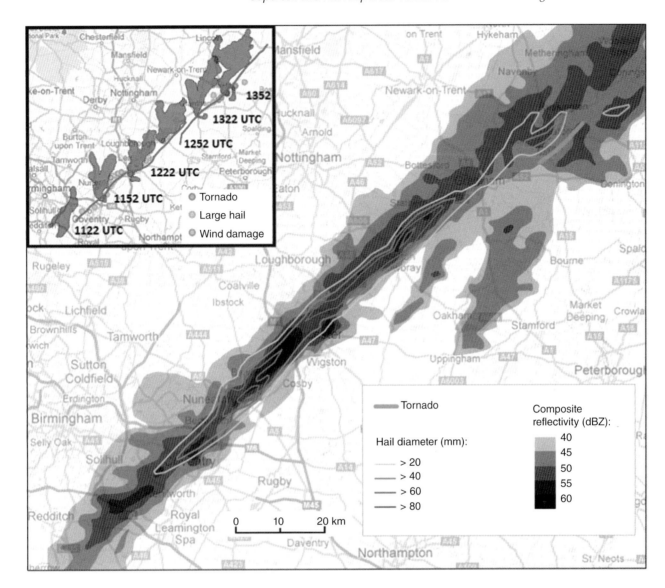

Figure 3.24 *Hail swathe over the Midlands associated with the Leicestershire–Lincolnshire supercell. Lime green line denotes the tornado track over Lincolnshire. Inset panel at top left shows the extent of the 45 dBZ reflectivity echo at 30 minute intervals (shaded), the radar-observed mesocyclone track (pink lines) and locations of individual severe weather reports (coloured circles). Background maps courtesy of Google Maps. Reproduced with permission from Clark and Webb (2013). © John Wiley & Sons Ltd. (See insert for colour representation of the figure.)*

data show that the Sancreed supercell was relatively small – the horizontal dimensions are around one-third of those on 28 June 2012, and one-half of those on 7 May 2012 (Figure 3.28). The smaller horizontal scale is commensurate with the shallower depth of the Sancreed storm – cloud tops were near 8 km, compared to 12–13 km on 28 June 2012. Notwithstanding the differences in scale, all three storms share a similar structure, including the deep, rotating updraught, the hook echo at the up-shear flank of the cell and the extensive band of increasingly light precipitation extending downwind, associated with the advection of precipitation well ahead of the storm core in the strong winds aloft. Despite the smaller scale of the Sancreed storm, reflectivity was no lower than in the larger storms on 7 May and 28 June 2012, demonstrating that relatively shallow, cool-season supercells are not necessarily less intense than their warm-season counterparts.

3.11   Concluding Remarks

In this chapter, we have provided a brief review of some of the types of weather systems in the United Kingdom and Ireland that produce severe convective weather, with emphasis on tornadoes and large hail. Having considered the typical morphology and life cycle of convective showers and thunderstorms, the concept of *storm mode* has been used to identify the key features and characteristic environments of damaging storms. A simple climatology has been examined to describe the temporal and spatial distribution of tornado events according to synoptic type. Although the spatial variations revealed by the climatology are instructive, their reliability is somewhat limited by the relatively short analysis period (10 years). Similar climatologies should be constructed over longer analysis periods in the future, when sufficiently long archives of radar data and surface analysis charts become available.

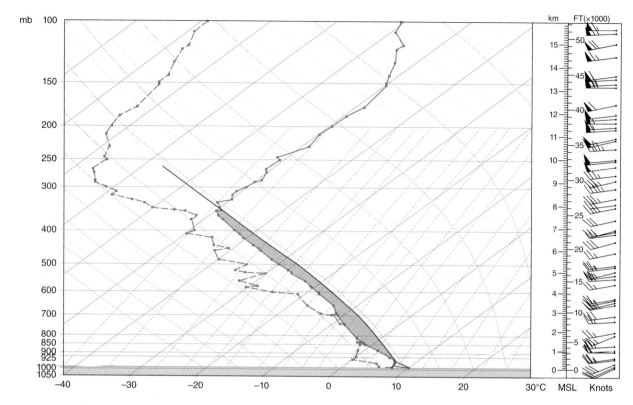

Figure 3.25 *Tephigram showing 0000 UTC 16 December 2012 Camborne radiosonde data, modified with 0930 UTC surface temperature and dew point temperature as observed at Camborne, which was immediately east of the storm at this time. Solid and dashed red lines show the environmental temperature and dew point temperature profiles. Blue line shows the parcel curve. The 'positive area', indicating the CAPE, is shaded orange. Adapted from a plot created using RAOB software. (See insert for colour representation of the figure.)*

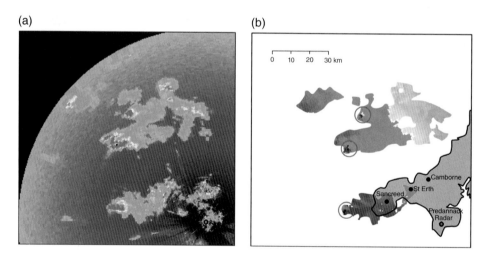

Figure 3.26 *Radar reflectivity (a) and radial velocity (b) at 0846 UTC 16 December 2012, showing three rotating storms near west Cornwall. Reflectivity and radial velocity scales are as in Figure 3.23. © Crown Copyright Met Office. (See insert for colour representation of the figure.)*

A key discrimination we have made is the division between *supercell* and *non-supercell* storms. Supercells are a form of highly organised convective storm possessing a deep, long-lived rotating core – a *mesocyclone*. Despite being generally shallower and less intense than their North American counterparts, supercells in the United Kingdom and Ireland are, nevertheless, associated with a number of different types of high-impact weather including heavy rainfall, large hail and tornadoes. As shown by two of the case studies, supercells may occur in both the warm and cool seasons of the year, the key parameters being the low-level instability and the vertical wind shear present in the storm environment. Despite the relative lack of deep instability in cool-season conditions, strong synoptic and mesoscale forcing, associated with mid-level jet streaks, may produce environments

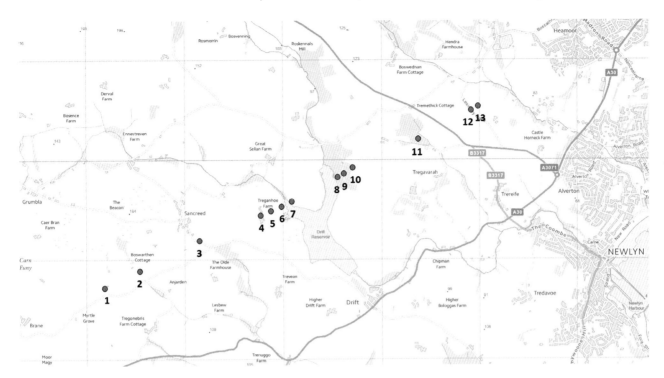

Figure 3.27 *Tornado damage track in the Sancreed area, following a site investigation by the author on 4 January 2013. Background map created using Ordnance Survey OpenData © Crown and database right 2015. Each blue square has dimensions 1 × 1 km. Contains Ordnance Survey data © Crown copyright and database right 2015. (See insert for colour representation of the figure.)*

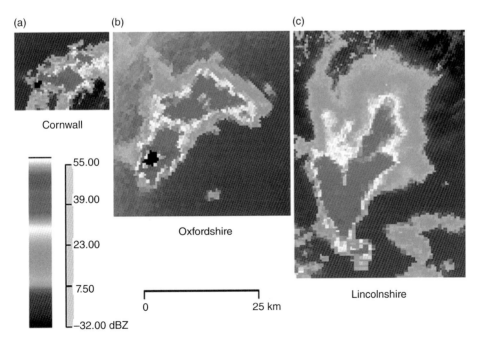

Figure 3.28 *Radar reflectivity imagery showing three examples of tornadic supercells that occurred within the United Kingdom in 2012. (a) 16 December 2012, (b) 7 May 2012 and (c) 28 June 2012. © Crown Copyright Met Office. (See insert for colour representation of the figure.)*

suitable for the initiation and sustenance of *high shear–low CAPE* storms. In such storms, embedded convection may have supercell-like characteristics, thus blurring our initial distinction between the supercell and non-supercell typologies.

Another type of *high shear–low CAPE* weather system which commonly produces severe weather in the United Kingdom and Ireland is the NCFR (see Section 3.10.1). Cold fronts exhibiting NCFRs have been responsible for some of the largest outbreaks of, mostly, weak tornadoes. Although they traverse the country relatively quickly, the intense precipitation strong winds and abrupt wind shear associated with NCFRs adds to the potential for disruption, for example, to transport infrastructure and aviation.

As shown by our case studies, modern meteorological observing systems allow the detection and observation of severe storms at high resolution in space and time. Whilst much of our understanding of supercell storms comes from studies in the United States, comparative efforts have been relatively rare in the United Kingdom and Ireland. However, the potential of weather radar to detect tornadic weather systems in the United Kingdom has been demonstrated by recent studies (e.g. Clark, 2011; Westbrook and Clark, 2013). Full deployment of Doppler and dual-polarisation radars in the UK will further enhance this effort, enabling more case studies and improving forecast capabilities. This will help fulfil the potential first suggested by early radar studies of severe convection in the United Kingdom (e.g. Browning and Ludlam, 1962).

In this chapter, we have not discussed at length the numerical modelling of high-impact weather. Operational forecast computer models now operate at grid lengths of order approximately 1–10 km and are thus practically capable of resolving organised convective storms (such models are termed *convection-permitting* or *convection-allowing*). Given these recent advances, an important area for future research is exploiting the synergy between high-resolution observations and modelling to further our understanding of the type of weather systems that produce unusual but high-impact weather, such as damaging thunderstorm winds and tornadoes (e.g. Smart *et al.*, 2012). Despite these technological advances, the work of recording, cataloguing and investigating, such events, as exemplified by TORRO, will continue to play an important role in scientific studies in the future.

References

AMS (2014) American Meteorological Society Glossary of Meteorology. Available online: http://glossary.ametsoc.org/wiki/MainPage. Accessed on 9 January 2014.

Bailey, N.S., Mortimore, K.O. (1984) The Teignmouth tornado of 26 January 1984. *J. Meteorol.* **9**: 281–287.

Barham, N. (2004) A flash flood assessment. An insurance perspective on the Boscastle floods of 16 August 2004. *J. Meteorol.* **29**: 334–339.

Bennett, L.J., Browning, K.A., Blyth, A.M., Parker, D.J., Clark, P.A. (2006) A review of the initiation of precipitating convection in the United Kingdom. *Quarterly Journal of the Royal Meteorological Society* **132**: 1001–1020.

Bolton, N., Elsom, D.M., Meaden, G.T. (2003) Forecasting tornadoes in the United Kingdom. *Atmospheric Research* **67–68**: 53–72.

Brown, R.A., Wood, V.T. (1991) On the interpretation of single-Doppler velocity patterns within severe thunderstorms. *Weather and Forecasting* **6**: 32–48.

Browning, K.A. (1964) Airflow and precipitation trajectories within severe local storms which travel to the right of the winds. *Journal of the Atmospheric Sciences* **21**: 634–639.

Browning, K.A. (2014) Letter: Large hail in supercell storms. *Weather* **69**: 27.

Browning, K.A., Foote, G.B. (1976) Airflow and hail growth in supercell storms and some implications for hail suppression. *Quarterly Journal of the Royal Meteorological Society* **102**: 499–533.

Browning, K.A., Ludlam, F.H. (1962) Airflow in convective storms. *Quarterly Journal of the Royal Meteorological Society* **88**: 117–135.

Browning, K.A., Pardoe, C.W. (1973) Structure of low-level jet streams ahead of mid-latitude cold fronts. *Quarterly Journal of the Royal Meteorological Society* **99**: 619–638.

Browning, K.A., Roberts, N.M. (1996) Variation in frontal and precipitation structure along a cold front. *Quarterly Journal of the Royal Meteorological Society* **122**: 1845–1872.

Buller, P.S.J. (1978) Damage caused by the Newmarket tornado 3 January 1978. *J. Meteorol.* **3**: 229–231.

Burt, S. (2005) Cloudburst upon Hendraburnick Down: The Boscastle storm of 16 August 2004. *Weather* **60**: 219–227.

Byers, H.R., Atlas, D. (1963) Severe-local storms: In retrospect. *Meteorological Monographs* **5**: 242–247.

Byers, H.R., Battan, L.J. (1949) Some effects of vertical wind shear on thunderstorm structure. *Bulletin of the American Meteorological Society* **30**: 168–175.

Byers, H.R., Braham, R.R. (1948) Thunderstorm structure and circulation. *J. Meteorol.* **5**: 71–86.

Carbone, R.E. (1982) A severe frontal rainband. Part I. Stormwide hydrodynamic structure. *Journal of the Atmospheric Sciences* **39**: 258–279.

Carbone, R.E. (1983) A severe frontal rainband. Part II. Tornado parent vortex circulation. *Journal of the Atmospheric Sciences* **40**: 2639–2654.

Carlson, T.N., Ludlam, F.H. (1968) Conditions for the occurrence of severe local thunderstorms. *Tellus* **20**: 203–226.

Clark, M. (2010) A survey of damage caused by the Monkwood, Hampshire, U.K. tornado of 3 November 2009. *Int. J. Meteorol.* **35**: 291–297.

Clark, M.R. (2011) Doppler radar observations of mesovortices within a cool-season tornadic squall line over the U.K. *Atmospheric Research* **100**: 749–764.

Clark, M.R. (2013) A provisional climatology of cool-season convective lines in the U.K. *Atmospheric Research* **123**: 180–196.

Clark, M., Pask, J. (2013) Site investigation in west Cornwall following the tornadoes of 16 December 2012. *Int. J. Meteorol.* **38**: 92–103.

Clark, M.R., Parker, D.J. (2014) On the mesoscale structure of surface wind and pressure fields near tornadic and nontornadic cold fronts. *Monthly Weather Review* **142**: 3560–3585.

Clark, M.R., Webb, J.D.C. (2013) A severe hailstorm across the English Midlands on 28 June 2012. *Weather* **68**: 284–291.

Davies-Jones R.P., Brooks H. (1993) Mesocyclogenesis from a theoretical perspective. *The Tornado: Its Structure, Dynamics, Prediction, and Hazards*. Geophysical Monograph **79**: 105–114. American Geophysical Union, Washington, DC.

Dessens, H. (1960) Severe hailstorms are associated with very strong winds between 6000 and 12000 m. *Geophysical Monograph* **5**: 333–336.

Doe, R.K. (2004) Extreme precipitation and run-off induced flash flooding at Boscastle, Cornwall, U.K. – 16 August 2004. *J. Meteorol.* **29**: 319–333.

Doswell, C.A. III, Brooks, H.E., Maddox, R.A. (1996) Flash flood forecasting: An ingredients-based methodology. *Weather and Forecasting* **11**: 560–581.

Elsom, D.M. (1983) Tornado at Bicester, Oxfordshire, on 21 September 1982. *J. Meteorol.* **8**: 141–148.

Golding, B., Clark, P., May, B. (2005) The Boscastle flood: meteorological analysis of the conditions leading to flooding on 16 August 2004. *Weather* **60**: 230–235.

Gray, M.E.B, Marshall, C. (1998) Mesoscale convective systems over the U.K., 1981–97. *Weather* **53**: 388–395.

Hand, W.H., Fox, N.I., Collier, C.G. (2004) A study of twentieth-century extreme rainfall events in the United Kingdom with implications for forecasting. *Meteorological Applications* **11**: 15–31.

Hobbs, P.V., Biswas, K.R. (1979) The cellular structure of narrow cold-frontal rainbands. *Quarterly Journal of the Royal Meteorological Society* **105**: 723–727.

Houze, R.A., Hobbs, P.V., Biswas, K.R., Davis, W.M. (1976) Mesoscale rainbands in extratropical cyclones. *Monthly Weather Review* **104**: 868–878.

James, P.K., Browning, K.A. (1979) Mesoscale structure of line convection at surface cold fronts. *Quarterly Journal of the Royal Meteorological Society* **105**: 371–382.

James, P.K., Browning, K.A., Gunawardana, R., Edwards, J.A. (1978) A case of line convection observed by radar using a high resolution colour display. *Weather* **33**: 212–214.

Kawashima, M. (2007). Numerical study of precipitation core-gap structure along cold fronts. *Journal of the Atmospheric Sciences* **64**: 2355–2377.

Keers, J.F., Wescott, P. (1976) The Hampstead storm – 14 August 1975. *Weather* **31**: 2–10.

Klemp, J.B., Rotunno, R. (1983) A study of the tornadic region within a supercell thunderstorm. *Journal of the Atmospheric Sciences* **40**: 359–377.

Knightley, P. (2006) Tornadogenesis across England on 28 July 2005. *Int. J. Meteorol.* **31**: 243–254.

Lemon, L.R., Doswell, C.A. III (1979) Severe thunderstorm evolution and mesocyclone structure as related to tornadogenesis. *Monthly Weather Review* **107**: 1184–1197.

Lewis, M.W., Gray, S.L. (2010) Categorisation of synoptic environments associated with mesoscale convective systems over the U.K. *Atmospheric Research* **97**: 194–213.

Ludlam, F.H., Scorer, R.S. (1953) Reviews of modern meteorology. 10. Convection in the atmosphere. *Quarterly Journal of the Royal Meteorological Society* **79**: 317–341.

Markowski, P., Richardson, Y. (2010) *Mesoscale Meteorology in Midlatitudes*. Wiley-Blackwell, Oxford.

Miller, M.J. (1978) The Hampstead storm: A numerical simulation of a quasi-stationary cumulonimbus system. *Quarterly Journal of the Royal Meteorological Society* **104**: 413–427.

Moller, A.R, Doswell, C.A. III, Foster, M.P., Woodall, G.R. (1994) The operational recognition of supercell thunderstorm environments and storm structures. *Weather and Forecasting* **9**: 327–347.

Morris, R. (1986) The Spanish Plume – testing the forecaster's nerve. *Meteorological Magazine* **115**: 349–357.

Mulder, K.J., Schultz, D.M. (2015) Climatology, Storm Morphologies and Environments of Tornadoes in the British Isles; 1980–2012. *Monthly Weather Review* **143**: 2224–2240.

Newton, C.W. (1960) Morphology of thunderstorms and hailstorms as affected by vertical wind shear. *Geophysical Monograph* **5**: 339–346.

Smart, D.J., Browning, K.A., (2009) Morphology and evolution of cold-frontal misocyclones. *Quarterly Journal of the Royal Meteorological Society* **135**: 381–393.

Smart, D., Clark, M., Hill, L. and Prosser, T. (2012) A damaging microburst and tornado near York on 3 August 2011. *Weather,* **67**: 218–223.

Trapp, R.J., Weisman, M.L. (2003) Low-level mesovortices within squall lines and bow echoes. Part II: Their genesis and implications. *Monthly Weather Review* **131**: 2804–2823.

Trapp, R.J., Stumpf, G.J., Manross, K.L. (2005) A reassessment of the percentage of tornadic mesocyclones. *Weather and Forecasting* **20**: 680–687.

Wakimoto, R.M., Wilson, J.W. (1989) Non-supercell tornadoes. *Monthly Weather Review* **117**: 1113–1140.

Webb, J.D.C., Elsom, D.M., Meaden, G.T. (2009) Severe hailstorms in Britain and Ireland, a climatological assessment and hazard assessment. *Atmospheric Research* **93**: 587–616.

Weisman, M.L., Klemp, J.B. (1984) The structure and classification of numerically simulated convective storms in directionally varying wind shears. *Monthly Weather Review* **112**: 2479–2498.

Weisman, M.L., Trapp, R.J. (2003) Low-level mesovortices within squall lines and bow echoes. Part I: Overview and dependence on environmental shear. *Monthly Weather Review* **131**: 2779–2803.

Westbrook, C., Clark, M. (2013) Observations of a tornadic supercell over Oxfordshire using a pair of Doppler radars. *Weather* **68**: 128–134.

Wicker, L.J., Wilhelmson, R.B. (1995) Simulation and analysis of tornado development and decay within a three-dimensional supercell thunderstorm. *Journal of the Atmospheric Sciences* **52**: 2675–2703.

4

Tornadoes in the United Kingdom and Ireland: Frequency and Spatial Distribution

Peter Kirk[1], Tim Prosser[2], and David Smart[3,4]

[1] Tornado and Storm Research Organisation (TORRO)*, Watford, UK,
[2] Tornado and Storm Research Organisation (TORRO)*, Doncaster, UK,
[3] Tornado and Storm Research Organisation (TORRO)*, London, UK,
[4] UCL Hazard Centre, University College, London, UK

4.1 Introduction

The cover of the very first volume of *The Journal of Meteorology* (Vol. 1, 1975–1976, October 1975) carried a simple map plotting the occurrence of tornadoes, funnel clouds and waterspouts in England, Wales and Scotland 'since January 1974' (see also Chapter 1). Within the covers, an article by Terence Meaden (Meaden, 1975), the then editor and founder of TORRO, outlined the purpose and aims of TORRO and its appeal to amateur observers and weather enthusiasts. The article went on to describe the foundation and work of the Tornado Division within TORRO, 'for the purpose of studying the occurrence, distribution, intensities, and characteristics of British and Continental tornadoes and associated phenomena'. Key to this work was the establishment of a database to form an ongoing record of tornado occurrences in the United Kingdom culled from the media, newspapers, the efforts of amateur and professional weather observers and, not least, reports from the general public. This database – continuously updated to the present day, extended back in time by diligent archive research and maintained in the form of a digital spreadsheet – now forms a unique record for the study of tornadoes in the United Kingdom and Ireland.

Until the foundation of TORRO, the recording and investigation of tornadoes in the United Kingdom and Ireland had been on an *ad hoc* basis. Reports appear in the meteorological journals founded in the latter half of the 19th century (e.g. *The Quarterly Journal of the Royal Meteorological Society*, *Symons's Meteorological Magazine*), but no systematic attempt was made to record and

* http://www.torro.org.uk/

catalogue tornadoes and tornadic damage. Indeed, few detailed investigations were carried out at all until the pioneering work of the meteorologist and climatologist Hubert Lamb, who investigated the damaging tornadoes on 21 May 1950 (Lamb, 1957). It is worth noting that Lamb himself was an enthusiastic supporter of TORRO's foundation and goals (Lamb, 1985).

In the post-Second World War years, further intermittent articles and papers appeared in the *Meteorological Magazine* (published by the Met Office) and *Weather* (published by the Royal Meteorological Society); however, reported events remained largely uncatalogued. With the advent of TORRO in 1974 and the instigation of the TORRO database, records became available which served as raw material for papers in *Weather* by Elsom and Meaden (1984) and in *The Journal of Meteorology* by Elsom (1985) and Reynolds (1995, 1999), the latter being the distillation of a university thesis on the subject (Reynolds, 1998). Since then, Peter Kirk has published further research in this field with Kirk (2007) and Kirk (2014). The first tornado climatology for Ireland was published by John Tyrrell (a director and former head of TORRO) (Tyrrell, 2003) after he started a systematic investigation in 1999 and examined historical reports back to 1950. A study by Holden and Wright (2004) used a simple statistical model to estimate the bias due to population and topographic factors in a 5-year climatology extracted from the TORRO data. Mulder and Schultz (2015) analysed a 33 year climatology, 1980–2012, using TORRO data. Their results are broadly in agreement with previous studies and the analyses presented later in this chapter.

In this chapter, having made some general comments on the nature and quality of the records, we examine two specific aspects of the database: firstly, the distribution of recorded tornadoes over time and, secondly, their spatial distribution across the United Kingdom and Ireland. The spatial analysis employs the technique of kernel density estimation (KDE) mapping to obtain a smoothed interpolation of the data and hence an estimate of the geographical density of events, that is, the number of events per unit area.

4.2   The TORRO Database

The TORRO database records information about every event accepted as being genuine in nature. Quality control involves examination and cross-checking of photographic evidence, eyewitness statements and reports from news media. Wherever possible, site investigations are undertaken to confirm the nature and extent of damage on the ground allowing verification of tornado touchdowns, length of ground track and assessment of the likely intensity of the tornado.

Key information entered into the database includes classification (including an indication of confidence in this classification); date; time; number of events; geographical location, with a grid reference; and the dimensions, direction and nature of the ground track. The latter allows an estimation of the intensity of the event according to the TORRO intensity scale (Meaden, 1985). Additional information includes, for instance, reports of hail and/or thunder, an evaluation and coding of the synoptic situation, fatalities and/or injuries caused and any source references. It should be emphasised that the database is a *work in progress* as, if and when new information comes to light, records are updated and events reassessed. The present investigations utilise a digital spreadsheet version of the database, allowing convenient manipulation, extraction and visualisation of the data. The 30-year period is a standard time frame in climatology, and we adhere to that convention, largely confining ourselves to the period 1981–2010.

The years 1981–2010 are particularly interesting as they cover a period of time when there has been a rapid evolution, even revolution, in the nature of the collection of the records. This is the 'digital era' after circa 1995 when personal computers and access to the Internet became more widely available to the public, thus enabling the development of, and widespread access to, digital news media, Internet forums and newsgroups. More recently, the advent of ubiquitous mobile phone cameras and social media, such as Twitter and Facebook, has further enhanced the ability of members of the public to communicate and share sightings. During this time, there also appears to have been a change in the public perception of tornadoes and severe weather hazards. The 30-year running mean of tornado reports, shown in Figure 4.1, generally increases with time in recent years. Initially, reports were mainly gathered from newspaper articles, but public awareness has gradually increased with time as was noted by Elsom *et al.* (1999). The large number of reports from the public for the major outbreak on 23 November 1981 was partly due to a TV appeal by Michael Hunt of Anglia TV and a press appeal by Mike Rowe. In the United Kingdom, significant events, such as the Selsey (7 January 1998), Birmingham (28 July 2005) and Kensal Rise (7 December 2006) tornadoes (Matthews, 1999; Kirk, 2006; Kirk, 2013), were reported in the national news media. More widely, television documentary-style coverage of natural hazards and disasters, such as series by the Discovery Channel (*Storm Chasers*), the American cinema film *Twister* (distributed by Universal Pictures, 1996) and the climate change debate have increased awareness of natural hazards and tornadoes specifically. When considering the changes in the frequency of tornado reports over time, the role of these confounding factors, increased ease of reporting and public awareness, needs to be borne in mind, even if it is not possible to objectively separate them from true climatological factors.

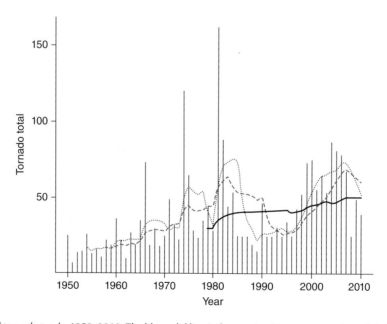

Figure 4.1   *UK and Ireland tornado totals, 1950–2010. The blue solid line is the running 30-year mean, the red dashed line the running 10-year mean, and the black dotted line the running 5-year mean. Reproduced with permission from Kirk (2014). © John Wiley & Sons Ltd. (See insert for colour representation of the figure.)*

An unfortunate result of the increase in public awareness is that media reports sometimes refer to incidents of strong straight-line winds, 'wind gusts', or land ('dust') devils as tornadoes. There are also instances of images being faked or doctored before being sent to the media or posted on web forums; therefore, a sceptical vigilance has to be maintained. As far as possible, TORRO scrutinises all reports carefully before entering them onto the database. Problems such as these underline the importance of site investigations and obtaining 'ground truth' for determining whether or not a definite tornado has actually occurred.

Tornadoes are likely to be under-reported over less populated areas. However, Doppler radars are able to detect areas of rotation within storms, including mesovortices (relatively shallow, small-scale rotating circulations as was shown for the 7 December 2006 tornado in Kensal Rise, London (Clark, 2011a)). At the time of writing, the UK Meteorological Office has a programme of installing dual-polarisation Doppler radars to provide coverage across the whole of the United Kingdom. There is the potential for further events, which would otherwise remain unknown, to be detected in this way. One such tornadic event was discovered west of Odstock Down, Wiltshire, on 3 November 2009 when several months later it was noticed that the radar imagery indicated the presence of mesovortices in the area. Investigations on the ground then found the track of a tornado (strength T1) along the path of one of the radar-detected circulations (Brown and Meaden, 2011). Obtaining a quantitative estimate of this under-reporting bias is difficult; however, Holden and Wright (2004) estimated that as few as 20% of all tornadoes were reported during the period of their climatology 1995–1999. Events which occur offshore close to the coast rarely do any significant damage and so are only reported if witnessed at the time. These events and those occurring during the nocturnal hours are likely under-reported.[1]

4.3   Tornado Frequency for the United Kingdom and Ireland: 1981–2010

4.3.1   Annual Number of Tornadoes and Tornado Days

Previous studies for the United Kingdom covering the period 1960–1989 found a mean number of tornadoes over land of 33.2 per year (Reynolds, 1999), hence the long-quoted figure for the incidence of

tornadoes in the United Kingdom of 33 on land per year, and a mean number of tornadoes over water, colloquially known as 'waterspouts', of 11.1 per year (Reynolds, 1998). However, some of these events were counted twice if they had crossed from land to water or vice versa and so cannot be added to each other to find a combined total. A combined mean total of 40.3 tornadoes per year for this period has been calculated by Kirk (2014). Elsom (1985) found that the annual decadal mean numbers for tornadoes in the United Kingdom had increased from 11 in the 1950s to 66 in the 1980s. Elsom and Meaden (1984) found a mean of 10 tornado days for 1960–1969 and a mean of 14 days for 1970–1979 for tornadoes over land for the United Kingdom. A 'tornado day' is defined as a day in which a tornado occurs, from 0000 to 2359 UTC.[2]

Tornado totals have been calculated for the UK by Kirk (2014) covering the period 1981–2010 (see Table 4.1), with 47.2 ± 10.5 tornadoes per year and 24.3 ± 3.3 tornado days per year. Excluding waterspouts there are 36.5 ± 10.1 tornadoes on 18.9 ± 3.0 days. For the UK and Ireland as a whole, there are 50.7 ± 10.9 tornadoes per year and 26.3 ± 3.8 tornado days per year. Excluding waterspouts there are 39.2 ± 10.3 tornadoes on 20.6 ± 3.5 days. It should be cautioned that the figures for Ireland show a sharp upturn from 1999 onwards, something not seen in the UK figures,

Table 4.1   *Tornado and tornado day totals, 1981–2010.*

Year	UK Tornadoes	UK Tornado Days	UK and Ireland Tornadoes	UK and Ireland Tornado Days
1981	162	19	162	19
1982	86	32	88	34
1983	43	24	45	25
1984	54	20	54	20
1985	24	14	25	14
1986	23	15	25	16
1987	24	19	24	19
1988	19	14	20	15
1989	15	14	16	14
1990	47	20	47	20
1991	23	11	25	12
1992	25	14	25	14
1993	30	24	31	25
1994	25	17	25	17
1995	31	22	35	26
1996	24	11	25	12
1997	43	26	43	26
1998	50	36	53	38
1999	67	36	74	42
2000	68	33	75	35
2001	47	26	56	29
2002	57	34	66	37
2003	49	26	54	30
2004	79	46	87	50
2005	75	40	81	42
2006	66	35	79	43
2007	60	32	68	37
2008	20	17	25	20
2009	44	25	49	29
2010	35	26	39	30

[1] Tornado totals collected by TORRO cover the United Kingdom (including the Channel Islands and Isle of Man due to their historical connections to the United Kingdom as Crown Dependencies) and the Republic of Ireland, out to a distance of 30 km offshore. Thirty kilometers has been chosen as a maximum distance that could be seen from land on a clear day. These totals include all confirmed and probable tornado reports but not *possible* tornado reports (i.e. events for which there is little or no supporting evidence). Where there is less certainty in information about a tornado such as track width or length, then the values in the database have been used being a most likely figure based on analysis of reports at the time. All tornado ratings referred to in this chapter are the maximum reached, or in the case of track lengths the total distance covered, during the lifetime of a tornado and are based on the International Tornado Intensity Scale (Meaden, 1985) which rates tornadoes on the wind speed (T scale), track length (L scale), track width (W scale) and track area (A scale). All mean figures appended by ± indicate a population mean using 95% confidence intervals.

[2] Universal Time Coordinate (UTC) is used throughout this chapter and for our purposes is equivalent to Greenwich Mean Time (GMT).

so there are likely events missed pre-1999 for Ireland. This effect may be related to the commencement of systematic investigations by John Tyrrell. The mean number of tornadoes for Ireland from 1981 to 1998 is 1.1 a year with a mean annual figure of 7.2 from 1999 to 2010. There is therefore more confidence in the mean annual figures for the United Kingdom than for the combined UK and Ireland figures over this period.

The number of tornadoes per year across the United Kingdom and Ireland is highly variable. During the period 1981–2010, this ranged from a low of 16 in 1989 to a high of 162 in 1981. The second highest yearly total is substantially lower than the 1981 total with 88 tornadoes recorded in 1982. Tornado day totals over this period ranged from a low of 12 in 1991 and 1996 to a high of 50 in 2004. The tornado day total for 1981 was just 19, reflecting the anomalously large number of reports in one outbreak on 23 November of that year (Meaden and Rowe, 1985; Rowe and Meaden, 1985).

4.3.2   Season and Month of Occurrence

Of the 1,504 tornadoes where the month of occurrence is known, 16.9% were recorded during winter (DJF), 14.9% during spring (MAM), 28.9% in summer (JJA), and 39.4% in autumn (SON). In terms of monthly totals, November was the most active month (Figure 4.2 and Table 4.2). The figure for November is, however, skewed with nearly half of all events (104) occurring on a single day on 23 November 1981. The other most active days in this period were 20 October 1981 with 29 tornadoes and 21 September 1982 with 26 tornadoes. For tornado days per month, a similar profile is found to that of tornadoes per month except the large outbreaks are filtered out (Figure 4.2 and Table 4.2).

23 November 1981 remains by far the most active day on record for the United Kingdom, with 104 tornadoes recorded on a squall line running ahead of a cold front. These occurred in a little over 5 hours in an area north of a line running from north Wales to south Essex and south of a line running from Blackpool to Hull. Of these, there were seven rated T0, 24 at T1, 47 at T2, 18 at T3

and three at T4 (Meaden and Rowe, 1985). Since the original paper was published on the event and as a result of the ongoing effort to reassess and update records, one tornado has since been discounted from this day (see Chapter 5).

Rowe (1985) found that in terms of outbreaks of 10 or more tornadoes on one day for the United Kingdom, there were 10 such events known between 1870 and 1984, nine of which were between 1966 and 1984 and all of which occurred between September and February. Since 1984, there have been 4 outbreaks of 10 or more tornadoes, namely, on 25 December 1990 (14 tornadoes), 30 December 2006 (11 tornadoes), 24 September 2007 (13 tornadoes) and 3 November 2009 (10 tornadoes), all of which were entirely confined to the United Kingdom. The most active days for Ireland only were 30 September 2001 and 1 January 2005 with four tornadoes each. The 30 September 2001 event was particularly notable in that four tornadoes occurred in close succession south-west of and around Mullingar, County Westmeath, with two of these tornado paths actually crossing over each other (Tyrrell, 2002).

Table 4.2 *Percentage of UK and Ireland tornadoes by month and tornado days by month, 1981–2010.*

Month	UK and Ireland Tornadoes	UK and Ireland Tornado Days
January	5.4	6.3
February	4.8	3.8
March	4.6	5.1
April	4.1	5.2
May	6.2	7.5
June	7.2	9.6
July	9.0	11.9
August	12.6	14.8
September	14.5	12.5
October	10.4	10.4
November	14.5	7.5
December	6.7	5.4

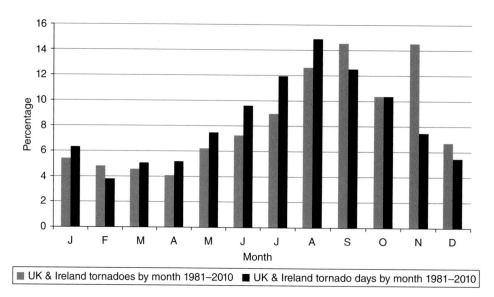

Figure 4.2 *Percentage of UK and Ireland tornadoes and tornado days by month, 1981–2010. Reproduced with permission from Kirk (2014). © John Wiley & Sons Ltd. (See insert for colour representation of the figure.)*

4.3.3 Hour of Occurrence

Of the tornado events in the database, there is a known hour of occurrence for 1013 of them. The most active hours of the day were 1400–1459 UTC at 11.4% and 1500–1559 UTC at 10.3% with a general peak from 1100 to 1759 UTC accounting for 57.7% of all tornadoes (Figure 4.3). The least active time of day was 2000 to 0459 UTC accounting for just 11.9% of known events. Due to the nature of tornadoes over water ('waterspouts'), these are only likely to be detected during daylight hours. All waterspout cases were reported between 0500 and 2059 UTC, with the exception of one case, the timing of which is open to doubt.

4.3.4 Intensities

Of the total number of 1521 tornadoes over this period, 824 have been assigned ratings based on the TORRO T scale (see also Chapter 1 and Appendix D) (see Table 4.3). The strongest tornado reached an intensity of T5/6 on 28 July 2005 at Birmingham in the West Midlands (Kirk, 2006). For an account of tornado extremes in the TORRO database, see Chapter 5.

Tornadoes of T0 strength are difficult to assess accurately due to the lack of damage caused. It is also difficult to rate tornadoes above T5 accurately due to the nature and complexity of the damage track.

4.3.5 Track Lengths

302 tornadoes have been assigned length ratings based on the TORRO L scale (see Table 4.4 and also Appendix D). Of the four L7s, the longest track was recorded for the 31 December 2006 T4/5 tornado with a length of 30.2 km from Ardmore, County Armagh, to Loanends, County Antrim (Tyrrell, 2008), in Northern Ireland. The second longest track was the 17 March 1995 Dunboyne, County Meath, tornado at approximately 29 km (Tyrrell, 2007) which is the longest track on record for the Republic of Ireland.

4.3.6 Maximum Track Widths

202 tornadoes have been assigned maximum width ratings based on the TORRO W scale (see Appendix D and Table 4.5), and of the three W8 tracks, two shared the record for the maximum width of 900 m (the highest value on record for the United Kingdom when two wider historical events are discounted). These were the 7 January 1998 Selsey, West Sussex, T3 tornado (Matthews, 1999) and the 4 July 2005 Burton Joyce, Nottinghamshire, T1 tornado (Brown and Meaden, 2010), although there is less certainty on the

Table 4.3 *Percentage of UK and Ireland tornado intensities, 1981–2010.*

TORRO Scale Intensity Rating	Percentage
T0	9.0
T1	25.6
T2	37.3
T3	21.5
T4	5.7
T5	0.8
T6	0.1

Table 4.4 *Percentage of UK and Ireland tornado track lengths, 1981–2010.*

TORRO Scale Length Rating	Percentage
L0	5.0
L1	7.9
L2	12.9
L3	27.8
L4	24.8
L5	14.2
L6	6.0
L7	1.3

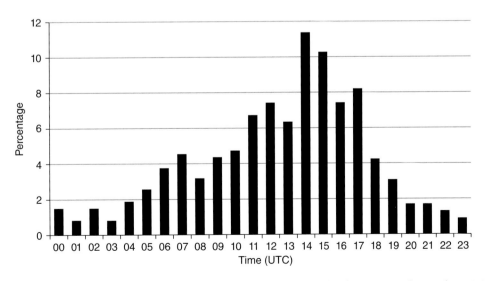

Figure 4.3 *Percentage of UK and Ireland tornadoes by hour, 1981–2010. Reproduced with permission from Kirk (2014). © John Wiley & Sons Ltd.*

rating for the latter event. The third W8 event was the 28 July 2005 Birmingham, West Midlands, tornado at 500 m. The widest tornado in Ireland was the 1 January 2006 Togher, County Meath, tornado at 460 m (Tyrrell, 2007), which is the widest track on record for Ireland.

There were no records of W0 tornadoes over this period, nor indeed in the whole database, and given these have maximum track widths of 2.1 m, they are very hard to detect and therefore easily overlooked. It is debatable as to whether a track of such low width is classifiable as being due to a definite tornado.

4.3.7   Directions of Travel

In all, 340 tornadoes have known directions of travel, given as the point of the compass they arrive from (Figure 4.4). Where track directions are known to have changed mid-event, then the initial direction at touchdown has been used. The most common direction

Table 4.5   *Percentage of UK and Ireland tornado maximum track widths, 1981–2010.*

TORRO Scale Width Rating	Percentage
W0	0.0
W1	1.0
W2	5.4
W3	21.3
W4	15.8
W5	28.7
W6	22.3
W7	4.0
W8	1.5

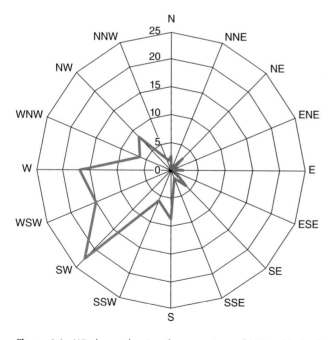

Figure 4.4   *Wind rose showing the percentage of UK and Ireland tornado by direction, 1981–2010. Reproduced with permission from Kirk (2014). © John Wiley & Sons Ltd.*

for tornadoes to move from is south-west, the prevailing wind direction for the United Kingdom and Ireland, towards the north-east (22.9%). Tornadoes moving in from the west were noted in 17.1% of cases and from the west–south-west in 15.0% of cases. Movement in the sector from the south clockwise round to the north-west makes up 84.7% of all reports. No tornadoes are reported as having moved in from the north-north-east. An analysis of the current complete database shows just 10 are known to have moved in from this direction with the last having occurred in August 1915 in Constantine, Cornwall, with none on record for Ireland.

4.4   Spatial Distribution of Tornadoes in the United Kingdom and Ireland

4.4.1   Simple Mapping of the Database

The distribution of tornado events from the complete dataset covering the years 1054–2013 is shown in Appendix C & colour insert Map C.1. This figure was composed by plotting the tracks of all known events over an image of Britain and Ireland derived from the NASA Blue Marble composite satellite image, with urban areas coloured based on a land-use atlas produced by the European Environment Agency in the year 2000. Stronger tornadoes are shown in a lighter shade and are plotted on top of lesser events. Note that the track widths are exaggerated, so this does not represent the actual land coverage.

Figure 4.5 shows the distribution of all recorded tornado tracks for the selected 30-year time period 1981–2010, and Figure 4.6 the geographical occurrence of *tornado days*, plotted on a modified 10 km grid, for the same period. Clearly, there is a concentration of events in England with a relative paucity of reports in the south-west, central Wales, northern England, southern and northern Scotland and western/south-western Ireland. On the other hand, relative concentrations of reports are found along the south coasts of England and Wales, south-west of London (with few reports from the London metropolitan area itself; see Section 4.7), the West Midlands, southern-central Scotland, the areas around Belfast and west of Dublin. A broad swathe of relatively high density appears to exist through England from the Wirral, in the north-west, to the East Anglia coast. Figure 4.6 suggests 'hotspots', that is, positive anomalies, along the south coast east of the Isle of Wight, the Suffolk coast around Ipswich, the south Wales coast, the area south-west of London, the West Midlands around Birmingham and the Wirral and Cheshire area in north-west England. There appears to be a general clustering of events in areas of high population density, for example, the London area and the south-east, the Bristol area and the south coast of Wales, the West Midlands conurbation and north-west England, suggesting a significant population bias in the data.

4.5   Issues with Mapping

As described previously, each record of a tornado that TORRO has collected consists of a description, various measurements (such as strength and path width), a location for the start of any

Figure 4.5   *Recorded tornadoes in the United Kingdom and Ireland, 1981–2010. Each event is plotted with a red marker. Reproduced with permission from Kirk (2014). © John Wiley & Sons Ltd. (See insert for colour representation of the figure.)*

damage track and (where known) a location for the end of any damage track. Although most of the tracks caused by stronger tornadoes were surveyed after the event, the majority of lesser events have only been inferred as being tornadic in origin from reports in the media or from information submitted to TORRO by third parties. As a result, it has not always been possible to give every event precise coordinates, and some may only be defined to an accuracy of 1 km or even, in a few rare cases, 10 km. These coordinates are mostly given as grid references, either using the Ordnance Survey National Grid reference system (OSGB) or the Irish Grid reference system (OSI). By convention,

these grid references refer to an entire grid square, rather than a point, with the size of the square defined by the accuracy of the reference. For the purposes of mapping, therefore, a 'best guess' for each event location was made as being the centre of the grid square given by the reference, rather than the south-western corner.

Although a straightforward map of tornadoes does show the general distribution, it does not give any direct information on the frequency of occurrence within any particular area, as this is something that is difficult to assess visually. It is useful, therefore, to calculate the 'density', or number of events per

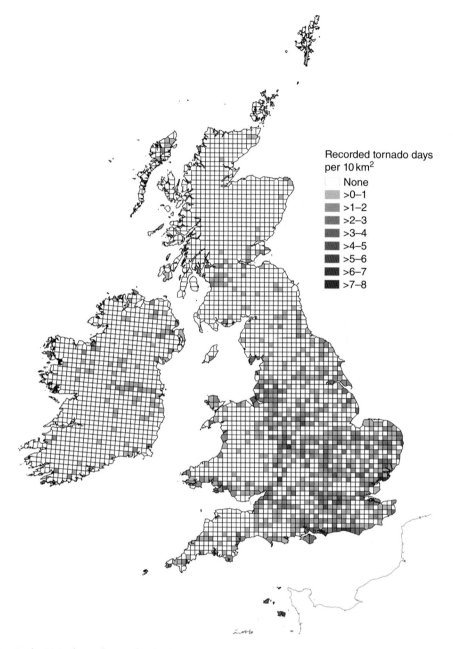

Figure 4.6   *Tornado days in the United Kingdom and Ireland, 1981–2010. © Tim Prosser. (See insert for colour representation of the figure.)*

Recorded tornado days per 10 km²
None
>0–1
>1–2
>2–3
>3–4
>4–5
>5–6
>6–7
>7–8

unit area, across the region being studied. This can be used to visualise better where tornadoes have occurred most often and allows some tentative hypotheses to be formed as to why this might be. A restriction does need to be made to the domain over which the density can be calculated. As mentioned previously, *waterspouts* do not leave any trace and can thus only be recorded by observation. A large proportion are, therefore, likely to be missed and to avoid this potential source of bias, only tornadoes which have tracked over land at some point in their path have been considered, and the domain for which the density is calculated is restricted to the land only. The density distributions reported here should be treated cautiously

as there may be a number of non-meteorological factors at work. In particular, only tornadoes which have caused damage to property are likely to be reported, and thus, urban areas will tend to attract more reports when compared to rural areas suffering a similar underlying density. An example of this may be the maximum in central England, best seen in Figure 4.7, which does coincide almost exactly with the conurbation around Birmingham. There is no obvious meteorological reason for Birmingham to be so afflicted, at least when compared with the rest of the Midlands, although of course it is entirely possible that further research may reveal a plausible cause.

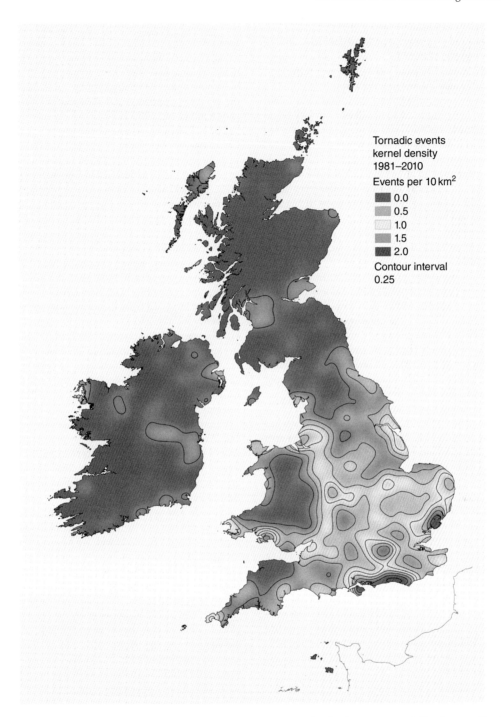

Figure 4.7 *Gaussian kernel density map of recorded tornado events in Britain and Ireland, 1981–2010. © Tim Prosser. (See insert for colour representation of the figure.)*

There is also a question as to whether there could be any bias caused by the availability of observers or surveyors, given that TORRO is a purely voluntary organisation. A particularly keen (or well-known) observer can increase the relative number of reports in their local area. This is a well-known problem in the natural history disciplines, for instance, when trying to ascertain species distribution. As mentioned previously, another source of bias or influence is the media attention given to some events.

Overall, the reader must bear in mind that issues with variable data quality and sample size present a challenge when interpreting data of this nature and require careful consideration; for a discussion of these points in relation to the much larger US tornado database, see Doswell (2007).

A traditional method of calculating density information across a defined area is to split the study area into smaller areas and count how many events fall within each division. This approach was used for previously published maps of British tornado frequency (Elsom and Meaden, 1984; Reynolds, 1999; Kirk, 2007), which counted the number of tornadoes recorded in each county between 1960–1982, 1960–1989 and 1980–2004, respectively. A density was then calculated by dividing the count by the known area of each county. A more common means of division is to use a regular grid to divide the study area into squares, usually based on a projected coordinate system (such as the Ordnance Survey National Grid). The grid pattern works well where the boundaries of the study area are relatively straight, but in this case, the boundary is a highly irregular coastline (remembering that we are only considering tornadoes over land). This can generate very small areas of land where the majority of a grid square is sea. If an event happens to occur within such a small area, it will produce an unrealistically high density. To avoid this problem, modification to the grid pattern is required, so that these small areas are merged into neighbouring ones.

In this study, a modified grid was generated using a method known as Voronoi tessellation (see Section Additional Information). On counting the number of events within each map division, it is preferable to count *tornado days* – the number of days a tornado has occurred – rather than just *tornadoes*. Closely spaced tornadoes on the same day will likely have been generated by the same storm or event, which will have affected any particular location only once, so this is a better measure of risk. Furthermore, without this reduction, a weak, intermittent track would generate a higher tornado count than a stronger, continuous track. This is obviously not an ideal representation. Once the number of tornado days has been calculated for each division, this can be divided by its area relative to a complete grid square to give a value in tornadoes per $100 \, \text{km}^2$.

Although the grid method works reasonably well in depicting the tornado distribution, it is subject to a number of difficulties. The most obvious problem is that the exact positioning of the grid lines can have a large effect on the maximum density, for example, by splitting (or not splitting) a closely spaced cluster of events in half. Another related problem is that the density recorded in each area takes no account of that in surrounding areas, so that a grid square with no tornadoes next to one with many records is assigned the same risk as one in a large area with no records, even though there could conceivably be an increased risk near to a known hotspot. One way to avoid these difficulties is to use *KDE* (see Section Additional Information), and this is the technique we have used to derive additional, more realistic spatial density estimates.

4.6 Kernel Density Mapping of Tornado Distribution

Our derived kernel density map is shown in Figure 4.7. Although we do not discuss the details here, the features in this map are robust to adjustment of the parameters used in the KDE analysis. As in the previous figures, the different treatment of the data reveals that the highest density of recorded tornadoes in Britain and Ireland is on the south coast of England, in particular the stretch from the eastern side of the Isle of Wight to Beachy Head, East Sussex. There are also maxima around the coast of Suffolk, to the south and west of London and in central England, as well

as a generally higher density in areas such as Cheshire, coastal Lancashire and southern Wales. The Channel Islands also appear to have a very high density although only three of the records there, out of a total of 15, have an intensity of greater than T1 and most may in fact be relatively weak tornadoes initially seen over the sea (as waterspouts) and with only short land tracks – thus, it is difficult to say if this is a particularly robust anomaly.

The overarching general concentration in England, that is, in the south-eastern quadrant of the United Kingdom, coincides with the climatological maximum in convective available potential energy (CAPE) as depicted in climate reanalyses (e.g. Romero *et al.*, 2007). CAPE is a measure of atmospheric instability which reaches its annual climatological maximum during the summer months. As has been shown, the frequency of tornadoes peaks in the late summer and early autumn (if we neglect the apparently anomalous month of November) (see Figure 4.2) when sea surface temperatures around the coast are at their annual peak. During the warm season of the year, the association of increased atmospheric instability with thunderstorm activity, either 'home grown' or imported from the near continent, and hence a greater frequency of tornadoes spawned by convective storms might explain some of this distribution. During autumn, the relatively strong north–south temperature gradient in the North Atlantic sector is associated with incursions of polar maritime air masses and the movement of cold fronts across the country, some of which turn out to be tornadic in nature (Clark, 2011b; see also Chapter 3).

Figure 4.8 shows the total annual density in Scotland, England and Wales broken down by the meteorological seasons of the year. The relation to the temporal distribution by month, presented previously in Figure 4.2, is clear with the most prominent geographical anomalies present in summer and autumn. The Birmingham, West Midlands, anomaly is most prominent during summer as is an apparent maximum to the south-west of London. We will not attempt to explain all the anomalies apparent in Figures 4.7 and 4.8. At present, we have no plausible explanation for the density maximum in proximity to the East Anglia coast, especially in the Ipswich area. It is possible that it is simply a 'recorder effect', due to one or more observers in the locality. A hypothesis that this may be related to sea breeze activity in the summer months is not supported by the breakdown into seasonal plots (Figure 4.8), as the occurrence of sea breezes peaks during the summer months. However, it may be that the much weaker anomaly running roughly northwards through East Anglia, then westwards into East Anglia, is related to the inland penetration of sea breeze fronts or a tendency for 'heat lows' to form during the summer months (see also Chapter 3). Here, we will look in more detail at two anomalies that have been identified and received attention previously in the literature.

4.7 The 'London Metropolitan' Anomaly

In a paper published in the journal *Monthly Weather Review*, Elsom and Meaden (1982), having investigated 150 years of data, pointed out an apparent deficit of tornado events within the inner London metropolitan area. They suggested that this effect was due to the suppression of tornadoes, from storms which otherwise would have produced them, by factors in the urban

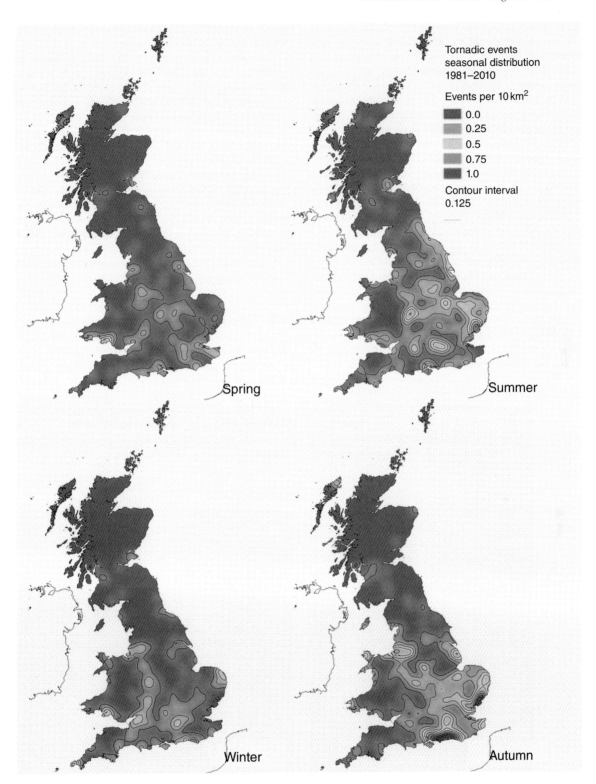

Figure 4.8 *Gaussian kernel density maps of recorded tornado events in Britain and Ireland, 1981–2010, by season. Spring, March, April and May; summer, June, July and August; autumn, September, October, November; winter, December, January, February. Note the change in scale from Figure 4.7. © Tim Prosser. (See insert for colour representation of the figure.)*

environment, for example, the increased roughness factor (a measure of the smoothness of the surface) and the urban heat island effect. Note that due to the prominence of London as a capital city with a large and dense population, we would expect any tornado events to be well recorded and documented, as emphasised by Elsom and Meaden. The possibility that weak tornadoes are suppressed in the inner London area does not preclude the occurrence of stronger events. A recent example was seen on 7 December 2006 when a T5 tornado tracked for 2.5 km across Kensal Rise in north-west London. This was the strongest tornado to hit London since a T7 tornado on 8 December 1954 tracked for 20 km from Gunnersbury, west London, to Southgate, north London (Kirk, 2013).

Data for our chosen 1981–2010 period also displays a weak local minimum in this area (see Figure 4.7). To investigate this further, in Figure 4.9a and b, we focus on the London area. Figure 4.9a suggests the positive anomaly is located over the south to south-western periphery of the modern London suburbs. Relative minima are located over north-east London and north Kent. For the current complete TORRO database, covering the years 1054–2013, the maximum is located in west London. In neither plot is there evidence for a deficit in the inner London area; in fact, the reverse appears to be true, especially so for the extended period of the complete database. Elsom and Meaden (1982) did not explicitly identify a local maximum in the area around London, although they commented on the higher observed frequency in the less densely populated suburbs and surrounding countryside. Figures 4.7 and 4.9a suggest that, in the 1981–2010 data, there is a prominent positive anomaly to the west and south-west of London covering a significant geographical area. This anomaly is most evident during the summer months.

Such an anomaly deserves further investigation to determine whether it is a real effect, for instance, partly associated with warm-season thunderstorm activity. If this is the case, a possible trigger for tornadogenesis could be the interaction of sea breezes originating from either the Thames Estuary or south coast or both, with convective storms initiated *in situ* to the west or south-west of London. Pedgley (2003) put forwards the idea that advection of the London heat island plume westwards was associated with the formation of deep convection west of London and implicated the inland penetration of sea breezes over southern England in the genesis of the Bracknell hailstorm on 7 May 2000. The sea breeze front has been identified as playing a possible role in the initiation of convective storms in the United Kingdom (see the review by Bennett *et al.*, 2006), although its significance overall is unclear. Detailed case studies and further climatological analyses are required to address this issue.

4.8   The Isle of Wight and South Coast Anomaly

The area in the vicinity and downwind (i.e. eastwards – downstream in the prevalent west or south-westerly winds) of the Isle of Wight, as far east as Beachy Head, East Sussex, has been recognised as a 'hotspot' for tornado and waterspout observations, both in the literature and from anecdotal reports. Matthews (2003) highlighted recent significant tornadoes in the area and assessed the possibility that the local topography had an influence on the genesis of vortices. He also pointed out the high population

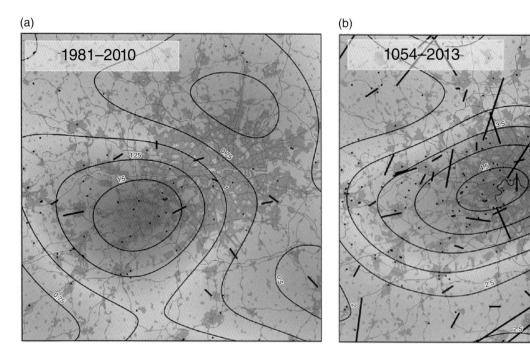

(a)   1981–2010

(b)   1054–2013

Figure 4.9   *Gaussian kernel density maps of recorded tornado events in the TORRO database for the London area. Density is shaded from red orange (high) to blue green (low); smoothed contours of density every 0.25 events per 10 km²; tornado events are plotted with points and lines against a modern urban shape-file background. (a) 1981–2010. (b) 1054–2013 (complete TORRO database). Created using OS OpenData™, contains Ordnance Survey data © Crown copyright and database right 2015. © Tim Prosser. (See insert for colour representation of the figure.)*

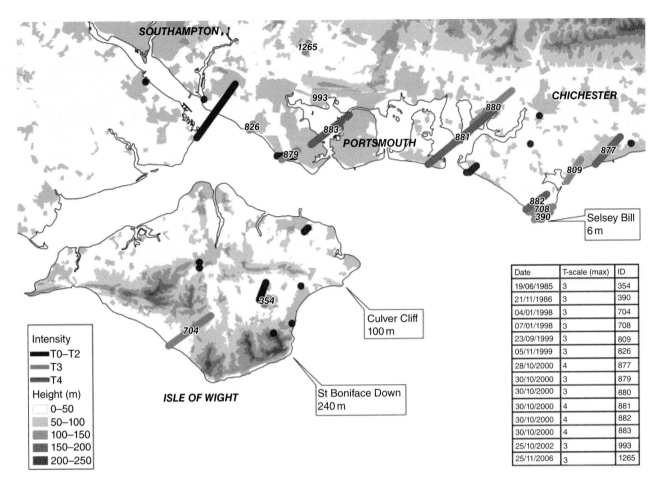

Figure 4.10 *1981–2010 tornado events in the vicinity of the Isle of Wight and Selsey Bill. Terrain height is shaded in grey according to the inset key (left), and urban areas are indicated by the transparent beige shading. Events are coloured according to strength (inset key, left). Event tracks are identified at their end points by their identification number in the TORRO database (inset key, right). Only events rated at T3 and above have been labelled. Created using OS OpenData™, contains Ordnance Survey data © Crown copyright and database right 2015. © Tim Prosser. (See insert for colour representation of the figure.)*

density in the area, which may influence the recording of events. This was followed up in a paper in *Atmospheric Research* (Meaden *et al.*, 2005) as well as the review paper on tornado forecasting in the United Kingdom by Bolton, Elsom and Meaden in the same journal (Bolton *et al.*, 2003).

The general mechanism invoked in these papers to explain the relatively high incidence in this region is the shedding of vortices downwind from the Isle of Wight. Such vortices are thought to be generated by airflow, under favourable atmospheric conditions, over the higher terrain of the Isle of Wight or promontories such as Selsey Bill on the West Sussex coast. The KDE analysis confirms that the Isle of Wight/south coast maximum is a robust anomaly; indeed, it is the largest, most intense anomaly apparent in the whole dataset (Figure 4.7). Zooming in on this region, Figure 4.10 shows an area immediately in the vicinity of the Isle of Wight and Selsey Bill, with the tracks of events from the 1981 to 2010 database overlaid. Visual inspection suggests that extrapolation of many of these tracks does indeed suggest a focus upwind near the island; however, the inference of a physical association remains largely speculation. As yet, the hypothesis that

vortex generation by the Isle of Wight plays an important role in tornadogenesis in the locality remains unsubstantiated by detailed meteorological studies or physical and/or numerical simulations.

4.9 Concluding Remarks

In this chapter, we have sought to describe the temporal and spatial frequency of tornado events as recorded in a subset of the TORRO database. The subset covers the most recent 30-year climatological period from 1981 to 2010. Allowing for caveats regarding recent trends in the reporting and recording tornadoes, the results for the time distribution analysis are broadly consistent with previous studies. Over this period, a mean number of about 47 tornadoes per year occurred on about 24 days per year for the United Kingdom and about 51 occurred on 26 days per year for the United Kingdom and Ireland combined. They were most common from July until October and were most frequent from 1100 to 1759 UTC. Most tornadoes were rated from an intensity of T1 to T3 and had a track length of L2 to L5 and a maximum width of

W3 to W6 on the relevant TORRO scales. The most common direction of travel was from the sector south clockwise round to north-west.

Our analysis for the spatial distribution across the United Kingdom and Ireland uses the KDE technique. Application of this method to the TORRO database presents its own challenges – principally the relatively small geographical area covered and the insular nature of the United Kingdom with its complex coastline. Despite limitations, we believe KDE will prove to be a useful method to analyse and visualise the data. Preliminary results, presented here, are consistent with what might be expected from climatological reasoning – tornadoes are most prevalent in the south-east region of the United Kingdom where the overall atmospheric instability is at its greatest during the warm season and the year-round surrounding sea surface temperatures are at their highest (see e.g. Holley *et al.*, 2014). These results are consistent with previous studies (e.g. Holden and Wright, 2004). The Isle of Wight/south coast maximum, previously identified and discussed in the literature, appears to be a robust feature in our analysis. However, we find no evidence for the inner London event deficit postulated by Elsom and Meaden (1982).

Previously unreported anomalies are also apparent, for instance, a marked maximum on the coast of East Anglia and a positive anomaly south-west of London. Further analysis is required to test the robustness and statistical significance of these features, and if they do prove robust, detailed studies will be required to try to elucidate their nature and origins. The TORRO database, collected by painstaking work by volunteers over many years, provides a unique resource for such studies.

Acknowledgements

The authors thank Mike Rowe for a review of a draft and assistance in clarification on a number of points. An anonymous reviewer is thanked for critical comments, helping to improve the clarity of this chapter. The work of Terence Meaden, Paul Brown, Mike Rowe and many others too numerous to mention in compiling the TORRO database over the years is gratefully acknowledged.

Additional Information

Kernel Density Estimation

This is a well-known method for extracting distribution information from geographical data and is commonly described as a 'heat map'. This method, instead of representing each event at a discrete location, smooths it out over the surrounding area based on a probability distribution, or kernel. The set of distributions for all the events are then added together to create a continuous overall density map. Kernel density estimation (KDE) has the advantage over simple counting in that there is no dependence on the exact position of arbitrarily chosen map divisions, but it does still have some parameters which must be chosen. The kernel type is not particularly important in determining the final result, and in this case, a Gaussian or normal kernel was chosen. For an in-depth discussion of these issues, the reader is referred to textbooks on geographical information analysis (e.g. O'Sullivan and Unwin (2002)).

The size of the kernel, often called the *bandwidth*, however, is a much more difficult parameter to choose. In the case of the Gaussian kernel, it is defined as the standard deviation of the probability distribution, and it determines the degree of smoothing which is applied to the data. A small bandwidth leads to less smoothing, in that the resulting density map is clustered around the actual events, whereas a large bandwidth smooths the density across a much wider area. The hard part of choosing a suitable value is finding one which does not show more detail than the distribution of events warrants – some clusters of tornadoes will be just chance – and yet does not smooth out any genuine local effects.

There are a number of mathematical methods described in the statistical literature for choosing the bandwidth. One commonly used method is *least squares cross-validation*, which attempts to minimise the difference between the actual estimated density based on the full set of data and densities estimated from various samples of the same data. For the filtered tornado data, this method suggests a bandwidth of approximately 14 km, and after some experimentation with different values, this value was selected.

We have additionally employed a 'clustering' algorithm to merge closely spaced tornadoes on the same day into single events, as they are clearly not independent. There are many algorithms available to perform clustering, but here, we have used a 'hierarchical clustering' method based on the separation of the mean position of the tornadoes within each cluster (e.g. see Maechler *et al.* (2013); R Core team (2013). Essentially, all the tornadoes are initially assigned to a single cluster each. If one or more pairs of clusters are closer than a specific target distance and the tornadoes within them occurred on the same day, then the two clusters closest together are merged. The merging continues on this basis until no clusters are closer than the target distance.

This method is comparable to those used in other tornado density studies. For instance, Dixon *et al.* (2009), conducting a study into the spatial distribution of tornadoes in the United States, used a similar method where any tornadoes closer than a specified distance were merged into a single event (picking the most damaging tornado as the primary event).

The estimate produced by the standard kernel density method has one major drawback, in that it produces a non-zero probability density over the sea. For example, a tornado occurring exactly on a straight section of coast would have only half of its kernel over land and the other half over the sea. As has been discussed, tornadoes over water have not been included in this dataset because they are very unlikely to produce verifiable damage and records are entirely reliant upon observation. The density estimate, therefore, should only apply to the land. This 'lost' density, if not corrected, causes a bias or error in the overall density estimate whereby the values close to the coast (but still on land) are lower than they should be. The literature provides no standard method to remove this bias, and indeed, it may be impossible to remove it entirely. There are, however, ways to improve the estimate. Here, the method of Diggle (1985) and Berman and Diggle (1989) has been used. This multiplies the density values produced by each kernel so that when the part over the sea is removed, the remaining values still sum to one. This ensures that the total sum of the density estimate is the same as the total number of events (post clustering).

Voronoi Tessellations

A set of seed points is specified, and for each seed point, a corresponding polygon consisting of all points closer to that seed than to any other is drawn. By defining the seed points as the collection of 10 km grid square centres which fall upon the land, a normal grid pattern is generated across most of the map. At the coastline, the grid is modified into irregularly shaped divisions which have a more even spread of area.

References

Bennett, L.J., Browning, K.A., Blyth, A.M., Parker, D.J. and Clark, P.A. (2006) A review of the initiation of precipitating convection in the United Kingdom. *Q. J. R. Meteorol. Soc.*, **132**: 1001–1020.

Berman, M., and Diggle, P. (1989) Estimating weighted integrals of the second-order intensity of a spatial point process. *J. R. Stat. Soc. Ser. B Methodol.*, **51** (1): 81–92.

Bolton N., Elsom D.M., Meaden G.T. (2003) Forecasting tornadoes in the United Kingdom. *Atmos. Res.*, **67–68**: 53–72.

Brown P.R., Meaden G.T. (2010) TORRO tornado division report: July 2005. *Int. J. Meteorol.*, **35**: 171–179.

Brown P.R., Meaden G.T. (2011) TORRO tornado division report: November-December 2009. *Int. J. Meteorol.*, **36**: 20–26.

Clark, M.R. (2011a) Doppler radar observations of mesovortices within a cool-season tornadic squall line over the U.K. *Atmos. Res.*, **100**: 749–764.

Clark, M.R. (2011b) A provisional climatology of cool-season convective lines in the U.K. *Atmos. Res.*, **123**: 180–196.

Diggle, P. (1985) A kernel method for smoothing point process data. *J. R. Stat. Soc.: Ser. C: Appl. Stat.*, **34** (2): 138–147.

Dixon, P., Grady, A., Mercer, E. Choi, J. and Allen, J.S. (2009) Tornado risk analysis: Is dixie alley an extension of tornado alley? *Bull. Am. Meteorol. Soc.*, **92** (4): 433–441.

Doswell, C. (2007) Small sample size and data quality issues illustrated using tornado occurrence data. *E-J. Severe Storms Meteorol.*, **2**(5). Retrieved 14 January 2014, from http://www.ejssm.org/ojs/index.php/ejssm/article/view/26 (accessed on 10 May 2015).

Elsom D.M. (1985) Tornadoes in Britain: Where, when and how often. *J. Meteorol.*, **10**: 203–211.

Elsom, D.M., Meaden G.T. (1982) Suppression and dissipation of weak tornadoes in metropolitan areas: A case study of greater London. *Mon. Wea. Rev.*, **110**: 745–756.

Elsom, D.M., Meaden G.T. (1984) Spatial and temporal distributions of tornadoes in the United Kingdom 1960–1982. *Weather*, **39**: 317–323.

Elsom, D.M., Reynolds, D.J., Rowe M.W. (1999) Climate change, the frequency of U.K. tornadoes and other issues raised in the media reporting of the Selsey Tornado, 7 January 1998. *J. Meteorol.*, **24**: 3–13.

Holden J, Wright A. (2004) U.K. Tornado climatology and the development of simple prediction tools. *Q. J. R. Meteorol. Soc.* **130**: 1009–1021.

Holley, D. M., Dorling, S. R., Steele, C. J. and Earl, N. (2014), A climatology of convective available potential energy in Great Britain. *Int. J. Climatol.*, **34**: 3811–3824.

Kirk, P.J. (2006) A Mammoth task: The site investigation after the Birmingham tornado 28 July 2005. *Int. J. Meteorol.*, **31**: 255–260.

Kirk, P.J. (2007) U.K. Tornado Climatology 1980–2004. *Int. J. Meteorol.*, **32**: 158–172.

Kirk, P.J. (2013) Kensal Rise, London tornado site investigation 7 December 2006. *Int. J. Meteorol.*, **38**: 244–252.

Kirk, P.J. (2014) An updated tornado climatology for the U.K.: 1981–2010. *Weather*, **69**: 171–175.

Lamb, H.H. (1957) *Tornadoes in England May 21, 1950*. Geophysical Memoirs, **12**: No. 99. London: HMSO.

Lamb, H.H. (1985) TORRO- and the importance of independent meteorological research. *J Meteorol.*, **10**: 180–181.

Maechler, M., Rousseeuw, P., Struyf, A., Hubert, M. and Hornik, K. (2013) Cluster: Cluster Analysis Basics and Extensions. From http://cran.r-project.org/ (accessed on 10 May 2015).

Matthews, P. (1999) The Selsey tornado of 7 January 1998: A comprehensive report on the damage. *J. Meteorol.*, **25**: 197–209.

Matthews, P. (2003) An investigation into the influence of the Isle of Wight on waterspout and tornado frequency and the effect of population densities on recording events. *J. Meteorol.*, **28**: 213–225.

Meaden, G.T. (1975) The tornado and storm research organisation- TORRO. *J Meteorol.*, **1**: 24–26.

Meaden, G.T. (1985) TORRO, The Tornado and Storm Research Organisation, The main objectives and scope of the network – Part A – Its formation and expansion. *J. Meteorol.*, **10**: 182–185.

Meaden, G.T., Rowe, M.W. (1985) The great tornado outbreak of 23 November 1981 in which North Wales, central and Eastern England had 105 known tornadoes in about five hours. *J. Meteorol.*, **10**: 295–300.

Meaden, G.T., Matthews, P., Bolton, N., Elsom, D.M., Gilbert, A., Reynolds, D.J., Rowe, M.W. (2005) Influence of an Island land mass on the frequency of waterspout and tornado formation in its vicinity. Example of the Isle of Wight with regard to waterspouts and tornadoes affecting the English South Coast. *Atmos. Res.*, **75**: 71–87.

Mulder, K.J., and Schultz, D.M. (2015). Climatology, Storm Morphologies and Environments of Tornadoes in the British Isles; 1980–2012. *Mon. Wea. Rev.*, **143**: 2224–2240.

O'Sullivan, D. and Unwin, D. (2002) *Geographic Information Analysis*. John Wiley & Sons, Hoboken, NJ.

Pedgley, D.E. (2003) The Bracknell hailstorm of 7 May 2000. *Weather*, **58**: 171–182.

R Core Team (2013) R: A Language and Environment for Statistical Computing. R Foundation for Statistical Computing, Vienna, Austria. From http://www.R-project.org (accessed on 8 May 2015).

Reynolds, D.J. (1995) Towards a climatology of severe local storms in the United Kingdom. *J. Meteorol.* **20**: 7–16.

Reynolds, D.J. (1998) Severe Local Storms in the United Kingdom: Climatology and Forecasting. PhD. Thesis. University of Wales, Swansea.

Reynolds, D.J. (1999) A revised tornado climatology of the U.K., 1960–1989. *J. Meteorol.*, **25**: 290–321.

Romero, R., Gayà, M., Doswell, C.A. (2007) European climatology of severe convective storm environmental parameters: A test for significant tornado events. *Atmos. Res.* **83**: 389–404.

Rowe, M.W. (1985) Britain's greatest tornadoes and tornado outbreaks. *J. Meteorol.*, **10**: 212–220.

Rowe, M.W., Meaden, G.T. (1985) Britain's greatest tornado outbreak. *Weather*, **40**: 230–235.

Tyrrell, J. (2002) Site investigations of a multiple tornado event in county Westmeath, Ireland on 30 September 2001. *J. Meteorol.*, **27**: 210–218.

Tyrrell, J. (2003) A tornado climatology for Ireland. *Atmos. Res.*, **67–68**: 671–684.

Tyrrell, J. (2007) Site investigation results of tornado reports in Ireland 2006. *Int. J. Meteorol.*, **32**: 266–270.

Tyrrell, J. (2008) The investigation of tornado events in Ireland, 2007. *Int. J. Meteorol.*, **33**: 197–201.

5

Tornado Extremes in the United Kingdom: The Earliest, Longest, Widest, Severest and Deadliest

Mike Rowe

Tornado and Storm Research Organisation (TORRO)*, Lymington, UK,

5.1 Introduction

The TORRO website currently contains a section with the title 'British and European tornado extremes'.[1] These extremes include earliest tornado, longest and widest tornado track, most intense (severest) tornado, most deadly tornado and largest tornado outbreak. These data have been extracted from the TORRO database of UK tornadoes. The database was originally compiled by the present writer from 1986 to 1994 and exhaustively checked and enlarged by David Reynolds. In recent years, it has been updated by Dr Terence Meaden and Paul Brown, who are now responsible for the monthly and annual summaries published in *The International Journal of Meteorology*. The database includes for each tornado the date, time, place, county, grid reference, severity and the direction, length and width of the path. Not all these parameters are known for every tornado, information on path direction, length and width being the most sparse – hence the importance of site investigations, which TORRO members have increasingly carried out in recent years. The section of the database up to 1960 is published in instalments in *The International Journal of Meteorology*.

Because of the interest attached to extremes, those given on the TORRO website have been widely quoted. However, the quality of the basic source material varies enormously, and this is one reason why some of the accepted extremes need revision. Other

reasons include the occurrence of new extreme cases (such as the Selsey tornado of 7 January 1998 with its exceptionally wide track) and the occasional discovery of new information about old cases, such as the Lincolnshire tornado of 1666 (formerly dated 1667). Furthermore, knowledge of the formation of tornadoes and other localised storms is always advancing, and this leads to particular cases being reanalysed. All dates before September 1752 have been converted from the Julian calendar ('Old Style') to the Gregorian ('New Style') unless otherwise stated.

5.2 Earliest Tornado

There is no doubt about the earliest tornado on record in the United Kingdom: it is a very severe case that occurred in London on 23 October 1091. A detailed account of this tornado is given by Meaden (1975), who also mentions a number of possible earlier cases, mostly in Ireland. Another account is given by Rowe (1999b). This tornado was described by two of the best 12th-century chroniclers, William of Malmesbury and John of Worcester. William's account in his *Gesta Regum Anglorum* ('The Deeds of the English Kings') is as follows (Figure 5.1):

A clash of conflicting winds, coming from the south-east, smashed over 600 houses in London on 17 October (23 October by the Gregorian calendar). Churches and houses, enclosures and walls were left in heaps. The violence of the winds produced a yet worse disaster: it raised the roof of the church of St Mary le Bow,

* http://www.torro.org.uk/

[1] http://www.torro.org.uk/site/whirlwind_info.php

Figure 5.1   *St Mary le Bow, London, 23 October 1091, an interpretation of how the tornado may have appeared. © Christopher Chatfield.*

and killed two people there. The timbers and beams were carried through the air – an amazing sight for those watching from afar, but a terrifying one for those standing nearby, lest they should be killed. Four timbers, 26 feet long, were driven into the ground with such force that scarcely four feet protruded. It was remarkable to see how they penetrated the hard surface of the public street, in the same arrangement as they had been placed by the craftsman's skill, until, being an obstacle to passers-by, they were cut off at ground level, since they could not be removed in any other way.[2]

The other account, from John of Worcester's *Chronicon ex Chronicis*[3] is shorter than William's but very similar to it. There are slight differences in the figures, though they agree on two deaths. For Scotland, the first recorded case was at Blairgowrie in September 1767 (Rowe, 1999a). It was found in the *Annual Register* for 1767 by Mr Frank Law, who sent the text to TORRO. The account reads:

> An uncommon phenomenon was observed on the water of Isla, near Cupor Angus [Coupar Angus], preceded by a thick dark smoke, which soon dispelled, and discovered a large luminous body, like a house on fire, but presently after took a form something pyramidal, and rolled forwards with impetuosity till it came to the water of Erick [Ericht]; up which river it took its direction with great rapidity, and disappeared a little above Blairgowrie. The effects were as extraordinary as the appearance. In its passage, it carried a large cart many yards over a field of grass; a man riding along the high road was carried from his horse, and so stunned with the fall, as to remain senseless a considerable time. It destroyed one half of a house, and left the other behind, undermined and destroyed an arch of the

new bridge building at Blairgowrie, immediately after which it disappeared.[4]

The situation in Wales is less certain. There is a very doubtful case for 24 May 1173. The *Brut y Tywysogion*[5] states that on Ascension Thursday '…there arose a most violent storm in the sky, of thunder and lightning, and whirlwind, and showers of hail and rain, which broke the branches of the timber, and threw the trees to the ground'. It will be seen that there is no conclusive evidence here of a tornado, and the location is not stated, so the event may not even have occurred in Wales. Slightly more definite is one on 16 September 1402, although the exact location of this is not known either. The *Annales Henrici Quarti*[6] reported that while the King was campaigning in Wales a severe storm arose which 'broke down, scattered and flattened the King's tent, and threw the king's lance with tremendous force and drove it into the royal armour'.[7]

The next Welsh tornado was in 1772. The newspaper report is taken from a letter dated 1 March, and the tornado – if that is what it was – is assumed to have occurred in February:

> People say that this place teems with more wonderful occurrences than any other in Wales…During a very deep snow, which fell in these parts, there happened in Llantysilio parish, in this neighbourhood, two whirlwinds, which did a great deal of mischief to a tenement belonging to Mr. Jones, of Bacha; by the violence of which two great stone walls were raised from the ground, and carried the distance of two large fields from the house near which they stood: all the apple-trees in the garden were rooted up, as were likewise two great gates and a piece of a mountain that was above the house, and tumbled down to the

---

[2] Hardy (1840), vol. 2, p. 505, M. Rowe's translation. There are more recent editions of this and several other chronicles quoted in this chapter.

[3] 'Chronicle of Chronicles', formerly attributed to Florence of Worcester; ed. Thorpe (1849), vol. 2, p. 29.

[4] *Annual Register*, 1767, vol. 10, Chronicle, pp. 127–128.

[5] 'Chronicle of the Princes', ed. Williams ab Ithel (1860), pp. 220–221.

[6] 'Annals of Henry IV', ed. Riley (1866), p. 343.

[7] For further details on both of these events see Rowe (1999b).

water side, above three fields distance off; and at another house a whole barn, a young colt, and two cows, were carried off a great way and destroyed, besides a great many other damages of the same kind in other places, insomuch that the inhabitants of these parts apprehended nothing less than the approach of the terrible day of judgment.[8]

TORRO researcher Paul Brown, who discovered this report, expressed reservations about it, suggesting that some of the description may be exaggerated. The present writer agrees, and without insisting that the whole story is a fabrication (TORRO knows of a damaging snowfall in Shropshire and north Wales on 2 February 1772), it is better to find another case to represent Wales's first definite tornado. That case occurred at Fishguard, Pembrokeshire, on 5 November 1798:

> On Friday, the 2d instant, there was so heavy a fall of rain at Fishguard, in Pembrokeshire, as to occasion the greatest flood known in the memory of man; and on Monday night about six o'clock, a violent whirlwind, taking a South-east direction, passed over that town into the sea, which, in its course, unroofed several houses, and forced in many windows; two women were dismounted, who were riding through the place, and, for the minute it lasted, the inhabitants thought of nothing but instant destruction.[9]

This chapter is concerned with events in the United Kingdom, but it is worth mentioning that the Republic of Ireland's first recorded tornado is earlier than any case known in the United Kingdom. It occurred at Rosdalla, near Kilbeggan, County Westmeath, on 30 April 1054 (Rowe, 1989a; see also Chapter 2, this volume). For Northern Ireland, the earliest known is near Gilford, Armagh, on 12 January 1751:

> The following Account comes attested by two Persons, who were both Eye-witnesses, and within twenty Yards of the Fire when it passed. – On the 1st of January last [Julian calendar], about 12 o'Clock at Noon, a Cloud was observed to arise off the Mountains of Morn [Mourne], N.W. from Newry, in the County of Down, in the North of Ireland; which Cloud came across the Country N.W. about 15 Miles, and when it was between Gilford and Tanderagee, about a Mile distant from each, there came out thereof a most surprizing Body of Fire, which appeared as if it reached from the Heavens to the Earth: It was about six Yards Square, and its Motion forwards so very slow, that any Person who had Presence of Mind might have gone out of its Way; its Course was direct, only in ten Yards it would have made an Angle, sometimes twenty, thirty, or forty Yards; it turned continually round like a Whirlwind, with a great Noise in the Air, like Cracking of Thorns in a Fire, and a prodigious Smell of Sulphur. The first Damage it did was about half a Mile from where it first broke out, which was to some large Trees in an Orchard, all which it split from Top to Bottom; all Trees and Hedges wherever it came shared the same Fate. Close by the said Orchard, it threw down a Corner of a Cowhouse, and within a hundred Yards of this it lifted the Roof off a very good Dwelling-House, threw down a Stone Partition, lifted up a Firkin of Butter, threw the Firkin to one Side of the Room, and the Butter to the other, broke half of a Linen Web that was in the Loom, and carried out a Glass Window, and not one

Person in the House was hurt; but about two Miles from this, it killed a Woman on the Highway. The Length this Body of Fire went across the Country was about twelve Miles to Loch [Lough] Neagh, where it is supposed it was extinguished by the great Quantity of Water there, for in its Course it lifted up all the Water wherever it came.[10]

One of the most interesting features of this account is the reference to 'a prodigious smell of sulphur'. A number of early tornado reports mention such a smell, but it is much rarer in modern-day accounts. The description of the tornado funnel as fire or smoke is also characteristic of early reports. A good example is the tornado at Great Malvern, Worcestershire, on 14 October 1761: 'It had the appearance of a volcano, and was attended with a noise as if 100 forges had been at work at once; it filled the air with a nauseous sulphrous smell; it rose from the mountains in the form of a prodigious thick smoak, and proceeded to the valleys, where it rose and fell several times, and at length it subsided in a turnep-field'.[11]

A few years later, a Mrs Dowsett, watching a tornado at Good Easter, Essex, on 18 August 1768, saw 'a great thick smoke arise out of the pasture fields…She heard a violent wind issue immediately "from under the smoke", (and "out of the ground", as she thought) which twisted the smoke up to some height, and went off violently'.[12]

5.3   Other Whirlwinds (First UK Record Only)

Many potential tornadoes do not reach the ground, and in that case, they are referred to as funnel clouds. Until recent years, they were severely under-recorded, because obviously they cause no damage. Our earliest known funnel cloud occurred at St Albans, Hertfordshire, on 26 May 1251, and is recorded in the *Chronica Majora*[13] of Matthew Paris, one of Britain's most famous medieval chroniclers. He described in great detail the effects of a severe thunderstorm, during which there appeared at St Albans, Hertfordshire, 'a torch like a drawn sword, but *plicabilis* [which here probably means "flexible; not solid"], and continuous thunder *cum murmure horribili* [probably "with a terrible roaring" rather than "murmur"]'.

Our earliest record of a waterspout concerns two spouts seen on the south coast of England in June 1233. Roger of Wendover, in his *Flores Historiarum* ('Flowers of History') mentions 'two huge snakes' fighting in the air, 'and after a long fight one overcame the other, and pursued the loser into the depths of the sea, and they were seen no more.'[14] C.E. Britton, in his important 1937 work *A Meteorological Chronology to AD 1450*, commented: 'This is probably a description of a display of aurora. Armies, ships or serpents fighting in the air is a very usual term in old writers to describe auroral appearances' (Britton, 1937). This is true enough, but it seems unlikely that this is a description of an aurora. We are not told that it happened at night, and the location was on the south coast of England, which could imply that the

---

[8] *Middlesex Journal or Chronicle of Liberty*, 10 March 1772.
[9] *Mirror of the Times*, 10 November 1798.

[10] *London Advertiser and Literary Gazette*, 12 March 1751.
[11] *Gentleman's Magazine*, October 1761, vol. 31, p. 477.
[12] *Annual Register* (1768), vol. 11, Chronicle, pp. 159–160.
[13] Greater Chronicles, ed. Luard (1880), vol. 5, pp. 263–264.
[14] Coxe (1842), p. 267.

Figure 5.2   *The south coast waterspouts of 1233. © Christopher Chatfield.*

phenomenon was to the south. Moreover, Roger's word *dracones*, though it can mean 'dragons' – which might suggest an aurora – can also mean simply 'snakes'; indeed, that is how Britton translated it. And two snakes in the air, seen from the south coast, are almost certainly waterspouts. Figure 5.2 shows Christopher Chatfield's reconstruction of the scene.

TORRO also collects records of the minor whirlwinds known as land devils or dust devils, which occur in fine weather, usually in the summer. What may be the earliest record of a land devil in Britain occurs in the *Miscellanies* of John Aubrey (1626–1697), better known for his gossipy *Brief Lives*. Aubrey received the information in 'a letter from a learned friend of mine in Scotland, dated March 35 [sic] 1695'. The friend had been informed by the witness, Mr Steward.

> When he was a boy at school in the town of Forres…he and his schoolfellows were upon a time whipping their tops in the church-yard before the door of the church; though the day was calm, they heard a noise of a wind, and at some distance saw the small dust begin to arise and turn round, which motion continued, advancing till it came to the place where they were; whereupon they began to bless themselves: but one of their number (being it seems a little more bold and confident than his companions) said, 'Horse and hattock with my top', and immediately they all saw the top lifted up from the ground; but could not see what way it was carried by reason of a cloud of dust which was raised at the same time. They sought for the top all about the place where it was taken up, but in vain; and it was found afterwards in the churchyard, on the other side of the church.[15]

If, however, we look for an account written by an eye-witness, and a precise date, this is the earliest, written by 'a very Sober and Credible Person', and published in the Royal Society's *Philosophical Transactions* for 1694 (p. 192):

[15] Aubrey (1721), pp. 158–160.

> On *Wednesday* last…there happened here, betwixt the hours of One and Two of the Clock in the Afternoon, a very terrible Whirl-wind amongst the Shocks of Corn, in that part of *Acrement Close*, which is in the possession of Mr. *Holt*, and took up into the Air about 80 or 100 Shocks, carried a great deal quite out of sight, the rest it scattered about the Field, or on the tops of the Houses or Trees thereabouts. I have seen of the Corn, which was carried a Mile distant from the Field; and it is reported by Persons of good Credit, that some was carried four or five Miles distant. The Whirl-wind continued in *Acrement Close* full half an hour; I myself, and several other persons saw at least three or four Waggon-Loads of Corn all at once whirled about in the Air.

The date of this was 11 August 1694, and the location Warrington in Northamptonshire.

5.4   Longest Tornado Track

The TORRO website quotes the longest British tornado track as 107.1 km from Little London (Buckinghamshire) to Coveney (Cambridgeshire) on 21 May 1950.[16] This track seems improbably long – it is over twice as long as any other on record in Britain – and this impression is strengthened by reading the detailed site investigation carried out by the distinguished climatologist Hubert Lamb (1913–1997), assisted by Mr J. Simmonds (Lamb, 1957). At the time Lamb was writing, it was usual to record all damage along a track as due to a single tornado, even if there were gaps of many kilometres. Thus, Dr C.E.P. Brooks, in his book *The English Climate* (1954), discussed the South Wales tornado of 27 October 1913 and then added: 'Between South Wales and Shropshire there was a break, represented only by severe thunder-storms, but the tornado appeared again between Ragdon (near

[16] The reference to Shipham, Norfolk, where the funnel cloud was last seen, is an error for Shipdham.

Church Stretton) and Shrewsbury, and again in Cheshire'. Today, no one would regard this as a single tornado, and the same applies to the May 1950 event, though the gaps in the track were admittedly shorter in the latter case. A comparable modern example is the outbreak of 11 tornadoes on 30 December 2006. These formed along a line extending from Winnersh (Berkshire) to Ludham (Norfolk), a distance of about 219 km. Brown and Meaden (2007) commented, 'In theory, almost all of them could be regarded as intermittent descents of the same vortex, but given that the distance between successive points of damage is typically 5–10 miles (8–16 km), we have decided to treat them separately'. A careful study of the data recorded by Lamb suggested that the 107-km-track tornado of May 1950 is best divided into a series of seven, none of which had a track length of more than 10 km. The most severe of these tornadoes reached force T5–6 at Linslade, Bedfordshire. Lamb reported, 'Most of the houses in two streets, Old Road and New Road, Linslade, were either unroofed or had their roofs badly damaged; and yard premises between the two streets were demolished. A heavier roof was carried away near the railway station, and cars parked in the station yard as well as an occupied horse-box were lifted and thrown about' (Lamb, 1957).

This site investigation by Hubert Lamb is a superb example of what can be deduced about a tornado by a very acute investigator. He noted:

All three tornadoes [meaning the main one, with the apparent 107 km track, and two in the same general area with much shorter tracks] passed through successive stages of regeneration, when the circulation developed down to ground level, followed by gradual weakening and decay, and eventually lifting off the ground again, until some fresh burst of energy occurred. Successive regenerations often took place at about 5-min. intervals. In several instances the regenerations were sudden and accompanied by sudden renewal or widening of the trail of damage. Nearly all the main bursts of energy, especially the more sudden ones, took place either immediately to the north of a ridge of hills or just on the crest, notably at points where the surface north-easterly winds ahead of the advancing whirl would suddenly increase in strength. At Bedford a more gradual regeneration took place over flat land...Over the south-facing slopes of ridges and over tightly grown coppices and avenues of trees situated across their paths, the tornadoes tended to break up. Evidence of the effect of ground barriers on the path and progress of the tornadoes was found at many points, particularly where hill slopes and dense coppices or woodland caused the path to zig-zag, though never departing by more than some 400 yd from its general line.

(Lamb, 1957)

If the 107-km track of this tornado is discounted, the next longest is 43 km for the tornado that tracked from Petworth (West Sussex) to Smallfield (Surrey) on 5 September 1958. However, the tornadic nature of the damage is not clear at several points along the track. There is also the possibility that the track continued to Oxted, which would increase its length to 54 km. The analysis of this case is complicated by the fact that there were two tornadoes in the area and separating them is difficult. The tracks worked out by the present writer for the TORRO database differ from those arrived at by Rowsell (1960). The TORRO tracks are almost parallel, whereas Rowsell has them crossing at about 40 degrees. Since Rowsell's tracks are based on a site investigation

(by Mr G.F.W. Clapp) carried out within a week of the storm, they carry considerable weight. However, in view of the uncertainties regarding this event, it is probably best not regarded as Britain's longest tornado track.

The tornado that passed from Swinstead (Lincolnshire) to Winthorpe (Nottinghamshire) – a distance of 40 km – on 8 October 1977 is a better candidate for Britain's longest-track tornado (Anon [Meaden], 1978; McLuckie and Taylor, 1978). Yet even here, there are problems. McLuckie and Taylor reported (p. 37): 'Damage is not continuous along the apparent track of the tornado. Instead, the evidence suggests that the tornado exhibited the characteristic "skipping" movement, only striking the ground at intervals and then retracting until sufficient energy had built up again for the tip to extend once more to ground level' (compare Hubert Lamb's similar comments about the 1950 tornado, quoted earlier in this section). Although McLuckie and Taylor show a single, continuous track on their map, the data in the TORRO archives – which includes a number of detailed press reports – show a 9-km gap in the path before reaching Grantham and a 13-km gap after it. Since there was a great deal of damage in Grantham itself, it is probable that these apparent gaps are due at least partly to lack of information. The tornado appeared near Swinstead at 1515 GMT and followed a very narrow path from about SSE to NNW, a relatively unusual direction for a tornado track in the United Kingdom. In Grantham (Lincolnshire), the damage was 'generally limited to a band no more than 15 metres wide, often with a very clear demarcation line between damaged and undamaged objects. According to one report, a garden shed disappeared completely but a greenhouse about 2 metres away was unscathed' (McLuckie and Taylor, 1978). An 11-year-old boy, Colin Scott, described the tornado as 'something like an enormous ice-cream cornet' according to a report in an unidentified newspaper.

One of the longest tracks in the database is (astonishingly, considering its early date) one of the best documented of all. The tornado occurred near Bexhill in East Sussex on 31 May 1729 and was the subject of a detailed site investigation by the cartographer Richard Budgen (1695–1731), who published his findings in a pamphlet, which included a large-scale map of the path. Budgen claims that the tornado's track extended from Bexhill Down to Smarden (Kent), which would indicate a length of 37 km. However, the latter part of the track was not well marked, was not mapped by Budgen and was parallel to but about 4 km to the west of the main track, probably indicating a second vortex. The main track, from Bexhill (East Sussex) to Rolvenden (Kent), was 25 km long and apparently continuous, but Budgen's map indicates that the tornado began as a waterspout over the English Channel, so 25 km may well be an underestimate. This was a severe tornado, reaching T4 at a number of points. A sample of Budgen's investigation:

At *Colliers-Green*, a House belonging to Mr *Richard Boys*, had the Chimney took off in the middle, all the Windows broke, and some of the Rafters, and the House uncovered. Near the House a Barn was blowed down, in which stood a Waggon that was turned bottom upwards; and two Dung-Carts were carried away in the Storm, of which they can find only some broken pieces about in the Fields. His Tenant, just as the Windows were drove in by a violent Impulse, was pushed against his Wife, and beat her down in the Chamber; a Child that sat in a Chair at the Feet of the Bed, was carried in his Chair and set in the Fire-Place. And the Gravel Stones, from the High-Way, and Glass from the Windows, were brought in with such Violence, as to stick in the Chairs, etc., like Shot discharged from a Fowling-piece.

(Budgen, 1730)

There are two more tornadoes with a reputed track length exceeding 30 km. On 19 August 1881, a T3 tornado covered 32 km from Upton to Elsham (Lincolnshire). There were six known damage points, fairly evenly spaced; at two other sites, a funnel cloud was observed, but it is not clear whether it touched down. The account suggests that there was damage before and after the known track (Anon [Symons], 1881). Finally, there is the 31-km track of the tornado that passed from Middle Winterslow to Apsley Copse, Wiltshire, on 1 October 1899. The distinguished meteorologist George James Symons (1838–1900)[17] carried out a thorough site investigation (Symons, 1900), from which it appears that this was the longest track in England where ground contact was continuous or nearly so. Symons was assisted by 'the extreme kindness of many residents on or near the track', including 'all the clergy through whose parishes it passed', and he continued, 'Owing to this kindness and hospitality he was able to almost live upon the track for several days, and to determine its direction and limits with a precision rarely attainable. The 6-inch. Ordnance map shows every barn and stable and clump of trees, and having joined up the entire series of maps, he took them with him, and either by tape or by azimuth instruments determined the track with all possible accuracy, generally within 100 ft'.

In Scotland, the longest tornado track is 11 km in the Newton Stewart area (Wigtownshire) on 20 January 1854:

WHIRLWIND. – A correspondent, writing us from Galloway, says that a squall of wind blew from the south for 10 minutes with terrific violence on the 20th ult., tearing up everything in its progress. At the farm of Auchlane, the property of Mr Stroyan, a house was unroofed, and other damage done. At Park, tenanted by Mr John Rae, a cattle-shed was nearly stript bare, three oat-stacks were upset, three trees torn up by the roots, and another snapt in the middle, two feet eight inches thick; a window was driven in, and a stable stript of its slates. At the Barr, tenanted by Mr Stroyan, it unroofed two houses, and upset five stacks. Newton-Stewart escaped with the blowing down of a few chimney cans, but at Penninghame House it committed fearful ravages, laying all waste. The blast seems to have travelled in a straight line.

(*Dumfries and Galloway Standard*, 1 February 1854)

A tornado (which was a waterspout for part of the time) passed from West Burra to Lerwick, Shetland, on 24 February 2000 (Anon [Rowe and Meaden], 2002). The track length was given as 12 km, but as there were no observations of the tornado between Trondra and Lerwick – a distance of 9 km – it should probably not be counted as Scotland's longest.

The longest track in Wales is 17 km, from Dyffryn Dowlais to Bedlinog, Glamorgan, on 27 October 1913. This very severe tornado, which killed five people, is considered in more detail

later in this chapter, as is the Belfast tornado of 2 September 1775, which had one of the longest tracks in Northern Ireland, recorded in the TORRO database as about 30 km. This almost equals the path length of the Wiltshire tornado of 1 October 1899 (though the latter is much more minutely documented), and it may be too short, since the tornado eventually passed over the sea. The tornado of 31 December 2006, which tracked from Ardmore, County Armagh, to Loanends, County Antrim, was the subject of detailed site investigations by Dr John Tyrrell, who concluded that the track was 30.2 km long (Tyrrell, 2007, 2008; see also Brown and Meaden, 2007). This is here regarded as Northern Ireland's longest tornado path, because of its much better documentation compared to the 1775 event.

5.5   Widest Tornado Track

The TORRO website gives the widest track as 0.75 miles (about 1200 m) at Fairlight, East Sussex, on 4 July 1946. However, the original report (Moon, 1946) is so brief and vague that the event is certainly best discounted. 'What was reported to have been a whirlwind' struck between 0030 and 0100 GMT. 'Damage to trees, etc. was particularly noted in a path about ¾ mile wide, with the church about the centre'. Despite the reference to a 'path' of damage, the event could well have been a microburst. The website also quotes a width of half to one mile (about 800–1600 m) at Fernhill Heath, Worcestershire, on 22 September 1810 (*Hereford Journal*, 3 October 1810). The very vague value for the track width does not inspire confidence in its accuracy. Another case which is probably best rejected concerns tree damage between Burton Joyce and Gonalston, Nottinghamshire, on 14 July 2005. The apparent track widened from 100 m at the start to 900 m at the end, but as Brown and Meaden (2010) state, the event could well have been 'some form of convective squall'.

A much better-documented case of an exceptionally wide track was the tornado at Selsey, West Sussex, on 7 January 1998. The damage was investigated by several TORRO researchers, most comprehensively by Peter Matthews (Matthews, 1999). He visited Selsey on 9 January and on three later occasions. He found a maximum path width of 900 m at the western end (where the vortex came ashore from the English Channel), narrowing to 600 m further east where it returned to sea. West Sussex Fire Brigade provided him with a large-scale map of Selsey showing the location of incidents reported to them; this showed the exceptionally wide track very clearly. The tornado struck at 2339 GMT during a severe thunderstorm. Observations of a funnel cloud were few but striking. A Mrs. Jinks saw and sketched a 'black wedge', while Mr. Alex Ralph saw 'a white cloud in the shape of a "triangle" whose width he estimated as ¾ mile (ca 1200 m) coming over the sea towards him'. Another witness, a Mr. Taylor, saw 'a blue light shaped like a "large blue eye" which lit up the sky towards the Isle of Wight. Below this was a swirling mass of cloud which extended to sea level and moved over the sea towards Selsey'. Besides the damage to a very large number of houses, there were some more unusual incidents. A Flymo lawnmower was lifted from a garage and deposited in the garden of a house over 100 m away. Part of a block of four garage units, the roof measuring some 12 m by 4 m,

[17] Symons was the founder of the British Rainfall Organisation (which revolutionised knowledge of rainfall in Britain and Ireland) and of the publications *British Rainfall* and the *Meteorological Magazine* – originally *Symons's Monthly Meteorological Magazine*.

took off from Gainsborough Road and landed in a garden in Beach Road, a distance of about 200 m, meaning that the structure must have been levitated over the intervening houses.

For Scotland, the widest track is probably that left by the 'typhoon of more than ordinary size' at Lumsden, Aberdeenshire, on 8 July 1875. 'The diameter of the moving current was about 500 ft [about 150 m]. It left a mark on grass, corn, and turnips, so that it could be easily measured. There was a cloud in the air whirling steadily with the wind, and reminded one of a tremendous whirlpool' (*Aberdeen Journal*, 14 July 1875). Another contender is the tornado at Yetholm Mains, Roxburghshire, on 12 May 1987. In a letter to TORRO, Mr. F.S. Roberton reported: 'We first of all had some thunder and then heavy hailstones and this was followed by a very strange high wind which seemed to follow a track from north to south only about 200 yards wide [approx. 182 m]. It took one tree out by the roots, and another very healthy chestnut had all the top removed including limbs 9 in. [23 cm] in diameter, and these seemed to have been lifted right off the top of the tree and deposited some distance away leaving the lower part of the tree undamaged'. Although the Yetholm width is greater, the Lumsden one appears to have been ascertained more accurately.

The widest track in Wales, like the longest, involved the tornado of 27 October 1913, which had a maximum width of 200 yards at Cilfynydd and about 1000 ft [about 300 m] at Edwardsville (Billett, 1914; Brown, 2013). A greater width was claimed for a 'whirlwind or tornado' at Rhyl on 8 July 1913, which affected 'a narrow strip a quarter of a mile wide' (about 400 m) according to *British Rainfall, 1913*. This value, however, sounds like a very approximate one. The record for Northern Ireland is apparently held by the Belfast tornado of 2 September 1775, which is discussed in the next section. The path 'scarcely exceeded half a mile in breadth'. However, as with many of these old accounts, TORRO's confidence in the accuracy of this figure is not great. A much better candidate is the tornado of 31 December 2006, which moved from Ardmore, County Armagh, to Loanends, County Antrim, already listed as having one of the longest tracks in Northern Ireland. A thorough site investigation by Dr John Tyrrell found a width of up to 300 m near Pigeonstown (Brown and Meaden, 2007).

5.6 Severest Tornado

Before examining the most powerful tornadoes known in the United Kingdom, it is necessary to be clear that the T scale is not an assessment of the tornado itself, but rather of the evidence that is known about it. This evidence is likely to be incomplete and may well be inaccurate, with exaggeration being more likely than understatement. There are two tornadoes that until recently have been graded T8 on the TORRO intensity scale. These are the earliest known British tornado, in London on 23 October 1091 (mentioned earlier) and the one at Southsea Common, Portsmouth, Hampshire, on 14 December 1810. The 1091 case is now given the more cautious assessment of T8(?) in the TORRO database. It is obviously impossible to tell how accurate the accounts of William of Malmesbury and John of Worcester are; these are not eyewitness reports, and both chroniclers were

probably writing over 30 years after the event. The 1810 tornado is described as follows:

> *Dec.* 14. The town and vicinity of *Portsmouth* were visited by a tornado, which passed in the direction of W.S.W. to S.E. and did very considerable damage. At South-Sea Common four houses were levelled to the ground, and as many more so much injured as to render it necessary to take them down; besides 30 others unroofed. At Haslar Hospital, and the Marine Barracks, chimneys were blown down, and the Government House and Chapel partly unroofed. The inhabitants of the houses facing the Grand Parade had not fewer than 100 panes of glass broken. The lead on the top of Messrs. Goodwin's bank was, by the irresistible power of this phenomenon of nature, rolled up like a piece of canvas, and blown from its situation.
>
> (*Gentleman's Magazine*, December 1810, vol. 80, part 2, p 583)

The damage at Southsea Common certainly suggests a very powerful tornado – the demolition of entire houses is very rare indeed in Britain – but attempts to find more detailed reports of this event in contemporary newspapers have not met with much success. The TORRO database therefore now rates the event more cautiously at T7/8. This tornado has been the subject of a number of errors. Rowe (1985) dated it 14 October instead of 14 December. At least one website gives the location as Plymouth, Devon.

Since it is often difficult to assess the strength of a tornado, it is worth asking whether any of those rated by TORRO as T7 could have been T8. Rowe (1985) examined the tornadoes then assessed at T6 or higher, but TORRO's rating of some of them has changed since then. The following discussion takes the cases in chronological order, except for the Lincolnshire tornado of 1666, which is reserved for extended treatment at the end of this section as TORRO now considers it the most severe tornado on record in the United Kingdom.

The first T7 case that we know of occurred at Nottingham on 11, 17 or 21 July 1558 (sources vary as to the precise date). The tornado is described by Raphael Holinshed in his *Chronicles of England, Scotland and Ireland* (1587). The text is given elsewhere in this volume (see Chapter 2). Holinshed does not give a precise date, but John Stow, in his *A Summarie of the Chronicles of England*, stated that it was 7 July, which is 17 July New Style (1598 edition: 293–294). In his *Annales* (1631 edition: 634), however, the date is given as 11 July (21 July New Style). The text of Stow's versions is almost identical to Holinshed's. The destruction of houses and churches, if correctly reported, would certainly suggest T7 (or more). Other references to this event are mostly derived from the Holinshed/Stow version at one or more removes, but there exists an independent account produced only a few months after the event (and, rather surprisingly, in Italy). It appears in a letter written by Gilbert Cousin at Padua in August or September 1558 (according to Richard C. Christie, who published the account in *Notes and Queries* in 1883). The original letter is in Latin; a translation follows:

> On the first day of July [Julian Calendar] there occurred a severe storm in England, near Nottingham. A countryman with his boy and four horses was busy ploughing his fields; he and three of the horses were killed by the violence of the wind and the size of the

hail. The boy and the fourth horse remained unharmed. Another man, who was loading his wagon with hay in the fields, seeing such a great storm near at hand, withdrew a little way, and immediately the wagon loaded with hay vanished, taken away by the force of the wind. A village called Suuentum [Sneinton] was almost totally destroyed, very many houses being torn apart. The whole church collapsed, leaving intact the holy font and the top part of the sanctuary. It pulled up many trees by the roots, and broke many in pieces. Lightning burnt a barn which was full of grain. Sheep, geese, chickens and vast numbers of small birds were killed by the size of the hail. The Duchess of Northumberland was an eye-witness of these events; and the citizens of Nottingham, questioned by Sir John Byron and some other members of the nobility, gave full testimony of the facts.[18]

Passing on to Stuart times and the reign of Charles II, the next T7 tornado occurred at Bedford on 29 August 1672. By this time, there was a brisk output of pamphlets detailing 'strange' events, and at least two covered the Bedford tornado. One of them described 'a horrible and unheard of tempest', which:

> …bore down two Houses in an instant, to the dreadful amazement of the Spectators that blessed be God, escaped maiming, yet knew not where to flie for shelter, but to run too and fro like persons amazed. In one of our Gardens it rent up the Onion and Reddish-Beds by the Roots, with an incredible violence, carrying them almost two Miles. It plucked up a large Apricock Tree by the Roots, and rent it from the Wall, to which it was nailed, and carried it over Houses and Hedges almost a quarter of a Mile…Wooburne [Woburn] also, as we are informed, felt something of this terrible Tempest, some Houses in that Town being levelled with the Ground by it. It is reported by Passengers upon the Road, that they see a great combustion in the Air, the Clouds as it were fighting one against another, in so much, that they thought at a distance the Town of Bedford was on a light fire.
>
> (Withnal *et al.*, 1672)

Once again, we have houses flattened, but there is perhaps an indication that at least some of the houses concerned were not built of brick or stone: 'It blew down several Houses at the further end of the Town and removed one House two yards out of its place and set the Threshold where the middle of the House was before'.

By the time we come to the Cowes (Isle of Wight) tornado of 28 September 1876, such events were being dealt with in considerable detail both in the press and in meteorological journals such as *Symons's Monthly Meteorological Magazine*, which, in the issue for October 1876 (vol. 11, pp. 121–124), reprinted long reports from the *Portsmouth Times* of 30 September and *The Times* of 2 October. The report in the *Portsmouth Times* covered the devastation in West Cowes in detail, while *The Times* described damage on the Hampshire mainland (regarded by TORRO as two additional tornadoes). With this tornado, there can be little doubt of its T7 status:

> A terrific whirlwind passed over Cowes early on Thursday morning, with most alarming and disastrous effects… The appearance of Cowes shortly after the occurrence was

astonishing…What was once the well known Globe Hotel had in a moment become a mere tottering ruin, the front being completely blown in, exposing the bed-rooms, in most of which lay heaps of bricks which had crushed through from the roof and walls. The Globe cottage attached to the hotel was similarly demolished, and in each house goods of all descriptions were demolished and in the wildest confusion…Eye witnesses with whom we conversed on the parade stated that the whirling in the air of the heavy things dislodged by the whirlwind was one of the most startling sights imaginable, and we have it from good authority that the yacht 'Palatine', owned by Lord Wilton…which was steaming up half-a-mile away in the offing, was strewn with barley blown from the stacks on shore, while a brick from one of the houses of the town, struck the forecastle deck with great force…At Gravel Lane we see a slate actually driven half-a-foot deep into a wooden window sill. Coming upon the terminus of the Newport and Cowes railway there is another scene of desolation; the engine-house, a large wooden building, is wrecked, four heavy carriages are blown over on their side, and the water-tank has overturned and smashed down on top of an engine. Down the Victoria-road… the dwelling of Mr. Maggs is nearly blown down…Proceeding into the open country we find a new phase of the affair…For a couple of miles across country big trees of long growth are torn up by the roots; fences and hedges are swept ruthlessly away; stacks are overturned; barns are raised [sic]; and last, though most important, cottages are unroofed and rendered untenantable…Broadfield's farm, occupied by Mr. John Davies, is a terrible scene of waste…The outbuildings of the farm are upset, and unfortunately three men – Leonard Drudge, Isaac Dunford, and George Parsons – were crushed beneath the falling barn. They were rescued as speedily as possible, and their injuries, though serious, are, we are informed, not of a dangerous character. At Place farm, occupied by Mr. Moor, there is a similar state of things, and taking the route of Tinker's-lane, passing several ruined labourers' cottages, the owners thereof pitifully bemoaning their fate, we came upon a summer residence of Mr. Redfern, who was fortunately away with his family when the whirlwind seized it and left it a miserable ruin with nothing but a staircase – leading to nowhere – left standing.
>
> (*Portsmouth Times*, 30 September 1876)

The South Wales tornado of 27 October 1913 is one of the best-known British tornadoes and sadly was one of the most deadly, with five fatalities. At the request of Mr. Clement Edwards MP, the Meteorological Office sent Mr. H. Billett 'to collect information on the spot that might throw light upon the meteorological conditions incidental to the occurrence'. The resulting report appeared in the series *Geophysical Memoirs* (Billett, 1914) and is the basis for a detailed assessment of the tornado by Paul Brown, published, to coincide with the centenary of the event, in *The International Journal of Meteorology* for September/October 2013 (Brown, 2013, and Figure 5.3). The tornado followed a 17-km track from Dyffryn Dowlais to Bedlinog along the Taff Valley, reaching its maximum strength at Abercynon and Edwardsville. At Abercynon, one row of thirteen houses had the roofs and joists completely removed, except for three or four houses in the middle of the row, and the partitions in the interior were smashed to bits. At Goitre Coed Farm, between Abercynon and Edwardsville, outbuildings with walls half a yard thick were severely damaged, 'the roof being quite gone'.

[18] Cousin (1562), vol. 1, p. 388, reproduced by Christie (1883). A virtually identical text is in Cousin's *Topographia Italicarum Aliquot Civitatum* ('Topography of Some Italian Cities'), Venice, 1559.

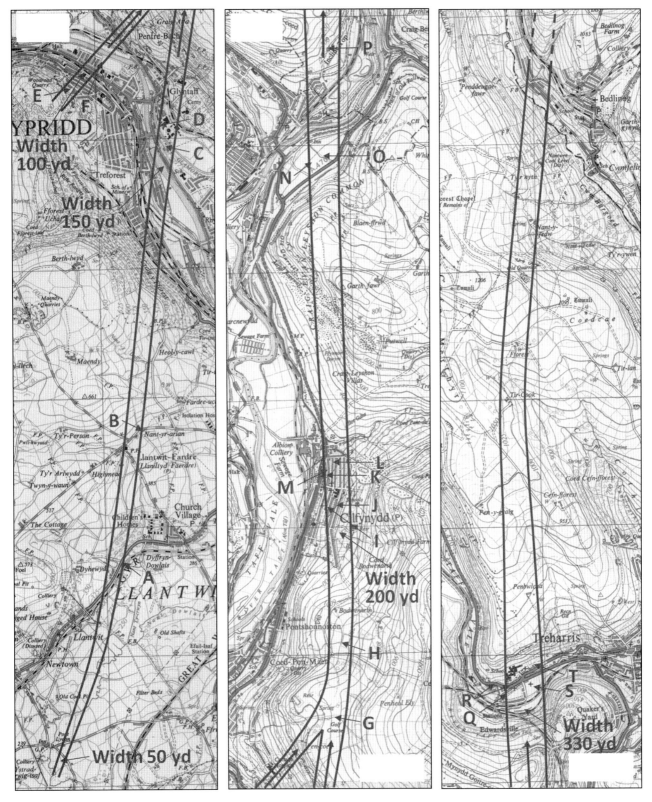

Figure 5.3   *The track of the South Wales tornado, 27 October 1913. For explanation of letters A to T, see Brown (2013). Reproduced with permission from Brown (2013).*

Mr. B.P. Evans of Edwardsville gave a graphic account of the arrival of the tornado:

> A few seconds before 5.50 p.m. we heard a noise resembling the hissing of an express locomotive. The sound grew rapidly in volume, at last resembling the rushing speed of many road lorries racing along. The oppressiveness that had been previously noticed increased, and the heat and air-pressure were pronounced during the rushing noise. We endeavoured to move out of the room to the passage for greater safety, because a hurried remark was made that the engines of these supposed passing loaded steam lorries had collided before the house, and were about to burst, when the panes of our window were broken by stones, tiles, slates, dried cement, and splintered timber. The missiles broke the Venetian blinds and struck the opposite walls. We made for the rear of the house, but all these windows were being bombarded also by small material and corrugated iron sheets. We could distinctly hear the chimney-pots fall on the roof, and material sliding off being dashed on the pavement and doorstep…After this crashing had ceased (this only lasted from 60 to 90 seconds), rain fell in torrents. The lightning set fire to the tar which had been sprayed some three weeks previously on the main Cardiff and Merthyr road, some 12 yards from our house door. A distinct smell of sulphur pervaded the air.
>
> (Billett, 1914)

Damage at Edwardsville was certainly at the T7 level; both the cemetery chapel and the Congregational chapel were wrecked. Several pieces of slate were found buried to a depth of 1.5 in. (3.8 cm) across the grain of trees.

The tornado at South Kelsey, Lincolnshire, on 25 October 1937, came to the notice of TORRO in 1984 and for many years was rated at T7 but has now been downgraded. Certainly, the newspaper photograph of the damage at South Kelsey (reproduced in Rowe, 1985) suggested a very severe tornado where a blacksmith's and a joiner's shop, both brick built, were destroyed. But the incident illustrates a frequent problem in assessing tornado strength, a situation where one piece of evidence suggests a much higher TORRO force than the rest of the evidence. At South Kelsey, the other damage mostly suggests T3, and the true maximum force is uncertain.

The most recent T7 tornado in Britain passed through London on 8 December 1954. Meteorologist Philip Eden published in *The Journal of Meteorology* a number of eyewitness accounts received after he had made a radio broadcast on the tornado (Eden, 1989) and to these were added a selection of reports received by the present writer as a result of a long-running press appeal for information on whirlwinds generally (Rowe, 1989b). The meteorology of the event was covered by Bull and Harper (1955) and Abbott (1955). The tornado was by far the most severe of six that formed along a line from Charmouth (Dorset) to London. It began at Chiswick where, at 1708 GMT, Gunnersbury station was unroofed, blocking and short-circuiting the track, and six people on the platform were injured. Rush-hour traffic was disrupted. Then at Acton, five houses had their end walls blown out or were unroofed, and many other houses were less seriously damaged. Over a dozen people were injured. It was reported that a car was lifted 15 ft [4.5 m] off the ground, but the present writer is unaware of the source of this claim. Buildings were also damaged further on at Golders Green, Hampstead Garden Suburb and Southgate. The *Acton Gazette* reported:

> In Acton Vale a house and a factory collapsed and an 80-ft [24 m] chimney was blown on to the Royal Standard Laundry, Bollo-lane. Trees blocked East Acton-lane, and in a line from Gunnersbury Station through Acton towards Willesden rubble lay thick on the roads…Mr. Bill Pierson, of 269, The Vale…said: "I was in an upstairs room when it began to shiver and shake. Then the roof was torn off. I have done 30 years in the Army and have served in two world wars but tonight was the first time I have been frightened"…One hundred yards away, in Mansell-road, four railwaymen rushed for telephones after their engine was stopped by trees blown down on to the line. Less than 50 yards from the engine a large storeroom of a lead factory belonging to James Gilder and Co., lay shattered over the ground and line.
>
> (*Acton Gazette*, 10 December 1954)

In contrast to these other T7 cases, the tornado that passed through Welbourn, Wellingore, Navenby and Boothby Graffoe in Lincolnshire on 23 October 1666 seems to have been underrated by TORRO, and we now consider that it is the most powerful tornado on record in Britain at T8/9. Thomas Short, writing in 1749, described:

> …a dreadful Storm of Thunder, accompanied with Hail, the Stones as large as Pigeon, or some like Pullets Eggs, followed by a Storm or Tempest, attended with a strange Noise; it came with such Violence and Force, that at *Welbourn* it levelled most of the Houses to the Ground; broke down some, and tore up other Trees by the Roots, scattering abroad much Corn and Hay. One Boy only was killed. It went on to *Willingmore* [Wellingore], where it overthrew some Houses, and killed two Children in them. Thence it passed on and touched the Skirts of *Nanby* [Navenby], and ruined a few Houses, keeping its Course to the next Town [Boothby Graffoe], where it dashed the Church Steeple in Pieces, furiously rent the Church itself, both Stone and Timber Work, left little of either standing, only the Body of the Steeple. It threw down many Trees and Houses. It moved in a Channel, not in great Breadth, or it had ruined a great Part of that Country. It moved in a Circle, and look'd like Fire.
>
> (Short, 1749)

Short gave his source as '*Clark's Exampl.*', which is perhaps Samuel Clarke's *A Mirrour or Looking-Glass both for Saints, and Sinners, Held forth in some Thousands of Examples* [etc.], the fourth edition of which appeared in 1671. A completely independent account, dealing with Welbourn only, was published by A.C. Hervey in the *Meteorological Magazine* (Hervey, 1919). This specifies the number of houses (44) which the tornado 'blew down' and confirms the death of the boy. The track was 'eleven or twelve score yards' wide, which is about 200 m. 'It had the Appearance of fire and a sulphrous smell' – another reference to a characteristic which, for whatever reason, tornadoes no longer seem to exhibit. This text was also published in a more complete form in the *Gentleman's Magazine* for April 1801 (vol. 71, part 1, p. 317). It is a church brief, that is, an appeal, read out in churches, for donations to help victims of a disaster. Samuel Pepys, for one, thought there were too many of them: 'To church, where we observe the trade of briefs is come now up to so constant a course every Sunday, that we resolve to give no more to them (Diary, 30 June [Julian calendar] 1661). From the Welbourn brief, we learn that the damage there amounted to 'six hundred and four pounds, fourteen shillings, and two pence, besides the building, amounting to eight hundred pounds more'.

A search of websites dealing with the villages affected by the tornado has turned up several short references. Thus, at Welbourn, 'of 80 stone houses only three were left standing, the timbers being so disposed that none could tell his own' (Welbourn Parish Council, 2005). No source is given for this, but the present writer has recently traced it to a letter in *Calendar of State Papers Domestic: Charles II, 1666–7* (vol. 175, no. 163). The letter was written at Coventry on 1 November by Mr Ralph Hope and adds that '…three or four persons were killed. Part of another town near Overston was blown down, and part of a church at Boothby; numbers of trees were torn up by the roots. At Denton, near Grantham, about the same time, happened a prodigious storm of hail, some stones 3 in. long, some like darts, bearded arrows, and other strange shapes'. The *Calendar*, which, it should be noted, gives a summary of each document rather than a verbatim transcript, has another letter on the tornado, written at 'Naneby' (Navenby) on 25 October by the rector, Dr Thomas Fuller:

> There has been a great storm of thunder and hail, some hailstones bigger than pullets' eggs, then a tempest of wind in which most of the houses at Wellbourn were thrown down, trees plucked up by the roots, &c; also some houses at Wellingore, the spire of a church in the next town dashed to pieces, &c; it caused such a terror that the people thought the world was at an end; had it been universal, and not come in a channel, the whole country would have been destroyed; some saw it coming like fire, moving in a circle, and yet it kept straight on; only three persons were killed.

And on 29 October, Luke Whittington wrote: 'There are strange reports of a hurricane near Lincoln'. Besides these references to the tornado, the *Calendar of State Papers Domestic* also has a good many notes of stormy weather generally during October 1666, partly because of its effects on shipping; the tornado clearly occurred in a disturbed spell. Thus, on 11 October, Hugh Acland wrote from Truro that 'The weather has been very stormy', and on 21st he amplified this: 'The weather has been very stormy, with much thunder'. On 25 October, R. Hope in Coventry reported, 'great rain…also thunder and lightning'. On 28 October, Silas Taylor said: 'The five days' storm has at last ceased, and no damage is done to the shipping'.[19] And on 1 November, Richard Watts at Deal stated: 'The wind has long been between south and west, with very much rain'. Samuel Pepys, in his diary, noted on the day after the tornado, '…a very foul morning, and rained; and sent for my cloake to go out of the church'.

Finally, another recently discovered account of this tornado is in a book published by Aurelian Cook in 1685 to celebrate the life and reign of Charles II, who died that year:

> And at a place called *Welbourn*, in *Lincoln*-shire, after a prodigious Thunder, with Hail-stones of a more than ordinary bigness, there followed such a Storm and Tempest, that its violence threw down most of the Houses to the ground, tore up Trees by the Roots, and

---

[19]This, along with a similar remark from Southwold in Suffolk on 30 October, may suggest that the depressions involved were deep but not exceptionally so.

dispersing several Ricks of Corn and Hay, passed to the next Village, called *Willington* [Wellingore], where it threw down firm Houses, and going forward to *Nanby* [Navenby], it fell so violently upon the Church, that it dash'd the Spire in pieces, and so tore and rent the Body of the Church, that it almost levelled it with the ground.

(Cook, 1685)

The evidence presented here strongly suggests that the damage caused by this tornado was the most severe on record in the United Kingdom, at any rate at Welbourn. Both the quality and quantity of evidence on this event are considerably better than for the London tornado of 1091 or the Portsmouth tornado of 1810. A minor point concerns the date: until recently TORRO thought that the event happened in 1667, the date given by Lamb in a list of British tornadoes which appeared as an Appendix to his *Geophysical Memoir* on the tornadoes of 21 May 1950. Lamb was clearly misled by Hervey, who gives 1667, but that is the date of the brief.

No T7 tornadoes are on record in Scotland or Northern Ireland. In Scotland, the Blairgowrie case of 1767, already dealt with, may well have reached T6. There is also a report (from an unidentified newspaper) of a presumed tornado at Cairnbaan, Argyllshire, on 28 September 1956, which '…tore the roof off an unoccupied bungalow and hurled it into a canal more than 500 yards away. Trees were uprooted and villagers ran for cover as the storm swept through the streets, but nobody was injured'. If accurate, this perhaps suggests T5, but more information is desirable. In Northern Ireland, there are three candidates for the severest tornado. The one on 31 December 2006, which tracked from Ardmore, County Armagh, to Loanends, County Antrim, has already been mentioned as possibly having both the longest and the widest path in Northern Ireland. Dr John Tyrrell, who investigated this event (with the help of Martin North, also of TORRO), reported: 'The impact of the tornado included numerous mature trees being uprooted or snapped (despite the fact that they were in their winter condition without any canopy resistance to the wind), sheds being lifted and moved, as well as a car being lifted and moved. A significant amount of damage was of T4 intensity, but marginally reaching T5' (Tyrrell, 2007). The area of most severe damage was near Crumlin, County Antrim, and was investigated and recorded by Martin North (Tyrrell, J., personal communication).

The tornado in the Belfast area on 2 September 1775 has also been mentioned for both the length and width of its path. Its severity may have been T4 at Duff's Hill, judging by the following account:

> At Shankhill bridge it commenced its work of destruction, by carrying off ten cocks of hay from the adjoining meadows, and also such corn as was cut; the reapers flying from the fields in the utmost terror. Keeping a north east course, it did considerable damage near Whitehouse; and entering the lower part of the parish of Carrickfergus, carried away all the hay and corn that were cut in the fields it passed over, having twirled them in the air in a most singular manner. Near lower Woodburn bridge, it tore several large trees out of root, and at the Windmill-hill some persons who were passing were lifted from the ground, and thrown into an adjoining ditch…Some houses were also injured; at Duff's-hill it entered the door of a house that was open, and carried away its rear, leaving the front standing. Crossing Kilroot and Braid-island, it seemed to gain

vigour. In the latter, it conveyed away a hay stack that was nearly completed, while the people who had been putting it up were at dinner; and at Larne lough, it lifted up the waters till they appeared like floating white clouds, and transported them to a considerable distance. Having touched a small part of Island Magee, where it did also much damage, it was at length lost in the channel.

This tornado was succeeded by vivid lightning, and most tremendous peals of thunder, accompanied with a heavy fall of rain and hail. The hail, or rather masses of ice, fell in a great variety of irregular shapes: several pieces measured upwards of 6 in. in circumference. The ground over which this hurricane passed scarcely exceeded half a mile in breadth.

(McSkimin, 1823)

Another T4 case occurred in Belfast on 26 September 1982 and was described by Dr Nicholas Betts in *The Journal of Meteorology* for March 1983. It was reported that '...a truck containing more than two tonnes of building material was lifted two metres above the ground and flung seven metres across the road. In the Limestone Road, large healthy trees were uprooted, and at the Dunmore Stadium, corrugated-iron high fencing was ripped up and flung 250 m through the air' (Betts, 1983). Interestingly, some eyewitnesses described the tornado as 'two spirals converging from different directions to form a single vortex'. The *Belfast Telegraph* of 27 September reported that more than 100 houses had been damaged, and Dr Betts stated that 30 houses 'had their roofs blown away' and 70 more were 'badly damaged'.

5.7   Largest Outbreaks

The first recognition that the United Kingdom occasionally experiences a considerable number of tornadoes in a single day came with the publication of R.E. Lacy's paper in *Weather* on the tornadoes of the years 1963–1966 (Lacy, 1968).

The most notable outbreaks found by Lacy were on 15 November 1966 and 1 December 1966, both with 13 tornadoes (the current totals for these two dates are 18 and 25 respectively). By far the largest outbreak occurred on 23 November 1981. It was described in detail by Rowe and Meaden (1985), who put the number of tornadoes reported at 102. A complete listing, presenting 105 tornadoes, was published by Meaden and Rowe (1985). Small adjustments were made in the TORRO database, reducing the number to 104 (*see* Rowe and Meaden, 2002), although a tornado near Stone, Staffordshire, dated December 1981 by Rowe and Meaden (2002), is probably attributable to the 23 November outbreak.

The 104 tornadoes affected a triangular zone stretching from Anglesey to the Humber estuary and Essex (Figure 5.4). There were five tornadoes on Anglesey alone; one of these, at Holyhead, was one of three that day to reach T4 (the others were at Stoneleigh, Warwickshire and near Southwell, Nottinghamshire); most of the others were fairly weak at T0–T2, and the few known track lengths were short, almost all under 3 km. The tornadoes were most numerous in the east Midlands and East Anglia. The apparent concentration in this area may, however, be partly due to the help given to TORRO by Michael Hunt, weather forecaster for Anglia Television, who made an appeal for information from viewers.

Descriptions of the tornadoes included a 'grey mass going down the lawn' (Wallasey, Merseyside), while at Grimsby (Lincolnshire), the tornado 'went past the window like a big top spinning'. A particularly interesting description came from Market Drayton, Shropshire, where a resident observed 'two gigantic hands of low cloud (fingers outstretched)' – presumably a reference to a number of funnel clouds. The characteristic tornado roar was often commented on, 'like an express train', as several witnesses put it. At Amlwch, Anglesey, a witness thought at first that the noise *was* an approaching train, until it increased so much that

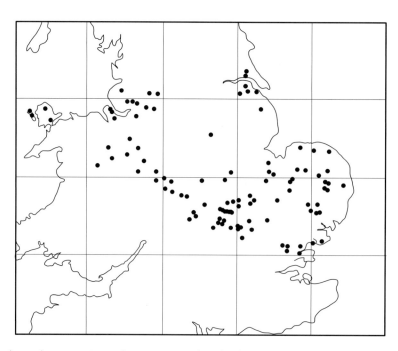

Figure 5.4   *Distribution of tornadoes on 23 November 1981. Reproduced with permission from Meaden and Rowe (1985). © Mike Rowe.*

the house shook, at which point 'I thought it must be an aircraft coming down'.

The second largest known outbreak occurred only a month before the largest one, on 20 October 1981. Turner, Elsom and Meaden (1986) recognised 31 tornadoes and gave full details of each one. The TORRO database reduces this total to 29 (*see* also Rowe and Meaden, 2002). This outbreak affected a smaller area than the one on 23 November, beginning in Somerset and stretching eastwards to Sussex and Kent (*see map in* Turner, Elsom and Meaden (1986)). No large outbreaks are known in Scotland or Northern Ireland.

5.8 Highest Death Toll

Fortunately, tornado deaths are extremely rare in the United Kingdom. The highest well documented death toll is five, at Abercynon and Edwardsville, Mid Glamorgan, in the T7 tornado of 27 October 1913 (Billett, 1914; Brown, 2013). Brown (2013) presented details of the deaths and added that the *South Wales Daily News* reported the number of fatalities as six, but without further information. One or two early tornadoes are claimed to have killed more than five people. A tornado hit Pillerton, Warwickshire, on 7 December 1222. 'The storm struck the buildings of a certain knight, crushing (*opprimens*) his wife with eight people of both sexes'.[20] The Nottingham tornado of 1558 is said to have killed a child and five or six men (see preceding text). A much better-known case is the storm at Widecombe, Devon, on 31 October 1638. Some very high figures have been quoted for the number of casualties in this event. Bonacina (1946) stated that 60 people were killed or injured, while Brooks (1954) gave 60 as the number of deaths. The question of the death toll was examined by Rowe (1980), who concluded that the number killed was possibly five to seven. It must be added that, although this was undoubtedly a very severe storm, its tornadic status can be questioned. The plane crash at Freckleton, Lancashire, on 23 August 1944, in which 61 people were killed, took place during a severe thunderstorm which appears to have included a tornado, but there is no evidence that the crash was due to the tornado. Finally, the TORRO website suggests that the collapse of the Tay Bridge on 28 December 1879, which resulted in the deaths of about 75 people, was probably attributable to 'two or three waterspouts which were sighted close to the bridge immediately before the accident' (*see* Botley, 1976). This is possible, but given that there was a severe gale blowing anyway the role of the waterspouts must remain speculative. In any case, there cannot have been conventional waterspouts in those weather conditions – possibly, they were eddy whirlwinds.

5.9 Concluding Remarks

It is clear from the data discussed that the severest tornadoes in the United Kingdom mostly occur in England: all those rated at T7 or higher were in England except the South Wales case of 1913. The

same is generally true for track length and width. The TORRO database contains 16 tornadoes with a track length of L7 or higher (over 21 km); all were in England except the two Northern Ireland cases (in 1775 and 2006) discussed earlier. Of 42 tornadoes with a track width of W7 or more (over 215 m), 36 were in England, with four in Wales and the two Northern Ireland cases just mentioned. Within England, there is no clear evidence of preferred areas for such events, though there are none north of Yorkshire.

Acknowledgements

The writing of this chapter would not have been possible without the collaboration of the many people who over the past 40 years have contributed information to TORRO, whether in the form of eyewitness accounts, site investigations or reports from earlier archival sources. The author is especially grateful to Paul Brown for supplying several early newspaper reports and the maps showing the path of the South Wales tornado of 1913 and to Christopher Chatfield for his impressive illustrations. The assessments given here by no means represent the last word on the subject, and further research will undoubtedly lead to improvements.

References

Abbott, S.G. (1955) The thunderstorms and tornadoes of 8 December 1954. *Weather*, **10**, 142–146.

Anon [Symons, G.J.] (1881) Three whirlwinds. *Symons's Monthly Meteorol. Mag.*, **16**, 133–136.

Anon [Rowe, M. and Meaden, G.T.] (2002) TORRO Tornado Division report for Britain and Ireland: January and February 2000. *J. Meteorol.*, **27**, 107–110.

Aubrey, J. (1721) *Miscellanies*, second edition. London: A. Bettesworth, J. Battley, J. Pemberton and E. Curll.

Betts, N.L. (1983) The Belfast tornado. *J. Meteorol.*, **8**, 76–80.

Billett, H. (1914) *The South Wales tornado of October 27, 1913*. London, His Majesty's Stationery Office, Geophysical Memoirs, no. 11.

Bonacina, L.C.W. (1946) The Widecombe calamity of 1638. *Weather*, **1**, 122–125.

Botley, C.M. (1976) Tornadic vortices and the Tay Bridge disaster. *J. Meteorol.*, **2**, 57.

Britton, C.E. (1937) *A Meteorological Chronology to AD 1450*. London, His Majesty's Stationery Office, Geophysical Memoirs, no. 70.

Brooks, C.E.P. (1954) *The English Climate*. London, English Universities Press.

Brown, P.R. (2013) The South Wales tornado of 27 October 1913. *Int. J. Meteorol.*, **38**, 196–205.

Brown, P.R. and Meaden, G.T. (2007) TORRO Tornado Division Report: December 2006. *Int. J. Meteorol.*, **32**, 243–248.

Brown, P.R. and Meaden, G.T. (2010) TORRO Tornado Division Report: July 2005. *Int. J. Meteorol.*, **35**, 171–179.

Budgen, R. (1730) *The Passage of the Hurricane, from the Sea-side at Bexhill in Sussex, to Newingden-Level, The Twentieth Day of May 1729, between Nine and Ten in the Evening*. London: John Senex.

Bull, G.A. and Harper, W.G. (1955) West London tornado, December 8, 1954. *Meteorol. Mag.*, **84**, 320–322.

Christie, R.C. (1883) The great storm near Nottingham in 1558. *Notes and Queries*, 6th series, vol. VIII, p. 304.

Cook, A. (1685) *Titus Britannicus: An Essay of History Royal: in the Life and Reign of His Late Sacred Majesty, Charles II. Of Ever Blessed and Immortal Memory*. London: James Partridg.

[20] Coxe (1842), p. 83.

Cousin, G. (1562) *Gilberti Cognati Nozereni Opera*. Basle: Heinrich Petri.

Coxe, H.O. (ed.) (1842) *Rogeri de Wendover Chronica sive Flores Historiarum*, vol. **4**. London: English Historical Society.

Eden, P. (1989) Memories of the West London tornado of 8 December 1954. *J. Meteorol.*, **14**: 416–418.

Hardy, T.D. (ed.) (1840) *Willelmi Malmesbiriensis Monachi Gesta Regum Anglorum atque Historia Novella*. London: English Historical Society.

Hervey, A.C. (1919) An old weather record. *Meteorol. Mag.*, **54**, 18.

Holinshed, R. (1587) *Chronicles of England, Scotland and Ireland*. Second edition, London.

Lacy, R.E. (1968) Tornadoes in Britain during 1963–6. *Weather*, **23**, 116–124.

Lamb, H.H. (1957) *Tornadoes in England May 21, 1950*. London, Her Majesty's Stationery Office, Geophysical Memoirs, no 99.

Luard, H.R. (ed.) (1880) *Matthaei Parisiensis Chronica Majora*, vol. **5**. London: Longmans *et al.*, Rolls Series.

Matthews, P. (1999) The Selsey tornado of 7 January 1998: a comprehensive report on the damage. *J. Meteorol.*, **24**, 197–209.

McLuckie, R. and Taylor, M.M. (1978) The Grantham-Newark tornado of 8 October 1977. *J. Meteorol.*, **3**, 35–39.

McSkimin, S. (1823) *The History and Antiquities of the County of the Town of Carrickfergus*. Second edition. Belfast: J. Smyth.

Meaden, G.T. (1975) The earliest-known British and Irish tornadoes. *J. Meteorol.*, **1**, 96–99.

Meaden, G.T. (1978) Eye-witness accounts of the Grantham-Newark tornado. *J. Meteorol.*, **3**, 40–42.

Meaden, G.T. and Rowe, M. (1985) The great tornado outbreak of 23 November 1981 in which north Wales, central and eastern England had 105 known tornadoes in about five hours. *J. Meteorol.*, **10**, 295–300.

Moon, A.E. (1946) Thunderstorms of July 4, 1946. *Weather*, **1**, 172–173.

Riley, H.T. (ed.) (1866) *Annales Henrici Quarti*, in *Johannis de Trokelowe et Henrici de Blaneforde…Chronica et Annales*. London: Longmans *et al.*, Rolls Series.

Rowe, M. (1980) The Widecombe calamity: how many died? *J. Meteorol.*, **5**, 315–317.

Rowe, M. (1985) Britain's greatest tornadoes and tornado outbreaks. *J. Meteorol.*, **10**, 212–220.

Rowe, M. (1989a) The earliest documented tornado in the British Isles: Rosdalla, County Westmeath, Eire, April 1054. *J. Meteorol.*, **14**, 86–90.

Rowe, M. (1989b) Further eye-witness accounts of the West London tornado of 8 December 1954. *J. Meteorol.*, **14**, 418–419.

Rowe, M. (1999a) The earliest known tornado in Scotland: Blairgowrie, September 1767. *J. Meteorol.*, **24**, 218–220.

Rowe, M. (1999b) 'Work of the Devil': tornadoes in the British Isles to 1660. *J. Meteorol.*, **24**, 326–338.

Rowe, M. and Meaden, G.T. (1985) Britain's greatest tornado outbreak. *Weather*, **40**, 230–235.

Rowe, M. and Meaden, G.T. (2002) British tornadoes and waterspouts of 1981. *J. Meteorol.*, **27**, 26–37.

Rowsell, E.H. (1960) Storms of 5 September 1958, over south-east England. *Meteorol. Mag.*, **89**, 252–257.

Short, T. (1749) *A General Chronological History of the Air, Weather, Seasons, Meteors, etc.* London: T. Longman and A. Millar.

Stow, J. (1598) *A Summarie of the Chronicles of England*. London: Richard Bradocke.

Stow, J. (1631) *Annales, or a Generall Chronicle of England*, ed E. Howes. London: Richard Meighen.

Symons, G.J. (1900) The Wiltshire whirlwind of October 1, 1899. *Q. J. R. Meteorol. Soc.*, **26**, 261–268.

Thorpe, B. (ed) (1849) *Florentii Wigorniensis Monachi Chronicon ex Chronicis*, vol. **2**. London: English Historical Society.

Turner, S., Elsom, D.M. and Meaden, G.T. (1986) An outbreak of 31 tornadoes associated with a cold front in southern England on 20 October 1981. *J. Meteorol.*, **11**, 37–50.

Tyrrell, J. (2007) Site investigation results of tornado reports in Ireland 2006. *Int. J. Meteorol.*, **32**, 266–270.

Tyrrell, J. (2008) The investigation of tornado events in Ireland, 2007. *Int. J. Meteorol.*, **33**, 197–201.

Welbourn Parish Council. Welbourn Parish Appraisal 2005. http://parishes.lincolnshire.gov.uk/Files/502/planappraisal.pdf (accessed 20 January 2014).

Williams ab Ithel, J. (1860) *Brut y Tywysogion; or, the Chronicle of the Princes*. London: Longmans *et al.*, Rolls Series.

Withnal, A. [also referred to as 'Mithnal'] *et al.* (1672) *A True relation of what happened at Bedford, On Munday last, Aug. 19 instant [Julian calendar], while Thundering, Lightning, and Tempestuous Winds tore up the Trees by the Roots [etc]*. London: Fra[ncis] Smith and R Taylor.

6

Site Investigations of Tornado Events

John Tyrrell

Department of Geography, University College Cork, Cork, Ireland

6.1 Introduction

Site investigations of local storm-damaged areas have engaged members of TORRO from its earliest days. There is no single right way to carry out a study of a storm's 'footprint'. As a result, there is ample scope for all manner of skills and experience to be used. TORRO has been an important vehicle for harnessing these and thereby improving our understanding of such storms, particularly tornadoes.

One of the first site investigations carried out in Britain and subsequently published was by Richard Budgen of an event in 1729 (Budgen, 1730). He was an outstanding cartographer who published a series of high-quality maps of the south coast region from 1723. Therefore, he was in an ideal position to map the details of the tornado that came off the sea near Bexhill, East Sussex, on 20 May 1729. His map showed meticulous detail as it recorded the 12-mile track crossing rivers, woods and villages, the damage swathe varying in width between 150 and 300 yards (137–274 m). His untimely death the following year at the age of 36 probably denied the scientific community other such maps.

Gradually, other studies of tornado events were published in the 19th century and early 20th century for the United Kingdom, such as the impressive 3-day survey of the tornado impacts in South Wales in 1913 (Billett, 1914). On mainland Europe, many similar studies of tornado damage tracks were conducted, no doubt influenced by the publication of guidelines for tornado damage surveys (Peterson, 1992). This culminated in the European-wide summary by Wegener (1917). He encouraged detailed measurements at many locations across a single swathe. In addition, as he had a particular curiosity about the effect of topographic features, he urged careful mapping of the track when it crossed these. This illustrates how a specific research priority can shape and possibly limit data gathering.

Political turbulence in Europe limited further developments. From the 1950s, ongoing progress in Britain was flagged with the publication of *The English Climate* (Brooks, 1954). In this, he described a number of tornadoes and mapped the tracks of 23 of them, thereby emphasising their dominant alignment from SW to NE. Generally, from the 1950s, new research largely ignored most previous work on tornado damage tracks, although there remained some awareness of the earlier meteorological studies.

TORRO was enthusiastic to go beyond the meteorology of tornadoes alone. But investigating tornado damage tracks was not one of the four aims initially set by TORRO's founders (Meaden, 1985). Nevertheless, it was clearly an approach intended to contribute significantly to the realisation of those aims. In fact, Rowe (1975) had stated that TORRO was formed 'to … encourage on site investigations', but this had still not been included in its statements of goals formulated in 1985. Despite this, most members realised the importance of investigating the damage tracks. So, from its earliest days, TORRO developed site investigation as a major methodology. Indeed, before TORRO was born, those who were to form the organisation had already started to develop their site investigation studies, as in the case of the tornado in Barton, Oxfordshire, in October 1966 (Meaden, 1977).

Site investigations of tornado events that came to the attention of TORRO were also carried out by local enthusiasts. No special preparation or training was required other than an attention to detail and a multidimensional curiosity. So, the investigation for the tornado that occurred on 12 January 1975 at Chipping Warden in Northamptonshire was carried out by a local resident, Rev. Robertson-Glasgow. The final summary of that event was produced from his information, local newspaper accounts and photographs (Meaden, 1976a). Such local involvement has always been important. The level of those contributions varied, but they were often important for two reasons. Firstly, there was the obvious practical consideration that where there were no other experienced investigators available they were the main source of detailed data for the event. Secondly, where a TORRO investigator did

Extreme Weather: Forty Years of the Tornado and Storm Research Organisation (TORRO), First Edition. Edited by Robert K. Doe.
© 2016 John Wiley & Sons, Ltd. Published 2016 by John Wiley & Sons, Ltd.

come from outside the affected area, local assistance opened up the area of investigation to the areas affected and to key eyewitnesses.

Over the years, the many weather enthusiasts across the country have channelled their energies into their own meteorological station to report weather conditions in their immediate area. Localised weather events such as tornadoes rarely pass over these or even fewer professional stations. So, despite many sharp eyes scanning the skies regularly each day, few have acquired site investigation experience that would develop from such encounters. Their primary purpose was to measure and record weather parameters at a particular geographical point. At best, it was a secondary matter to record other extreme weather events in their wider local region.

By 1994, 20 years on from TORRO's birth, although many tornadoes had been recorded, it was apparent that site investigations had been carried out by relatively few enthusiasts. There had been many site visits. But it is the attention to recording the detailed impacts of tornadoes in a systematic way that differentiates a site visit from a site investigation. A site visit does have value in briefly viewing how the event occurred and possibly reporting selected impressions of the effects. A site investigation, on the other hand, is a systematic recording of impact details, both physical and non-physical. As a result, the track dimensions are defined, the relative intensity of the event is established, together with the timing, duration and speed of the vortex. In addition, there are human consequences that are noted where this is possible.

6.2 Getting Started: How Site Investigations Come About

All site investigations begin with a report of severe local wind damage where a tornado is suspected. This comes from people who have been affected. How that report travels to the site investigator filters the information and may influence the focus of the site investigation. Generally, there have been four main pathways for this information:

1. A local resident may contact TORRO directly, normally via its website reporting tool. The reporting tool has attempted to define a number of key parameters that will assist in any site investigation follow-up as well as providing archive information that will enable it to be entered into its database in a meaningful and comparative way.
2. A report to the local media, usually a radio station or local newspaper. They will decide whether to use it, depending on how 'reportable' it is. Much detail is normally dropped, but information about significant damage retained. Local media vary enormously in their response to such reports. Whether they reach the site investigation stage may depend on personal initiatives within media organisations rather than particular policies guaranteeing follow-up and verification.
3. Through the web's network of weather-related special interest groups. These have increased in recent years and have attracted many reports of possible tornadoes. The reports are not always first-hand, and the content can be very slight.

Two major barriers in using these are the establishment of the poster's identity or location and the increase in hoax reports. Nevertheless, many reports have proved to be worth further investigation, and subsequent site investigations have followed.
4. Meteorological organisations receive reports of possible tornadoes, some of which are forwarded to TORRO for follow-up. None of the organisations in the United Kingdom have a formal policy to do this, but for a long time, many professional meteorologists have been aware of TORRO's activities, and not only have they ensured that key reports are passed on, but some of them have extended their interest – indeed, their intense enthusiasm – to become part of TORRO enterprise themselves. But overall, the information from this source can be rather general at first and may provide much less focus for a starting point of a site investigation.

When the reporting process has put the investigator in touch with local eyewitnesses, the prospect for an effective site investigation is enhanced significantly. Making such contact is now relatively easily done through the Internet. Investigators have identified making this connection as one of the particular challenges in preparing to respond to a report. Local people mostly know where the damage occurred, who has been affected and the impacts that would not be readily apparent to a stranger. It has been found that very few have known details for the entire track. Exceptionally, briefings have been provided by emergency services. However, investigators have also been frustrated by restrictions imposed by emergency personnel, as in North London in 2006 (Brown and Meaden, 2007).

Investigations often start where the report originates. This could be anywhere along the track – but rarely at the beginning or the end. These tend to be located late on, for often the damage and other impacts are least and less evident at these extremities. They are mostly overlooked by media reporting that focuses upon the quick, big story. Indeed, people tend not to report impacts that seem to be slight compared with what quickly goes into the public domain. It is part of the site investigation to seek these out. This can be particularly challenging when a tornado has varied in intensity during its life so that severely damaged areas alternate with areas that hardly seem to be impacted at all. It is estimated that a definitive start and end to the damage track is established in just over 60% of investigated cases.

6.3 Site Investigation Methods

The site investigation seeks to record all information that will assist in confirming whether a tornado was involved or not. The tornado funnel is not always seen, especially at night. In these and other circumstances, wind impacts, particularly the damage pattern, have been used successfully to differentiate between 'straight-line' winds and those of a tornado vortex. Thus, the details of what has been damaged or destroyed and its location, the alignment of the debris and where it was finally located with reference to its original position before the storm are all significant.

Damage patterns are only part of the investigation. To identify what has happened also requires eyewitness information about their experience of the storm and in particular the passing of the tornado. This may range from weather sequences to sounds and sensations. Many of these details go beyond the determination of the tornado's intensity. But they explore other important dimensions of the event.

Mapping is a key in establishing and recording the location, direction, width and variations in the tornado's track. From the early years, printed maps at scales of 1:25,000 to 1:10,000 have been particularly useful. Maps and vertical photography now accessible on the web have considerably enhanced the range of base maps for plotting local detail of damage and debris. Sketch mapping has frequently been added to these since the precise map requirements are unknown at the outset of the investigation. Nearly all investigators have produced an annotated map as a result of their investigation.

Questionnaires have been most productive when the investigator and witness complete it together. Often, unexpected information emerges in the process. But this takes time and time is often limited when the investigator is a volunteer operating on a limited time budget. Questionnaires have also been productive when residents are absent (Elsom, 1989). But without person-to-person interaction, a high response is rare. Questionnaires have been very productive in yielding timing details which, when added to other replies, can be converted to worthwhile assessments of the tornado's forward speed and possible variations. They have also established the weather sequences immediately before, during and after the tornado, particularly with regard to the sharp transitions that occur and whether the funnel was rain-wrapped, as has often been the case. In addition, non-physical details such as noise sequences, pressure effects, personal injury and reactions are all part of the tornado experience that would never be recorded if information gathering was by observation alone.

For the eyewitness, the use of a questionnaire has turned a conversation into an interview and placed the information into a formal research context. This has sometimes given confidence in sharing important information that would not otherwise emerge. At times, there is a fear of the media and how their personal information will be treated. It comes as a reassurance to some that the investigator is not a media person but is part of a formal research organisation.

A photographic record of a tornado occurrence was for many years primarily made by the researcher. This still remains an important way the site investigator records tornado damage. One reason for this is that the tornado's intensity is measured by its estimated maximum wind speed, which is normally assessed from the most severe damage. A record of the damage is, therefore, an important part of the event documentation.

Today, the potential of mobile phones and other digital products means that images of a tornado vortex as well as its effects may abound. However, tornadoes are unexpected phenomena, and available close-up images of individual tornadoes still appear to be rather limited. Since a tornado passes a single spot in a very short time and people are normally concerned about safety and protection issues, photographing what is happening is not a priority. Nevertheless, a wealth of images of the tornado's damage and debris may be unearthed during an investigation.

Field sketching of the funnel, cloud features and surface effects creates a familiar perspective of what had been seen. Constructing the sketch with an eyewitness involves sharing information that often yields a great deal more detail about the event than would otherwise be achieved. The earliest attempt to use this technique for a tornado funnel appears to have been made in 1638 (Rowe, 1975), and a fine engraving from a sketch of a funnel near Maidenhead, Berkshire, was published in 1853 (Chatfield, 2004). However, field sketches often get overlooked when case material is finally published.

Sketching captures the very different shapes, sizes and contexts of tornado funnels. Such detail is only available from local eyewitnesses. Occasionally, they are able to sketch what they see as the event unfolds. Whirling debris in the tornado funnel was the particular focus of the sketch made of the Birmingham tornado of 1931 (Meaden and Chatfield, 2009); but the Gosport, Hampshire, tornado of August 2001 was sketched to emphasise how 'ribbons of condensation appeared to branch downwards like talons, rapidly rotating around each other and converging at the base' (Gilbert, 2002); while an outline sketch by an artist in Selsey shows a wide wedge-like funnel in full contact with the sea surface approaching from the Isle of Wight direction (Matthews, 1999b). In the case of the 2011 Craigbrack tornado in Northern Ireland, a sketch was created with the eyewitness (Figure 6.1) to capture the tornado's dimensions and track after it had caused considerable structural damage to a farm (Tyrrell, 2012). Sometimes sketching is more effective than words in conveying detail. Types of funnel shapes and how they changed during the event, debris details, whether the rotation was cyclonic and even whether there was more than one vortex involved have been recorded in this way. So, in the Calshot, Hampshire, tornado of November 1999, a witness sketched the shape of the funnel confronting him from 20 m, with the paths of corrugated steel roof sheets rotating in the funnel (Matthews, 2000).

The site assessment of tornado intensity: From early historical work identifying tornadoes, it became apparent that historical sources were more likely to report severe tornadoes rather than less powerful ones. As the need to make distinctions grew, the development of an intensity scale was considered a necessity. This was developed and tested privately from 1972 and was circulated within TORRO from 1974. It then became an important tool for investigations and focused this work on particular effects that indicated tornado wind speeds (Meaden, 1976b).

The intensity scale proved to be a very valuable tool for site investigation work. Based on the Beaufort scale of wind speeds, which was taught throughout the UK's educational system, its rationale was readily known. Care was taken not to use it too rigidly. Its ability to make distinctions on the basis of wind speed and its effects was easily understood and readily adaptable in later years. Its ability to separate tornado intensities at the weaker end of the spectrum was a particular benefit.

Applying this scale focused attention on the most severely affected parts of the tornado's track and less on its beginning and end, although many investigations tried to locate these as well. Tree and roof damage have been the dominant forms of tornado impact on the landscape, by a long way. As a result, the use of a tornado classification scheme that uses such damage features to estimate vortex wind speeds is widely applicable. Nevertheless, the application of the scale is not easy, and apparent mismatches

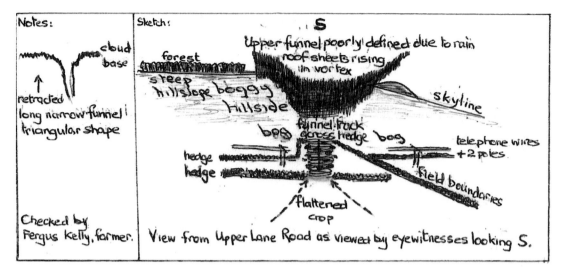

Notes:

cloud base

retracted
long narrow funnel
triangular shape

Checked by
Fergus Kelly, farmer.

Sketch:

S

Upper funnel poorly defined due to rain

roof sheets rising in vortex

forest

steep hillside boggy
hillside

skyline

bog

funnel track across hedge   bog

telephone wires + 2 poles

hedge

hedge

field boundaries

flattened crop

View from Upper Lane Road as viewed by eyewitnesses looking S.

Figure 6.1   *A sketch of the Craigbrack tornado from eyewitness information, October 2011. John Tyrrell (2012).*

between the tornado intensities applied from tree damage and house damage have had to be resolved. One positive feedback from site investigations led to the further modification of the intensity scale by expanding the indicators in each of the classes, particularly as building types and materials changed and other expressions of modernisation developed (see Chapter 1).

6.4   Site Investigation Outcomes: The Growing Understanding of UK Tornadoes

As membership spread across most counties of the country, TORRO became uniquely positioned to carry out site investigations. The urgency to do this grew as the extent of tornado frequency became apparent, far beyond the capacity of a single institution to respond to. The accumulation of investigation results from every part of the country has resulted in a significant major advance in understanding the tornado hazard. It has not been possible to investigate all reported tornadoes, but TORRO records show that site investigations have varied between 6 and 41% of known tornado events per year. As a result, particular features of the tornado event have been explored.

Core winds: The core of the tornado at ground level is an area of violent inflow from 360 degrees, converging towards the axis of the tornado, where the lowest atmospheric pressure at the ground surface will occur. Strong uplift occurs around that point. The pressure gradient into the core is always strong, and as it arrives, the pressure change is very sudden. Site investigations occasionally find a record of these features, as in the case of a barograph trace located within 500 m of the tornado core of the Bosham, Hampshire, tornado of October 2000 (Gilbert, 2001). This trace made a near-instantaneous fall and rise of 4 hPa. Garden and rooftop anemometers have been valuable for determining inflow wind strengths, as at Hill Head, Hampshire, in November 1999, when 135 km/h was logged just metres away from the core. From such a detail, other wind characteristics were calculable (Gilbert, 2000).

Witnesses caught within the vortex circulation have spoken of winds suddenly changing direction, but not noticeably decreasing in their violence. This would require the witness to be close to the core, as in small-diameter tornadoes. At Retford, Nottinghamshire (August 1997), meteorological instruments recorded wind direction changes over the full 360 degrees (Fowler, 1997). The strength of this wind shear is itself potentially damaging, especially where structures have already been weakened by the initial assault of the wind. A stark effect of violent winds on opposite sides of the vortex was evident when walls collapsed into a road, but on opposite sides, at Lower Earley, Berkshire, in June 1998 (Graham, 2000). In contrast, witnesses caught in the core of a wider and slower-moving tornado had a different experience. In 1969, as the Bodmin, Cornwall, tornado arrived, '…the noise was terrific … then there was dead calm … then the terrific noise and tumult again' (Rowe and Meaden, 1985).

Detailed site investigations have shown that even with smaller tornadoes, the most significant destruction is to larger-sized structures and objects. This is a result of the way wind pressure around the core has been exerted, because it is wind pressure that causes most of the damage rather than stresses exerted by atmospheric pressure. So roofs and even walls often go, but the bread bin or microwave in the kitchen stay on the worktop, because the windflow has been diverted. Recording this level of detail for individual buildings has been very patchy but worthwhile when achieved.

Site investigators frequently receive statements linking the core pressure drop with the sensation of popping ears. Severe cases are likely to be due to such an effect, as when excruciating ear pain lasting for 20–30 seconds was felt by a couple as a tornado passed at Pagham, West Sussex, in September 1999 (Gilbert, 2000). Many other less painful occasions have been recorded.

'Explosive' effects: Site investigations in the United Kingdom have produced many cases where damage has been caused to buildings by tornadoes and the debris has been thrown outwards. This has been described in some cases as an 'explosive effect'. Its destructiveness and violence merits such a description, but the evidence has also given rise to alternative explanations.

The severe drop in pressure within the core of the tornado has the potential for creating violent air movements, due to the strong suction that occurs as a result. An interesting case is reported for the West Cornwall tornado of December 2012 where a farm building 'exploded' outwards at opposite ends, leaving the middle section intact (Clark and Pask, 2013). A similar effect can occur within a building, as when particularly strong suction resulted in a closed bolted internal garage door leading into a house being broken open when a pair of screws were forced out and a second pair loosened (Meaden, 2001).

An alternative understanding of such explosivity has given a dominant role to the vortex winds rather than atmospheric pressure. The mechanism involved is derived from the wind exerting a direct positive pressure on the outside of the building. This results in some areas of accelerated airflow which creates a suction effect. This is a different suction effect compared with what would result from the low-pressure core of the tornado. Also, there is still a relative outward pressure from the building itself to be added. The end effect is to give the appearance of an 'explosive' failure (Buller, 1985).

Clearly, there is a possibility that both effects contribute to the damage from time to time. Not all tornado tracks have evidence of an 'explosive' failure. Where this does occur, the appropriate interpretation might well depend on the orientation of the building with respect to the passing tornado and its core. But where both windward and leeward sides of a structure fail at the same time and the debris is projected outwards, it is quite likely that the former process is at work. It will also depend where the funnel core is located at the time of the 'explosive' effect. Often site investigations are inconclusive with regard to this detail.

Timing and forward speeds: Timing details from eyewitnesses and residents have helped to match events with recorded meteorological data and products such as radar images of storm cells. The timing of the start of the tornado is particularly helpful. It has helped to clarify how a vortex developed and moved, whether it is from a gust front at the forward edge of the outflow from the parent storm cell, whether it occurs between interacting adjacent storm cells or whether it occurs at the downdraught–updraught interface. The latter appears to be the commonest position in the case of an isolated storm, the vortex normally being positioned on the right flank of the intense radar echo, as demonstrated by the site investigation for Long Stratton, Norfolk, in December 1989 (Elsom, 1993). Exact timing details are rarely determined, but during the Selsey, West Sussex, tornado of January 1998, a clock battery popped out with the pressure change and immediately stopped the clock (Meaden, 1998).

The forward speed of tornado funnels can vary enormously. This can be a significant detail to relate to the nature of the damage effects and the lifespan of the tornado funnel. The rapid forward movement of tornadoes limits the orientation of the debris on the ground. With the usual cyclonic rotation of the funnel, the debris is mostly flung forward and to the right. This effect was noted in the Farnborough, Hampshire, tornado of September 2007 which had a forward speed of 50 knots (Brown and Meaden, 2007) and the South Ham, Hampshire, tornado of June 1996 of 64 km/h (Matthews, 1999a), while the Monkwood, Hampshire, tornado of 2010 investigation had a forward speed estimate of 90 km/h and threw debris forward to the left (Clark, 2010).

In contrast, it has been found that a vortex may have a speed as slow as 6 km/h, as at Pagham, West Sussex, in September 1999 (Gilbert, 2000). This would appear to be exceptionally slow moving. From a small sample, more typical forward speeds have been 25–35 km/h, as in Reading in June 1998 (Graham, 2000) and Atherfield (January 1998), Isle of Wight (Burberry, 1998). Both slow- and medium-speed tornadoes have tended to produce debris fields with alignments in a wide variety of directions, enabling the investigator to more readily confirm the presence of a rotating vortex.

The lifespan of a tornado is affected by the relative speed of the parent cloud that spawned it. It is thought that this often accounts for the relatively short lifespan of UK tornadoes. The faster-moving parent cell causes the trailing funnel to stretch and fragment (i.e. to 'rope out'). An example appears to have been the cluster of short-lived tornadoes in the Solent in October 2000, where the funnels were travelling at approximately 80–90 km/h beneath a parent storm cell moving at 110 km/h (Gilbert, 2001).

Multiple tracks: Site investigation work has been key in identifying tornadic storms that have spawned more than one tornado. It is the natural assumption that a single tornado would be responsible for a damage track, even if it appeared to be fragmented, particularly if the event is a stand-alone one. For example, at the end of one tornado track in East Anglia, mapped for some 16.8 km (10.5 miles), the investigator was informed of more damage only 2.4 km (1.5 miles) away (Briscoe, 1997). But this proved to be a separate event that partly ran parallel to the first one, occurring at the same time (Figure 6.2). Such parallel tracks have occurred very close to one another, as when two tornadoes struck Bosham village, Hampshire, in Chichester Harbour (Figure 6.3). These were a mere 600 m apart during an outbreak of six tornadoes over a 25 km (15.5 miles) stretch of coastline (Gilbert, 2001).

Coordinated site investigation work has also confirmed that on occasions, widespread multiple outbreaks have occurred within the United Kingdom. Then, multiple individual tornadoes were reported, and site investigation work was as extensive as possible. A particularly well-researched event of this kind was in October 1981 when at least 31 tornadoes were confirmed as certain or probable from site investigations or, where site visits were not possible, by questionnaires (Turner et al., 1986). Site investigations have confirmed similar large-scale outbreaks and have given new insights into cold fronts and triple-point environments (where a cold front intersects a warm front).

Within a broader damage track, definable smaller tracks have sometimes been noted, leaving smaller, narrower trails winding their way with the main track but not necessarily following it in its entirety; indeed, they may be extremely limited in their extent. These are usually caused by suction vortices which rotate slowly around the fringe of the tornado core. They occur in the more intense tornadoes, but TORRO has only been able to document a small number of cases (Elsom, 1985).

Intensity variations: Right from the early beginnings of its site investigation work, TORRO investigators noted that damage tracks sometimes appeared to be fragmented and discontinuous. Matching this evidence were accounts from eyewitness that described how they watched the funnel of a tornado seeming to be in touch with the ground only intermittently (often described as 'skipping'). Others reported that it appeared to be growing and then shrinking.

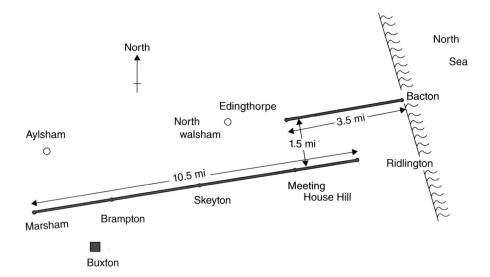

Figure 6.2 *Parallel tornado tracks of the Edingthorpe and Marsham tornadoes, February 1997. Charles Briscoe (1997).*

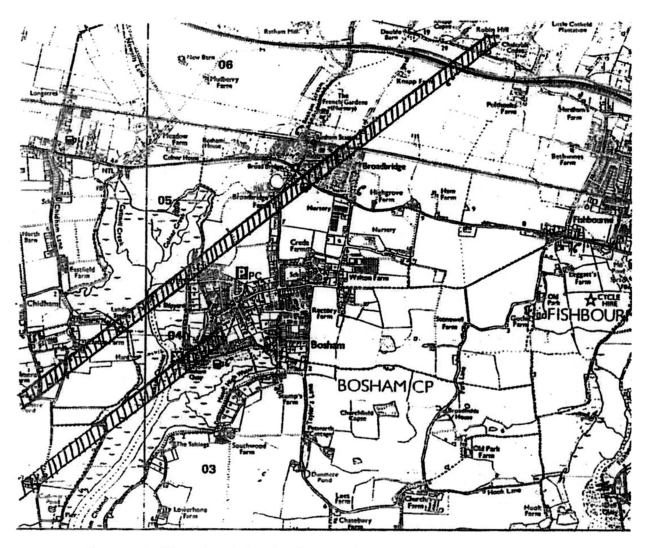

Figure 6.3 *Parallel tornado tracks through Bosham in October 2000. Reproduced from Gilbert (2001).*

Damage maps have shown such gaps are often on slopes that face the advancing tornado funnel. One such case was the Annesley, Nottinghamshire, tornado of August 1985 (Osborne, 1985). It advanced from the west through Annesley village, creating a relatively short track of approximately 2.5 km (1.5 miles). It then passed over two west-facing slopes without inflicting damage but did so before and after each of them. This was confirmed by both damage mapping and eyewitness accounts. Similar cases have been identified, as at Overbury Park (February 1990) in the county of Hereford and Worcester where the funnel was reported as lifting temporarily on approaching rising ground (Wood and Mortimore, 1991) and at Dullingham in East Anglia (November 1991) where buildings on a stretch of rising ground were undamaged (Pike, 1992), while the Southampton tornado of November 1998 was a slight variant of this since the severest damage was inflicted on the downhill side of the terrain, suggesting that in this instance the funnel was dragged strongly downwards (Matthews, 1999c).

This pattern seems counter-intuitive. But the evidence from site investigations shows that it is real. Normally, the expectation would be that a wind-facing slope would be the most vulnerable to damage. This is so with straight-line winds and is, therefore, common to popular experience. But all that is required of the tornado for this effect is that either the approaching funnel lifts more rapidly than the slope incline or that the vortex rotation declines sufficiently for a short period to reduce its damaging impacts. In either case, a discontinuous track is created.

However, from time to time, the contrary pattern of damage also occurs, where the severest damage is on slopes facing the tornado. In the case of a north Derbyshire tornado in 1995, the tornado tracked from the SSE across eastward flowing tributary streams of the River Rother. The investigation recorded notable damage on south-facing slopes and crest ridges but much less on north-facing slopes, even in wooded areas which normally would be severely damaged along a tornado track (see also Prosser and Hill, 2006).

Discontinuities in the damage track are due to more than topography. They have occurred in many cases without any significant topographic influence. Among those factors that have been found to give rise to this effect in built-up areas are changes in alignments of streets, changes in house design and variations in the quality of building materials. Similar variation occurs in rural areas with regard to the condition of trees, the age of woodlands and forest areas, their stability with regard to their rooting systems and many other conditions. However, seemingly 'missing tracks' are sometimes revealed. When the ground survey for the Little Stretton, Shropshire, tornado of September 2007 found such a gap on farmland, an aerial reconnaissance was able to discern the track crossing the ploughed fields (Brown and Meaden, 2008).

It is unfortunate that the term 'skipping' has been adopted by many to describe this effect because it implies that it is a function of vortex behaviour alone, namely, its vertical rise and fall. Where there is considerable variation in topography and types of surface such as in the United Kingdom, the effect appears to be relatively frequent. Site investigations clearly establish it. But increasing numbers of case studies are showing that it is a function both of the details of the terrain and the dynamics of the tornado causing them to weaken and intensify several times over.

More systematic site investigation case studies are necessary before the complexities of this relationship are established.

Where there are discontinuities in the damage track, there have still been other significant effects of the tornado that have been continuous along the route of the vortex, even when the physical damage is reduced. For example, in many cases, it is apparent that they have still been exposed to significant amounts of flying debris and local people have still been subject to the many psychological traumas associated with the passing of tornadoes.

Topography: Topographic influences upon tornadoes in the United Kingdom appear to be considerable. Site investigations have revealed a significant matching between topographic detail and the tornado track, from the start to the end of the track. With a countryside as topographically diverse as the United Kingdom, many of these associations are becoming increasingly apparent. But so many possible factors are involved in each case that a very large population of cases is needed in order to confirm suspected processes. One feature that has emerged from the few computer simulations that have focused on this issue is that the tornado's intensity and structure can be modified as a result of the fluctuating inflow into the tornado core close to the surface. This varies significantly with the details of the topography, so generalisations are very difficult without numerous case studies (Lewellen, 2012).

Terrain roughness has emerged as one possible factor in the initiation of tornadoes. There are two elements to this. Physical variations in altitude and relief may create turbulence, convergence and uplift. But surface vegetation contributes to terrain roughness as well, and this has a similar effect, as, for example, where forested areas are known to have a much rougher surface than agricultural land. The relief required to do this may be relatively modest, as with the Stroud, Hampshire, tornado of January 1999. This developed on the easterly slope of Cold Hill, which is part of a SW–NE ridge, from where the tornado formed and from which the tornado tracked eastwards (Gilbert, 1999).

A further process that has been established from site investigations is the way a steep valley may appear to steer a tornado/waterspout. An excellent case was that of the Barmouth waterspout from Cardigan Bay in September 1984. After it crossed the coast (and thus temporarily became a tornado), it entered the Mawddach estuary where the funnel at the surface was then steered away from the parent storm cell by high ground to its north and south (Figure 6.4). The site investigation established that it then appeared to accelerate as it was pulled away from its parent cell, at the same time thinning as it stretched, until it finally dissipated (Smith and Harper, 1986). The investigation was mostly by photography and interviews as little of the vortex passed over land.

Coastal regions are fertile areas for possible topographic effects on tornado and waterspout development. One such region that has been relatively well documented by TORRO with many site investigations is along the south coast of England, where the influence of lee coasts, and the Isle of Wight in particular, has been very significant (Matthews, 2003). This effect has contributed to the long history of tornadoes impacting upon a relatively densely populated coastline.

Noise: The sound of a tornado appears to be at the heart of the tornado experience for residents and others in its path. During night-time events, personal memories are dominated by it. But

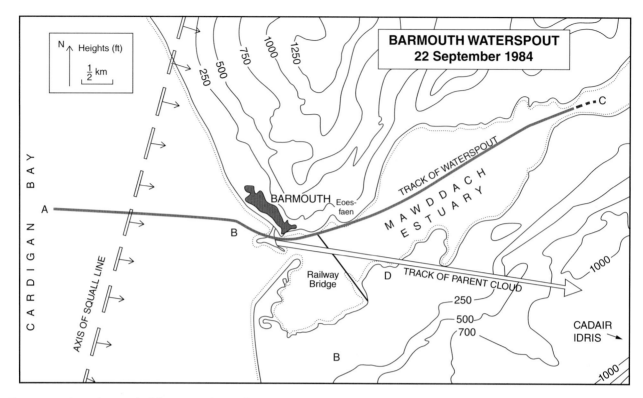

Figure 6.4 *The surface track of the Barmouth tornado – waterspout (A–B) constrained by the surface terrain to diverge from its upper parent storm cell (B–C), September 1984. Smith and Harper (1986).*

even during the daytime, it is a source of great anxiety, fear and even terror. The site investigator encounters this frequently when gathering personal accounts of the event. The subsequent report usually expresses this as a simple statement to the effect that most residents commented on the noise created by the tornado, but such brevity belies the intensity of the experience.

Rarely have eyewitnesses said that there was simply a sound of a strong wind. Instead, many analogies are used to compare similar sounds with that of the tornado. The sound of a tornado has most frequently been reported as a 'roar', often referring to trains in a tunnel or jets flying at rooftop height and even crashing in the ground nearby, the explosion of a massive bomb, a massive shredder or simply a terrific noise, as in a horror movie. The sound is often described as deafening, terrifying, fearsome, frightening and as never having been heard before. All this points to a new, unique experience in the life of those affected, one which they feel inadequate to describe in a way that captures its immensity.

Investigators have found that the reactions people had to this were instinctive. 'We rushed to get away'. 'I just fell to the floor'. 'I thought we've got to get out of here'. Investigators report that many of those affected saw themselves in a life-threatening situation and retreated from it in a way that would afford the greatest protection to themselves and their families. Yet it was always brief: the noise went as quickly as it came. Some investigations report a duration of about 70 s, others as brief as just 3 or 4 s. Its onset was mostly sudden, and its cessation was equally as abrupt, adding to the dramatic effect. Eyewitness accounts indicate that a tornado has to be very close before it can be heard. Very occasionally, a more gradual build-up of noise as well as its eventual

reduction is reported. But it still produced similar reactions in most cases. Even the roar of inflow winds themselves have been noted as distressful, as in an infant school during the passage of the Leeds and Harrogate tornado of September 2006 (Prosser and Hill, 2006).

When the site investigation encounters such reports, the really significant part of the record is not so much the physical sound, as scientifically interesting as that is. Rather, it is the trauma it imposes upon the people who experience it. This often lasts for a considerable period of time. This sometimes results in lilapsophobia (the fear of tornadoes and hurricanes) which is expressed as an ongoing fear that any storms will become severe. This develops from such a negative experience of severe weather and may occur in both children and adults. Indeed, where site investigations have been conducted some time after the tornado has occurred, an investigator may come across this condition from time to time. Comments such as 'I am still afraid of high winds' are commonly made by those who have had a tornado encounter. However, lilapsophobia extends far beyond this and is related to fears and insecurities that are quite debilitating. Because of the well-known negative effects of ramping up the dramatic elements of such events, particularly on children, investigators invariably listen and record and do not prompt nor enlarge in order to extract harrowing details.

The sound of a tornado is unique. But it has many components. Some sounds are produced as a result of the shear effects in the atmosphere above and others by the vertical movements of turbulent, unstable air. But at the surface, these combine with the noise of structures being torn apart and debris being cast far and wide. So, attempts to scientifically define these sounds by analysing recordings made in the path of tornadoes have concluded that the audible sound waves are too variable to make helpful generalisations.

Damage and debris: Damage detail is at the heart of the evidence for interpreting the event. Trees provide excellent evidence and are found at most investigated sites. Since many trees are surrounded by relative open space, the particular debris from individual trees has often been identified when the scattering has not been very extensive. As a result, the branch fragments often demonstrate the rotation of the vortex winds, being orientated in multiple directions (Figure 6.5).

A further very common effect on trees noted in site investigations occurs where twisting of individual snapped trees is clearly evident (Figure 6.6). When the tornado track has been much wider than the diameter of the tree crown, it has been suggested that these have been caused by either small vortices only a few metres wide within the tornado or a very severe shear within the tornado's funnel. The twisting produces large vertically elongated break surfaces and splinters of wood.

Details of roof damage have been particularly helpful for interpreting an event. Roof tiles that are intact on one side of the street have been mostly stripped from the other. The leeward side of roofs tends to be most damaged. Even when damage is not evident from the road view, investigators have found that access to their rear reveals severe damage (Meaden, 1998). Garage roofs

Figure 6.5 *Tree branches from a black poplar tree thrown to the NW through N to SE by the Leeds–Harrogate tornado, September 2006. Tim Prosser and Louise Hill (2006).*

are very common and have been indicative, since non-tornadic winds tend to fold these over while a tornado rips them off (Figure 6.7). It has been found that much structural damage is often out of sight from external examination and requires local knowledge (Clark and Pask, 2013).

The debris from the damage is usually widely scattered. An advantage gained from an investigation some time after the event has been the retrieval during that period of much scattered debris. Debris scattering is invariably much more extensive than the tornado track itself. This debris field therefore defines the wider destructive impact of the tornado. It is many times the area of the damage track alone. The lifting and transport mechanisms that are responsible are not resolved by the site investigation. It has been found that most of the debris is carried over short distances, but exceptionally it is strewn for a kilometre or more, as at Lower Loxley, Yorkshire, in 2004 (Brown and Meaden, 2009c), West Cornwall in 2013 (Clark and Pask, 2013) and Kings Langley, Hertfordshire, in 1979 (Buller, 1979).

The debris field has frequently included remarkable effects. These are remarkable because they are quite unexpected and initially may seem to defy logic. These have normally occurred within an inner zone of the debris field. Examples include bales of hay sucked through a skylight onto a shed roof at Stone in Staffordshire (Rowe and Meaden, 2002); a chipboard deeply embedded in a wooden fence panel in Northamptonshire (Brown and Meaden, 2009c); concrete blocks landing on the roof of a neighbouring barn at Winslow, Buckinghamshire (Rowe and Meaden, 2002); a garage roof found several days later in a garden over 500 m away at Hoghton, Lancashire, as well as a caravan being destroyed by large, heavy wooden stakes rotating and then embedded into their side (Hall, 2005); roof tiles swept from a nearby roof underneath a parked car without inflicting a scratch on it, in Crewe, Cheshire, in 2004, and where at the time of the investigation, a garden shed had still not been found (Brown and Meaden, 2009a); and a roof tile going through the front door glass and then through the timber back door in Chipping Warden, Northamptonshire, in 1975 (Meaden, 1976a). Some of the debris has been extremely large. It includes whole mobile homes torn from their anchorage (Meaden, 2001) and whole barns (Brown and Meaden, 2010).

By drawing attention to certain types of damage, site investigators have marked up matters that may be of concern to developers and planners, particularly in urban areas. In this regard, it is important to stress that this damage may not arise from a direct hit by the vortex alone but also from the extreme inflow winds that cover a wider area around the vortex as well as the projectiles in the debris field. Common vulnerable structures noted in investigations include modern petrol stations, bus shelters, advertising hoardings, brick walls and garden structures.

Weather sequences: Eyewitness accounts frequently comment on the rapid changes in weather condition both before and after the passing of the tornado. Investigators have recorded different sequences. But the new conditions are always sudden and violent. Frequently, people note that storm clouds either arrive or develop quickly, becoming very dark, black and heavy. Thunder and lightning may occur, but this is far from being universal. A representative sequence is captured by those caught by the Coventry tornado of August 1975 (Burt, 1976). One resident noted that

Figure 6.6 A hornbeam tree twisted by the Leeds–Harrogate tornado, September 2006. Tim Prosser and Louise Hill (2006).

Figure 6.7 Garage roofs ripped off, their doors twisted open and objects missing after the Farnborough tornado, September 2007. Sarah Horton (2007).

At 6.15 p.m. the sun was shining brilliantly. At 6.30 p.m. the sky was overcast and the wind started to ... make a noise like someone sucking in air through rounded lips. As I ran to the tool shed, large drops of rain started to fall. I had intended to run back to the house but this was impossible as by now, 6.31 p.m., the wind speed had increased to 100 mph [the nearby airport recorded a sudden increase from 8 to 90 kts and was shaking the control tower making the staff hide in fear]. It was darker and the rain had given way to large misshapen hailstones ('more like chunks of ice', at the airport) which covered the lawn to 30 mm in a few seconds ... At 6.35 p.m. the wind dropped and the hail gave way to heavy rain, which ceased at 6.55 p.m., when the sun shone again.

In this case, as in many others, the intense rain or hail hides the tornado funnel and, at best, can only be seen as it approaches in the form of a curtain of mist racing across the fields.

But there are also cases when the weather seems to stop and a calm or even an eerie silence dominates everything. Even normal natural sounds seem to be missing, of birds and other animals. Even humans have said that feelings and sensations appear to change at this point. So at Retford in Nottinghamshire, calm conditions preceded and then followed the 90s of tornadic fury (Fowler, 1997); at Winkhill in Staffordshire, 'the rain stopped and it went very quiet, then the wind picked up again very strongly and suddenly we saw this tornado' (Evans, 1997); while investigations in Norfolk (Briscoe, 1997) and Selsey, West Sussex (Matthews, 1988), found people remarking on the quietness occurring after the tornado.

Such sequences relate to the tornado's position within the storm. If, as is often the case, the tornado is in the updraught area of the storm, the preceding intense precipitation (of rain or hail) located in the downdraught will be absent. The transition may be marked by a sudden eerie calm before the violent winds of the tornado abruptly arrive. But this has also been related to rapid pressure changes and the physiological reactions this induces, creating uncertainty and anxiety, a possible biological early warning system.

Non-tornadic winds: These are mistakenly reported from time to time as tornadoes because of the severity of the wind and the localised damage they have caused. However, in carrying out an investigation to confirm such reports, it has become clear that no tornado has occurred, despite filtering the report information to reduce such occasions to a minimum. So, a report of damage from intense winds during torrential rain at the Llandeilo Show, Dyfed, in August 1992 was finally identified as a severe squall as a result of a site investigation (Rowe and Meaden, 1994).

Also, land devils, eddy whirlwinds and various straight-line winds such as microbursts have each been found to be responsible for some of the effects reported, rather than a tornado. A site investigation near York in August 2011 showed surface damage patterns of a microburst as well as a tornado and so added to the relatively meagre published case studies for such phenomena in the United Kingdom (Smart *et al.*, 2012).

Extreme straight-line winds are well capable of moving debris from damage sites over considerable distances. Thus, a long, narrow trail of debris that has been moved by the wind is not conclusive that a tornado was involved. But heavy debris such as tiles, bricks, planks, etc. can only be carried upwards if there are extremely strong updraughts. If this uplift is to any significant height, it is unlikely to be the work of anything other than a tornado.

Funnel lightning: A number of electrical effects associated with a tornado have been reported during a small minority of site investigations, particularly in descriptions of the funnel. This is not a constant feature of tornadoes, since it is noted in only less than 5% of TORRO's published reports. But the observations have led to a debate over their significance.

Tornadoes often occur in thunderstorms, although many do not. When they do, lightning discharges may occur within the storm system. These have been reported both before the funnel has developed and after it has passed but rarely as it passes. However, the occasional suggestion that the funnel itself may be electrified has been based on observations such as made by a resident of Congresbury, Somerset, in January 1993 who, looking out of the window as the tornado passed, 'saw a big flashing ball of light followed by another white light. There were sparks coming out of it' (Weeks, 1995). Specifically within the funnel, a 'red glow' was reported in the South Ham, Hampshire, tornado of June 1996 (Matthews, 1999a). As well as lights, evidence of scorching has also suggested some type of internal electrical activity (Matthews, 1985). But generally, electrical effects receive little comment unless it is something remarkable, or lightning strikes occur, as during the Coventry tornado of August 1975 (Burt, 1976).

However, it is the primary role of the site investigation to establish what happened rather than to explain it. Scientific observation does not reject information because it does not fit into current theory. Nor do the observers have to provide an explanation, although they are frequently in a reasonably good position to do so.

6.5 Site Investigation Experience

The experience of the investigator is part of the site investigation itself. The investigator enters a localised disaster zone and encounters a situation that may be quite outside all previous experiences – certainly as far as the first investigation is concerned. This is particularly intense in urban environments where there are far more people and a dense concentration of houses and other vulnerable structures. Initially, this can be bewildering and overpowering for the investigator, despite any previous training, background and learning about tornadoes.

Site investigations have been possible because of the commitment and determination of the TORRO members involved. Enthusiasm for finding out more about extreme weather events has energised many of the investigations and has helped to apply different combinations of personal aptitudes and experience with a variety of training backgrounds, all involving forms of investigation, enquiry and research. So, investigators have reported that previously acquired skills in interviewing, fieldwork, mapping, recording, evaluating information of all kinds, investigative procedures from different professions, understanding complex engineering questions and even a well-developed 'common sense' have all been brought together and applied in different combinations to the investigation.

But there were several significant challenges that investigators have recognised at the outset of their task. A primary one has been to decide when to carry out the investigation. A desire to be on site as quickly as possible, especially in urban areas where damage is repaired and debris gathered very quickly, had to be matched against the time available. Some of this had to be used in preparing maps and equipment and establishing local contacts, all of which made for a time-effective investigation. The use of a questionnaire has also been found to be time effective, rather than depending on unstructured interviews that overly depend on the investigator's instinct to guide and develop. Frequently, return visits at a later date added significant information, when access was easier, the awareness of debris distribution was more common and people were less traumatised. Indeed, even later, first investigations have been found to be worthwhile. In addition, it was always more successful when the weather was suitable!

In some ways, an investigation embraces more people than those on-site because they have access to the experience and knowledge of other TORRO members to assist in the evaluation of the evidence that is gathered at any one particular site. Immediate access to such a diverse pool of experience and knowledge is unique compared with the readily available capacity of an educational or research institution. A significant proportion of investigations have used this resource, particularly by investigators doing their first site investigation. An example of the unique online support that was available for an investigator shows how significant this is for building up a quality database for these events. In the exchanges below on TORRO forum, 'Inv' is a first-time investigator, 'Memb 1–Memb 7' are the TORRO members, and 'F' is the location of the report investigated:

Memb 1: text of a news report is posted verbatim on the forum without comment.

Memb 2: 'Pictures were just on News 24 … they showed' (some details are then given).

Inv1: 'I've been over to F this lunchtime (spent about 2.5 hours there). Got lots of photos, talked to witnesses. I think I found the start and end of the path of damage. … Now I need to get the photos up here'.

Inv1: Presents the first report, describing observations step by step, with a link to a map.

Memb 3: 'Had my doubts listening to the radio this morning, but no doubts about the photos – def tornado'

Inv1: Now gives the web location of the rest of the many photos taken.

Memb 3: Gives a commentary on these, highlighting details that support it being a tornado.

Inv1: 'I hope I didn't miss anything important. I did *really* enjoy doing it. I like piecing all the picture together and also talking to people in the street. They were all amazing … There were lots of small fascinating things … (examples given). It's also amazing how rumour spreads from one street to another. A few people in different streets mentioned they had heard that someone had seen a funnel actually split in two. Where I think that came from was from the chap who saw the garage roof split off in two directions after it had hit the house'.

Memb 4: 'Good work I – I am sure that is tornado damage there. It's quite an experience, eh?'

Inv2: 'I have just visited the site … this was tornado damage for sure' (some details given and photos posted).

Memb 5: 'Great work – certainly looks fairly like tornado damage – the path is very interesting'.

Memb 6: 'Wonderful stuff … an excellent job for a first SI. We usually leave it to the investigator to assess the strength etc and there are guidelines'. (details given).

Inv1: 'From that I'd say some of the damage was T2. Particularly these points' (details given).

Memb 7: That path is interesting … think the squall line was overtaking the tornado, forcing it to turn to the right? (discussion follows).

For most investigators, each event has proved to be a remarkable learning experience. Learning about tornadoes goes far beyond what appears in a textbook. It has been commented that reading does not prepare you for what a tornado does and what it leaves behind. Despite being relatively well informed, investigators have declared themselves surprised by the apparent ferocity of even the smallest tornado. To some, this was beyond any preconceived idea they had. Surprise that there were no fatalities was matched by remarkable stories from people who appeared to have escaped death by the narrowest of margins. This takes some handling immediately after an event, and listening skills become most important. However, it is part of the openness that many people have shown to investigators after the event, which has led to a great deal of important detail about an event being passed on.

Other sources of surprise to investigators are small items that remain intact amidst major destruction, the precision with which the damage track may be defined, the often sporadic nature of the damage, the role of debris in causing damage, embedded items that almost defy logic, lifting effects (from roofs to cows to people) and the distance large tree branches can be thrown by the smallest tornadoes – indeed the intensity of small tornadoes is a common surprise to investigators. As a result of all this, the investigation is often a significant learning experience for the investigator.

A survey of learning experiences from carrying out site investigations has identified the following as the most common:

- There is no such thing as a simple tornado vortex. Model representations fall short of the reality.
- Tornado events do not readily conform to set seasons, occurring during a night in December as well as a hot summer's day.
- They affect any one dwelling for an extremely short period of time. This may vary from a mere few seconds to a few minutes at most.
- Their noise is unique and overwhelming.
- Single tornadoes may have quite variable damage tracks, but sometimes, more than one tornado must be involved.

6.6 Concluding Remarks

The site investigation is one of the principal foci for a voluntary organisation such as TORRO; it is where the major contribution can be made by a significant number of its members and associates. Together with those who have sent in many of the reported events over the years, the investigators have added significantly to our knowledge of tornadoes and related severe weather events across the country.

One of the major outcomes from the investigations over the years has been to establish that wind fields are much more complex than is often assumed and certainly as represented in textbook models. If a tornado track was always straight, continuous and long, there would still be a significant challenge in establishing the local expression of its major characteristics. But this is not so. It may well be that some first-time investigators have entered the disaster zone with some of those assumptions, but they soon have to let them go and let the damage, debris and other impacts speak for themselves.

Collecting eyewitness evidence has added immeasurably to the quality of the investigation's results. Where it has been possible to carry this out, it has proved to be the most time consuming aspect of an investigation. Indeed, it sometimes requires a return visit to complete. But over the years, the cooperation of local people has been remarkably effective, particularly when a well-structured interview has been prepared. It is clear that many eyewitnesses do not come forward initially with their information for many complex reasons.

The role of the site investigator is evolving. Over the past years, site investigations have evolved as more has been learnt about tornadoes and as lessons from previous on-site experiences are absorbed. For many of these events, the knowledge acquired has become more readily accessible with the associated web discussions and subsequent published reports. As site investigations continue into the future, there are increasing opportunities for the results of this work to be of assistance in many ways, such as the development of forecast techniques and testing of meteorological models as well as profiling vulnerabilities for local authorities, planners, social scientists and others.

Many of the outcomes of the site investigations have been unpredictable for the investigators involved, just as the events themselves were unpredictable for the local communities on the receiving end of the tornado. The severity of 'small' tornadoes, the unpredictable threat of flying debris from small shards of glass to whole roofs, the fear and long-term trauma they cause are just some of the consequences for which the investigator acquires a new level of awareness. All these are products of even small tornadoes. Since these dominate the country's tornado profile, TORRO investigators have built up considerable experience in recording them and distinguishing them from other types of severe winds.

Site investigations have become opportunities for investigators to explore their scientific curiosity about severe weather in a specific local event on behalf of the general TORRO community. Indeed, it is often one of the reasons for becoming a part of it. There are still many unanswered questions about tornadoes in the UK environment. Many of the answers to these questions will emerge from continuing detailed site investigation work with closer links to the physical and social sciences.

Acknowledgements

The following site investigators who have kindly shared their experiences for this chapter are gratefully acknowledged, namely, Chris Bell, Ian Brindley, Charles Briscoe, Paul Domaille, Anthony Gilbert, Sarah Horton, Nicky Lambert, John Mason, Harry McPhillimy, Tim Prosser, Tim Sharp and Chris Warner.

References

Billett, H. (1914) *The South Wales Tornado of October 27, 1913: With a Note on the Remarkable Pressure Oscillations Observed on August 14, 1914.* HMSO, London.

Briscoe, C.E. (1997) Tornadoes in North Norfolk, 18 February 1997. *J. Meteorol.*, **22** (220) pp.212–214.

Brooks, C.E.P. (1954) *The English Climate.* The English Universities Press, London (2nd edition, 1964).

Brown, P.R. and Meaden, G.T. (2007) TORRO tornado division report: January to April 2007. *Int. J. Meteorol.*, **32** (323) pp.317–320.

Brown, P.R. and Meaden, G.T. (2008) TORRO tornado division report: September – December 2007. *Int. J. Meteorol.*, **33** (327) pp.96–102.

Brown, P.R. and Meaden G.T. (2009a) TORRO tornado division report: January to May 2004. *Int. J. Meteorol.*, **34** (341) pp.237–244.

Brown, P.R. and Meaden G.T. (2009b) TORRO tornado division report: July 2004. *Int. J. Meteorol.*, **34** (342) pp.282–287.

Brown, P.R. and Meaden, G.T. (2009c) TORRO tornado division report: August and September 2004. *Int. J. Meteorol.*, **34** (343) pp.312–319.

Brown, P.R. and Meaden, G.T. (2010) TORRO tornado division report: January – May 2009. *Int. J. Meteorol.*, **35** (352) pp.317–322.

Budgen, R. (1730) Transcript of The Passage of the Hurricane from the Sea-Side at Bexhill in Sussex to Newingden Level 20 May 1729 between Nine and Ten in the Evening. AMS6044 National Archives, East Sussex Records Office.

Buller, P. (1979) Tornado damage in the home counties, 24 June 1979. *J. Meteorol.*, **4** pp.235–240.

Buller, P. (1985) Structural damage caused by tornadoes in the United Kingdom. *J. Meteorol.*, **10** (100) pp.221–226.

Burberry, M. (1998) The 11.45 Atherfield to Cridmore tornado (the cauliflower express) Sunday 4 January 1998. *J. Meteorol.*, **23** (230) pp.189–192.

Burt, S.D. (1976) The Coventry tornado of 5 August 1975. *J. Meteorol.*, **1** (11) pp.342–346.

Chatfield, C. (2004) War of the elements: tornadoes of the 19th century, part II, *Convection*, **5** TORRO membership publication, pp.7–9.

Clark, M. (2010) A survey of damage caused by the Monkwood Hampshire, U.K. tornado of 3 November 2009. *Int. J. Meteorol.*, **36** (353) pp.291–297.

Clark, M. and Pask, J. (2013) Site investigation in West Cornwall following the tornadoes of 16 December 2012. *Int. J. Meteorol.*, **38** (379) pp.92–103.

Elsom, D.M. (1985) Tornadoes in Britain: where, when and how often. *J. Meteorol.*, **10** (100) pp.203–211.

Elsom, D.M. (1989) Tornado at Didcot, Oxfordshire, on 23 November 1984 and its relationship to 'cold air funnels'. *J. Meteorol.*, **14** (138) pp.109–114.

Elsom, D.M. (1993) The thunderstorm gust front as a trigger for tornado formation: the Long Stratton tornado of 14 December 1989. *J. Meteorol.*, **18** 175 pp.3–12.

Evans, D.E. (1997) Staffordshire tornado May 1997. *J. Meteorol.*, **22** (220) p.227.

Fowler, C. (1997) Tornado at Retford, 31 August 1997. *J. Meteorol.*, **22** (223) pp.345–347.

Gilbert, A. (1999) Tornado at Stroud, Hampshire, 26 January 1999. *J. Meteorol.*, **24** (240) pp.214–217.

Gilbert, A. (2000) Tornado at Pagham, West Sussex, 12 September 1999. *J. Meteorol.*, **25** (247) pp.82–88.

Gilbert, A. (2001) The multiple tornado outbreak of 30 October 2000 around the Solent, U.K. – the site investigation. *J. Meteorol.*, **26** (262) pp.281–291.

Gilbert, A. (2002) Tornado at Gosport in Southern England, 9 August 2001: author's sighting and eyewitness report. *J. Meteorol.*, **27** (270) pp.197–204.

Graham, E. (2000) Thunderstorm and tornado at Reading, 13 June 1998. *J. Meteorol.*, **25** (252) pp.231–237.

Hall, S. (2005) A tornado site investigation at Hoghton, Lancashire, 2 May 2005. *Int. J. Meteorol.*, **30** (303) pp.338–346.

Lewellen, D. (2012) Effects of Topography on Tornado Dynamics: A Simulation Study. AMS 26th Conference on Severe Local Storms https://ams.confex.com/ams/26SLS/webprogram/Paper211460.html (accessed on 9 May 2015).

Matthews, P. (1985) Lightning inside a tornado? *J. Meteorol.*, **10** (104) pp.375–376.

Matthews, P. (1988) The Selsey tornado. *J. Meteorol.*, **13** (133) pp.355–357.

Matthews, P. (1999a) Tornado at South Ham, Basingstoke, 7 June 1996. *J. Meteorol.*, **24** (239) pp.172–178.

Matthews, P. (1999b) The Selsey tornado of 7 January 1998: a comprehensive report on the damage. *J. Meteorol.*, **25** (240) pp.197–209.

Matthews, P. (1999c) A 'giant hand' sweeps the trees as a tornado hits Southampton. *J. Meteorol.*, **24** (243) pp.358–363.

Matthews, P. (2000) The fifth of November – a day to remember at Park Gate. *J. Meteorol.*, **25** (254) pp.355–361.

Matthews, P. (2003) An investigation into the influence of the Isle of Wight on waterspout and tornado frequency and the effect of population densities on recording events. *J. Meteorol.*, **28** (280) pp.213–225.

Meaden, G.T (1976a) Tornadoes of 12 January 1975 in England. *J. Meteorol.*, **1** (6) pp.187–193.

Meaden, G.T. (1976b) Tornadoes in Britain: their intensities and distribution in space and time. *J. Meteorol.*, **1** (8) pp.242–251.

Meaden, G.T. (1977) The Tornado in Barton, Oxfordshire 16 October 1966. *J. Meteorol.*, **2** (16) pp.103–106.

Meaden, G.T. (1985) TORRO, the tornado and storm research organisation: the main objectives and scope of the network, *J. Meteorol.*, **10** (100) pp. 182–193.

Meaden, G.T. (1998) Selsey tornado, the night of 7-8 January 1998, *J. Meteorol.*, **23** (226) pp.41–48.

Meaden, G.T. (2001) Site study of the T4 Bognor Regis tornado of 28 October 2000. *J. Meteorol.*, **26** (265) pp.3–14.

Meaden, G.T. and Chatfield, C.R. (2009) Tornadoes in Birmingham, England 1931 and 1946 to 2005. *Int. J. Meteorol.*, **34** (339) pp.155–162.

Osborne, J. (1985) Tornado at Annesley, Nottinghamshire, 16th August 1985. *J. Meteorol.*, **10** (104), pp.372–374.

Peterson, R.E. (1992) Early studies of tornado damage. *J. Meteorol.*, **17** (165) pp.10–16.

Pike, W.S. (1992) Two East Anglian tornadoes of 12 November 1991 and their relationship to a fast-moving triple point of a frontal system. *J. Meteorol.*, **17** (166) pp. 37–50.

Prosser, T. and Hill, L. (2006) Site investigation of the Leeds and Harrogate Tornadoes of September 17, 2006. Unpublished manuscript.

Rowe, M. (1975) A history of tornado study in Britain. *J. Meteorol.*, **1** (1) pp. 20–24.

Rowe, M. and Meaden, G.T. (1985) British tornadoes and waterspouts of the 1960s part 5: 1967–1969. *J. Meteorol.*, **10** (98) pp.112–123.

Rowe, M. and Meaden, G.T. (1994) TORRO tornado division report: January to April 1993. *J. Meteorol.*, **19** (197) pp.98–100.

Rowe, M. and Meaden, G.T. (2002) British tornadoes and waterspouts of 1981. *J. Meteorol.*, **27** (268) pp.26–36.

Smart, D., Clark, M., Hill, L. and Prosser, T. (2012) A damaging microburst and tornado near York on 3 August 2011. *Weather*, **67** pp.218–223.

Smith, J.P. and Harper R.D.M. (1986) The Barmouth waterspout: 22 September 1984. *J. Meteorol.*, **11** (105) pp.9–14.

Turner, S., Elsom, D.M. and Meaden, G.T. (1986) An outbreak of 31 tornadoes associated with a cold front in Southern England on 20 October 1981. *J. Meteorol.*, **11** (106) pp.37–50.

Tyrrell, J. (2012) Site investigations of tornado reports in Ireland, 2011. *Int J. Meteorol.*, **37** (367–376) pp.70–81.

Weeks, A.H. (1995) The Congresbury tornado of 12 January 1993. *J. Meteorol.*, **20** (197) pp.94–98.

Wegener, A. (1917) *Wind-und Wasserhosen in Europa*. Friedrich Vieweg & Sohn, Braunschweig, 301pp.

Wood, F.E. and Mortimore, K.O. (1991) Damaging tornado at Overbury in the County of Hereford and Worcester: 2 February 1990. *J. Meteorol.*, **16** (157) pp.84–86.

Part II
Thunderstorms and Lightning

7

Epic Thunderstorms in Britain and Ireland

Jonathan D.C. Webb

Thunderstorm Division, Tornado and Storm Research Organisation (TORRO)*, Oxford, UK,

7.1 Introduction

Figures 7.1 and 7.2 present summaries of two private records of thunderstorms; these span 60 years and 30 years respectively. Both observers currently report to TORRO and their station records in north-east Bristol and in North Yorkshire also include some outstanding events (Williams, 2000; Cinderey, 2005). It may be noted that while increased automation of weather stations has enabled more collection of instrumental data from remote areas, it has reduced the availability of 24-hour 'audiovisual observations' including thunder.

Methods of summarising thunderstorm observational data are presented in Table 7.1. Widespread reports of thunder/lightning can sometimes be attributable to thundery showers occurring widely but with gaps in the 'overhead activity'. If there are reports of widespread overhead thunder, this indicates highly organised convection such as Mesoscale Convective Systems (MCSs) or (a subdivision of the latter) thundersquall lines (Pike, 1994b; 1999).

Showery situations are more likely to be replicated on several successive days than are situations favouring major thunderstorm systems, though even with the former, subtle day by day variations in the distribution of convective showers must be expected. Prichard (1987) and Grant (1995) both refer to the unusual sequences of widespread thunderstorm days from 22 to 26 June 1980 and from 12 to 22 May 1983. June 1982 was the most thundery calendar month in the United Kingdom with an average of 11% of stations reporting thunder each day (see also Chapter 8).[1]

Widespread overhead thunder can be identified by (a) observers' reports of overhead thunder, the latter defined by TORRO as within 5 km or 'close'; (b) the distribution of lightning strikes showing a high-density 'whiteout'; and (c) strong evidence,

albeit not direct proof, from an MCS anvil cloud shield on infrared satellite imagery (Figure 7.3). The criteria used by Gray and Marshall (1998) for the classification of a thunderstorm cluster as an MCS includes (i) concurrent observations of thunder and lightning over an area extending 100 km or more in at least one direction; (ii) precipitation, some heavy, also occurring simultaneously across an area over 100 km across in at least one direction; and (iii) a continuous, consolidated anvil cloud shield which confirms the extensive thunder is not merely from widespread individual storms. The most extreme type of MCS (in terms of magnitude) is the Mesoscale Convective Complex (MCC) for which Maddox (1980) used the additional parameter of an anvil shield at or below −52°C covering an area over 50,000 km² and persisting for at least 6 hours.

Thunderstorm severity is also difficult to define precisely. It can be based on the peak intensity (lightning flash frequency) of a storm or on the total number of discharges. The main risks of personal injury and damage to buildings and infrastructure derive from cloud-to-ground strokes though aerial lightning strikes are significant hazards for aircraft. Moreover, ground strikes vary in power; considerable damage and disruption can be associated with a single powerful cloud-to-earth discharge – usually a positive strike. Positive lightning discharges (Rakov, 2003) are able to propagate continuously without 'steps', whereas the more common negative discharges cannot propagate continuously so the propagation is 'stepped'. Public (and observers') perception of storm severity will also be influenced by the intensity of precipitation. The international definition of a severe thunderstorm does not refer to electrical activity, but it is a definition which implies the need for organised convection. It is defined as a storm with one or more of the following: a tornado, wind gusts over 50 kt or hail 20 mm diameter or more.

For TORRO dataset purposes, the guideline for logging 'significant lightning incidents' is that these are cases where, for example, a person or people are struck (directly or indirectly), houses or other buildings are struck, trees are struck and damaged

* http://www.torro.org.uk/

[1] This percentage will almost certainly have been understated since only a minority of stations maintained a 24-hour weather watch.

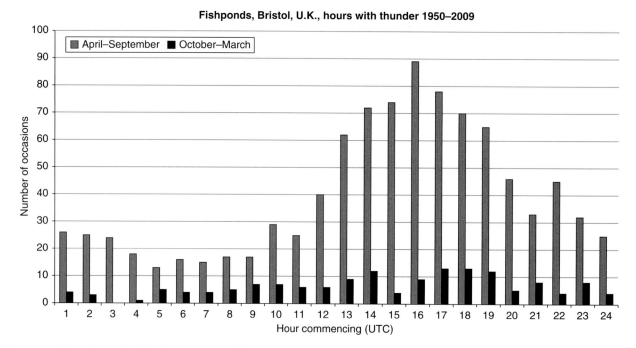

Figure 7.1 *Diurnal record of thunderstorms in north-west Bristol, England, 1950–2009. Observations by Bryan A. Williams. (See insert for colour representation of the figure.)*

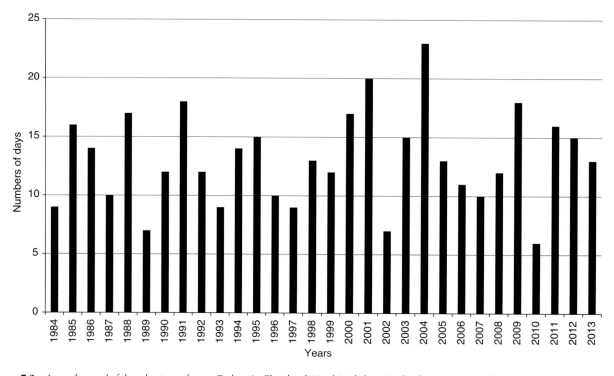

Figure 7.2 *Annual record of thunderstorm days at Carlton-in-Cleveland, North Yorkshire, England, 1984–2013. Observations by Mike Cinderey.*

or strikes cause an electricity supply or telecommunications cut over a large area. TORRO data for the recent 25-year period from 1988 to 2012 has an average number of 166 significant lightning incidents per year in Britain and Ireland. The highest annual total was 378 in 2006, and the highest event total was 124 on 10–11 July 1995. However, it should be noted that the reported incidents

are likely to be only a proportion of the actual cases. Considerably higher incident totals have occurred in some earlier 20th-century events described in the following pages.

The following selection highlights those events which were outstanding for both the intensity and geographical extent of thunderstorm activity. Extreme electrical activity has been the principal

Table 7.1 *Methods of summarising thunderstorm observational data.*

Criteria, Statistic	Comments, Methodology	Advantages	Caveats
Days with thunder heard	Any days on which thunder is heard between 0000 and 2400 UTC	Most easily available data, simple definition (in theory!)	Ideally requires a 24-hour watch on the weather. Audibility of more distant thunder very dependent on background noise with quiet rural locations likely to record the most realistic results. No indication of how active a storm is. Other issues with this statistic are discussed by Bob Prichard in Chapter 8.3.1 (see also Prichard, 1985 and 1987).
Days with overhead thunder	Definition used by TORRO is within 5 km (3 miles) or reported 'close'	Much less subject to background noise	Requires a consistent definition and consistent application of it. Does not indicate how active the storm is
Hours of thunder	The methodology usually involves the total number of hours (0000–2400 UTC) during which thunder was heard at least once. Alternatively, the total duration of thunderstorms, based on the official criteria that a thunderstorm ceases 10 minutes after the last thunder	More indication of significant events	Requires close '24-hour watch'. No direct relationship to how active a storm is
Number of lightning damage incidents	Compiled from observers' reports, news reports, etc.	Indication of the impacts of electrical activity	Can be heavily influenced by one major storm episode, dependent on reporting of events (tends to be biased towards populated areas). Reports need careful quality control as incidents may be incorrectly attributed to lightning in news stories
Days of widespread thunder	Widespread reports of thunder/lightning can be attributable to thundery showers occurring widely (with gaps in the 'overhead activity'). If there are reports of widespread overhead thunder, this indicates very organised convection such as Mesoscale Convective Systems (MCSs) or (a subdivision of the latter) thundersquall lines	Potentially, the best indicator of how thundery a specific month or year is	Consistent definition can be difficult because of changes to available station reports. Does not distinguish between showery situations (with distant thunder reported at many locations) and instances of widespread overhead thunder associated with a large storm system such as an MCS. Not necessarily indicative of severity
Lightning strikes recorded	As recorded by sferics networks	Not dependent on audibility of thunder	Can be heavily influenced by one major storm episode, spurious data still possible

selection criterion although most such episodes also produced some extreme rainfalls, damaging hail and destructive local winds. One event, on 5 September 1958, was so outstanding for both electrical activity and hail that it appears in both relevant chapters of this book. The selection is restricted to the past 150 years (since regular meteorological journals first appeared) although some accounts of earlier 'epics' can be found elsewhere, for example, the storms of 1 August 1846 (Webb, 1996b). Although none of the events selected below are from the latest decade, Chapters 9 and 14 confirm that more recent years have had their share of extreme convective events, albeit sometimes localised in nature. Moreover, slow moving multicell thunderstorms (e.g. on a convergence zone) can produce prolonged and intense electrical activity at a rather more local level. In the year of TORRO's launch, a thunderstorm of this nature produced up to 9 hours of continuous and often spectacular activity across the Bristol area overnight on 15–16 June 1974 (Pook, 1974; Mortimore, 1975; Williams, B., pers. comm.).

Because of the issues of realistically classifying thunderstorm severity (some noted previously), there is inevitably an element of subjectivity in this selection of events. Most of them have been partly or predominantly night-time events, not entirely coincidental as these provide, arguably, nature's most impressive and awesome spectacle. Most of these events have also, sadly,

been associated with lightning fatalities although, fortunately, changing lifestyles have significantly reduced the risks over the past 100 years as discussed in Chapter 10 on lightning impacts.

7.2 Selected Epic Thunderstorm Events

7.2.1 2–3 August 1879

These thunderstorms occurred overnight and on the 'cold' side of a shallow depression which was drawing extremely hot air northwards over the near continent. The violence of the storms is likely to have been associated with destabilisation of a plume of very warm, moist air at mid-levels (i.e. at an elevated rather than 'boundary' level). The NCEP reanalysis chart (Figure 7.4) indicates very strong directional wind shear, quite similar to the cases of the elevated[2] 'south coast storms' of 5 June 1983 and the elevated 'Wokingham storm' supercell of 9 July 1959 (see also case studies in Chapter 9).

[2] Elevated convection is convection occurring within an elevated layer, that is, a layer in which the lowest portion is based above the earth's surface and boundary layer. Elevated convection occurs independently of surface heating and usually involves the release of instability at medium levels by general forced ascent ahead of a frontal zone and/or upper trough.

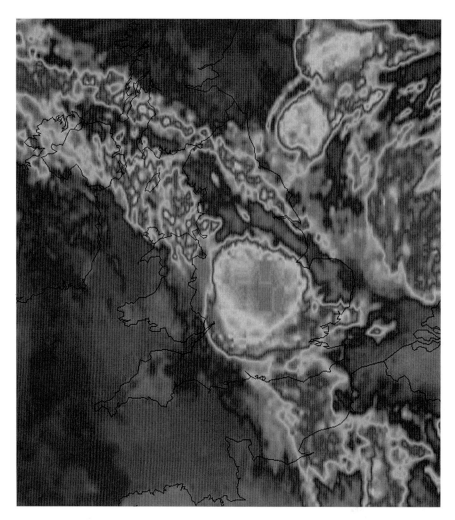

Figure 7.3   *Infrared satellite image, 1330 GMT, 22 July 2006. A Mesoscale Convective System (MCS) moved NNE across England accompanied by widespread overhead thunderstorm activity (see Webb and Pike, 2007). Cloud top temperatures (deg C): green −30 to −40°C, green/yellow −40 to −50°C, yellow −50 to −58°C, orange −58 to −60°C, red below −60°C. Image courtesy of Robert Moore. © EUMETSAT 2006. (See insert for colour representation of the figure.)*

The most outstanding incident of the 1879 episode was the west London ('Richmond') hailstorm at about 0200 GMT, which caused massive destruction of glass from Reigate, Surrey, north to Ealing and north-west London; a full survey was presented in *Symons's Meteorological Magazine* (Symons, 1879). A severe hailstorm also affected the Bury St Edmunds area of Suffolk. At Weybread (Suffolk), a correspondent reported that thunder and lightning were incessant from 0300 to 0345 GMT with intense rain, followed at 0345 GMT by hailstones the size of small oranges which wrecked glasshouses. At Barrow and Bevington (Suffolk), hail penetrated slated shed roofs, and many windows were broken. The church at Wells-next-the-Sea in Norfolk was struck by lightning and seriously damaged.

The storms were also outstanding for spectacular lightning displays, reported from *British Rainfall* observers at places as far apart as Haverfordwest in Pembrokeshire and Costessey, Norwich, and there were many reports of lightning damage. In the Cambridge area, the second of two violent storms lasted from 0100 to 0430 GMT and was accompanied by lightning discharges

at the rate of 56 per minute with incessant thunder; Grantchester Mill, Cambridge, recorded 100 mm of rain overnight. *The Times* reported the very vivid lightning and near incessant thunder peals at Leicester and noted that 'a large number of the day excursionists remained at Leicester station for several hours, being afraid to venture into the streets' (Prichard, 1993).

7.2.2   5–10 June 1910

A very thundery period affected England and Wales between 5 and 10 June 1910, the storms being especially prolonged and violent across the Thames Valley and the South Midlands on the 7th and 9th. Southern Ireland was also severely affected at times, especially overnight on the 5th–6th. A large anticyclone covering northern Scotland and southern Scandinavia receded north-east and declined, while a shallow area of low pressure over Biscay drifted slowly into southern Britain (source *Daily Weather Report*). Despite a good deal of cloud, temperatures rose to 26–28°C in places between the 7th and 9th.

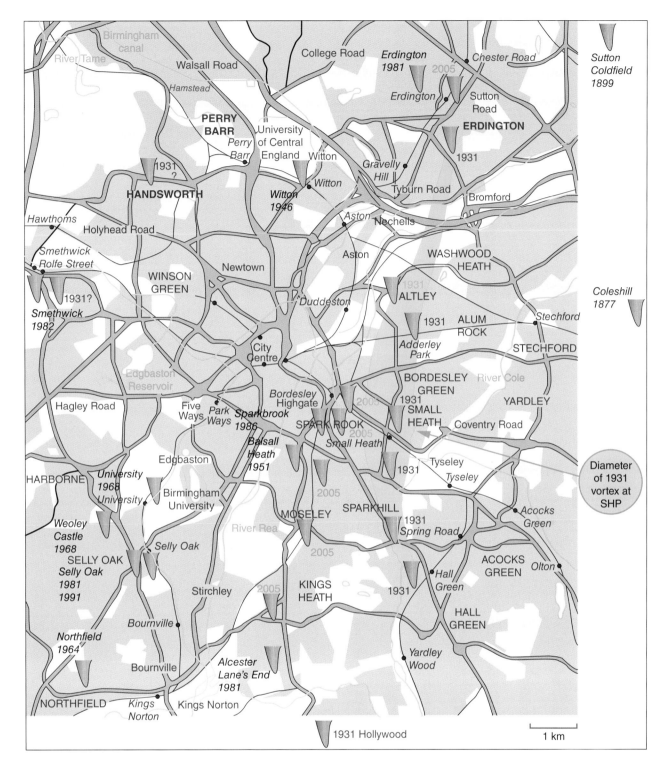

Figure 1.9 *The map shows the tornado track of 28 July 2005 through the city of Birmingham. Maximum intensity was T5/6 on the International T-Scale. Sites of tornado damage in preceding years are indicated too. Chief investigators were Ian Brindle, Matthew Capper and Peter Kirk. Details are given in papers by Knightley (2006), Kirk (2006), Pearson (2006), Smart (2008) and Meaden and Chatfield (2009).*

Extreme Weather: Forty Years of the Tornado and Storm Research Organisation (TORRO), First Edition. Edited by Robert K. Doe.
© 2016 John Wiley & Sons, Ltd. Published 2016 by John Wiley & Sons, Ltd.

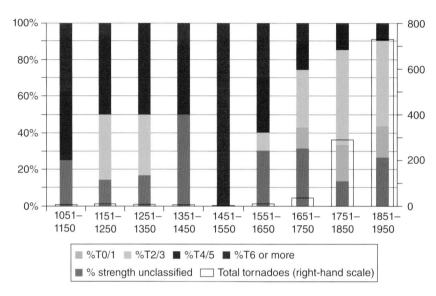

Figure 2.2 *Cumulative percentages of tornadoes by T-strength for each 100-year period from AD 1051 to 1950 (left-hand scale). Also showing totals of all tornadoes in the same 100-year periods (right-hand scale).*

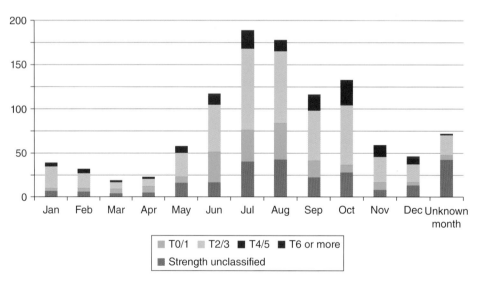

Figure 2.3 *Monthly totals of tornadoes by T-strength from AD 1054 to 1950.*

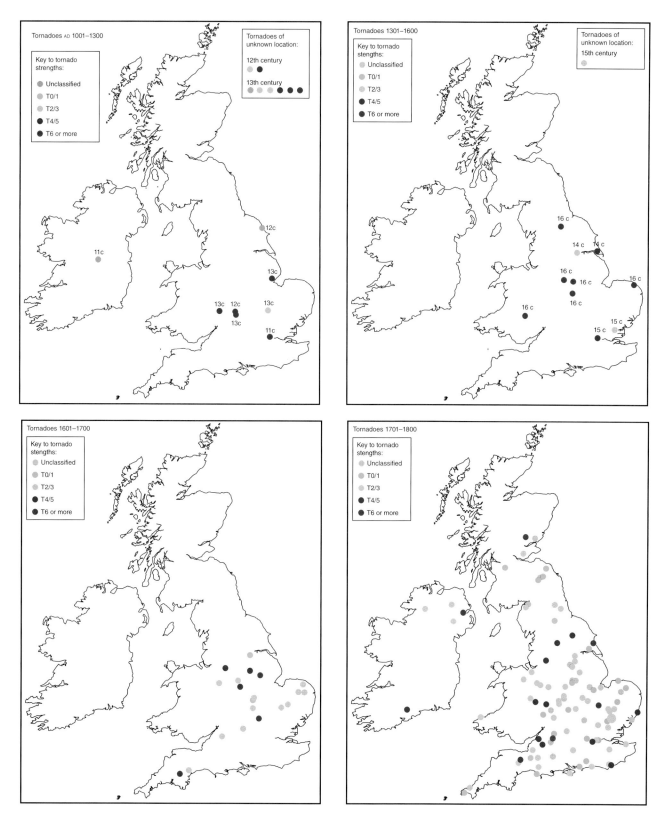

Figure 2.4 (a) Geographical distribution of known tornadoes by strength for 11th–13th centuries (see text for the meaning of the upper right inset). (b) as in Figure 2.4a but for the 14th–16th centuries (see text for the meaning of the upper right inset). (c) as in Figure 2.4a but for the 17th century. (d) as in Figure 2.4a but for the 18th century.

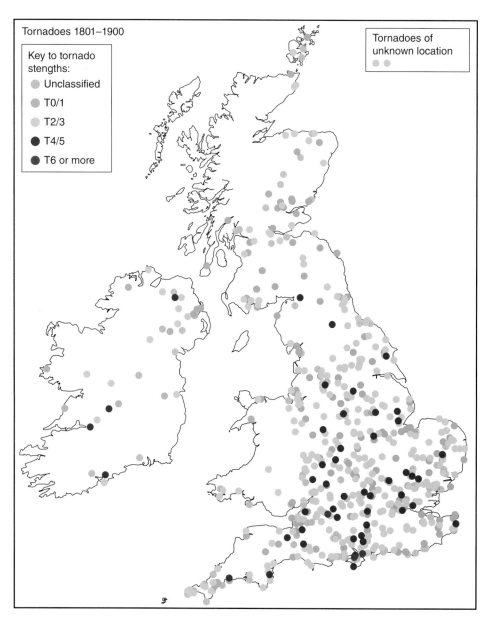

Figure 2.4 *(Continued) (e) as in Figure 2.4a but for the 12th century (see text for the meaning of the upper right inset). Contains Ordnance Survey data © Crown copyright and database right 2015.*

(a)

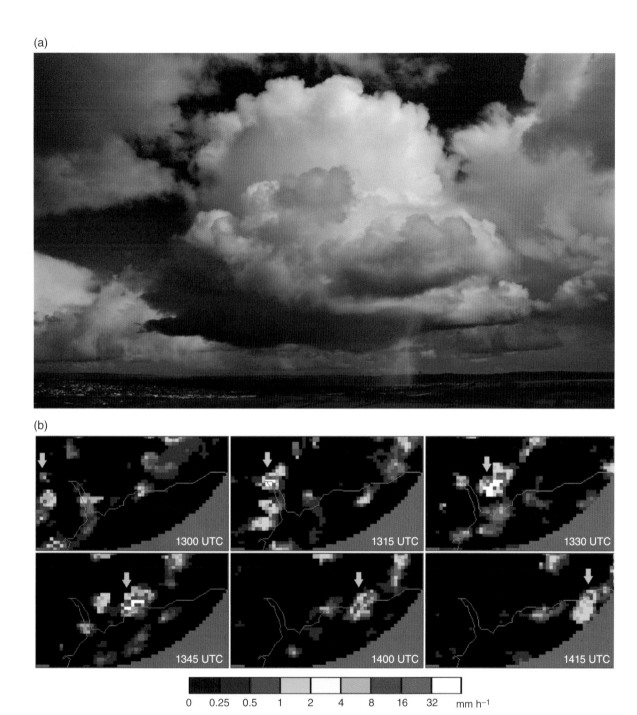

(b)

1300 UTC

1315 UTC

1330 UTC

1345 UTC

1400 UTC

1415 UTC

0   0.25   0.5   1   2   4   8   16   32   mm h⁻¹

Figure 3.2 (a) Example of single-cell convection near Exeter, Devon, at 1310 UTC on 10 April 2012. The view is towards the east and the cell is about 9 km distant. At the time of the photograph, the cell is just entering the mature phase of development, as evidenced by the newly developed narrow shaft of precipitation visible under the base of the cell. © Matt Clark. (b) Sequence of 1-km resolution radar rainfall imagery between 1300 and 1415 UTC on 10 April 2012, showing the development and dissipation of the photographed cell (marked by the grey arrow in each panel). © Crown Copyright Met Office.

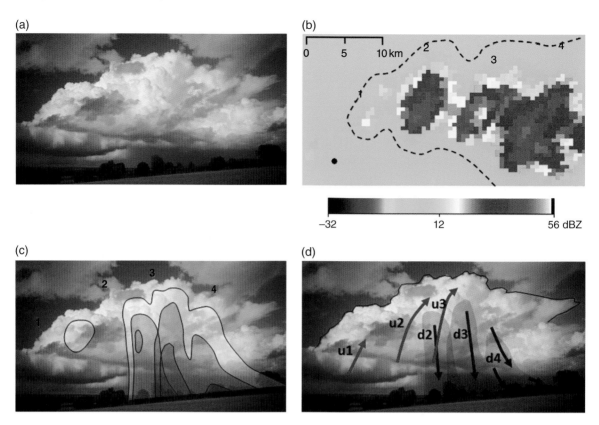

Figure 3.3 *Multicellular convection near Radstock, Somerset, on 29 July 2013. (a) Photograph of the storm at 1805 UTC, as viewed from the south-west. © Matt Clark. (b) Radar reflectivity at 1806 UTC. Black circle shows approximate location from which the storm was photographed. © Crown Copyright Met Office. (c) Idealised structure of the precipitation areas associated with each cell. Yellow, orange and pink shades denote reflectivity greater than 20, 30 and 40 dBZ, respectively. (d) Idealised schematic of updraughts (red arrows), labelled u1-u3, and down-draughts (blue arrows), labelled d2-d4, within each cell comprising the multicell storm. Cell (1) is in the developing stage, cell (2) is in the early mature phase, and cell (3) is in the mature phase (note that it is partially obscured by cell (2) in panels (a), (c) and (d)) and is associated with the heaviest precipitation at ground level. Cell (4) is in its weakening phase.*

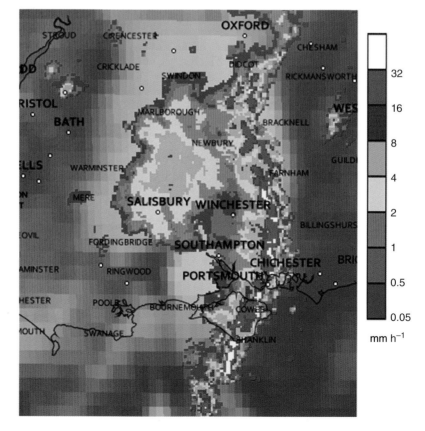

Figure 3.4 *Radar rainfall rate and satellite infrared image of a mesoscale convective system exhibiting linear morphology over southern England at 2000 UTC 22 October 2013. The system was moving towards the north-east. Panel shows data over an area of width 150 km. © Crown Copyright Met Office.*

(a)

(b)

(c)

(d)

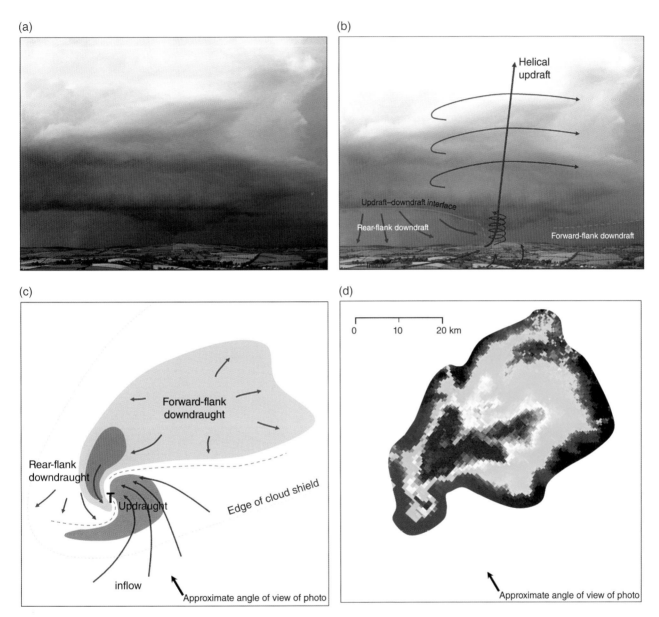

Figure 3.5 *(a) Photograph of a supercell thunderstorm near Newtownards, Northern Ireland, on 31 July 2006. The view is from the south-east. Photo © Martin North. (b) Annotated view of the same storm, with idealised representation of the main airflows within the storm. Red arrows indicate inflow and updraughts. Blue arrows indicate downdraughts. © Martin North. (c) Idealised schematic of a supercell, showing the updraught and downdraught regions, and storm-relative low-level winds. Dark grey shading denotes typical location of large hail. Light grey shading denotes location of smaller hail and rain. 'T' indicates location of tornado, where present (after Lemon and Doswell (1979)). (d) Radar reflectivity image of a large hail- and tornado-producing supercell which occurred over the Midlands on 28 June 2012. Reflectivity scale is given in Figure 3.3. © Crown Copyright Met Office.*

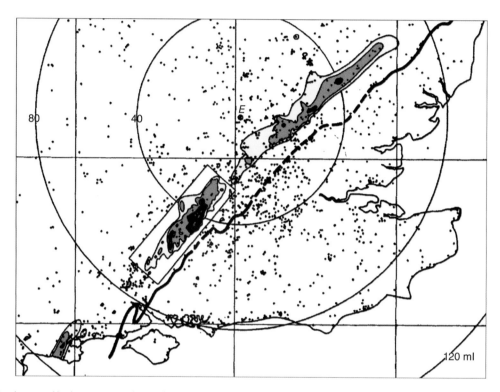

Figure 3.6 *Distribution of hailstone size in the 9 July 1959 'Wokingham' storm. Yellow, orange, black and pink shaded areas denote hail diameters exceeding 0.64, 1.27, 2.54 and 4.13 cm, respectively. Dots indicate individual hail observation points (442 also lie within the rectangle surrounding the most severely affected area). The thick black line shows the radar-observed locus of the right flank of the main storm. Browning and Ludlam (1962) was the first study of a supercell thunderstorm, paving the way for the recognition of supercells as a distinct class of severe thunderstorm. Adapted from Figure 3.2 of Browning and Ludlam (1962). Reproduced with permission of John Wiley & Sons Ltd.*

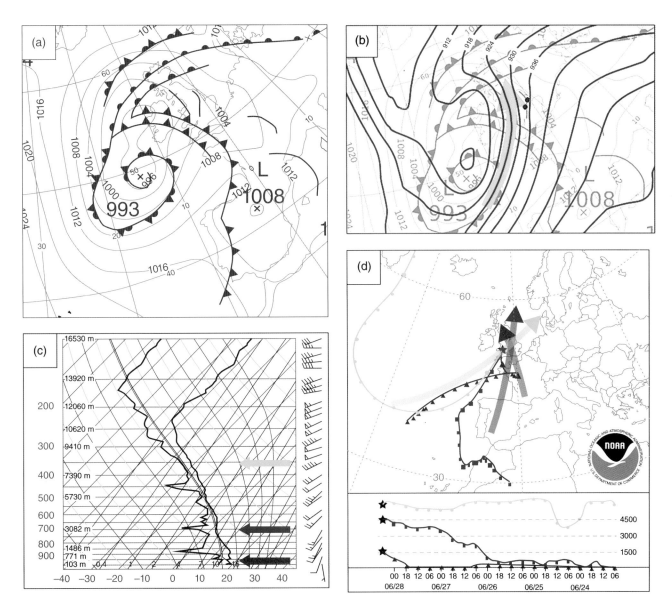

Figure 3.10 *Schematic illustrating the Spanish Plume synoptic set-up over the United Kingdom on 28 June 2012. (a) Met Office surface analysis chart for 1200 UTC 28 June 2012. Isobars are drawn at 4 hPa intervals and fronts are drawn using standard symbols. Crown Copyright Met Office. (b) 300 hPa geopotential height at 1200 UTC 28 June 2012 (blue contours; contour interval is 60 m; contour labels indicate the height, in tens of metres, of the 300 hPa pressure surface). The blue arrow indicates the axis of strongest winds at 300 hPa. The red dots indicate locations of the two tornadoes that occurred within ±2.5 hours of analysis time. (c) 0000 UTC Nottingham radiosonde data, illustrating conditions typical of the warm sector and associated plume. Bold black lines indicate the environmental temperature (right-hand line) and dew point temperature (left-hand line) profiles. Red, blue and green arrows indicate the approximate heights of the end points of the trajectories marked in (d). Courtesy of University of Wyoming. (d) Trajectories of air at 1500 m (red), 3000 m (blue) and 6000 m (green) above ground level over the 5-day period ending 0900 UTC 28 June 2012. Arrows indicate idealised trajectories of airflows at these heights in a typical Spanish Plume set-up. © Courtesy of NOAA Air Resources Laboratory.*

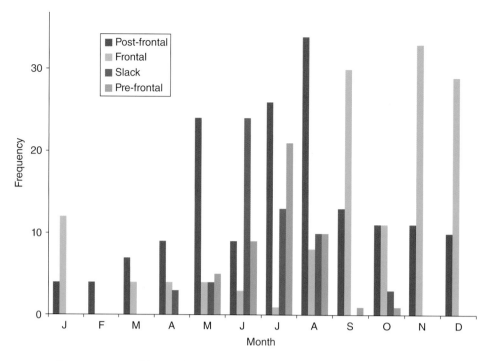

Figure 3.14 Monthly tornado totals by synoptic type during the period 2003–2012.

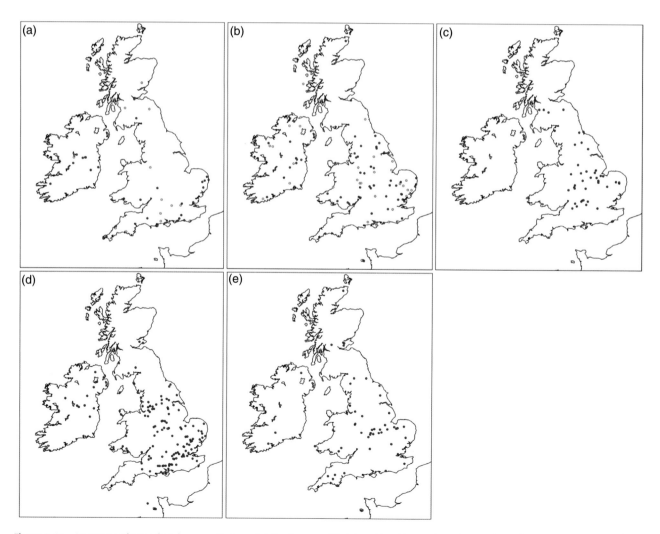

Figure 3.15 Locations of tornadoes by synoptic type. (a) Cool-season (October–March) post-frontal tornadoes, (b) warm-season, post-frontal tornadoes, (c) pre-frontal tornadoes, (d) tornadoes associated with frontal bands and (e) tornadoes in slack pressure patterns.

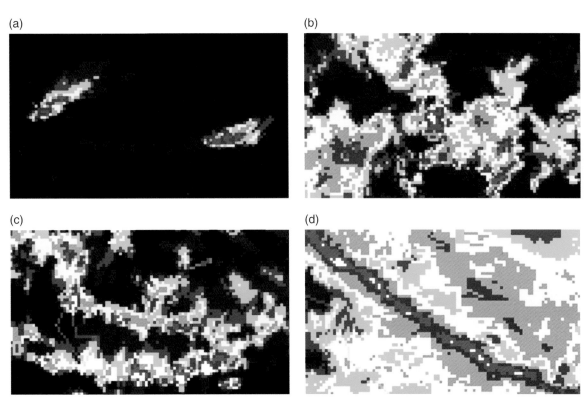

Figure 3.16 *1-km resolution radar rainfall imagery showing examples of four morphologies of convection: (a) cell, (b) cluster, (c) broken line and (d) line. Radar rainfall rate scale is given in Figure 3.2. © Crown Copyright Met Office.*

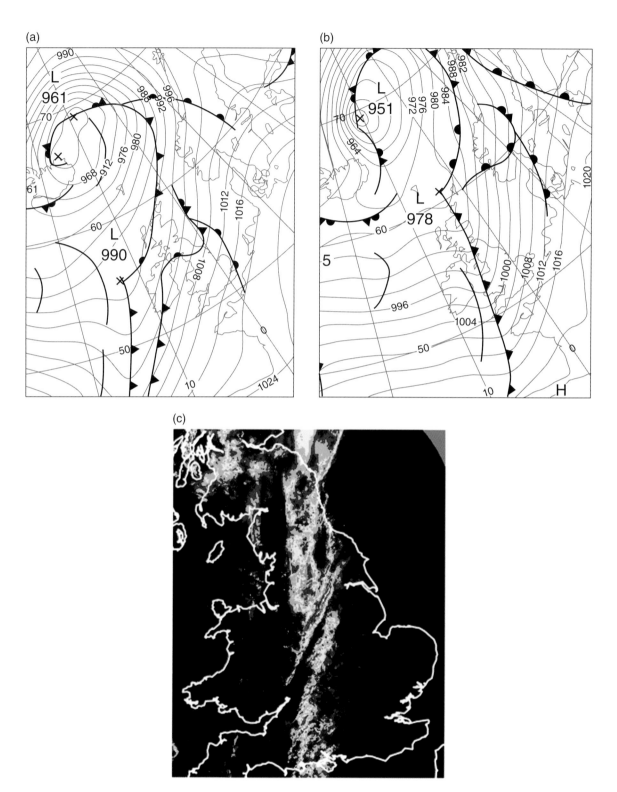

Figure 3.17 Surface analysis charts for (a) 0000 UTC 29 November 2011 and (b) 1200 UTC 29 November 2011. (c) Composite radar rainfall data at 1500 UTC 29 November 2011, showing the narrow cold frontal rainband. Radar rainfall rate scale is given in Figure 3.2. © Crown Copyright Met Office (2011).

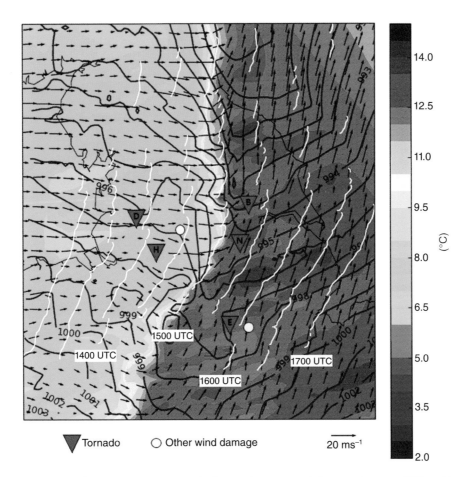

Figure 3.18 *Surface analysis of wind (arrows), temperature (shading) and pressure (solid contours) at 1500 UTC 29 November 2011. Analysis has been constructed using 1-minute resolution data from the Met Office's network of automatic weather stations. Locations of NCFR segments at 30-minute intervals between 1300 and 1700 UTC are indicated by the white solid lines. Tornado locations are indicated by inverted triangles. Letters indicate location of tornado report. D = Darwen, Lancashire; H = Hyde and Heaton Moor, Cheshire; N = New Rossington, South Yorkshire; B = Breighton, East Yorkshire; E = East Leake, Leicestershire. Non-tornadic wind damage locations are indicated by the yellow circles.*

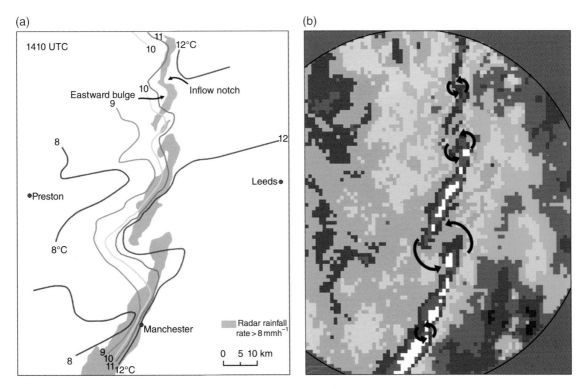

Figure 3.19 *A closer view of the segment–gap structure of the 29 November 2011 NCFR over northern England. (a) Contoured surface temperatures and (b) radar rainfall rate at 1-km resolution. Scale is as in Figure 3.2. Grey shaded areas in (a) denote regions of radar rainfall rate greater than 8 mm h⁻¹. Black arrows in (b) indicate the probable locations of cyclonic vortices along the line. © Crown Copyright Met Office.*

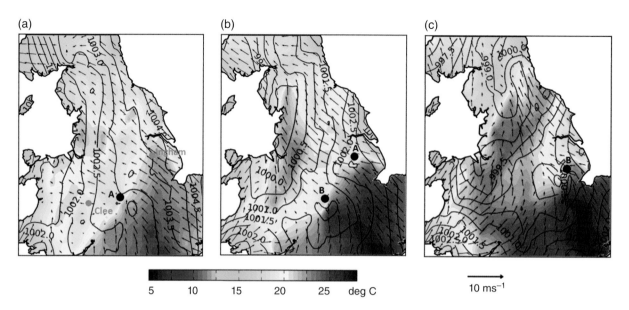

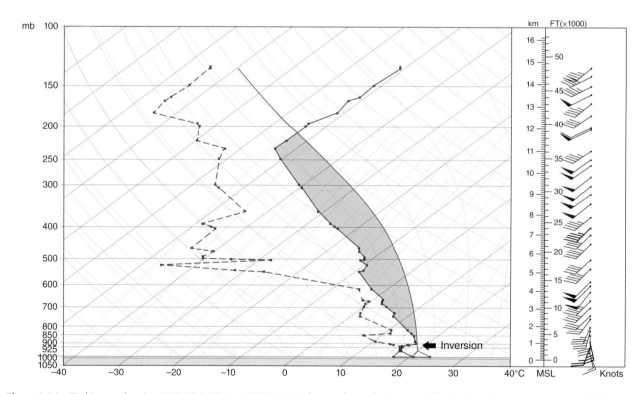

Figure 3.20 *Analysis of surface temperature (shaded), pressure (solid contours) and winds (arrows) for (a) 1000 UTC, (b) 1200 UTC and (c) 1400 UTC 28 June 2012. 'A' denotes the location of the first surface-based supercell of the day, which produced the tornado in Leicestershire. 'B' denotes the location of the second surface-based supercell, which produced the swathe of large hail through Leicestershire and Lincolnshire and the tornado in Lincolnshire. Reproduced with permission from Clark and Webb (2013). © John Wiley & Sons Ltd.*

Figure 3.21 *Tephigram showing 1200 UTC 28 June 2012 Nottingham radiosonde data, modified with surface temperature and dew point temperature as observed immediately south-east of the Leicestershire–Lincolnshire supercell. Solid and dashed red lines show the environmental temperature and dew point temperature profiles. Blue line shows the parcel curve. The 'positive area', indicating the CAPE, is shaded orange. Adapted from a plot created using RAOB software.*

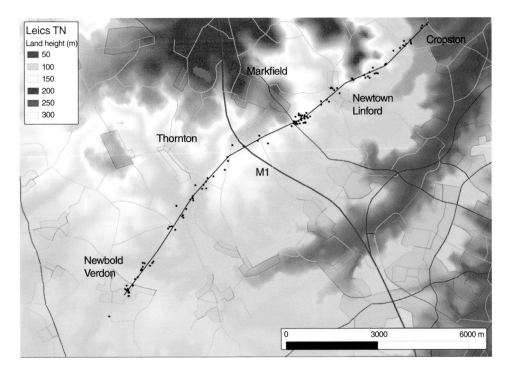

Figure 3.22 Damage track of the T4 Leicestershire tornado of 28 June 2012, as revealed by a site investigation conducted by TORRO member Tim Prosser. Black dots indicate individual instances of damage, and the black line is the estimated track of the centre of the tornado circulation. The colours indicate the altitude of the terrain (see key, inset). Contains Ordnance Survey data © Crown copyright and database right 2015. © Tim Prosser.

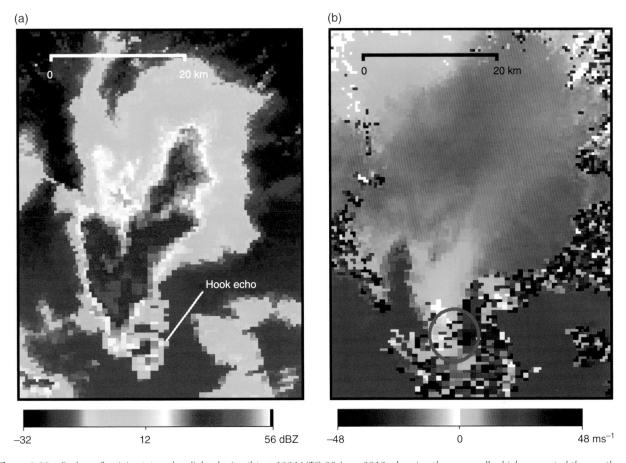

Figure 3.23 Radar reflectivity (a) and radial velocity (b) at 1321 UTC 28 June 2012, showing the supercell which generated the swathe of large hail in Leicestershire and Lincolnshire and the tornado in Lincolnshire. The velocity couplet, indicative of storm rotation, is marked by the pink circle in panel (b). © Crown Copyright Met Office.

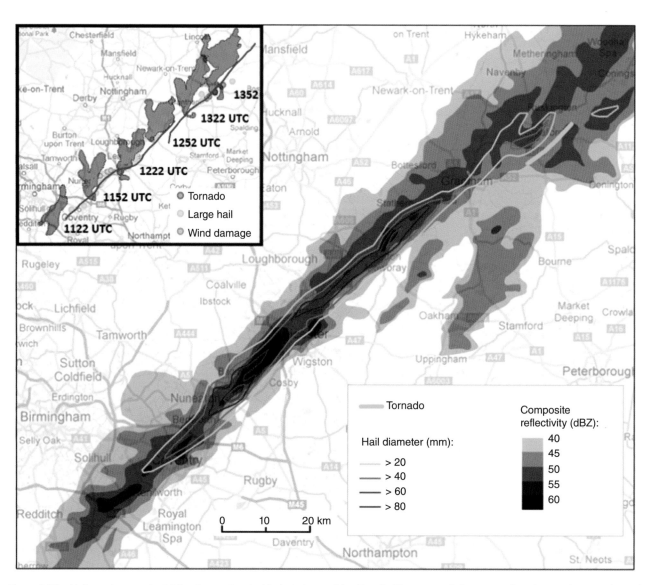

Figure 3.24 *Hail swathe over the Midlands associated with the Leicestershire–Lincolnshire supercell. Lime green line denotes the tornado track over Lincolnshire. Inset panel at top left shows the extent of the 45 dBZ reflectivity echo at 30 minute intervals (shaded), the radar-observed mesocyclone track (pink lines) and locations of individual severe weather reports (coloured circles). Background maps courtesy of Google Maps. Reproduced with permission from Clark and Webb (2013). © John Wiley & Sons Ltd.*

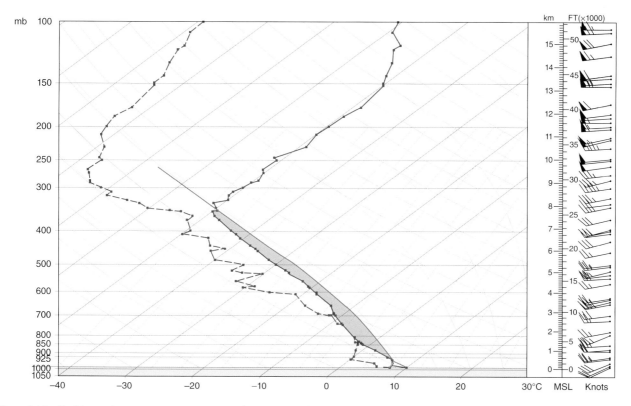

Figure 3.25 *Tephigram showing 0000 UTC 16 December 2012 Camborne radiosonde data, modified with 0930 UTC surface temperature and dew point temperature as observed at Camborne, which was immediately east of the storm at this time. Solid and dashed red lines show the environmental temperature and dew point temperature profiles. Blue line shows the parcel curve. The 'positive area', indicating the CAPE, is shaded orange. Adapted from a plot created using RAOB software.*

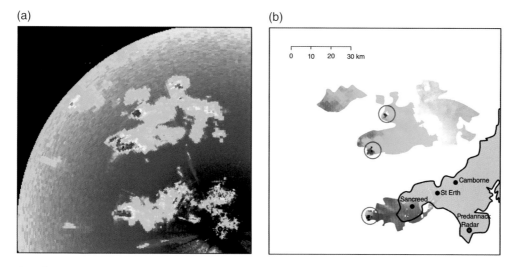

Figure 3.26 *Radar reflectivity (a) and radial velocity (b) at 0846 UTC 16 December 2012, showing three rotating storms near west Cornwall. Reflectivity and radial velocity scales are as in Figure 3.23. © Crown Copyright Met Office.*

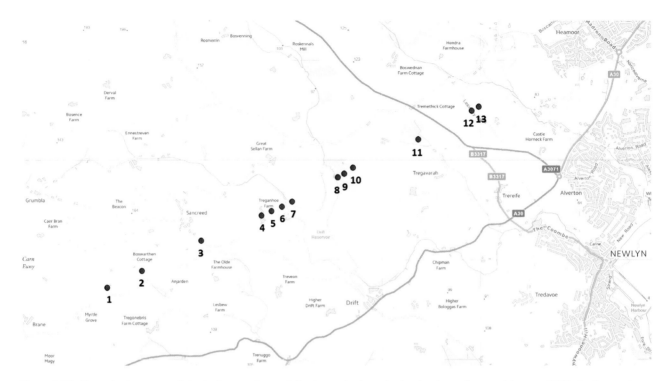

Figure 3.27 *Tornado damage track in the Sancreed area, following a site investigation by the author on 4 January 2013. Background map created using Ordnance Survey OpenData © Crown and database right 2015. Each blue square has dimensions 1×1km. Contains Ordnance Survey data © Crown copyright and database right 2015.*

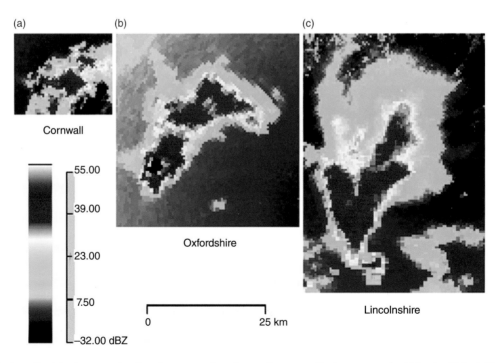

Figure 3.28 *Radar reflectivity imagery showing three examples of tornadic supercells that occurred within the United Kingdom in 2012. (a) 16 December 2012, (b) 7 May 2012 and (c) 28 June 2012. © Crown Copyright Met Office.*

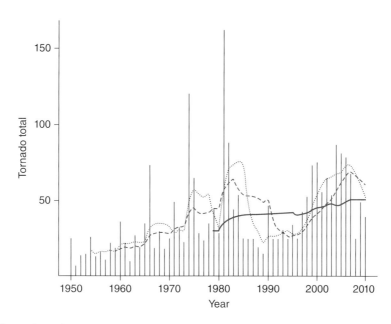

Figure 4.1 UK and Ireland tornado totals, 1950–2010. The blue solid line is the running 30-year mean, the red dashed line the running 10-year mean, and the black dotted line the running 5-year mean. Reproduced with permission from Kirk (2014). © John Wiley & Sons Ltd.

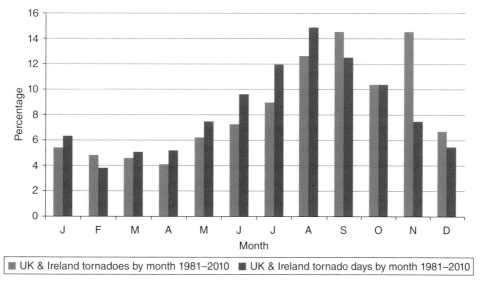

Figure 4.2 Percentage of UK and Ireland tornadoes and tornado days by month, 1981–2010. Reproduced with permission from Kirk (2014). © John Wiley & Sons Ltd.

Figure 4.5 *Recorded tornadoes in the United Kingdom and Ireland, 1981–2010. Each event is plotted with a red marker. Reproduced with permission from Kirk (2014). © John Wiley & Sons Ltd.*

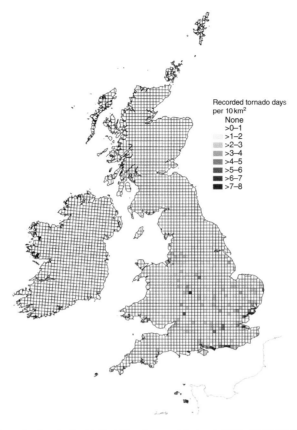

Recorded tornado days
per 10 km²
None
>0–1
>1–2
>2–3
>3–4
>4–5
>5–6
>6–7
>7–8

Figure 4.6 *Tornado days in the United Kingdom and Ireland, 1981–2010. © Tim Prosser.*

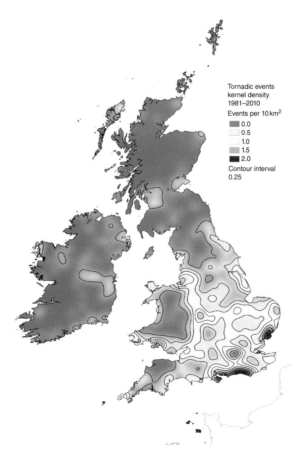

Tornadic events
kernel density
1981–2010

Events per 10 km²
0.0
0.5
1.0
1.5
2.0
Contour interval
0.25

Figure 4.7 *Gaussian kernel density map of recorded tornado events in Britain and Ireland, 1981–2010. © Tim Prosser.*

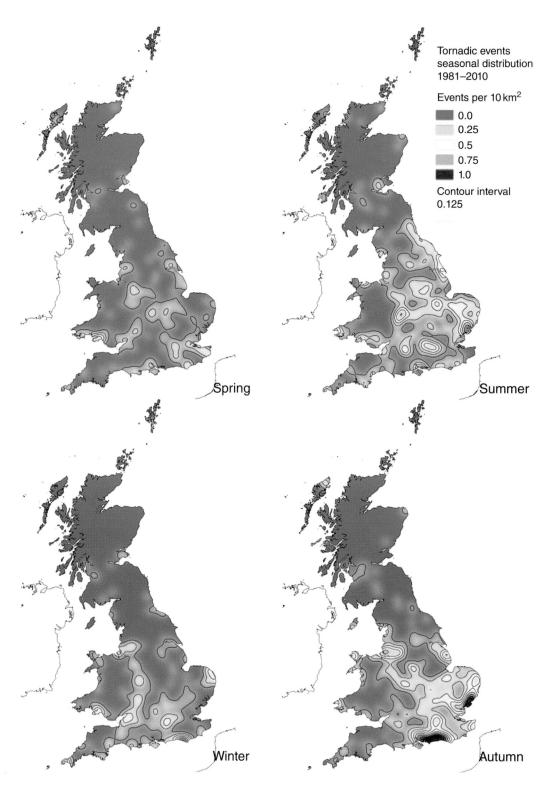

Tornadic events
seasonal distribution
1981–2010

Events per 10 km²

	0.0
	0.25
	0.5
	0.75
	1.0

Contour interval
0.125

Spring

Summer

Winter

Autumn

Figure 4.8 *Gaussian kernel density maps of recorded tornado events in Britain and Ireland, 1981–2010, by season. Spring, March, April and May; summer, June, July and August; autumn, September, October, November; winter, December, January, February. Note the change in scale from Figure 4.7. © Tim Prosser.*

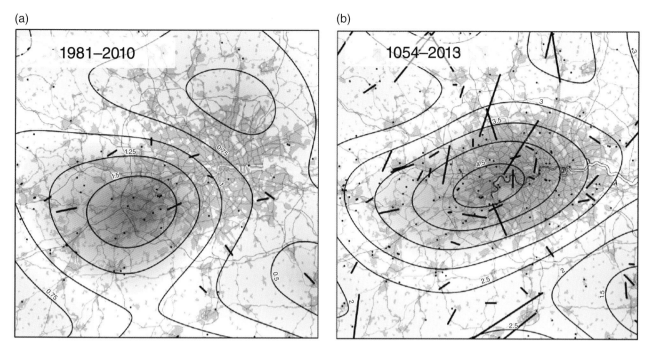

Figure 4.9 Gaussian kernel density maps of recorded tornado events in the TORRO database for the London area. Density is shaded from red orange (high) to blue green (low); smoothed contours of density every 0.25 events per 10 km²; tornado events are plotted with points and lines against a modern urban shape-file background. (a) 1981–2010. (b) 1054–2013 (complete TORRO database). Created using OS OpenData™, contains Ordnance Survey data © Crown copyright and database right 2015. © Tim Prosser.

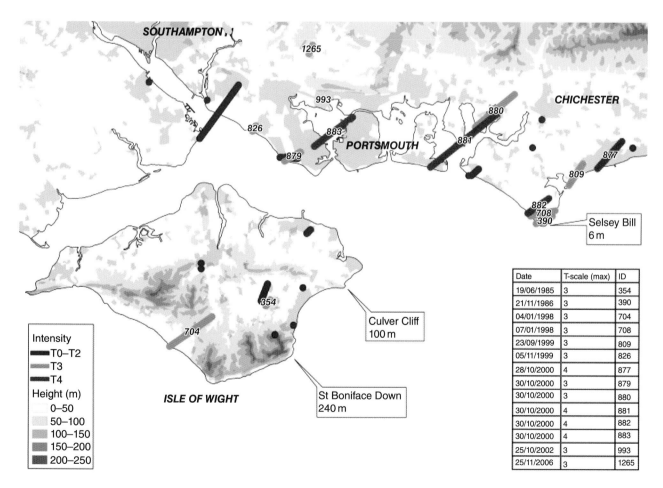

Date	T-scale (max)	ID
19/06/1985	3	354
21/11/1986	3	390
04/01/1998	3	704
07/01/1998	3	708
23/09/1999	3	809
05/11/1999	3	826
28/10/2000	4	877
30/10/2000	3	879
30/10/2000	3	880
30/10/2000	4	881
30/10/2000	4	882
30/10/2000	4	883
25/10/2002	3	993
25/11/2006	3	1265

Figure 4.10 1981–2010 tornado events in the vicinity of the Isle of Wight and Selsey Bill. Terrain height is shaded in grey according to the inset key (left), and urban areas are indicated by the transparent beige shading. Events are coloured according to strength (inset key, left). Event tracks are identified at their end points by their identification number in the TORRO database (inset key, right). Only events rated at T3 and above have been labelled. Created using OS OpenData™, contains Ordnance Survey data © Crown copyright and database right 2015. © Tim Prosser.

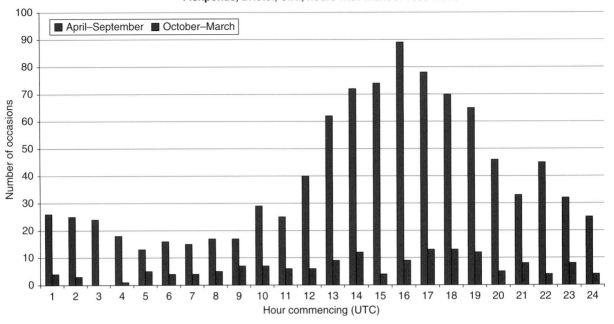

Figure 7.1 *Diurnal record of thunderstorms in north-west Bristol, England, 1950–2009. Observations by Bryan A. Williams.*

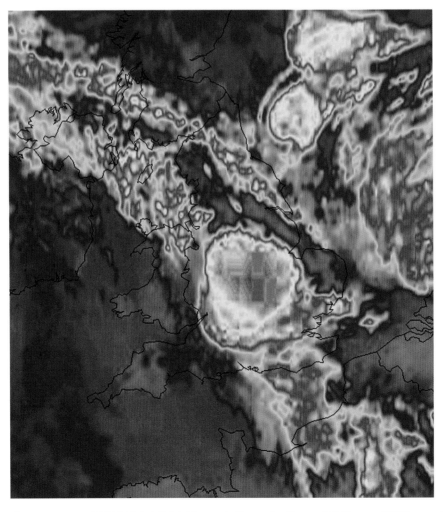

Figure 7.3 *Infrared satellite image, 1330 GMT, 22 July 2006. A Mesoscale Convective System (MCS) moved NNE across England accompanied by widespread overhead thunderstorm activity (see Webb and Pike, 2007). Cloud top temperatures (deg C): green −30 to −40°C, green/yellow −40 to −50°C, yellow −50 to −58°C, orange −58 to −60°C, red below −60°C. Image courtesy of Robert Moore. © EUMETSAT 2006.*

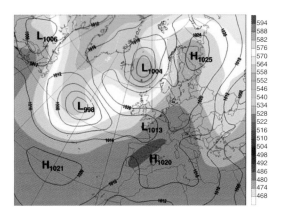

Figure 7.4 *NCEP reanalysis chart for 3 August 1879, 0000 GMT. These charts show surface pressure (msl) isobars (solid black lines), 500 hpa heights in dam (shading, see legend on right) and 500–1000 hpa thickness in dam (broken white lines). Courtesy of NOAA ESRL/PSD reanalysis project (see Compo et al., 2011). Meteocentre, http://meteocentre.com/.*

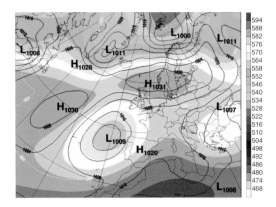

Figure 7.6 *NCEP reanalysis chart, 0000 GMT, 8 June 1910. Courtesy of NOAA ESRL/PSD reanalysis project (see Compo et al., 2011). Meteocentre, http://meteocentre.com/.*

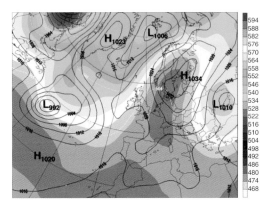

Figure 7.8 *NCEP reanalysis chart for 29 August 1930, 0600 GMT. Courtesy of NOAA ESRL/PSD reanalysis project (see Compo et al., 2011). Meteocentre, http://meteocentre.com/. © NOAA.*

Figure 7.19 *Lightning over Tydavnet, County Monaghan, Ireland, night of 25–26 July 1985. © Donal McEnroe.*

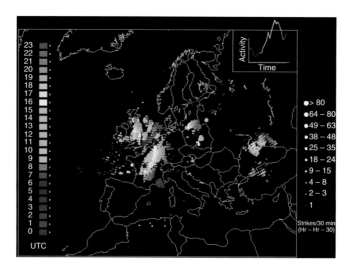

Figure 8.11 Lightning distribution and intensity on 18 August 2004, as compiled by © Wetterzentrale from the Met Office's lightning detection network.

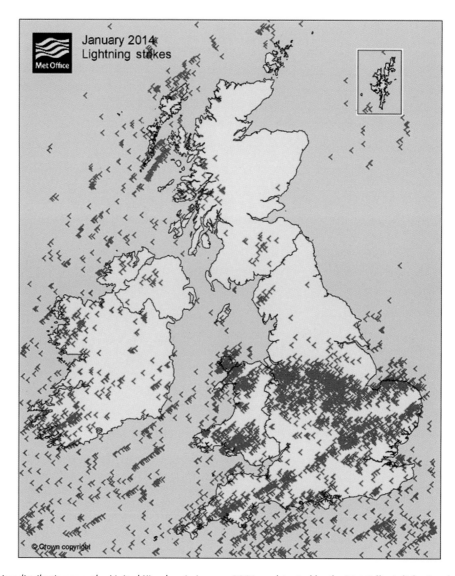

Figure 8.13 Lightning distribution over the United Kingdom in January 2014, as detected by the Met Office's lightning detection network. Met Office, UK. © Crown copyright.

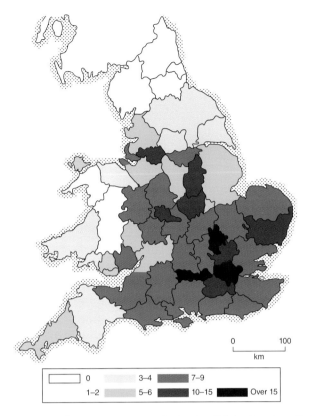

Figure 9.10 *Geographical distribution of storms of H2 or more severity, England and Wales. Number of hailstorms (per 1000 km² per 100 years) reaching or exceeding H2 intensity, for counties of England and Wales, 1930–2009 (legend above).*

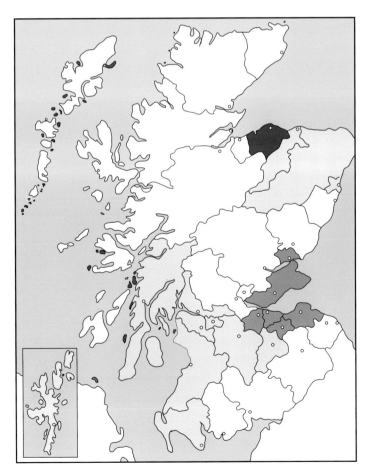

Figure 9.11 *Geographical distribution of H2+ hailstorms in Scotland by counties, Number of hailstorms (per 1000 km² per 100 years) reaching or exceeding H2 intensity, for lieutenancy areas of Scotland, 1930–2009 (white=0–1, light blue=2, mid blue=≥3, dark blue=5).*

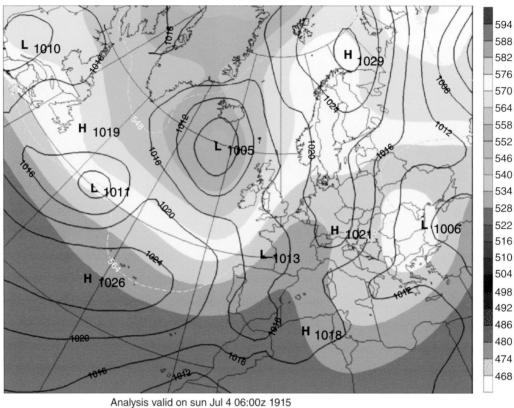

Mean sea level pressure (hPa)
500–1000 hPa thickness (dam)
500 –hPa Heights (dam)

Analysis valid on sun Jul 4 06:00z 1915
20th century reanalysis – meteocentre.com

Figure 9.20 *NCEP reanalysis chart, 4 July 1915, 0600 GMT. Courtesy of NOAA ESRL/PSD reanalysis project (see Compo et al., 2011), Meteocentre.*

Figure 10.2 *Lightning above power lines, Cardiff. © Chris Cameron-Wilton.*

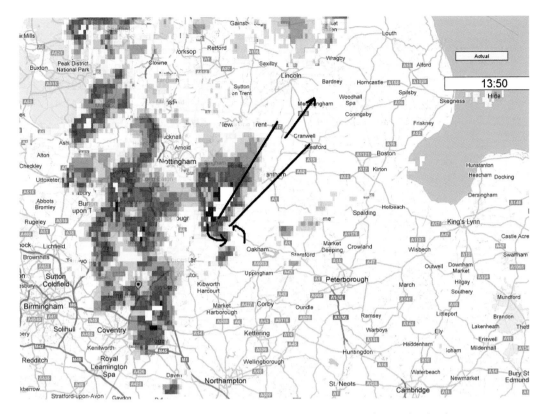

Figure 12.2 *The storm of 28 June 2012. The hook echo shows the approximate location of strong low-level rotation (arrows), and the black lines show the approximate track the storm will take. A tornado remains possible with this storm, along with large hail. Image courtesy of MeteoGroup, UK Ltd. (2012), www.raintoday.co.uk. Data from UKMO.*

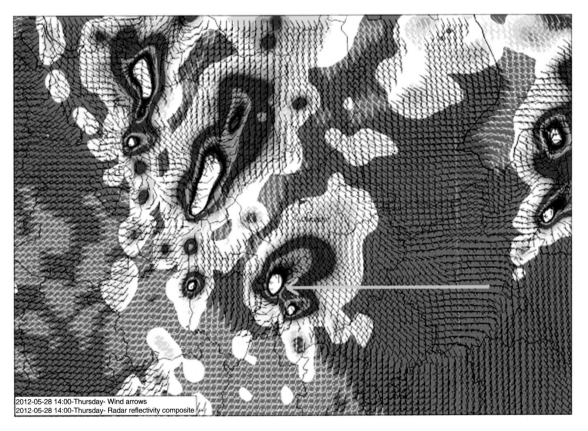

2012-05-28 14:00-Thursday- Wind arrows
2012-05-28 14:00-Thursday- Radar reflectivity composite

Figure 12.3 *A simulated supercell thunderstorm. This image shows the output of the MG, incorporating WRF® model for 1400 GMT on the 28 June 2012. Image courtesy of MeteoGroup, UK Ltd. (2012). WRF® is a registered trademark of the University Corporation for Atmospheric Research (UCAR).*

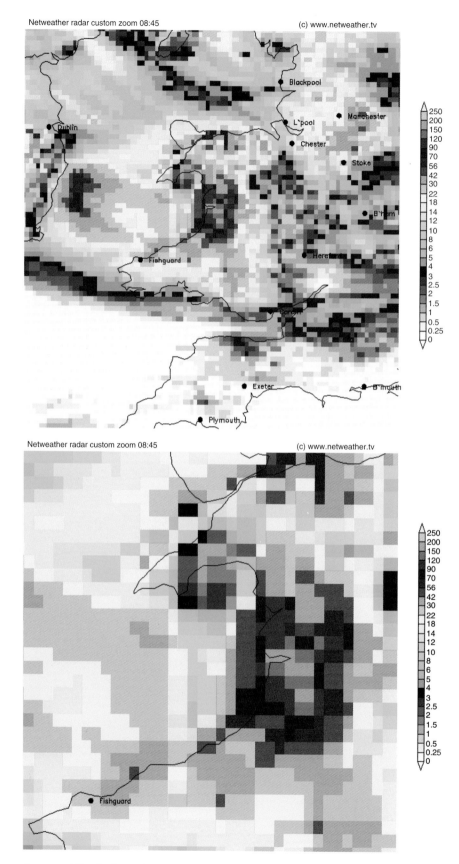

Figure 14.4 *Left: cumulative rainfall radar for Wales on 8 June 2012. Right: close-up on region affected showing up to 144–192 mm locally (left image) Image supplied. © NetWeather.*

Figure 14.8 *Some extreme summer rainfall events by area (km²) affected, for the United Kingdom. For further information, see British Rainfall 1912, 1922, 1948 and 1968.*

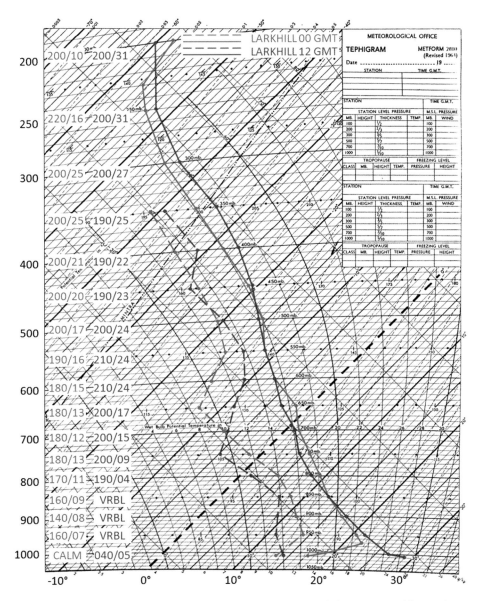

Figure 14.15 *Larkhill tephigrams (dry bulb and dew point temperatures) and upper winds for 0000 GMT (blue) and 1200 GMT (red) 29 May 1944. Drawn from data published in the Upper Air Section of the Daily Weather Report. © Paul Brown.*

Tornado Intensity - TORRO Scale

Unknown
Waterpouts
0 1 2 3 4 5 6 7 8 9

Map C.1   Tornadoes in the United Kingdom and Ireland 1054–2013 *This map shows the distribution of tornado events from the complete TORRO dataset covering the years 1054–2013. The figure was composed by plotting the tracks of all known events over an image of the United Kingdom and Ireland derived from the NASA Blue Marble composite satellite image. Urban areas are coloured based on a land use atlas produced by the European Environment Agency (year 2000). Stronger tornadoes are shown in a lighter shade and are plotted on top of lesser events. Note that the track widths are exaggerated so this does not represent actual land coverage. This map was produced by Tim Prosser.*

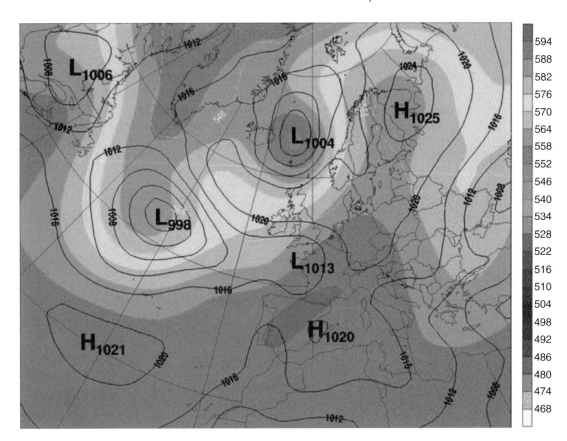

Figure 7.4 *NCEP reanalysis chart for 3 August 1879, 0000 GMT. These charts show surface pressure (msl) isobars (solid black lines), 500 hpa heights in dam (shading, see legend on right) and 500–1000 hpa thickness in dam (broken white lines). Courtesy of NOAA ESRL/PSD reanalysis project (see Compo et al., 2011). Meteocentre, http://meteocentre.com/. (See insert for colour representation of the figure.)*

Overnight 5th–6th, violent thunderstorms affected an area extending west–north-westwards from Sussex across to Salisbury Plain and on to south Wales. At Brighton, thunder persisted for 8½ hours from 2200 GMT to 0630 GMT. Surface charts on 7–8 June (Figure 7.5) showed a light to moderate north-easterly airflow across the Midlands, while the 500-hPa reanalysis charts (Figure 7.6) indicate a large cut-off vortex situated north-west of Iberia with south-west England within its circulation. Increasingly humid air edged northwards into southern England. Unusually prolonged thunderstorm activity, with widespread lightning damage, occurred in a broad swathe from Surrey west–north-westwards to Herefordshire and southern Wales on the 7th. The most severe downpours and local flooding occurred in the evening. The north-east surface wind may well have been under-cutting the advancing warm, humid air as happened in a more recent instance of a prolonged thunderstorm event in the area in May 1993 (Pike, 1994a). Indeed, Mill (1911) noted that the most severe storm downpours occurred almost simultaneously throughout this swathe between 1930 and 2330 GMT on 7 June 1910, a feature indicative of a convergence zone.

The *Henley Standard* reported that '…the rumbling noise of the thunder up the Thames Valley prevailed all day'. In Oxfordshire, Churchill school, near Chipping Norton, recorded a daily rainfall of 108 mm. At Stow-on-the-Wold (Quar Wood), most of the 90 mm of rain recorded fell in 2 hours in the evening, and in nearby valleys, 'houses were flooded to the tops of doors on ground floors'. The Evenlode Vale was reported to be in flood for miles. At Kingham (Fowler, 1913), the great thunderstorm began at 1930 GMT with incessant thunder and lightning from 2000 until 2330 GMT. The brass vane on the manor house was struck and crumpled. At the railway level crossing between Kingham and Churchill, the crossing keeper was opening the gate for a train at the very moment that a tall elm tree was struck by lightning 20 yards away; this was one of several trees in the area which were struck and had bark ripped off. At Swerford, also near Chipping Norton (Mill, 1911), the observer reported seven separate thunderstorms (three directly overhead) during this extraordinary day, culminating in a prolonged and 'alarming' storm from 2045 to 2330 GMT; thunder was heard during 12 hours of the day. Further south at Wantage, three periods of violent overhead electrical activity occurred: around noon, from 2000 to 2100 GMT, and from 2200 to 2300 GMT when a large barn was struck and set alight.

There were many damaging lightning strikes in Oxfordshire. Deddington parish church was struck with damage to the pinnacles, while houses were struck in Kingham, Lyneham, Churchill, Deddington, Hook Norton, South Newington, Ducklington (Witney) and Boars Hill (Oxford). At Chipping Norton, two hayricks were struck and set on fire, and farm livestock were killed by lightning in neighbouring villages. In north Gloucestershire, lightning dislodged the chimney of a house at Oddington and struck and fired a group of farm buildings at Moreton-in-the-Marsh.

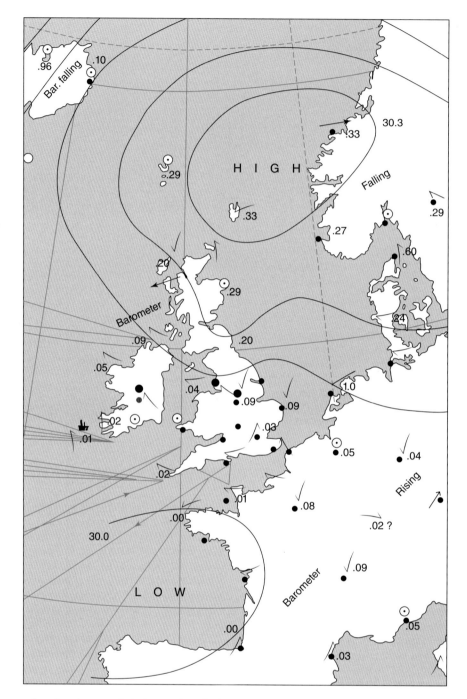

Figure 7.5 *Surface weather chart (pressure in inches, and winds) for 0700 GMT, 8 June 1910. Adapted from the Daily Weather Report of the Met Office, UK. © Crown copyright.*

At Coleford in the Forest of Dean, a chimney was destroyed, and a man sustained eye injuries when evidently receiving an electric shock via cutlery and his spectacles. A ploughman was struck and killed at Badminton, while many livestock were killed in the Tewkesbury/Evesham area. Further afield, three houses and a hotel were damaged by lightning in Windsor (Berks), while lightning hit three houses in Swansea.

The widespread storms on 9 June were notable for extreme rain and hail falls (Webb, 2011a). Pressure was quite uniform on the morning of the 9th, and the area of warm, humid, stagnant air will have been conducive to the development of convergence zones. The *Monthly Weather Report* indicated a very shallow low centre tracking east–north-east into the eastern English Channel on 9th. The 500-hPa reanalysis chart for 1200 GMT on the 9th showed the cut-off cold vortex south-west of the British Isles within the axis of a deep upper trough just west of these islands. This corresponds closely to the scenario of the 'modified Spanish Plume' described by Lewis and Gray (2010).

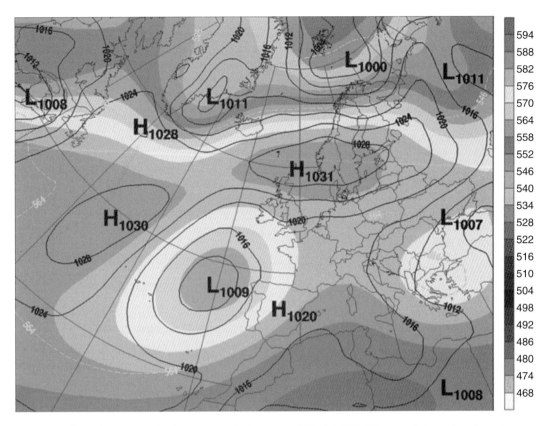

Figure 7.6 *NCEP reanalysis chart, 0000 GMT, 8 June 1910. Courtesy of NOAA ESRL/PSD reanalysis project (see Compo et al., 2011). Meteocentre, http://meteocentre.com/. (See insert for colour representation of the figure.)*

Storms of historic intensity occurred in the Reading and Oxford areas. At Caversham, Reading, torrential rain commenced at 1215 GMT, changing to intense hail between 1225 and 1300 GMT. At Peppard, 74 mm fell from 1200 to 1330 GMT. At Wheatley, the deluge commenced at 1242 GMT, while at nearby Waterstock, the storm began at 1300 GMT with a 'hurricane' followed by the hailstorm which lasted from 1315 to 1415 GMT; rain continued until 1530 GMT though most fell before 1500. At Pyrton Hill, meteorologist W.H. Dines wrote: 'thunder was first heard soon after noon and was practically continuous from 1230 to 1600 GMT, very severe from 1400 to 1500 GMT. Nearby, three trees were struck by lightning. The storms appeared to move slowly from south to north' (Mill, 1911).

Two independent rain gauges at Wheatley, Oxfordshire, recorded over 125 mm of rain on the 9th: School House with 132 mm and Holton Cottage with 127 mm, the former including a carefully measured 110 mm in 1 hour. However, both readings had caveats since the former was from a non-standard gauge, while the latter was from a gauge evidently in an unorthodox exposure (Mill, 1911). The nearby standard gauge at Waterstock recorded 100 mm, but it had become completely choked by hail (indeed overflowed), and it is estimated that the true fall was appreciably higher and close to that of the Wheatley recordings. The consistency between the Wheatley readings and the support derived from neighbouring rainfall stations (see rainfall footprint in Mill, 1911) do indicate that the peak recorded values at Wheatley were very close to the truth. Moreover, referring to the

Hampstead storm of 14 August 1975, Tyssen-Gee (1975) indicated that about 30% of hard hailstones of 20 mm diameter were likely to bounce out of a standard rain gauge (marble size hailstones were reported to have stripped foliage in Wheatley on 9 June 1910).

The flood and damage at Wheatley is described in the *Henley Chronicle* of 17 June 1910. Flowing currents were 45 to 90 cm deep in most streets of the village, the water carrying earth and debris washed down from the surrounding small hills as well as many household goods and live ducks and geese! The main street 'became a raging torrent within 30 minutes' with many houses flooded to depths of 45–90 cm. One lady was rescued from drowning. In Crown Street, where water was 180 cm deep, tenants in one dwelling escaped via a bedroom window; water even damaged a Dutch clock hanging near the ceiling of the house. A wall was washed away at the west end of the village, a 6-ft (180-cm) gulley was scoured in the meadow turf, and a hayrick weighing nearly 2 tonnes was shifted several yards by the water. The village post office was inundated, cutting communications. The *Oxford Journal Illustrated* (1910) published photographs showing the peak flood tide marks in Crown Street (with a resident pointing to one at 180 cm), the displaced hayrick and damage to roads which had been under 30–60 cm of debris. In Waterstock, 'a deluge of hail and water raced through the village and the houses' while a large elm tree was uprooted by the force of water.

The various observations of hail all indicate hailstones of 20–35 mm diameter: at Caversham (Reading), marble to large walnut size (circa 20–35 mm diameter); Assendon (Henley), half an ounce (corresponding to about 32 mm diameter); and Waterstock, walnuts (20–30 mm across). The hail wrecked glass-houses and frames in all three aforementioned localities, and in Waterstock the hailstones were piled in broad mounds 90–120 cm deep near obstructions and lay nearly 10-cm level depth on open ground. Hail shattered the glass dome on Waterstock House. Plants were reduced to bare stalks, and the complete devastation of vegetation and glass in Waterstock Rectory garden is described in a letter in *British Rainfall* 1910 which was also reproduced in *The Times*.

Numerous incidents of lightning damage were reported, some involving serious injuries. In Caversham, a female teacher was struck in her home at the time the local post office was hit, and she suffered scorched hands and loss of movement. In Wheatley, another lady was struck and lost the use of her arm for several days. In Farthinghoe, north Oxfordshire, a man was struck in a horse-driven cart; he was thrown out of it as the startled horse fled. The famous windmill at Brill, Buckinghamshire, was struck with one of the sails torn off and shivered into pieces. The violence of the rain and hail storms on 9 June 1910 (with intensities of at least H4 on the TORRO International Hail Scale) can be attributed to destabilisation of a very humid air mass by surface heating, local convergence and general ascent ahead of a major upper trough.

7.2.3   7–10 July 1923

The first half of July 1923 was notable for a spell of classic 'hot and thundery' weather. Prichard (1977, 1999) assessed the London thunderstorm overnight 9–10 July 1923 as the most severe, in terms of electrical activity, to have affected the UK capital in the 20th century. The previous day had seen an exceptional flash flood event in the Carrbridge area of Inverness-shire in north-east Scotland as the first wave of heat broke in wide-spread thunderstorms (McConnell, 2000).

The surface chart for 0700 GMT on the 10th, from the *Daily Weather Report* (DWR) (Figure 7.7), shows a conspicuous trough extending northwards across eastern England. The United States Weather Bureau (USWB) synoptic chart for 1300 GMT on 10th showed a corresponding frontal zone extending from western France through the 'spine' of England to the northern Irish Sea within this broad surface trough. A strong upper ridge covered much of Europe. A deep upper trough was disrupting to the west of Ireland, and the 500-hPa airflow over England was south–south-easterly, light in the east but increasing further west. The severe outbreak of storms from the late evening of the 9th occurred on the leading edge of very hot tropical continental air returning slowly westwards. Storm cells appear to have been repeatedly generated at mid-levels and drifted north–north-west in a swathe from Sussex to south Lincolnshire. More than 100 mm of rain fell in Sussex with 116 mm at Rottingdean and 103 mm at Seaford.

A brontometer at Chelsea recorded 6924 flashes between 2200 GMT on the 9th and 0400 GMT on the 10th with the lightning frequency peaking at 47 discharges per minute during the storms from 2300 to 2400 GMT. At Wadhurst, Sussex, 1600–1700 flashes per hour were observed at the height of the storm. *The Times* reported that through the night 'in many houses lights were to be seen and it was obvious that many people were too

frightened to go to bed'. The London fire brigade reported that lightning started fires in 11 separate London districts. A tramcar on the Victoria embankment was struck and set alight. In Kentish Town road, a manhole and 4 m of pavement were torn up while lightning sparked an explosion and fire in a house in Brockley. A lead manufacturer's premises in Isleworth was struck by lightning and severely damaged. Two houses in Bromley and four houses in Caterham, Surrey, were struck by lightning as were several buildings in Mitcham. A large residence at Walton-on-the-Hill, Surrey, was struck and completely destroyed by the resulting fire. In Horley, two cottages and the roof of the local Conservative Club were damaged by strikes. Further afield, in Luton, a factory and a farmhouse were fired by lightning. At Farcet near Peterborough, lightning struck a farm and fired and destroyed four cottages and a hayrick.

Another brief but violent outbreak of thunderstorms affected south-east England later on the 10th when Eton College chapel was damaged by lightning. Subsequently, even hotter air spread back westwards with temperatures reaching 35°C in Reading and Bristol on the 12th.

A rather similar, spectacular 'warm front' thunderstorm, albeit of rather shorter duration, lit up the skies across London and surrounding counties on the night of 14–15 July 1945 (Douglas and Harding, 1946).

7.2.4   17–18 July 1926

On the 17th, a trough over south-west Ireland extended south-east to Brittany, while surface winds were south-easterly. A depression deepened rapidly over north-west France and moved northwards. This unusually 'dynamic' breakdown of a July heatwave resulted in very widespread and spectacular thunderstorms with numerous incidences of large hail (especially in the west) and exceptionally heavy rainfall (with 154 mm at Abergwesyn, Cardiganshire on the 18th), often accompanied by strong winds. Seaton, Devon, reported a 12-hour period of continuous thunderstorm activity (*Monthly Weather Report*). At nearby Axminster, thunder was continuous for 13 hours from 2200 GMT on 17 July to 1100 GMT on the 18th, and 71 mm of rain fell in 3 hours.

Seven severe category hailstorms occurred, all affecting south-west England and Wales. Severe hail caused damage in Devon from Hartland in the north-west to the Dorset border near Lyme Regis in the south-east. Irregular hail up to 70 mm (average dimension) was observed at Combe St Nicholas near Chard (Somerset) where severe thunderstorm activity was reported for 8 hours from 0200 to 1000 GMT on the 18th. Pieces of ice which fell near Gwennap in West Cornwall in the early afternoon of the same day measured 100 mm by 50 mm by 25 mm (Meteorological Office, 1927).

7.2.5   28–29 August 1930

A notable late summer heatwave saw temperatures exceeding 32°C on three successive days, 27–29 August, over England with 34°C in London on both the 28th and 29th. Thunderstorms affected western areas overnight on the 26th–27th as the hot spell began, and vivid lightning accompanied storms across the south-east on the night of the 29th; 63 flashes of sheet lightning per minute were observed in south-west London at 2330 GMT (*Monthly*

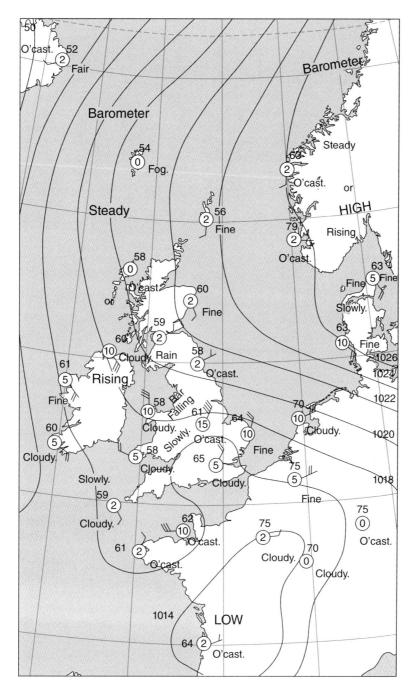

Figure 7.7   *Surface analysis for 0700 GMT, 10 July 1923. Adapted from the Daily Weather Report of the Met Office, UK. © Crown copyright.*

Weather Report). However, the most severe thunderstorm activity occurred in a zone across northern England and south and east Scotland from around midday on 28th to noon on 29th. Here, the strongest vertical wind shear is evident. A returning warm front is shown straddling western parts of England, the northern portion of a quasi-stationary frontal zone extending north across Portugal and Biscay (USWB). A trough extended further northwards over northern England and south-east Scotland (Figure 7.8).

In Dundee, the first violent storm lasted from 2100 to 0130 GMT, with the observer noting that the maximum lightning

frequency reached 30 discharges per minute (the average was about 12 per minute). A correspondent to the *Dundee Courier* likened the storm to the one he had witnessed in London on 10 July 1923 (Campbell 1984). The storm was followed by a very misty morning which gradually darkened due to the thickening cloud which preceded a second severe storm at noon. Between 2136 and 2200 GMT on the 28th, 30 mm of rain fell at Leuchars met station (Fife), and the total for the 24-hour period to 1800 GMT on 29th was 52 mm. At Leuchars, thunder was recorded in 12 separate hours between 1700 on the 28th and 1600 GMT on the

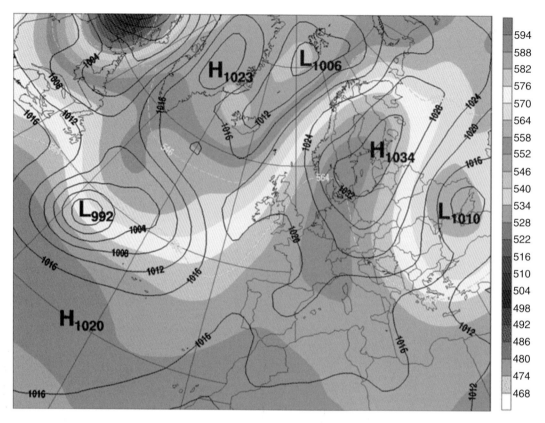

Figure 7.8   *NCEP reanalysis chart for 29 August 1930, 0600 GMT. Courtesy of NOAA ESRL/PSD reanalysis project (see Compo et al., 2011). Meteocentre, http://meteocentre.com/. © NOAA. (See insert for colour representation of the figure.)*

29th. Heavy thunderstorms with hail were recorded in the station register at 2330 GMT on the 28th and 1100 GMT on the 29th. The University Hall at St Andrews was set on fire by lightning; fortunately, no students were in residence, and the fire was put out quickly. Further north up the eastern coastal counties of Scotland, lightning fired the rail station at Buckie, Banffshire, and struck and severely damaged a school near Stonehaven with the bell tower destroyed. Dunbar also recorded two fierce thunderstorms, late on the 28th and on the morning of the 29th, the former being accompanied by 30-mm-diameter hailstones. In Arbroath, the chimney stack of the infirmary was struck by lightning, and a large portion of copestone was dislodged. There were incidences of ball lightning observed in both Perth and Dundee. *The Times* reported that Edinburgh was plunged into darkness at midday with street lamps and shop lights switched on. Several houses in Haddington and Dunbar (both East Lothian) were struck by lightning.

A notable hailstorm swept across Peeblesshire on the morning of 29th as reported in *St Ronan's Standard* which provided a vivid description. This included an account of the approach of a huge arc-shaped cloud, day suddenly turning into night with headlights of vehicles flashing, and then 'throughout could be heard the continual breakage as window after window fell to the huge lumps of ice'. Indeed, hundreds of west-facing windows in Peebles were broken, paint was scraped off buildings, and sandstone walls were pitted. Near Glentress Forest, hailstones 38 mm across were measured during the storm which produced continuous thunder and lightning.

Further south, the Nottingham Evening Post carried vivid accounts of the thunderstorms and reported that on the night of the 29th, 'Thunder rolled and roared in ceaseless agony and vivid lightning lit up the sky and silhouetted tall buildings and church towers'. Over Hornsea, East Riding of Yorkshire, 'the whole dome of the heavens was at times lit up with flashing violet light across which huge, long zig-zags sported on their courses' (*Hull Daily Mail*). A 'terrific electric hailstorm' caused severe damage to glasshouses at Aldborough, while lightning damaged two adjoining houses in Grimsby. At Downholme, North Yorkshire, raging thunderstorms sent a wall of water through the village carrying away stone walls and ripping up roads. A similar extreme flash flood was reported in the Goyt Valley in the Peak District, while hailstones at least the size of pullet's eggs were observed at Helmshore, Lancashire.

7.2.6   18–22 June 1936

A depression moved very slowly north-east across Biscay, with an east to south-easterly airstream over Britain and Ireland (Figure 7.9) and maximum temperatures exceeding 30 °C in places. This was a classic 'modified Spanish Plume' scenario. Major upper trough disruption resulted in a cut-off cold vortex becoming slow moving over Biscay and producing a very backed south-easterly airflow at 500 hPa. The USWB frontal analysis indicates that a warm front pushed slowly north over the English Channel on the 19th to lie across Wales and the Midlands at

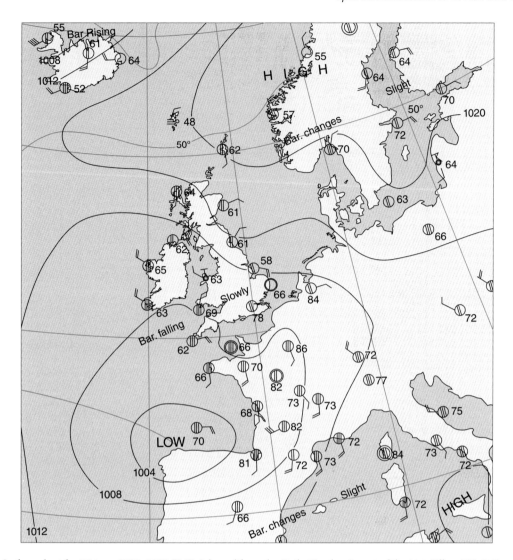

Figure 7.9 *Surface chart for 20 June 1936, 1800 GMT. Adapted from the Daily Weather Report of the Met Office, UK. © Crown copyright.*

1300 GMT on 20th. A very pronounced surface trough pushed north-east on 21st, coincident with the leading edge of an occlusion on the contemporary DWR chart (Figure 7.10).

Thunderstorms were exceptionally widespread and severe on the 19th, 20th and 21st with a remarkable 600 to 700 lightning damage incidents reported in Britain and Ireland. Indeed, the Thunderstorm Census Organisation (TCO) Annual Report for 1936 records that the 3 days 19th–21st all had overhead thunderstorm activity covering more than half of England and Wales; the areal coverage for the three days was 82, 70 and 52% with significant activity also on the 18th and 22nd. During June 1936, the National Grid reported 76 power system breakdowns due to lightning, the highest monthly incidence between 1934 and 1947. Twelve severe hailstorms were reported on 20th–21st during the peak of the 5-day thundery spell. The observer at Ross-on-Wye reported two thunderstorms on each of the 3 days, while at Kew Observatory, eight separate thunderstorms were recorded between the 18th and 21st.

The most persistent and severe thunderstorm activity on the 19th occurred across southern coastal counties. In Portland,

Dorset, it was reported that vibrations from the thunderclaps cracked many windows. Five houses were struck by lightning in Brighton, and at nearby Shoreham, a 20-minute hailstorm beat paint off a cross-channel aeroplane. Manston, Kent, recorded 8 hours of thunder between 1400 GMT on the 19th and 0200 GMT on the 20th, and in Broadstairs, a violent hail squall broke windows and hurled a boat 30 yards.

On the 20th, activity was especially severe from just west of London across Salisbury Plain, Somerset and the south-west Midlands to south and mid-Wales. A breathtaking lightning display over north Somerset was witnessed by south Bristol residents during the late evening of the 20th; slightly earlier, their area had been affected by a severe overhead storm with hailstones around 30 mm across (Weeks, A., pers. comm.). At Avonmouth, lightning struck a ship blasting the topmast onto the other side of the dock, and in Portishead, a hotel was struck and then flooded. The *Western Mail* published striking photos of cloud-to-ground lightning over Cardiff during the late evening. Two houses were struck and fired in Gloucestershire, at Notgrove in the Cotswolds and Coleford in the Forest of Dean. At Crickhowell near Brecon,

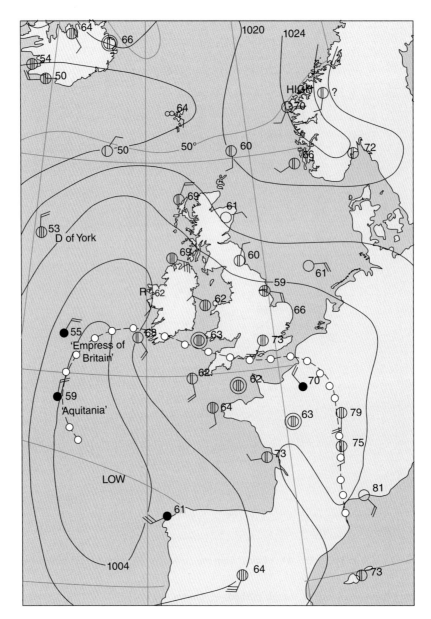

Figure 7.10   *Surface chart for 21 June 1936, 1800 GMT. An occlusion is drifting northwards into southern England preceded by a pre-frontal trough, focussing severe thunderstorm development. Adapted from the Daily Weather Report of the Met Office, UK. © Crown copyright.*

almost continuous thunder rolled from 2015 GMT to after midnight. From 2325, storm activity was overhead with 'flash after flash of blinding intensity which left eyes aching and thunder which shook windows' (Sandeman, 1936). Four trees were struck nearby at Kincoed.

On the 20th, observers at two official Met Office climatological stations witnessed hail of, respectively, 32 mm diameter at Malvern and 38 mm diameter at Horfield, Bristol (*Monthly Weather Report*, June 1936). Hops were severely damaged by hail in the Ledbury area. Egg-sized hail was reported from Chepstow, while it was reported that golf ball-sized hailstones broke every window at Felinfach Post Office near Aberaeron (Ceredigion) where severe flash floods occurred. Indeed, 101 mm of rain was recorded on the 20th at Aberaeron where a 'tidal

wave' swept away trees, part of a wool factory including machinery, a small footbridge, hedges and a huge boulder.[3]

On the 21st, thunderstorms became especially severe and damaging during the evening, just ahead of the occluded front – from the Thames Estuary north-westwards across the Midlands to north Wales. Daily rainfalls on the 21st included 101 mm at Nuneaton (Warks) and 112 mm at St Albans (Herts). In Leicestershire, violent local winds uprooted trees at Tugby, and windows were cracked by hail at Hallaton. The first of two severe thunderstorms in the Chester area was accompanied by 25-mm-diameter hailstones and gusts to 52 kt at Sealand (Flower, 1936). The storms in the St Albans area were the subject of a letter in *The Times*

[3] 101 mm of rain was recorded on the 20th for the 24 h to 0900 on 21st.

from the Met observer at Rothamsted (Harpenden). Here, 80 mm of rain fell, mostly in two cloudbursts during the prolonged thunderstorm from 1500 to 1930 GMT. There was almost continuous thunder and hailstones 20–25 mm diameter which extensively damaged local market gardens. St Albans Abbey was struck by lightning, and over 300 telephone lines attached to the town's exchange were put out of action. Lightning also struck five houses and a factory in the town. Five sections of the local railway were washed away.

7.2.7 The West Country Thunderstorm of 4 August 1938

A strong contender for the title of Britain's greatest thunderstorm of the past 150 years is the colossal storm which struck the southwest of England in the early morning of 4 August 1938, at the height of the holiday season there! Severe storms also affected south Ireland and later extended to many other parts of Wales and England. This account has been compiled from reports of observers affiliated to the TCO, whose activities were taken on by TORRO from 1982 (Meaden, 1984; Mortimore, 1990). These observers recorded thunderstorm times and details as well as

logging incidents of lightning and flood damage. Use has been made of accounts in the *Exeter Express & Echo, Devon and Exeter Gazette* and *Mid-Devon Advertiser*. Reports from *British Rainfall* observers have also been included (*see also* Webb, 2013).

7.2.7.1 General and Synoptic Background

The chart for 0700 GMT on 4 August (Figure 7.11) showed an anticyclone over Scandinavia, and a shallow depression centred over the Celtic Sea. The previous day saw temperatures above 27°C in southern England, with maxima of 30°C on the south coast at Bournemouth and Southsea (where there was shelter from the light north-easterly wind) and over 32°C in north-west France. The 500-hPa reanalysis showed an upper trough extending right down to Iberia, and on the 1200 GMT NCEP chart, this had developed a cut-off 'cold' vortex at 500 hPa south-west of Ireland. The situation corresponded closely to the 'modified' Spanish Plume scenario discussed by Lewis and Gray (2010). Strong ascent ahead of the extending upper trough will have been a major factor in storm development. The surface chart for 1300 GMT (see *Met Mag*, Sept 1938) plots the occluded front from, approximately, Newquay to Start Point to Cherbourg, while

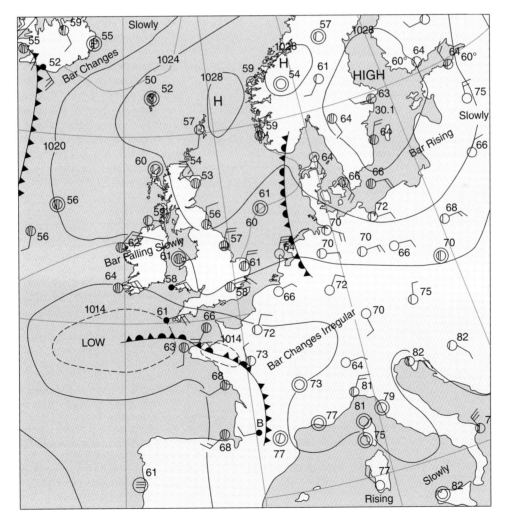

Figure 7.11 *Surface synoptic chart, 4 August 1938, 0700 GMT. Adapted from the Daily Weather Report of the Met Office, UK. © Crown copyright.*

the NOAA/NCEP reanalysis data indicates a belt of high precipitable water (over 24 kg/m^2) extending from Biscay across south-west England, south-west Wales and southern Ireland. This is likely to correspond to the axis of the 'plume'.

Prolonged thunderstorm activity affected south-east Ireland overnight 3rd–4th. At Berehaven, Cork, a severe thunderstorm episode lasted for 12 hours from 1700 GMT on 3rd to 0500 GMT on 4th. There were many lightning-related incidents in Ireland, especially livestock fatalities and haystack fires.

7.2.7.2 Morning Thunderstorms in Devon and Cornwall

Table 7.2 presents a selection of observations relating to the England 'west country' storms. In west Cornwall, the thunderstorms were exceptionally violent in the Penzance/Mounts Bay area. The storm was overhead at 0100 GMT when a lightning strike cut power to the Marazion area for 6 hours. The telephone exchange at Land's End was also knocked out by lightning at 0100 GMT. At Newlyn, around 6000 flashes of sheet lightning were observed as well as 'six flashes like red bars from cloud to sea' (Glasspoole, 1938, quoting C. Harpur). In nearby Gulval and Zennor, marble-sized hail smashed windows. The *Penzance Evening Tidings* reported that a resident of Newmill measured a hailstone 38 mm long. In Gulval, eight houses were damaged by lightning in four separate incidents.

At Torquay, Devon, the borough meteorologist waded through knee deep water to reach the station in Abbey Park. The observer's full account of the storm was quoted in the *Torquay Herald and Express* and summarised in the station's climatological return to the Meteorological Office:

'Distant thunder was heard at midnight. At 3 am (0200 GMT) residents were awakened by a violent rumble of thunder preceded by a brilliant flash, and others followed'. Initially just a few spots

of rain fell. Then 'between 0400 and 0415 GMT a heavy shower with hail occurred, after which the rain almost ceased, but from 0517 the rain greatly increased again in intensity…by 0600 GMT the rain was torrential and continuous with vivid streak, fork and sheet lightning appearing in all directions giving wonderful luminous effects, while the consistent loud crackling of thunder was like repeated gunfire'.

From 0612 to 0800 GMT, 108 mm of rain was recorded with 57 mm falling between 0612 and 0708 GMT and heavy rain continuing to after 0830 GMT (the highest 2 hour fall was 114 mm). Later in the morning, 'the clouds gathered again' with further heavy rain falling from 1000 to 1100 and 11.30 to 1300 GMT, bringing the total rainfall from 5 a.m. to 2 p.m. (0400 to 1300 GMT) up to 6.39 in. (162 mm). During the overhead phase of the storm, two cloud types were noted: 'one coming from the south-east and the other, a lower type of nimbostratus, from the north-east'. The observation is consistent with an elevated storm system but with sufficient lower-level moisture for a low-level 'seeder' cloud. Similar observations were made during a more recent 'plume' event on 26 May 1993 when up to 129 mm of rain fell during a quasi-stationary, albeit narrower, thunderstorm system across central southern England when there was, likewise, an 'undercutting' low-level north-easterly wind (Pike, 1994a). In both cases, the plume of warm air was subject to forced ascent both from the north-east and from the post-frontal air mass to the south-west.

Autographic records (from Dines tilting syphon recording rain gauges) were kept at both Torquay and Paignton. The hyetogram (rainfall chart) for Torquay is reproduced in Webb (2013). At Paignton, 69 mm of rain fell in only 45 minutes from 0556 to 0641 GMT. The Torquay observer also remarked that 'at the height of the storm hailstones as large as small walnuts broke the glass roofs of a number of conservatories'. The meteorological observer at nearby St Marychurch reported hail the size of large sugar lumps lying 10 cm deep like snow.

Table 7.2 *Observations of thunderstorms at selected locations, 4 August 1938 (TCO or BR observers).*

Location	Lat/Long	Periods of Thunder (GMT)	Notes
Penwithick, St Austell (Cornwall)	50:37N 4:78W	0200–0530, 0630–0930	R^2 + a few large hailstones, nearby house struck and severely damaged by lightning
Paignton (Devon)	50:44N 3:56W	0000–1045, 1125–1300	Observer awoke at 0308 h to almost continuous tl 4 miles W, SW, hail at 0610 h when fresh NE wind arose, lightning flashes mainly 'vertical'
Torquay–Abbey Park (Devon)	50:46N 3:54W	0000–1100, 1130–1300	Overhead by 0600 h with TLR^2
Exeter –Heavitree (Devon)	50:72N 3:50W	0404–0830, 0955, 1020–1200, 1225–1335	tl <3 miles by 0436 h, Overhead TL 0504–0532 h, heavy close TL 0819–0830 h, contin overhead TL 1225–1300 h with R^2, 4 incidents of houses struck and damaged <½ mile
Moretonhampstead (Devon)	50:66N 3:76W	0315–1400	Frequent overhead TL 0315–0930, R^2
Copplestone (Devon)	50:81N 3:75W	0230–1400	Overhead 0540 h, 44 mm rain, 3 telegraph poles struck ½ mile north, Morchard Road station telephone box disabled by strike
Ottery St Mary (Devon)	50:75N 3:28W	0330–1030, 1230–1345	Heavy TLR 1230–1345 h
Axminster (Devon)	50:78N 3.00W	0200–1400	To west, all AM, then violent overhead TLRH 1230–1315 h, further distant tl to SE at 2300 h
Minehead (Somerset)	51:21N 3:47W	0524–0822, 0945–1112, 1305–1405	1 mile 0554 h, overhead 1107 h (nearby house struck), <3 miles 1332 h, 2 houses struck and severely damaged (huge ball of red flame observed)

It is evident that some onset times refer to when the observer was first woken by the storm.
H, heavy hail; R, heavy rain; R^2, violent rain; tl, thunder and lightning; TL, heavy thunder and lightning.

In Exeter, thunder had been audible from around 0200 GMT.… 'Towards five o'clock (0400 GMT) people were awakened by somewhat louder rumblings of thunder… to the westward were masses of thunder clouds whose peaks, reaching almost to the overhead, were visible over the dawn streaked sky. As an orange hue suffused the sky to the east, the thunder clouds in the west, now alive with brilliant lightning zig-zagging over the Newton Abbot area, became illuminated with an eerie glow. Between six and seven o'clock (0500–0600 GMT), as the thunder grew louder, low clouds began to move in from the south-east towards the storm centre. These quickly developed themselves into a violent thunderstorm which enveloped the city. Many householders were seriously alarmed, dogs set up their short, sharp barks which tell of fright, and lights appearing in many a window told of the little comfort of civilisation which people call upon in times of stress. Many of those who thunder terrifies hid in cupboards or other dark places. Many people complained of shocks while using telephones at the height of the storm and there were some cases of fainting'. It may be noted that the preceding account in the *Exeter Express and Echo* agrees very well with the notes of the TCO observer in Heavitree, Exeter (Table 7.2).

It was likewise described as a night of terror in mid-Devon. At Moretonhampstead, the police described how the extent of damage was unclear as most people were too frightened to leave their homes. One resident described the storm as like having the Victoria Falls, a howitzer artillery battery and a searchlight tattoo in one's own garden! At Crediton, homes shook as if they would fall apart. In Okehampton, sleep was impossible, and many terrified people dived into the darkest cupboard and stayed there! At Hedgebarton, near Bovey Tracey, 5.86 in. (149 mm) of rain fell between 0415 and 0800 GMT accompanied by continuous lightning, with a total of 6.45 in. (164 mm) by the afternoon. At Widecombe-in-the-Moor (Dartmoor), the thunderstorm raged from 0330 to after 1130 GMT with lightning a 'non-stop searchlight'. The activity was overhead for 3 hours with intense rain and hail. A lane was torn up into channels up to 6 ft deep over a distance of 2 miles.

Another wave of storms affected the region between 1130 and 1400 GMT; this time, the severest activity was rather further east over the Exe, Otter and Axe catchments. At Honiton, the observer recorded that 48.5 mm of rain fell between 1130 and 1300 GMT. Douglas (1938) noted that the individual storm activity moved north–north-westwards but the general movement of the storm zone was to the north-east. The observed cell versus storm motion showed development of new cells on the right flank of the storm system, which is consistent with conceptual model of growth in an organised multicell storm (in an environment of veering winds with height; see Chapter 3).

7.2.7.3   *Lightning and Flood Damage*

There were scores of lightning damage incidents across Devon, many personally witnessed by TCO observers. At Topsham, lightning blew out a gas fire grate, hurling a man across the room, also knocking tiles off his home and the neighbouring house. In Monks Road, Exeter, lightning blasted the whole gable off a house, the explosion filling the street with smoke. In Heavitree, Exeter, lightning blew a large hole in the roof and caused internal plaster to fall off. In Burnthouse Lane, Exeter, lightning demolished a chimney and brought down the ceiling of a house. In a field at Starcross, a stack of harvested oats was set on fire by lightning. In Chudleigh Knighton, a farmhouse was struck and set on fire, while a roof was damaged by another direct hit in Newton Abbot. A telephone kiosk in Moretonhampstead was struck and exploded in fire, and the local TCO observer reported telephones were struck and wrecked in both her own house (where woodwork was set on fire) and a neighbour's where the instrument was blown to pieces. Another TCO observer at Penton (A Montague) reported that 70 telephones (including his own) were severely damaged in country areas around Crediton.

In North Devon, lightning wrecked the chimney of a house at Chawleigh and fired a Dutch barn near South Molton. At Mortehoe, near Woolacombe (Figure 7.12), lightning twice struck the Fortescue Hotel, first setting a wireless set on fire and then hitting a chimney; in the second incident, a guest suffered burns via gas piping. In Ilfracombe, the thunderstorm lasted around 12 hours, and it was reported that many people fainted. Lightning damaged a house and brought down a tree. Extensive power cuts in North Devon caused problems for food traders when refrigerators failed.

A gamekeeper at Tamerton near Plymouth had a lucky escape when lightning snapped a telephone wire which then fell on him, while a chimney was demolished by a strike at Plymstock. There was a courageous rescue by fishermen from Paignton after two men sent a distress call when lightning struck their small yacht. The lightning had torn the sails to ribbons, leaving the vessel tossing around in the turbulent seas of Torbay. In the Dartmouth area, a tree was struck and hurled into the river Dart. On the other side of the Dart estuary, near Dartmouth, terrified cattle broke chains in their stalls and ran amok in a field overturning churns of milk. In Teignmouth, lightning 'could be seen playing on carriage doors of passing trains'.

By the following day, 5th, the Cornwall Mutual Insurance company at Truro had been notified of claims in respect of 96 animals killed by lightning, the most in a single storm in its (then) 60-year history. Claims were made for a total of 157 animals killed by lightning in Devon and Cornwall in the storms of 1 and 4 August combined.

In fact, 65 incidents of significant lightning damage were reported across Devon during a 12-hour period on the 4th with a total of 175 such incidents across England and Wales in the 24-hour period. There were 222 incidents over Britain and Ireland between 1200 GMT on 3rd and 1200 GMT on 5th (and a further 32 on the 5th after 1200 hour). The 65 incidents in Devon on 4 August 1938 included 40 cases with buildings struck (35 of these private dwellings). Of these 40 incidents, 33 (83%) caused significant damage to at least one house. There were nine incidents involving people struck in the Devon storms; seven occurred indoors, a reflection of the early morning activity with (fortunately) relatively few people outdoors. The outdoor incidents affected two soldiers at a gun range at Dalditch camp and a farm worker milking cattle at Bishops Nympton (North Devon). There were a total of 31 incidents of lightning damage in Cornwall.

Overall across England and Wales on 4 August 1938, 18 people were struck in 14 incidents and 125 houses and other buildings were hit by lightning. TORRO statistics for recent years indicate an average of 29 people struck per year over the 25 years from 1988 to 2012 (see chapter 10); however, such comparisons need to be considered in the context of fewer people being exposed to lightning in daily activities nowadays.

Figure 7.12 *Lightning striking the sea west of Woolacombe, North Devon, on 4 August 1938, at 0445 GMT. Richard C.J. Sanders, first published in* British Thunderstorms *1934–1937, Thunderstorm Census Organisation (TCO), Huddersfield 1947 (Bower 1947). © TORRO.*

An area of around 500 km² received over 100 mm rain (Eden, 2008). In Torquay, part of the railway embankment was washed away, and further up the line near Newton Abbot, a shed was carried 100 yards (91 m) by the force of floodwater. Near Kingskerswell station, residents of a row of cottages improvised planks to walk precariously from their bedrooms to the station platform, while the local constable rushed to Torquay to obtain a boat! Cars were stranded everywhere, and Fleet Street, Belgrave Road and Avenue Road in Torquay resembled raging rivers, and in the Upton area of the town, buses were submerged to their rooftops. In some homes, the force of floodwater fractured gas pipes. Many shops in Torquay and Paignton were inundated. The post office in Newton Abbot was flooded up to the sorting benches, while the telephone exchange at Ashburton was likewise swamped and put out of service. A portion of a single track railway was washed away at Tower Hill station. Many bridges were washed away on Dartmoor including a footbridge at Peter Tavy, part of a bridge over the river Dart at Dartmeet and another bridge at Princetown. At Haytor large boulders were displaced by the flood, while at Becky Falls, Manaton, 'huge trees were uprooted by the raging torrents which carried wooden bridges away like matchsticks'. At Bickington, Devon, the river Lemon overflowed; three bridges were swept away as well as a bowls pavilion with its chairs. Buckland Bridge, Ashburton, was swept clean away. In Cornwall, extensive flooding affected premises in Par, Bodmin and Launceston.

The storm zone, ahead of the front, continued to push slowly north-eastwards. Temperatures on the 4th reached 28°C at Southsea and 27°C at Rothamsted (Herts) and Kew (London). The storms continued to be very electrically active with widespread lightning damage across south Wales, Wiltshire, Gloucestershire and the Midlands: storms were especially fierce across Leicestershire with 34 incidents of lightning damage here. The worst incident of the day occurred when a young cyclist was struck and killed by lightning on exposed downland at Marden near Devizes, Wiltshire. Between Dofen and Llanelli (Carmarthenshire), a young woman was struck and badly burned while walking. In Dorset, there were incidents of boats being overturned at both Poole and Swanage; fortunately, the occupants were rescued.

Later in the evening, another area of active thunderstorms moved north from France across east Sussex, Kent and Essex (Douglas, 1938). Indeed, the very thundery period (which began on 1 August 1938) persisted for a further week with the upper-level low and associated cold pool drifting close to southern England. Daytime storms were especially widespread and severe on the 12th with numerous severe hailstorms.

7.2.7.4 Organisation of the Storms in the South-West

While the 1938 event was more than two decades before the advent of satellite imagery to confirm cloud top data, the persistence of overhead thundery activity (i.e. discharges within 3 miles) across a very large area for 6 hours or more is strongly indicative of an MCS, indeed possibly an MCC. Douglas (1938) noted the very unusual occurrence of overhead thunderstorms simultaneously across much of Devon and Cornwall, and this is very evident in the observations of TCO observers. For instance, in the hour to 0600 GMT, overhead thunder and lightning was reported at Brixham, Torquay, Exeter, Moretonhampstead, Copplestone, Lifton (near Launceston), West Down (near Braunton) and Minehead.

It is very likely that three separate, albeit 'overlapping', MCSs crossed the south-west on 4 August 1938, each one a little further east. At 0600 GMT, surface winds were east–north-easterly 11 and 14 kt at Boscombe Down and Bristol respectively. Contemporary upper air analyses on 4th (*Daily Aerological Record*) were from aircraft observations and mostly confined to the lowest 2 km. However, at 1100 GMT, RAF Cranwell, Lincolnshire, did record winds at 6 km (about 500 hPa) as 180°/8 kt. At 1700 GMT, Lympne (Kent) reported 170°/14 kt at 4 km (700 hPa). Although some distance from the main region under discussion, these observations are consistent with the NCEP reanalysis chart at 0000 GMT and with a slow north–north-westward drift of storms. There was therefore significant directional wind shear between the surface and 500 hPa but little speed shear. This moderate shear was consistent with confirmed observed hail up to a maximum of 20–25 mm diameter (H3 on the TORRO H scale). The degree of shear was sufficient to support the development of MCSs (through the merger of multi-cellular thunderstorms).

7.2.8   5 September 1958

An old, rippling front appeared on the 0600 GMT DWR chart, from approximately Bournemouth to Dublin; this was dropped from the 1200 GMT chart. A depression was slow moving well to the west–south-west of the British Isles, while pressure was high over eastern Europe. The general surface wind flow was south-easterly. However, by 1800 GMT, a shallow surface low was evident across the English Channel, ahead of which low-level winds backed east–north-east (Figure 7.13). On the 500-hPa contour chart, a cold vortex was evident west of Ireland (Figure 7.14). Winds veered with height over central south England (ref the Larkhill ascent); they were SSW 25–35 kt at mid-levels and 60–70 kt near the tropopause.

Some thunderstorms had been observed across the North Midlands in the morning, for example, at Watnall, Nottingham. At Loughborough (Leics), the local press reported that, during a short thunderstorm around 1000 GMT, lightning struck a factory building, wrecking a chimney and dislodging masonry; a gas fire was hurled off a wall and struck a lady's toe. Early in the afternoon, an isolated storm developed for a time in the Bedford area (Ludlam and Macklin, 1960). This was probably the same storm

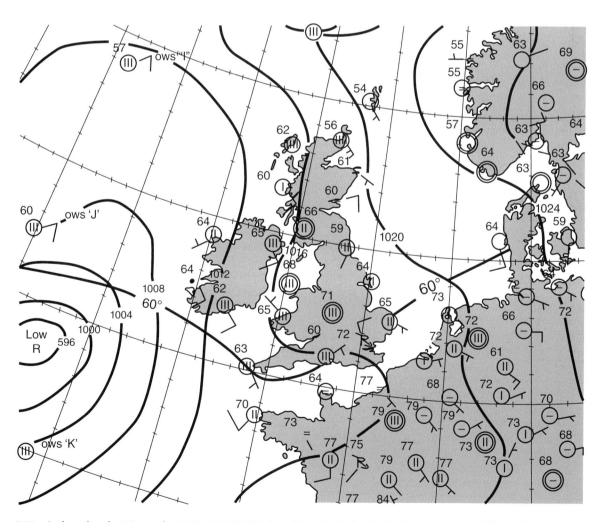

Figure 7.13   *Surface chart for 5 September 1958, 1800 GMT. Adapted from the Daily Weather Report of the Met Office, UK. © Crown copyright.*

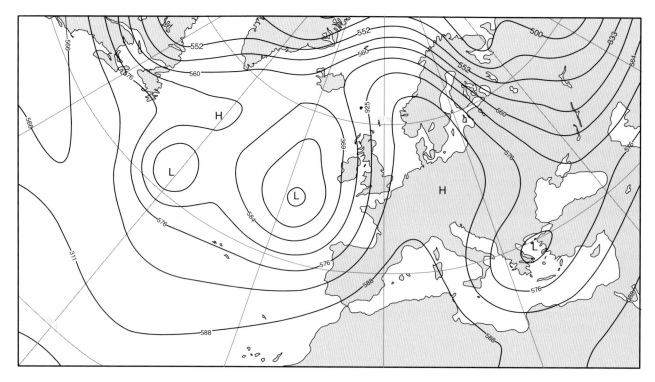

Figure 7.14   *500-hPa chart for 5 September 1958, 1800 GMT. Courtesy of ECWMF/ERA Interim Analysis. © ECMWF.*

system which was observed from north Essex (see below) and which deposited 26 mm in 40 minutes at St Neots (Hunts) from 1450 GMT (Meteorological Office, 1963).

During the late afternoon and evening, a large area of severe thunderstorms developed and moved north-east across London and south-east England. Some extreme rainfalls occurred as the storm area grew, probably into an MCS, as the 'Horsham hailstorm' (see Chapter 9) merged with further severe storm developments in the Brighton area. 131 mm of rain fell in 2 hours at Knockholt, Kent, while at Sidcup 63.5 mm fell in only 20 minutes, still a national record for that duration. The TCO observer at Dartford, Kent, reported the thunderstorms lasting from 1845 to 2215 GMT and overhead from 2000 to 2110 GMT. 70 mm of rain fell in 105 minutes accompanied by hailstones up to 25 mm across.

Lightning counters at Tatsfield and East Hill recorded just over 5000 flashes and 7000 flashes, respectively, while a pen record from Lasham produced an estimated 9575 strokes. Even allowing for some discharges triggering double counts, Horner (1960a, b) estimated that about 2000 discharges occurred in the hour from 1800 to 1900 GMT. Indeed, actual observations confirm this extremely intense electrical activity. A TCO observer in east Dulwich, south London, noted that an east–south-easterly wind at 1330 GMT had backed to north-easterly by 1730 GMT. The first thunder was heard at 1840 GMT, and he remarked that 1900 lightning discharges were observed between 1900 and 2000 GMT with peaks of 36 flashes per minute at 1942 and 1955. A local ground strike damaged telephone lines and fuses in several houses. Elsewhere, lightning struck the chimneys of two houses in Brighton. The Horsham area felt the impact of the electrical aspect of the storm too, lightning blowing three holes in the roof of a grocery shop in Oakhill Road and also hitting a substation at

Petworth resulting in the loss of power to over 7000 homes in Midhurst, Slinfold, Rudgwick, Billingshurst and Wisborough Green. Elsewhere in the south, lightning struck the television aerial of a house in Slough. At Petworth, lightning hit a large chimney cowl on the church roof, smashing tiles.

Further north, the following account of the storm was received from an observer at Rickling, north-west Essex (Banks, J.C., pers. comm. – note all times adjusted to GMT). At 1200 GMT, skies were mainly cloudless if slightly hazy with an ESE wind; however, the tops of three mature cumulonimbus heads were visible distant to the west (probably the Bedford storm noted earlier). At 1300, skies were evidently cloudless but hazy. At 1400, high cirrus and cirrostratus were invading the sky from the south-west with the sun turning watery. There followed cirrocumulus and altocumulus castellanus, the latter developing, typically, very smooth bases beneath the white towers. At 1500, a slightly darker roll of cloud appeared in the west, which was followed quite soon by the first flash of lightning. Quite rapidly, lightning became frequent (up to 20 flashes per minute) with muffled but continuous thunder: 'like emptying a sack of potatoes on a wooden floor'! At 1600, people working on top of a corn stack refused to continue for safety reasons. By 1830 GMT, thunder was continuous with one lightning stroke lasting 5 seconds. One farm worker evidently experienced an electric shock from the starting handle of an Austin 7 car. At 1930, the air breaker on a combine harvester started glowing, as did a field gate latch and several posts in an adjacent meadow. About a minute later, the combine harvester was struck by lightning starting a small fire.

Meanwhile, only a few spots of rain fell before 2000 GMT. At 2000, the first significant rain fell. Around this time, a remarkable lightning flash was observed to the north: 'a flash performed a loop, then blew out with two descending sparks like a distress flare'. Rain became heavy for a while, amounting to 15.5 mm before a veer of the wind to south-westerly heralded a clearance, albeit with spectacular lightning still visible as the storm receded north.

The occurrence of intense lightning activity even away from the precipitation cores is also evident in the observations from the Slough area where many flashes were seen to streak upwards and outwards from the storm anvil, spreading across clear skies with extensive branching (Prichard, 1999). Near Wisbech in Cambridgeshire, the thunderstorm lasted for 4 hours, from 2015 to 0015 GMT on 6th. Lightning struck and killed cattle 5 miles to the north-east and set a corn stack alight 5 miles to the south-east. At Holt, Norfolk, vivid lightning with distant thunder were observed from 1945 GMT with overhead activity between 2025 and 2030 GMT and 'close' discharges continuing until 2057; at 2026 GMT, one lightning flash 'appeared to break up into several parts which fell like sparks through the air' (quite similar to the description from north Essex noted previously). In all, some 1900 lightning flashes were noted. Moderate rain fell from 2016 to 2030 GMT.

British Rainfall 1958 has further accounts of the storms including the tornado and wind damage at Horsham and Gatwick and maps of the distribution of rain and hail.

Overnight, thunderstorms were also reported as far west as Bristol, Birmingham (Elmdon), Shawbury (Shropshire) and Manchester (Ringway). Photographs of both cloud-to-ground lightning strokes (at Mottingham) and upward branching lightning (Mill Hill) were published in the 1958 *Meteorological Magazine* (Anon, 1958). Further discussion of this event can be found in chapters 8.2 and 9.10.11.

7.2.9 22–24 June 1960

This massive thunderstorm episode has been described by Pike and Webb (2004) and was notable for an exceptionally large numbers of thunderstorm hours recorded at individual stations. An upper vortex became slow moving over the Celtic Sea and south-west England (Figure 7.15). A shallow surface depression drifted northwards into the Irish Sea with the associated occluding frontal system becoming complex as a new triple point depression formed over the western English Channel; additional small frontal wave lows developed as the system moved very slowly north-east overnight (see hourly charts drawn by W. S. Pike in Pike and Webb, 2004). There was very strong convergence (forcing) along the front at all levels below 500 hPa. This was a case where thunderstorms very quickly grew in intensity and areal extent around dusk (especially on the 22nd). Radiative cooling of unstable medium-level cloud tops after dark was identified as a possible contributory cause of the explosive destabilisation of the plume although evidence for this factor is 'circumstantial'. As in 1938, the initial and most severe phase of the 1960 event was characterised by individual storm cells moving north–north-west, while the general movement of the storm zone was slowly north-eastwards. The front became quasi-stationary over East Anglia early on the 24th.

Three major storm phases were identified. Thunder was heard continuously for 8 hours in the English western Midlands overnight on the 22nd–23rd and in a total of 14 hours (within a 24-hour period) at Wittering near Peterborough. Very extensive lightning, flood and (more locally) hail damage occurred with

two direct human fatalities and numerous livestock casualties. There were 125 known reports of lightning damage across England. Pike and Webb (2004) noted that higher livestock casualties occurred prior to the widespread use of metal and celcon barns to shelter animals. The *Bournemouth Evening Echo* reported that the traditional instinct of residents to awake, leave their beds and go downstairs during a violent overnight storm proved fortunate for five residents of a thatched cottage at Corfe Castle in Purbeck which was struck by lightning, fired and burnt to a shell. As a variation of the theme of daytime storms turning day into night, the *Leicester Mercury* described the spectacular thunderstorm as 'turning night into day'.

A lightning counter at Douai Abbey School, Woolhampton in Berkshire, recorded more than 7000 discharges in the 5 hours from 0145 to 0645 GMT. Duns Tew manor, Oxfordshire, received 113 mm of rain of which 101 mm fell in 5 hours from 0415 GMT.

7.2.10 8–9 August 1975

The very hot 'high summer' of 1975 was noteworthy for several outstanding thunderstorm events (Webb, 2011b). The most extensive thunderstorm episode occurred on 8 August, the hottest day for many (Figure 7.16), and a full account of developments on this dramatic day has been given by Prichard (1976). Grant (1995) noted that this was the day with the most extensive UK thunderstorm activity during the whole period 1972–1994 inclusive. An old rippling frontal zone lay across the Irish and Celtic Seas, while shallow low pressure developed across northern France and the western English Channel and drifted north. Very hot tropical continental air brought maximum temperatures of up to 34°C in several places in the Midlands and southern England.

Prichard (1976) commented that by late evening, '…the curtain comes down with a mass of thunderstorms stretching from Northern Ireland across Wales and the west country to central southern and southeast England with other areas of activity over Cornwall, Cumbria and southwest Scotland' (see also chapter 8.4).

At Shanklin, Isle of Wight, strong blasts of very warm air were felt as the first lightning storm made landfall just out of audible range around 2000 GMT. Very rapid development of new cumulonimbus cells to the south followed, and by 2100, the sky was ablaze with lightning to seaward. Between 2130 and 2230 GMT, a violent thunderstorm passed overhead with lightning reaching a peak frequency of 45 discharges per minute (counted), continuous thunder for 25 minutes and torrential rain driven by a force 8 squall (Flitton, 1976). Power was cut off, and a cliff shelter became crammed with drenched holidaymakers!

The most prolonged thunderstorm activity of all was across Cornwall, near the slow moving frontal zone. *The Journal of Meteorology* reported thunder at Penryn, from 0755 to 1315 GMT, 1735 to 1856 GMT and 2101 to 2300 GMT, that is, in 11 separate hours. At Constantine, thunderstorms lasted from 0900 to 1400 GMT and 1800 to 2340 GMT (COL, 1975), while at St Mawgan Met Office thunder began at 0725 and lasted until just after 1600 GMT with a further storm around midnight. There was extensive disruption of electricity and telephone services, and three houses were struck and damaged in Truro.

The evening also brought a severe thunderstorm on the Clyde coast and to the west of Glasgow which lasted for nearly 3 hours (Blackshaw, J., pers. comm.).

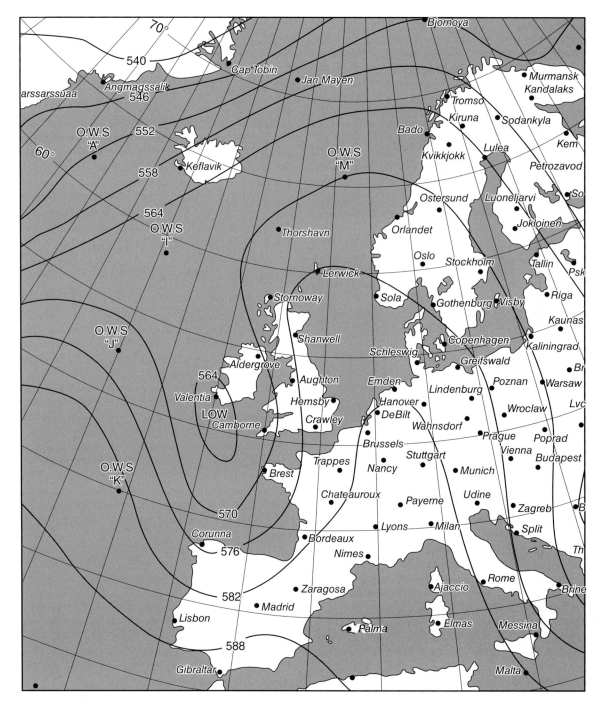

Figure 7.15 *500-hPa analysis 23 June 1960, 0000 GMT. Adapted from the Daily Aerological Record of the Met Office, UK. © Crown copyright.*

7.2.11   13–14 June 1977

Low pressure was moving slowly north-east over northern France, while an associated upper cold vortex was slow moving over western France carrying the thunderstorms west–north-westwards with the associated mid-level winds. Only the far south-east of England had a taste of the very hot air which was over the near continent, Folkestone recording 26°C.

This overnight outbreak was especially violent across south-east England (see also chapter 8.4) although the storms remained active as they moved north-west across Wiltshire and the western Midlands during the morning, reaching as far north as Merseyside and Manchester in the early afternoon of the 14th. Prolonged and brilliant lightning was associated with many incidences of damaging lightning strikes. 63 mm of rain fell in 2 hours at Souldrop (Beds) and 57 mm in 48 minutes at Biggin Hill (Kent). Throughout

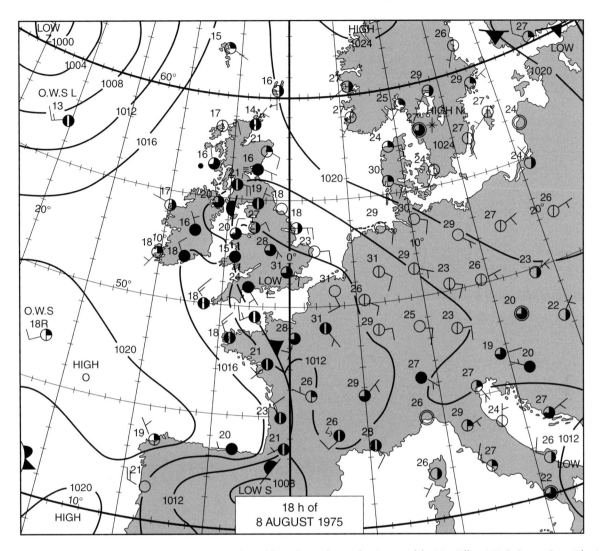

Figure 7.16 *Surface chart, 8 August 1975, 1800 GMT. Adapted from the Daily Weather Report of the Met Office, UK. © Crown Copyright, 1975.*

Kent, London and Essex, almost continuous lightning and window-rattling thunder was reported. In Shenfield, Essex, lightning struck a row of six terraced houses, all of which were evacuated, while two other houses were hit and damaged in Stanford-le-Hope. More than 100 electricity transformers were damaged by lightning in Essex. Lightning cut power to 10,000 homes in Sittingbourne (Kent), and in Rochester, lightning split a chimney. Other casualties of the lightning included a sausage factory which was set on fire in Ipswich and the Met Office computer at Bracknell which was knocked out by a lightning-induced power surge. Eight houses were also struck in the Birmingham area, and three men received electric shocks when a supermarket was hit in Solihull.

7.2.12   25–26 July 1985

This was one of Ireland's most violent and extensive thunderstorm events on record. While many severe thunderstorm episodes across Ireland originate as a north-westward extension of a plume event over south-west Britain (e.g. 3–4 August 1938), this 1985 event was a case of a huge MCS developing and moving

north–north-east over central and eastern Ireland (Figure 7.17). At 1200 GMT on 25 July, a sharp upper trough was located west of Ireland, and this extended down to west of Portugal. The surface analysis is shown in Figure 7.18.

At Kilkenny met station, a series of thunderstorms passed overhead between 1445 GMT on 25th and 0615 GMT on the 26th, all tracking south to north. Frequent strokes to ground were observed, and at 0215 GMT, an adjacent strike resulted in several instruments and items of office equipment being damaged beyond repair with both telephone and printer sockets being blown to bits.

At Mullingar, thunderstorms were observed during 12 separate hours on the 25th, and a house was struck and fired at Knockatee, 4 miles away. At Knockroe, Monaghan, the first period of thunderstorm activity occurred from 1652 to 1904 GMT (overhead just after 1730). A much longer period of thunderstorms began at 2055 GMT and persisted through to 0650 GMT on the 26th (Skeath, H., pers. comm.). There were several periods of intense, even extreme overhead activity (Figure 7.19), for instance from 2245 to 2320 GMT when the buzzing sound of St Elmo's Fire was witnessed shortly before an electricity pole

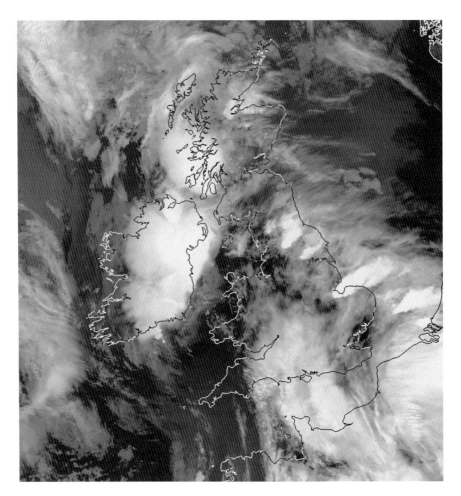

Figure 7.17   *I/R satellite image, 26 July 1985, 0400 GMT. A huge thunderstorm system and anvil shield covers central and eastern Ireland.*
© *University of Dundee.*

400 m away was struck and fired by lightning. At Cookstown, Co Tyrone, thunderstorm activity persisted from 1530 GMT on the 25th right through to 0800 GMT on the 26th. Thunder was reported to be continuous from 2308 to 0030 GMT with 30–40 flashes per minute observed around 2330 GMT. In the Republic of Ireland, 295,000 customers lost power, while in Northern Ireland, 161 transformers were damaged by lightning. Among many other incidents, lightning ignited a gas pipe at a factory in Cork and partly destroyed an arched bridge at Dyfarm in Co Kilkenny. Mortimore (1986) described the extensive lightning and hail damage, highlighting the huge numbers of livestock casualties attributable to lightning. There was very extensive damage to underground telephone cables at Drogheda, Co Louth, after lightning hit overhead lines and travelled to the underground network (D. McEnroe, personal communication).

The storms also affected south-west Scotland late on the 25th, while thundery rain was widespread over south and central Scotland on the 26th. At Elderslie, west of Glasgow, the thunderstorm broke at 2015 GMT with very frequent and close electrical activity from 2045 to 2115 GMT; from 2115 to 2330 GMT, lightning continued to flash almost constantly although the storm receded to the north-west and became distant (J. Blackshaw, personal communication). At Inchlaggan, thunderstorms persisted from 1925 GMT to midnight, and at Loch Goilhead, Argyll, from

1840 GMT to midnight. Many people were rescued from vehicles stranded in floodwaters around Loch Lomond, while a ceiling collapsed when lightning blew apart the roof of a house in Kirkintilloch. In the Edinburgh area, the storms cut power to 35,000 homes on the 26th. At the tail end of this storm system on the 27th, a devastating lightning strike at Grantown-on-Spey (Inverness-shire) blew a 27-m fir tree apart. The discharge also tunnelled under a lawn, blowing out cabling and fuse boxes, before sucking out two windows; turf debris also chipped the window and door of a nearby bungalow.

7.2.13   24 May 1989

On 19 May 1989, an extraordinary local thunderstorm event deposited 193 mm of rain in 2 hours at Walshaw Dean near Halifax in West Yorkshire. This storm, which occurred near a slow moving frontal zone, was the subject of several published papers (e.g. Acreman, 1989) before the reading was officially accepted. An upper trough persisted for several days just west of these islands. Cornwall and southern Ireland experienced violent thunderstorms in the early hours of the 21st, causing extensive electricity supply and telecommunications failures. There were also severe local storms with large hail in southern Ireland on 23rd with power and telephone service outages again, in Co Cork and Co Limerick (Meskill, D - pers comm).

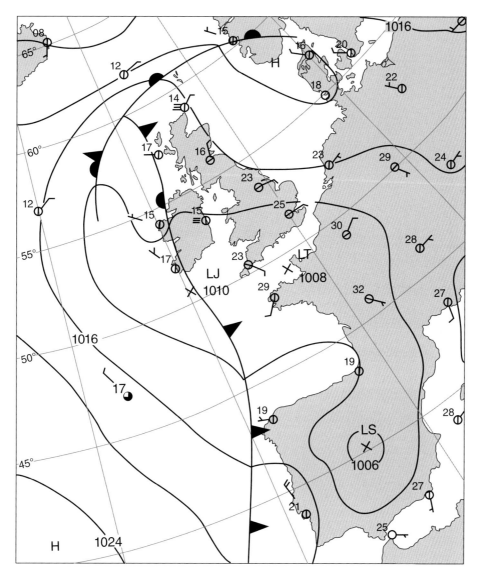

Figure 7.18 *Surface chart 25 July 1985, 1800 GMT. Adapted from the Daily Weather Summary of the Met Office, UK. © Crown copyright).*

England and Wales experienced an exceptionally widespread and severe outbreak of thunderstorms on the 24th. There was a very slack surface easterly airflow with a hint of a trough extending north from low pressure over Iberia. Temperatures peaked at around 27°C.

One area of thunderstorms, a 'repeat' of an outbreak on the 23rd, moved north from France across central southern England, the West Midlands and east Wales from dawn to mid-morning.

Then soon after midday, there was explosive 'home-grown' development of storms to the west and south-west of London (Figure 7.20) with this development later 'rippling' westwards to Dorset and East Devon. The area of storms grew into a large MCS as it moved north into the Midlands. A violent thunderstorm in the Oxford area dumped 68.5 mm of rain in about 2 hours at Sandford on Thames and hail 10–20 mm diameter fell in the east of the city. Elsewhere, a microburst occurred near Farnborough airfield in Hampshire where 60 mm fell in under 2 hours (Waters and Collier, 1995). Hailstones 23 mm diameter fell at Beaufort Park near the (then) Met Office headquarters at Bracknell, while 68.5 mm of

rain also fell at Chineham near Basingstoke. An especially severe hailstorm occurred in the Taunton area of Somerset with hailstones 40–50 mm diameter causing structural damage at Blagdon Hill in the Blackdown Hills. A daily rainfall of 110 mm was recorded at Swallowcliffe near Salisbury.

The *Northamptonshire Chronicle and Echo* published a photo of extreme day darkness and flooded roads in Wellingborough. Two houses were struck and damaged in nearby Finedon as was a chemical works in Wollaston. The intensity of the storm was confirmed by personal observations at Deanshanger, where almost total darkness descended and triggered street lighting, during a storm which was accompanied by 12 close cloud-to-ground lightning strokes per minute at the peak (R. Peverall, personal communication). A lightning strike damaged houses on opposite sides of a street in Newport Pagnell. At Keyworth, Nottinghamshire, the observer also noted that the severe thunderstorm caused day darkness which switched on street lights for 30 minutes – until lightning hit a local substation and all power was cut (Hodgson, J. – *pers. comm*). There were at least 37 incidents of

Figure 7.19 *Lightning over Tydavnet, County Monaghan, Ireland, night of 25–26 July 1985. © Donal McEnroe. (See insert for colour representation of the figure.)*

Figure 7.20 *I/R satellite image, 24 May 1989, 1243 GMT. Note the residual storms (from the overnight MCS) over the north-west Midlands and the fresh thunderstorms exploding to the west and south-west of London. © University of Dundee.*

lightning damage this day, to add to nine in the previous day's storms across the West Midlands. All this occurred in one of the finest Mays of the 20th century!

7.2.14   8–9 August 1992

A full description and analysis of this event has been provided by Pike (1994b). A complex depression was moving northwards into southern England while at 500 hPa there was a southerly airflow ahead of an extended upper trough. This 'plume' event featured a thundersquall line on the leading edge of the (second) main storm complex, accompanied by near continuous lightning; 12–20 discharges per minute were recorded by observers in Oxfordshire and Berkshire with even more extreme electrical activity across the East Midlands. At least 35 buildings (public and private) were struck by lightning. Figure 7.21 shows the second (main), large storm area centred over The Wash.

Storms moving north from France from 2006 GMT onwards produced an astonishing lightning display at Guernsey Airport, with more than 60 visible flashes per minute accompanied by virtually continuous rumbling thunder. Later, at Great Gaddesden, Hertfordshire, lightning intensity also approached 60 discharges per minute from 2330 to 0000 GMT. Other observations by TORRO observers included the following (inc times of thunder):

Mortimer (Berks)	2045–2210 GMT 2230–2330 GMT 0100–0230 GMT	Intensity regularly reached 15–20 discharges per minute
Woodlands St Mary (Berks)	2045–0330 GMT	12 discharges per minute
Crowmarsh (Oxon)	2050–2300 GMT 0110–0130 GMT	Lightning frequency increased to 20 discharges per minute by 2300 with TLR[2]h
Berkhamsted (Herts)	0030–0245 GMT	Overhead with an average of 40 flashes per minute at peak
Stony Stratford, Milton Keynes (north Bucks)	2230–0200 GMT 0200–0300 GMT	Over 30 lightning discharges per minute, including many cloud-to-ground strikes
Fleet (Lincs)	2315–0330 GMT	Overhead with 26–28 flashes per minute
Ruddington (Notts)	0130–0500 GMT	Overhead, by 0300 GMT, lightning frequency reached 30 flashes per minute with the whole sky illuminated from horizon to horizon

Figure 7.21   *I/R satellite image, 9 August 1992, 0349 GMT. Note the two large storm systems, the second covering the North Midlands.* © *University of Dundee.*

7.2.15   24 June 1994

This was one of the few MCS clusters over Britain which approached MCC status. Throughout the afternoon and evening, a cold front edged slowly across south-west England, while very warm air was advected northwards across England and Wales (Figure 8.10); temperatures reached 28°C widely (29.3°C at Heathrow), while at 1800 GMT, Paris reported 32°C. Dew points exceeded 20°C during the afternoon, for example, 22.5°C at Mortimer, Berkshire, where the dry bulb maximum was 29.1°C. Scattered thundery activity affected the Irish Sea and adjacent coastal areas during the afternoon.

Between 1530 and 1630 GMT, vigorous 'home-grown' thunderstorms erupted over the western Midlands. These storms were steered north–north-eastwards in the strengthening 500-hPa wind flow ahead of a deep upper trough which was approaching the British Isles from the west. The storms were fed by the low-level inflow of hot air provided by the moderate south-easterly winds. A severe hailstorm affected an area just west of Wolverhampton between 1600 and 1630 GMT. TORRO's David Reynolds noticed a greenish colour to the sky across the affected area, and he later carried out a site investigation. Golf ball-sized hailstones severely damaged three garden centres near Pattenham, Staffordshire. Greenhouse plants were strewn with broken glass (see photograph by D. J. Reynolds in Webb, 1996a), while parked cars were dented and perspex roofing was peppered with holes. Widespread lightning damage, including numerous livestock casualties, occurred across the northern Midlands and West Yorkshire.

Much further south, an organised cluster of thunderstorms had developed near the north Brittany coast, associated with a sea breeze converging on the surface south-easterly wind (Young, 1995). This system developed into a classic MCS as it drifted north–north-eastwards towards south-east England. Meanwhile, a broad pre-frontal trough of low pressure, associated with a plume of very moist air, was swinging north-east into central southern England. The approaching cold front had already set off some late-afternoon thundery showers in Dorset and Wiltshire. The arrival of the gust front, ahead of the MCS, intensified storm development and expanded the storm system on the leading edge, with a crop of extremely vicious storms crossing Berkshire (Knightley, 1994), east Oxfordshire and north Buckinghamshire between 1800 and 2000 GMT.

Wokingham (Berkshire) was hit by a fierce squall during which hail 10 mm diameter accompanied winds of up to 41 kt. During a ferocious storm at Saunderton Lea, near Princes Risborough (Bucks), torrential rain was accompanied by pea-sized hailstones which were lashed horizontally by a gale-force wind, amidst continuous thunder and lightning. A few miles north-west, a 'freak' storm was reported from Brill where hail the size of large peas covered the ground to a depth of 4 cm. By now, the leading edge of the MCS was crossing south-east England, accompanied by excessive electrical activity and violent squalls. Lightning struck a yacht in Chichester marina. At Brighton, 11 to 12 discharges per minute were observed from 1820 to 1850 GMT, while at Broadstairs 12–30 strikes per minute were noted during 3 hours of thunderstorm activity from 1935 to 2240 GMT. Indeed, nearly all observers in Sussex, Kent, London and Essex described electrical activity as intense or very

intense. Hail 13 mm diameter accompanied winds of an estimated 50 to 80 mph in the Gillingham area; further south, Herstmonceux met station recorded a gust of 54 kt (63 mph). Large hail and violent squalls also affected the Brentwood area of Essex. 52 mm of rain fell in 2 hours at Billericay, Essex. Another observer at Bottisham, Cambridgeshire, witnessed 20–25 lightning discharges per minute from 2055 to 2115 GMT.

Grant (1995) noted that 47% of UK climatological stations reported thunder on 24 June 1994; between 1972 and 1994, the only more 'thundery' day was 8 August 1975 (57%). In the London area, this event was also associated with a huge surge in asthma-related problems reported to medical general practitioners or accident and emergency units; this prompted several research papers in medical journals.

7.3   Concluding Remarks

While the plain statistic 'day with thunder heard' often relates to brief thundery showers, all areas of Britain and Ireland occasionally experience one of nature's greatest spectacles: prolonged, intense and vivid electrical storms. Epic events such as those described earlier can bring, across a large area, all the meteorological hazards associated with thunderstorms, including lightning strikes, flash flooding, severe hail and violent squalls.

Acknowledgements

Special thanks are due to Bob Prichard and Keith Mortimore who have both been very prominent in thunderstorm data collection, associated TORRO research and published reports and articles over many years. Sincere thanks to Mark Beswick, Joan Self and Steve Jebson, at the *National Meteorological Library and Archive* in Exeter, for copies of DWR charts and some climatological returns and daily registers and also to Neil Lonie, *Dundee University*, for satellite images. Sincere gratitude is also expressed to all TORRO colleagues and members for their reports, likewise officials and observers of the TCO (founded by Morris Bower in 1924) which TORRO incorporated in 1982; also many invaluable observations from members of the *Climatological Observers Link* (COL) and, not least, the *British Rainfall Organisation* (founded by George Symons in 1860).

References

Acreman, M. (1989) Extreme rainfall in Calderdale, 19 May 1989. *Weather*, **44**, 438–446.

Anon. (1958) Lightning with upward branching, lightning flashes. *Meteorol. Mag.*, **87**, 369, 379.

Betts, N. L. (2003) Analysis of an anomalously severe thunderstorm system over Northern Ireland. *Atmos. Res.*, **67–68**, 23–34.

Bower, S. M (1947) British Thunderstorms 1934–1937, Thunderstorm Census Organisation (TCO), Huddersfield 1947.

Browning, K. A. and Ludlam, F. H. (1962) Airflow in convective storms. *Q. J. R. Meteorol. Soc.*, **88**, 117–135.

Campbell, S. (1984) Dundee's dark day. *Weather*, **40**, 259–260.

Cinderey, M. (2005) The North Yorkshire-Teesside storm of 10 August 2003. *Weather*, **60**, 60–65.

COL (August 1975) Climatological Observers Link (COL) Bulletin.

Compo, G. P., Whitaker, J. S., Sardesmukh, P. D., Matsui, N., Allan, R. J., Yin, X., Gleason, B. E., Vose, R. S., Rutledge, G., Bessemoulin, P., Bronniman, S., Brunet, M., Crouthamel, R. I.; Grant, A. N., Groisman, P. Y., Jones, P. D., Kruk, M. C., Kruger, A. C., Marshall, G. J., Maugeri, M., Mok, H. Y., Nordli, O., Ross, T. F., Trigo, R. M., Wang, X. L., Woodruff, D. and Worley, S. J. (2011) The twentieth century reanalysis project. *Q. J. R. Meteorol. Soc.*, **137**, 1–28.

Douglas, C. K. M. (September 1938) The thunderstorms of August 1938. *Meteorol. Mag.*, 195–202.

Douglas, C. K. M. and Harding, J. (1946) The thunderstorm of the night of 14–15 July 1945. *Q. J. R. Meteorol. Soc.*, **72**, 323–331.

Eden, P. (2008) *Great British Weather Disasters*. Continuum Books London.

Flitton, S. (1976) Unusual warm air blasts before a violent thunderstorm. *J. Meteorol.*, **1**, 199–200.

Flower, W. D. (1936) Severe thunderstorms in the Chester district and over the Dee Estuary. *Meteorol. Mag.*, **71**, 141–142.

Fowler, W. W. (1913) *Kingham Old and New Studies in a Rural Parish*, B. H. Blackwell Oxford. Available at: http://www.archive.org/details/kinghamoldnewstu00fowluoft (accessed 20 January 2011).

Glasspoole, J. (September1938) The thunderstorm rains of August 1–12, 1938. *Meteorol. Mag.*, 202–204.

Grant, K. (1995) Thunderstorm days and hail days in the United Kingdom 1972–1991. *J. Meteorol.*, **20**, 281–297.

Gray, M. E. B and Marshall, C. (1998) Mesoscale convective systems over the U.K. 1981–1997. *Weather*, **53**, 388–396.

Horner, F. (1960a) The thunderstorm of 5 September 1958 – lightning discharges and atmospherics. *Meteorol. Mag.* **89**, 257–260.

Horner, F (1960b) The design and use of instruments for counting local lightning flashes. *Proc. Inst. Electr. Eng.*, **107**, 321–330.

Knightley, R. P. (1994) Thunderstorm of 24 June 1994 at Reading. *J. Meteorol.*, **19**, 362–363.

Lewis, M. W. and Gray, S. L. (2010) Categorisation of synoptic environments associated with mesoscale convective systems over the U.K. *Atmos. Res.* **97**, 194–213.

Ludlam, F. H. and Macklin, W. C. (1960) The Horsham hailstorm of 5 September 1958. *Meteorol. Mag.*, **89**, 245–251.

McConnell, D. (2000) The Carrbridge cloudburst of 1923. *Weather*, **55**, 407–415.

Maddox, R. A. (1980) Mesoscale-convective complex. *Bull. Am. Meteorol. Soc.*, **61**, 1374–1387.

Meaden, G. T. (1984) The early years of the thunderstorm census organisation. *J. Meteorol.*, **9**, 310–313.

Meteorological Office (1927) British Rainfall 1926. HMSO, London, pp. 51–52 and 62–66.

Meteorological Office (1963) British Rainfall 1958. HMSO, London, Part 3, pp. 15–21.

Mill, H. R. (1911) *British Rainfall 1910*. Edward Stanford, London, Part 2, pp. 116–120.

Mortimore, K. O. (1975) Thunderstorm over Bristol. *Weather*, **30**, 379–380.

Mortimore, K. O. (1986) The severe Irish thunderstorms of 25/26 July 1985. *J. Meteorol.*, **11**, 299–306.

Mortimore, K. O. (1990) Thunderstorm climatological research in Great Britain and Ireland: a progress report and aims for future study. *Weather*, **45**, 21–27.

Pike, W. S. (1994a) The remarkable early morning thunderstorms and flash flooding in Central Southern England on 26 May 1993. *J. Meteorol.*, **19**, 43–64.

Pike, W. S. (1994b) Fifth TORRO Conference: Tornadoes and Storms 3. Part 1. The thunderstorms of 8–9 August 1992. *J. Meteorol.*, **19**, 304–321.

Pike, W. S. (1999) The Thunder-squall line of 29 May 1999. *J. Meteorol.*, **24**, 237–246.

Pike, W.S. & Webb, J. D. C. (2004) On the classic overnight thunderstorms of 22–23 and 23–24 June 1960 in the British Isles. *J. Meteorol.*, **29**, 277–291.

Pook, C. A. (1974) Heavy thunderstorms over Bristol, 15/16 June 1974. *Weather*, **29**, 390–392.

Prichard, R. J. (1976) The thunderstorms of 8 August 1975 *J. Meteorol.*, **1**, 160–162.

Prichard, R. J. (1977) London's violent thunderstorm of 9–10 July 1923 – was it the worst this century? *J. Meteorol.*, **2**, 236–238.

Prichard, R. J. (1985). The spatial and temporal distribution of thunderstorms. *J. Meteorol.*, **10**, 227–230.

Prichard, R. J. (1987). The incidence of widespread thunderstorms in England and Wales: 1946–1985. *J. Meteorol.*, **12**, 83–86.

Prichard, R. J. (1993) Thundering through the years. *J. Meteorol.*, **18**, 126–127.

Prichard, R. J. (1999) Seventh TORRO conference: Tornadoes and storms 5. Part 2. London's top ten thunderstorms in the twentieth century. *J. Meteorol.*, **24**, 354–358.

Rakov, V. A. (2003) A review of positive and bipolar lightning discharges. *Bull. Am. Meteorol. Soc.*, **84**, 767–776 Available at: http://dx.doi.org/ 10.1175/ BAMS-84-6-767 (accessed on 10 May 2015).

Sandeman, R. G. (1936) Unusually severe thunderstorms in Breconshire. *Meteorol. Mag.*, **71**, 143.

Symons, G. J. (August 1879) The thunder and hail storm of August 2nd–3rd. *Symons Meteorolo. Mag.*, **97–113**, 125–128.

Tyssen-Gee, R. (1975) Hampstead deluge of 14 August 1975 [+ Editor's note]. *J. Meteorol.*, **1**, 6.

Waters, A. J. and Collier, C. G. (1995) The Farnborough storm – evidence of a microburst. *Meteorol. Appl.*, **2**, 221–230.

Webb, J. D. C. (1996a) TORRO hailstorm division, annual summary: damaging hail in Great Britain and Ireland 1994. *J. Meteorol.*, **21**, 89–103.

Webb, J. D. C. (1996b) The hailstorms of 1 August 1846 in central and eastern England. *Weather*, **51**, 413–419.

Webb, J. D. C. (2011a) Violent thunderstorms in the Thames Valley and south Midlands in early June 1910. *Weather*, **66**, 153–155.

Webb, J. D. C. (2011b) The great summer heatwaves of 1975 and 1976 in the U.K., and some violent storms. *Int. J. Meteorol.*,**36**, 255–261.

Webb, J. D. C. (2013) The phenomenal West Country Thunderstorm of 4 August 1938. *Int. J. Meteorol.*, **38**, 206–218.

Webb, J. D. C and Pike, W. S (2007) Thunderstorm squall associated with a mesoscale convective system, 22 July 2006. *Weather*, **62**, 270–275.

Williams, B. A. (2000) Fifty years of thunderstorms in north east Bristol 1950–99, *J. Meteorol.*, **25**, 161–165.

Young, M. V. (1995) Severe thunderstorms over southeast England on 24 June 1994: a forecasting perspective. *Weather*, **50**, 250–256.

8

Thunderstorm Observing in the United Kingdom: A Personal Diary of Days with Thunder 1953–2013

Bob Prichard

Claygate, Surrey, UK.

8.1 Introduction

It is often one of life's mysteries how and why we follow particular interests. I have a copy of the *Observer's Book of Weather*, inscribed with an aunt's note wishing me a happy seventh birthday in January 1955. So my interest in the subject, unique in my family, was well established by that date. In fact, there are weather events in 1953 that I still have memories of, but I do not know whether they started my interest in the subject or fed what was a very early hobby – perhaps a mixture of both. If a friend of my father had not called on us at our Leigh-on-Sea (Essex) home on the Sunday afternoon of 1 February 1953, the date of the catastrophic North Sea flood, and told of what he had seen from the road bridge at Benfleet looking across a flooded Canvey Island, would my hobby of, and later career in, the weather have taken off?

That first book on the weather fed my interest, but in fact, it contains many errors (and I do not just mean that it is now out of date). I do not remember the Lynmouth flood disaster of August 1952, but there is a plate in the book showing what purports to be a colour painting of that storm cloud from a distance. That made quite an impression on me – and on others, for at least until recently it adorned part of a wall of one of the rooms of the headquarters of the Royal Meteorological Society at Reading. But whatever storm it claims to show, it is not the Lynmouth one, which was not an isolated storm surrounded by clear, blue sky, as depicted in the book, but was embedded within a large rain area around a depression in the western English Channel. It was many years later before I realised that this depiction of one of British meteorology's most dramatic events was false, reinforcing the maxim that you should not trust everything you read – including much that has been written about thunderstorms and other severe weather events.

I do not know where I picked up the falsity that thunder is caused by two clouds 'colliding', but I think it was not that uncommon a heresy when I was young – and it was reinforced for me in an odd manner one evening when I was eight or nine. It was a warm summer's evening and I was lying on my bed with the curtains open. I watched as two clouds came closer to each other, and as they merged, there was a clap of thunder! I even invented a game where I moved old rugs around our garden lawn and went 'bang' as they 'hit' each other. I enjoyed that game until, one day, I noticed a neighbour watching me from her upstairs room.

8.2 Early Observations

My earliest memory of a thunderstorm, perhaps also a formative one, I have dated, after some research using NOAA's excellent reanalysis charts, is that of 22 August 1953 (Figure 8.1).[1] I recall watching the annual Southend carnival in Leigh Broadway when a thunderstorm passed overhead. I was whisked out of the deluge to sit on a (clean!) fishmonger's slab! It took me over 50 years to track down the date, but the chart for this day shows a cool, showery westerly which would fit the bill.

During the 1950s, I noted some thunderstorms I observed and kept patchy weather records. I started a full study of the weather in 1960 and have a complete record from wherever I was living from April that year. That record includes an account of every storm I have observed (or that occurred at wherever was my home at the time when I was away for whatever reason), and there are now over a thousand in my files. Virtually all the storms were observed at various locations in England.

[1] Further charts are available via www.wetterzentrale.de/topkarten/fsreaeur.html

---

Extreme Weather: Forty Years of the Tornado and Storm Research Organisation (TORRO), First Edition. Edited by Robert K. Doe.
© 2016 John Wiley & Sons, Ltd. Published 2016 by John Wiley & Sons, Ltd.

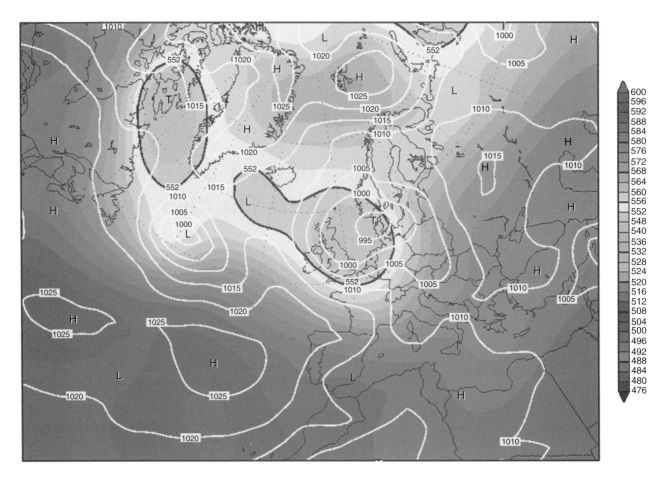

Figure 8.1 *Surface isobaric chart and 500 hPa chart for 0000 UTC on 22 August 1953. Courtesy of NOAA/NCEP reanalysis project.*

At some stage fairly early in the record, I devised my own seven-point scale of severity ('1' the most severe) which I have used, unchanged, ever since. There is always the likelihood of a 'drift' in assigning an imprecise scale over all these years, but I do look back over my record of earlier storms from time to time, and of those I recall the scale seems to work consistently. If anything, by my current 'standard', I was a little harsh in my assessments in earlier years, and I have upgraded by one scale number some events. I find it interesting that I can remember quite well many of the storms in the 1960s – but not so many from the 1980s to 1990s. I do not think that this necessarily reflects any decline in events or interest, but rather that life gets busier as you get older and some hobbies take a lower priority to other activities. My scale is certainly coarse and imprecise, but I have never come across a perfect scale. The standard classification that meteorological observers use tends to bring rainfall intensity into consideration, though it is meant to refer to frequency of lightning. The Beaufort letters 'TLR' that are featured in publications like the Met Office's *Daily Weather Report* and the code ('97' as it used to be) for a heavy thunderstorm could often actually mean a thunderstorm with heavy rain – not the same thing at all. But the definition based on frequency of lightning is not really satisfactory either: I would always rate a storm with a few fierce, close flashes as more severe than one with many undramatic mostly cloud-to-cloud flashes. Indeed, one of my relatively few

'1'-rated storms, on 3 April 1963 (discussed later), gained this because of one damagingly dramatic flash out of just two flashes in the storm. Broadly, I suppose one could say my scale is linked to the impact of the storm (but only its electrical impact). The purist may criticise it, but I repeat that I do not believe any scale is perfect. This account details most of the storms to which I gave my highest ratings (mostly 1 or 2: Table 8.1).

8.2.1   Thunderstorm Observing

Thunderstorm observing does bring many frustrations. You can go for months without seeing one and then have an engagement which prevents you watching the next one that comes along: I still try not to book summer activities (like attending concerts) that may mean I will miss a storm – though in recent summers there has not been much to miss. One frustration has eased, though. In my childhood and teen years, I would often hear sferics[2] on the radio or 'atmospherics' as they were then called ... where did that name come from? As they got louder, I would look forward to a storm – but nothing happened. Nowadays, sferics displays on the internet at least mean you do not waste time and 'nervous' energy waiting for something that may never happen. In south-eastern

---

[2] A sferic is a broadband electromagnetic impulse that occurs as a result of a natural atmospheric lightning discharge.

Table 8.1 *The Prichard thunderstorm severity scale.*

Scale Number	Description
7	Only noticed by fluke or diligence
6	Noticed by most in a position to do so, but with little comment
5	The first scale number for a definite, if weak, storm
4	Generally noticed and remarked upon, but pretty 'run of the mill'
3	'That was a good storm'
2	'Wow'!
1	'Almost unbelievable'

England, sferics are frequently picked up on radios from storms on the near European continent that are not coming our way. This may happen on many a summer day – it certainly seemed to in my young days. My preferred listening frequency is at the extreme top end of the medium wave scale, about 550 m – a silent frequency where you soon got used to the likely distance of any lightning strokes you might pick up. It is now my only use of a radio set and is useful for the more isolated or developmental storms that might occur before the time-delayed sferics appear on the internet maps.

From my earliest records until 1971, my home was at Leigh-on-Sea, near Southend, in Essex. There were several notable thunderstorms during that period. My recording of these was patchy during the 1950s, but there were two in mid-July 1955 that I still recall. The first was on the 14th, a hot, sultry day on which there were six deaths through lightning in England, including that of a soldier in a tent at Bisley (Surrey) and two people leaning on a metal fence at the Royal Ascot race meeting (47 racegoers were reported as suffering shock or injury from this event). I gave this day's storm at Leigh a '4', as I have no evidence it was especially severe, but I do recall seeing lightning from it. Three days later, early on the Sunday evening of 17 July, I saw a thundercloud on the south-eastern horizon and noted that a thunderstorm was 'raging' by 10.30 p.m. – but the cloud only covered the southern half of the sky and no rain fell for some time. Thunder was still rumbling at 6 a.m. next morning. I also noted that the storm was not forecast (how often have I noted that over the years – and still do). The *Daily Weather Report* stations of Guernsey, Felixstowe and Watnall were the only ones to report thunder with rain, but lightning was widely reported as far north as Manchester. A boy was killed by lightning in a tent at Ramsgate (Kent), where nearly 90 mm of rain fell. The synoptic pattern showed high pressure over northern Britain and low pressure over France.

I have no records of storms from 1956 to 1957, but during the summer of 1958, no fewer than three merited scale readings of 1 or 2 (this compares with none at all throughout the 4 years 2010–2013). The classic one was that of 5 September 1958, which was probably the most severe I have ever experienced (although it has a rival in the one of 4 June 1982 at Loughton). There had been thunder on 7 days in the previous fortnight, including a lively storm on the morning of 28 August (that day's storms were expertly studied by Pedgley (1962): reading about this one now, and with a memory of it, perhaps I might have rated it '2' rather than the '3' I gave it at the time). There was a warm and very humid southerly airflow over the country on 5 September (Figure 8.2). I recall seeing the first

flashes to the south from halfway up a tree in our garden at 7.20 p.m. By 8 p.m., the sky was almost continuously alight, and soon there was a barrage of thunder. My grandfather reported that dozens of flashes of lightning were occurring simultaneously all over the sky for about an hour. I regret to state that this storm seriously frightened me, and I stayed under the bedclothes! Not my finest hour! My father was trapped by floodwater on his drive home from London and did not get home until well after midnight – and of course, there was no means of communication from cars in those days. Parts of Surrey and Sussex were even more severely hit, with reports of tornadoes and giant hail, while nearly 130 mm of rain fell in about 2 hours at Knockholt (Kent). I 'only' recorded 24 mm at Leigh-on-Sea.

8.3 Thunderstorms of the 1960s

The next severe storm I recorded was on 23 June 1960 (Pike and Webb, 2004). Two further storms received a scale mark of 2 that summer, the most interesting of the three occurring on the afternoon of Friday 15 July, during a maths lesson at my school, Westcliff High School for Boys. A storm with intermittent thunder and lightning approached slowly from the west from 3 p.m., and at about 3.30 p.m. there was a tremendous cloud-to-ground stroke followed by a deafening clap of thunder about 4 seconds later. Under its centre, there was a report of lightning seeming to dart across a room, but no damage was reported. There were several more hefty cracks of thunder over the next few minutes in what I noted as an unusually severe storm for the area in a westerly airflow. Later that summer, thunder was heard on 9 days in August at Leigh-on-Sea, as well as 3 in September.

An oddity was that a 2-rated thunderstorm occurred on 4 May in both 1961 and 1962, but it is to 3 April 1963, already alluded to, that I turn for the next noteworthy event, again observed from Westcliff High School for Boys. My write-up of this described it as 'one of the most spectacular and unusual storms in the period covered' (up to 1968 at that point) – albeit that there were only two flashes and bangs. The surface weather chart looks innocent (Figure 8.3), with a slack northerly airflow over eastern England, but an upper trough was digging southwards across the area. The morning's forecast, though, was for a rather warm day with sunny periods as an anticyclone was expected to push north-eastwards across the country. It failed to do this, and showers developed over eastern Britain; some were accompanied by thunder. In the Leigh area, clouds built steadily during the day, and a very dark cloud came over from 3 p.m. Rain began around 3.30 p.m., and from 4 to 4.45 p.m. there was much heavy hail, which settled to a depth of 2–3 cm on grass. Black cloud kept rolling across the sky from the north and east, but the north-western and extreme southern portions of the sky remained brighter for over an hour. Then, around 4.30 p.m., came a tremendous flash of lightning and clap of thunder. A geography teacher at the school reported the lightning ran down a wire in his classroom and was accompanied by a cloud of smoke: he was advised by a sceptical head of science 'not to have so many sugar lumps in his tea'! The smoke was surely not an illusion, though! What was without doubt was that the same flash (or a branch of it) tore holes in the roofs of two semi-detached houses half a mile to the west of the school. Ten minutes later, there was a further flash about a mile to the south.

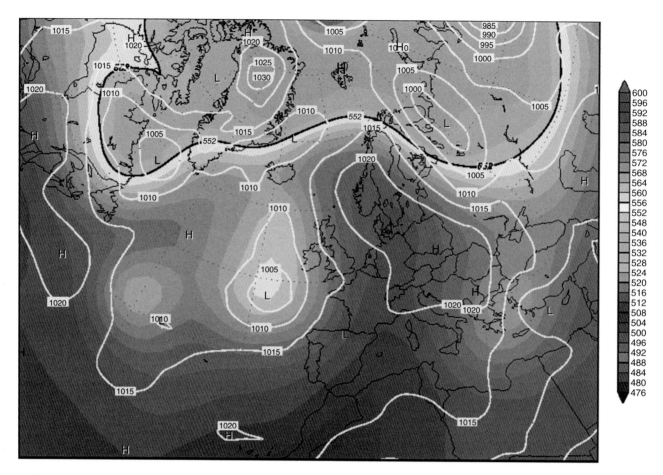

Figure 8.2   *Surface isobaric chart and 500 hPa chart for 0000 UTC on 6 September 1958. Courtesy of NOAA/NCEP reanalysis project.*

Rain, torrential at times, continued until 6 p.m. and intermittently for the rest of the evening. The rain and hail were accompanied by considerable particles of soot. 19 mm of rain was recorded at the Southend weather station, about 3 miles to the south-east of the school; I only recorded 11 mm at my home, not far from where the house was struck, but speculated that much of the hail may have bounced out of my plastic-funnelled rain gauge.

This storm does illustrate the 'shock factor' of a very close discharge, especially when, as in this case, it is the first flash of a storm. Quite a few reports of 'balls of lightning' and other unusual effects are doubtless 'only' extremely close, damaging flashes. Most of us do not see many of these; I have probably seen fewer than ten strokes at such close quarters in my fifty years of observing. So they are liable to be written up in flowery prose, as perhaps I did here. But, as I noted earlier, I would still rate a one- or two-flash storm of close, damaging lightning as more severe than a storm with mostly cloud-to-cloud flashes occurring several times a minute: I gave this one a 1 rating. A couple of 2-rated storms followed that year, on 13 June and 17 August.

Four storms merited a 2 rating in 1964, the last two of these on 18 July – though the events of the previous day are perhaps of more interest. The chart on that Friday morning showed an anticyclone centred over Norfolk, with a 'respectable' central pressure of 1025 hPa. Despite its presence, there were a few thunderstorms across the south-east, including Leigh, that morning.

I noted that the day had started sunny, and although there was a darkening of the sky, there was no recognisable advance of a thundercloud. It was a strange day: the first brief storm, with heavy rain, occurred around 10 a.m., and then much of the day was sunny, but three further rumbles of thunder were heard at well-spaced intervals in the late morning and early afternoon. Lightning damaged houses in north and east London. A thundery low was in evidence by the afternoon of the 18th, and that day brought us heavy thunderstorms around 9 a.m. and 7 p.m.

I left school in July 1965 and then worked for a year as a 'scientific assistant' at Kew Observatory in south-west London – observing the weather and tabulating the records. This was, in effect, my 'gap year' (I don't think the term was yet in use), recommended by my headmaster, and what a valuable year it proved, introducing me to the skills of weather observing, codes, etc. in a small office staffed largely by others as keen as I was (which in my later forecasting career I was to find was rather a rarity). It was marked by a few 'good' thunderstorms, though the most severe was observed from Leigh when I was off duty, on the night of 10/11 June 1966. After a hot, sunny afternoon, a large area of thunderstorms moved north in association with a depression over Biscay. I first saw lightning at 9.46 p.m., and this soon became very frequent as the storm system approached into a previously clear sky. Several storms passed overhead or nearby during the next five and a half hours (although there was only 6 mm of rain).

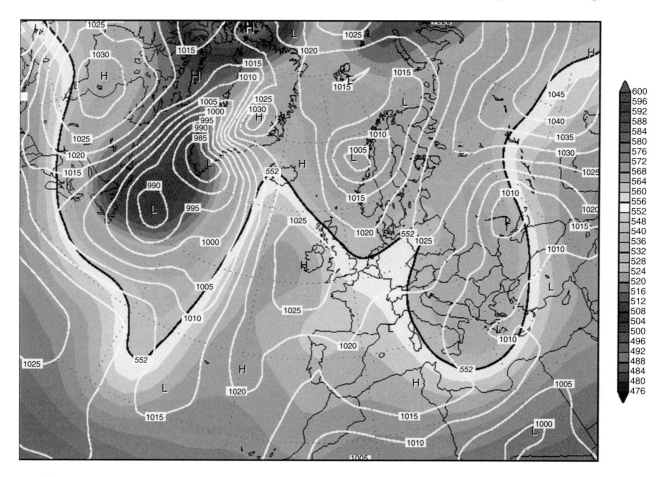

Figure 8.3 *Surface isobaric chart and 500 hPa chart for 0000 UTC on 3 April 1963. Courtesy of NOAA/NCEP reanalysis project.*

At 2 a.m., I noted 'three deafening cracks a second or so apart that sounded like hundreds of fireworks bouncing from rooftop to rooftop'; this followed several big bangs that got progressively closer, but after it, there were only distant rumbles, though also frequent lightning for another hour or so. This merited a 1 rating, and I recorded three 2-rated storms during the rest of that summer, including the last ones I observed during my time at Kew on the afternoon of 30 August.

I had an 'interesting' afternoon on Saturday 30 July, the day England won the World Cup. I was on duty on my own (the usual weekend pattern) but wanted to watch the match. Kew had a live-in caretaker and his wife, with whom I was on good terms, and I asked if I could possibly watch the match with them. Unfortunately, they had visitors and turned down my request. When I arrived for my shift at 1 p.m., the off-going observer offered to take me to watch it at his house in north London and then bring me back. I accepted his kind offer. It was a showery day, and there were heavy showers at his location, and at Wembley, but no thunder was apparent. When I got back to Kew, I had to 'hindcast' the 4 p.m. observation and, using the rain gauge chart and thermograph, was quite pleased with my efforts as I phoned it in to London Weather Centre with the 7 p.m. report. A little later, I saw the caretaker. 'Good thunderstorm we had this afternoon', he reported! So I had to send a correction: I don't think the supervisor there had much difficulty working out what had happened.

There were no repercussions, though I understand my boss did find out. It may sound irresponsible, but we were not an 'important' site: our nearest airfield, Heathrow, was only 5 miles away and had its own observers. Kew finally closed in 1980, its historicity finally succumbing to some rather obvious economic facts. Incidentally, the picture on my colleague's television failed 15 minutes from the end of the World Cup Final, so I never saw the winning goals!

Between October 1966 and June 1970, I was a student at Keele University (north Staffordshire) and enjoyed plenty of lively storms, with quite a few during the holidays at Leigh-on-Sea. Keele is particularly prone to 'Cheshire Gap' showers, some thundery – those that owe their presence and often quite surprising intensity to an airflow that has had a very long sea track before being squeezed through the North Channel and then encountering further convergence as they come off Liverpool Bay and approach the Peak District. But it also had its fair share of land-based storms. In May 1967, thunder was heard there on 10 days, an equal personal 'best' for me. At least a couple of those were distant rumbles that I noted and reported to a previously unaware observer. The Keele met station was run then by Professor Stanley Beaver, Head of Geography, who had set it up when the university started in 1950. Geography was not one of my subjects, but I soon developed a very good rapport with the 'Prof'. My first 2-rated storm at Keele came on the evening of 11 May 1967 and was followed next day by one reported to me from Leigh-on-Sea. In the summer, for the second consecutive year, there was a fierce lengthy night-time storm at Leigh, on 22/23 July (I think we regarded them as quite normal summer events then, but they have scarcely been seen in recent years).

That summer, I took a temporary clerical job in an office that was no more than 200 yards east of London Weather Centre in High Holborn (central London). On the afternoon of 10 August, a severe storm hit London's East End from about 3.30 to 5 p.m. From inside the office I was in, I heard several rumbles of thunder and saw one flash of lightning, and there was heavy rain for about 20 minutes. Imagine my surprise when, checking through all the old observation books when I later worked at London Weather Centre, I noticed that no thunder had been reported from there that afternoon! At that time, their office was in a basement with no windows (clever that!), but they did do hourly observations from the roof and also had a weather shop fronting onto High Holborn. This is, therefore, a useful point for a short diversion from my chronology.

8.3.1   The 'Days with Thunder Heard' Statistic

This is probably one of the least useful of all met statistics: making comparisons between different parts of the country is fraught with problems. If 24-hour reporting met stations can miss nearby storms (and there were two even more blatant examples during my time at Manchester's Ringway airport in the early 1970s), what chance has the enthusiast with work, family and hobbies to balance? Night-time thunder is an obvious difficulty, and noise is another problem. How often does a plane go over, car or lorry pass, someone in the house call out to you, as you strain to hear a distant rumble having picked up a loud sferic on the radio? Nevertheless, the keener amateur observers probably give us the best records: it is the 'professionals' who are prone to let us down. The observer at Southend met station told me in the early 1960s that he only reported thunder when it was 'nearby' and ignored distant rumbles. TORRO has introduced a definition of 'nearby thunder' (within about 5 km), but this is obviously rather arbitrary and only likely to be of any use to the keener observer. In recent years, metars (weather observations) at UK civil airfields (where most observers are now air traffic staff, not met personnel) have used 'VCTS', a vague term that may be of use to aircrew but is a further complication to accurate statistics since experience shows that it does not necessarily mean thunder has been heard at the site (although it should do).

The '5-km' definition brings to mind a related issue. Most of us grew up with the 'knowledge' that we could work out how far away a storm is by counting the seconds between flash and rumble. This is useful for distant storms but becomes less accurate the nearer they come. Five seconds represents 1 mile (3 seconds: 1 km) is the rule – but the flash may be a mile directly above you in the cloud, not a mile distant down the road.

Furthermore, as we seek to find which parts of the country are the most (or least) thundery, what are we measuring? Nearby hills may generate a lot of distant storms, giving you several days of thunder but few of any consequence (south Lancashire comes to mind), while some other areas may get thunder quite infrequently but are prone to storms of frightening severity when they do come (south Dorset, perhaps).

8.3.2   Back to the 1960s

1968 remains the most eventful year of weather observing I have experienced (Prichard, 1992), not least – but far from the only reason – because it had several notable thunderstorms. At Leigh-on-Sea, two consecutive April evenings (17th and 18th) brought lively thunderstorms northwards from France in association with waves on a cold front. So frequent was the thunder and lightning that I am surprised that I managed to note so much detail (at least two flashes a minute) while still watching the storm.

The summer of 1968 brought three exceptional, widespread thunderstorm events. The first, at the beginning of July (Figure 8.4), affected western and northern regions, and I caught the early stages of this prolonged outbreak before leaving Keele for Leigh in the early afternoon of 1 July. The late morning storms were violent enough, but those of the next morning were clearly even more severe, with torrential rain, heavy hail and a gust of 56 kn. The next severe outbreak, on 10 July, affected much of southern Britain, especially the West Country, but was quite weak at Leigh. It was an interesting example of an unforecast development (of which there have been many, seemingly especially some of the most severe, over the years). Pressure was high over the British Isles on the 9th, while very hot air was moving north into southern Europe. In the evening, there was a sudden fall of pressure over north-west France. While sitting in the garden in the balmy warmth, I noticed a rapid development of cirrus which then advanced rapidly across the sky in a north–north-easterly direction. By late evening, there were numerous sferics on the radio. We had just a little thunder soon after daybreak but, more impressively, much lightning (but no thunder) on the next night: my notes say that 'some of the lightning was purple'.

The events of mid-September 1968 have been widely chronicled (see, e.g. Jackson, 1977). Over the weekend of 14/15 September, I recorded over 100 mm of rain at Leigh, and there were lengthy thunderstorms from the Saturday afternoon through to late on the Sunday morning. A depression lay just to the south of England, and the seasonal warmth of the southern North Sea was a key factor in the night-time events in the north-easterly airflow. At 10 p.m., there were isolated sferics. By 10.15 p.m., there were one or two loud ones – and 5 minutes later came the first flash of lightning, with the first thunder at 10.27 p.m.; a lively storm soon followed. There was extensive flooding in the locality (and across much of south Essex, south London and north Surrey). It is of interest, and follows a perceived trend of reduced thundery activity in recent years (which we might hope will have been reversed by the time this book comes out), that on 24 August 2013 the Southend area had a deluge that can be compared with this one – but without thunder.

There were 7 days of thunder at Leigh in September 1968, with further severe activity on the afternoon of 23 September in a showery westerly. Back at Keele, the next day of thunder was not until 24 April 1969 – a rumble in a shower about an hour after I had returned for the summer term. Another lively May followed, culminating in severe storms on the 29th and 30th; I wrote up the latter one in an article for *Weather* (Prichard, 1970). It was a storm that developed over south Cheshire in a slack airflow in mid-afternoon and expanded, rather than moved, into part of north Staffordshire, including Keele, where there was heavy rain and hail, and overhead thunder and lightning, for about an hour.

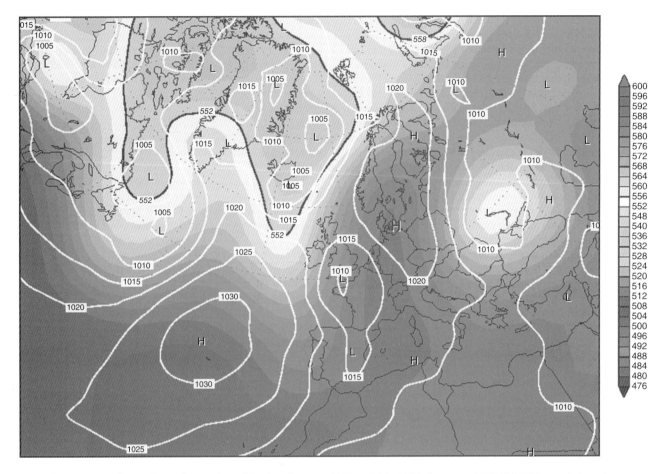

Figure 8.4 *Surface isobaric chart and 500 hPa chart for 0000 UTC on 2 July 1968. Courtesy of NOAA/NCEP reanalysis project.*

No rain fell barely 2 miles to the east and south of Keele. There was yet another severe storm at Keele on the evening of 14 June, as well as a lively one next evening.

8.4 Thunderstorms of the 1970s

Thunder occurred on 3 days at Leigh-on-Sea in February 1970 (in my absence), quite a feat for the normally quietest month of the year. The very localised storm in the early evening of 22 February, just after the passage of a small secondary depression from the west, was close and noisy. There was yet another clap of thunder during a heavy snow shower around 8.15 a.m. on 2 March. My final term at Keele saw several interesting storms, which fortunately managed not to interfere with revision and final exams! On 13 May, in an otherwise cloudless sky, a small patch of cumulus appeared to the north of Keele around 12.30 p.m. It built steadily *in situ* into a cumulonimbus by 2 p.m.; as it developed further, it expanded across about half of the sky, and among the distant rumbles came a few bigger bangs around 3.30 p.m. The overhead development faded soon after 4 p.m. – leaving a decaying cumulonimbus in much the same place it had first appeared. It finally decayed around 5.30 p.m. There was a fair amount of thunder in June too, and I observed one storm on the hot afternoon of 9 June in a train as I returned from an interview for a post in the Met

Office from Bracknell to London Waterloo. After a night of lightning, a 2-rated storm affected Keele in the late afternoon of 11 June, with thunder at about 10–15-second intervals for 80 minutes. Manchester caught the worst of this storm; two boys were killed by lightning, and hailstones the size of golf balls were reported in the city centre, cracking several windows. My final Keele storm was an active night-time one, with much lightning, on the night of 19/20 June as an old front over western Britain was unexpectedly reactivated.

My Met Office career, and move away from Leigh-on-Sea, began as a forecaster at Manchester's Ringway (as it was then known) airport on 10 May 1971, and I lived at nearby Cheadle Hulme until 1975. Thundery activity during this period was less marked than during the Keele years, but was not without its highlights, although overall it proved to be a rather frustrating 4 years for a thunderstorm observer. The first storm of note that I observed at Cheadle Hulme came on 31 July 1972 as the centre of a slowly filling depression tracked south-east over Cheshire. A series of cells drifted west or north-west off the hills to the east to give an evening of frequent thunder and lightning and spells of heavy rain. On 9 August, in a showery south-westerly airflow, a storm passed directly over Cheadle Hulme shortly after 1 p.m., giving a spell of violent rain and hail with a fair amount of thunder and lightning for about 30 minutes. Ringway airport lies 4 miles south-west of Cheadle Hulme; the first two flashes I observed were in that direction, and all the rest were no more than 2 miles to the north-east of my site – but the professional met observers and forecasters at Ringway did not notice this storm (and they had an office with a very good view from west to east via south)!

We should never forget that there is always an element of risk in watching, or being out in, thunderstorms. On 30 July 1973, a weakening ridge of high pressure covered the British Isles, but the air aloft was cold. Thunderstorms developed over many central areas of England in the afternoon, and one drifted north-east a few miles to the south and east of Cheadle Hulme. It was clearly missing us, with no threat of rain, and I walked to a nearby field to get a good view of the lightning which was flashing about every 5 minutes. After about 20 minutes, the storm centre was well away to the east, so I turned my back on the field to walk home. There was then an enormous crack of thunder, followed by a lengthy rumble. Lightning must have gone to earth just across the field: my emotions were rather mixed – annoyance that I had not seen the flash, but concern that it was clearly an earth (cloud-to-ground) strike not very far from where I had been standing. There was further thundery activity not far to the east later in the afternoon, without any rain – but I stayed indoors! Serious flooding was reported from Glossop, about 15 miles to the east.

And so to another storm that was not noticed by the met staff at Ringway. On 16 June 1974, a weak area of low pressure covered the British Isles. After some overnight thundery activity, there

was a sunny spell in the early afternoon. Cumulus quickly developed, and one patch rapidly grew into a large cumulonimbus which gave a violent thunderstorm well to the west of Cheadle Hulme. I noted at least 30 rumbles of thunder and several flashes of lightning 4–7 miles distant to the west and south-west (so not far from Ringway).

Unfortunately, the worst storm during my 4 years at Cheadle Hulme, on 4 July 1975, came when we were visiting my parents at Leigh-on-Sea. Pressure was low over the continent, with a weak north-easterly airflow over northern England (Figure 8.5). It was difficult to piece together the development (which may have been triggered by a sea breeze from Liverpool Bay converging on the north-easterly). At least 36 mm of rain fell, giving serious flooding in and around Cheadle Hulme (but only 3 mm at Ringway), and reports spoke of frequent close thunder and lightning for about 30 minutes. And something similar happened 9 days later – when I was on duty at Manchester Weather Centre and missed it! I did eventually get to see a fierce storm – but from near Belfast's Aldergrove airport where I worked as a forecaster during August 1975. On the 8th, very hot air covered Britain, and plenty of thunderstorms developed within it. A cold front lay down the Irish Sea, and it was no more than pleasantly warm over Northern Ireland. Impressive castellanus and floccus cloud drifted north over the sky in the afternoon, and storms followed in the evening. I had a good view of their approach from the grounds

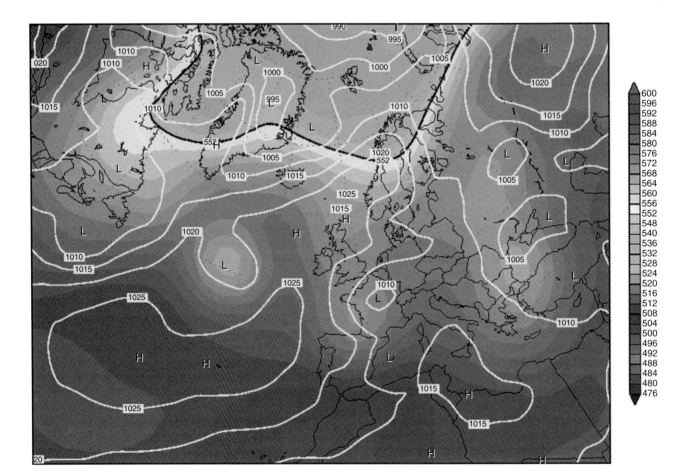

Figure 8.5   *Surface isobaric chart and 500 hPa chart for 0000 UTC on 4 July 1975. Courtesy of NOAA/NCEP reanalysis project.*

of the large farmhouse where I was lodging. For some time, the storms skirted us to the east, but they slowly edged nearer, with several earth strokes as darkness fell. I retreated inside, and a storm passed overhead with torrential rain around 10 p.m.; the thundery activity continued until midnight.

I moved to Epping (Essex) in December 1975 and began working at London Weather Centre. But the next 'big storm' was at neither venue. During the very hot weather of early July 1976, we were staying with friends at Middle Barton, 15 miles north of Oxford. A series of storms affected Oxfordshire and Buckinghamshire during the afternoon and into the night of 3 July. As viewed from Middle Barton, although these were medium-level storms, there were some very loud cracks of thunder from several cloud-to-ground earth strikes, both in the early afternoon and early evening. In mid-July, I attended the Royal Meteorological Society's Summer Meeting at the University of East Anglia in Norwich and viewed a lively night-time storm early on 16 July, though it appeared that the storm was fiercer at Epping. Four days later, I was, at last, 'at home' for a major storm. On the afternoon of 20 July, a series of thunderstorms moved north-east over south-east England in connection with a small wave on a cold front that had crossed the country from the north-west. Between 3 and 5 p.m., there were at least 120 claps/rumbles of thunder, many of them close, with several earth flashes; 28 mm of rain fell. This outbreak of thunderstorms had not been forecast.

It may be noted that I have reported here on several severe storms in the 'drought' summer of 1976 – it was certainly not without thunder. As the weather became more generally disturbed in the autumn, thunder occurred at intervals at Epping, including on two successive December evenings (6th and 7th) in a showery south-westerly airflow. February and March 1977 were quite lively too, and there was a remarkable outbreak on 20 March – the third consecutive day with thunder. A depression was sinking southwards into France, with a gentle north-easterly airflow over southern England. Large cumulus developed into cumulonimbus by midday, and thunder and lightning broke out from 12.45 p.m. Until 2 p.m., it was mostly a little way to the north and west. Heavy hail, rain and snow were falling by then, but it was fairly quiet for a while. At 2.10 p.m., I was minded to open the back door to fetch a broom to try and clear hail from a gutter – just as I turned the handle, there was a 'swish' from adjacent point discharges with an immediate, simultaneous flash and colossal bang. I decided to stay indoors. The next 10 minutes brought 6 more very close discharges, after which there was a 15-minute interval before the next, overhead, flash. Heavy hail (with a maximum diameter of about 7 mm) covered all surfaces, and the temperature fell to 2.2°C (from 8.4°C). The rain gauge reading was 19 mm.

Another fierce storm struck Epping on the evening of 12 May 1977 in a showery trough extending from a depression over the North Sea, while early on 14 June, a 1-rated night-time storm struck as part of a massive storm system that moved north-west from the south-east coast as far as Merseyside and north Wales – having developed in hot Belgian air and gained enough momentum to keep going well into very cool air. At Epping, the storm was approximately overhead from 3.10 to 3.45 a.m., with lightning every 2 or 3 seconds, and thunder was virtually continuous from 3.15 to 4.30 a.m. The storm system did not completely clear until after 8 a.m., though I noted that

there were very few earth strokes. 23 mm of rain fell at Epping, but 37 mm 4 miles north at Harlow where it provided an early baptism of fire for our 3-day-old twins in the hospital where they were born!

I moved to Loughton, 4 miles south-west of Epping, on 2 May 1978, and was greeted by storms on both the 4th and 5th. I stayed there for 26 years, so was able to develop a good understanding of the local weather characteristics. The first 2-rated storm came on 4 June; in fact, it was the second afternoon in four when a storm developed over north London as cold frontal troughs drifted north-east. These storms gained intensity as they moved out towards the Epping Forest ridge, and on the 4th, there was thunder and lightning about every 5 seconds for over 20 minutes with some enormous bangs. Early on 13 December 1978, there was a lively early morning storm in a vigorous showery south-westerly airflow, after a little thunder on the previous afternoon. Thunder and lightning occurred about every minute for a while around 4 a.m., with a short spell of torrential rain and hail and a severe squall. There were several close, very loud discharges.

A thundery May 1979 ended with an impressive storm on the evening of the 30th along a slow-moving front that was separating hot continental air from cool, dull weather over much of England. There was thunder and lightning about every 5 seconds for about 40 minutes, with several earth strokes and flashes within a mile.

8.5 Thunderstorms of the 1980s

There was no scarcity of thunder over the next few months (in June 1980, it was reported on 10 days at Leigh-on-Sea – beating my 'record' of 9 days there in August 1960). However, it was 29 July 1980 (Figure 8.6) before I next observed a heavy storm at Loughton, as a cold front moved north-east into a very humid easterly airflow. After a sunny afternoon, it became overcast and very oppressive by 6.30 p.m. There was lightning every 2 or 3 seconds from 7.33 to 8 p.m., with 22 mm of rain from 7.30 to 8.10 p.m. The storm passed overhead, giving over 500 flashes in all, but only two or three fierce earth strokes. I had just written my monthly weather column for the *Loughton Review* free newspaper commenting on how storm centres seemed to be missing us, with a consequent lack of rain; I rewrote it later that evening.

There was an impressive thunderstorm in central London on the afternoon of 9 July 1981, which my work pattern that day fortunately enabled me to watch closely. Hot, stagnant air covered the country, and there were thunderstorms in various districts throughout the day. Between 2 and 3 p.m., the sky took on a grey-yellowy hue over London, as large cumulus, and altocumulus, covered the sky and haze and smoke converged below the cloud cover. The temperature was 26°C. By 3.15 p.m., the sky was very dark, still with a yellow tinge, to the west–south-west, and soon a shower of huge raindrops began at London Weather Centre. After about 5 minutes of heavy rain, the first flash was seen about a mile to the west–south-west. From 3.20 to 4 p.m., there was frequent, close thunder and lightning (up to about 6 flashes per minute), intense rain, a spell of large hail and a very squally wind for a few minutes (with a gust of 42 kn). At the peak of the rain, around 3.30 p.m., visibility looking west along High Holborn was reduced to about 50 m; 58 mm of rain fell in 50 minutes. There was a rapid improvement shortly after 4 p.m., with the storm

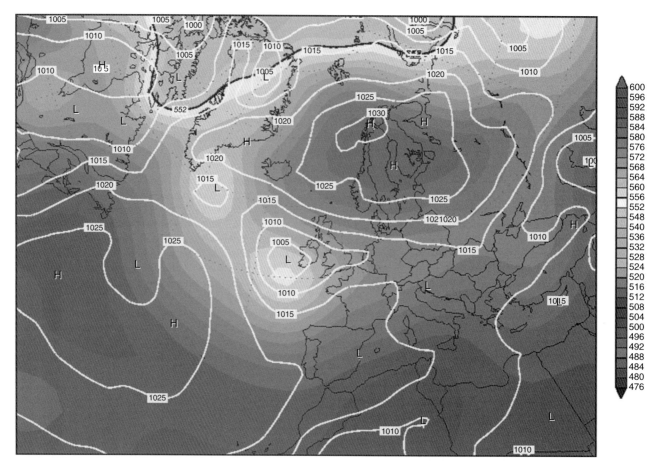

Figure 8.6 *Surface isobaric chart and 500 hPa chart for 0000 UTC on 30 July 1980. Courtesy of NOAA/NCEP reanalysis project.*

cloud well away to the east–north-east by 4.15 p.m. Exactly 4 weeks later, another day of widespread thunderstorms in hot, stagnant air gave central London a battering on 6 August – not once, but twice with a fierce mid-morning storm giving 35 mm of rain in an hour and frequent, often very close, thunder and lightning. Then for about 30 minutes from midday, it was almost as dark as at night; during the following 90 minutes, there was considerable thunder and lightning (up to 10 discharges per minute for much of the time, some of them close) with heavy rain at times.

There were 27 days with thunder at Loughton in 1982 and a personal record of 32 days in 1983 (equalled at Claygate, Surrey, in 2014). This was the most active period of thunderstorms in my records. It was heard on 10 days in June 1982 (Prichard, 2012), and on the 4th I witnessed probably the most severe storm of my life. Under slack low pressure, the morning was mostly sunny, and the temperature rose to 29.5°C. Large cumulus began to develop before midday, with the first reports of thunder from Hertfordshire around 2.30 p.m. At Loughton, thunder began rumbling around 3.30 p.m. and continued for the next hour or so as a developing storm drifted north to the west. Towards 5 p.m., there was a fair amount of close lightning – for a time without much rain – so I decided I would remain under a shop canopy I had reached near my home in my journey from work rather than risk going out into the open. Other folk seemed

quite unperturbed by the forked flashes and big bangs, but one flash during this period struck and set fire to a house 0.25 miles to the north. It did appear, though, that Loughton had missed the worst of this storm. Suddenly, though, a new cell went up on the south-eastern edge of the Hertfordshire storm, developing just to the south of Loughton and drifting north, being continually fed by the hot, sunny air not far to the south and south-east. The next few minutes brought frequent flashes to earth all around, torrential rain, with hail for a time, and a gusty wind near the end of the storm. Between 5.20 and 6.30 p.m., 69 mm of rain fell out of a storm total of 79 mm. About 200 houses in the area were flooded, some up to first-floor level and fire tenders were called from all over the county. Telephones were put out of action, and floodwater blocked some roads. The storm highlighted the risk of flash flooding in the area and led me into discussions with the District Council on how to improve the forecasting of them – an involvement which was to end bizarrely 5 years later, as will be described in the next section. Sharp thunderstorms brought further local flooding on other days during the rest of the month, with a particularly heavy deluge on the afternoon of the 25th only just stopping in time to prevent further serious flooding: fire brigades across the county had already been summoned, and some houses were protected by sandbags. I recorded 182 mm of rain that month. And it was not over yet: 36 mm fell in a series of thundery downpours on 15 July.

There were 9 days of thunder at Loughton in May 1983, culminating in a most impressive lengthy night storm on the last night of the month, which I witnessed from London Weather Centre. Here, thunder and lightning was almost continuous from 1 to 2 a.m., while the fiercest storm arrived from the south around 3.45 a.m. to give torrential rain and almost continuous thunder and lightning for some 30 minutes with a number of intense discharges, sometimes two or three a minute. Loughton was similarly affected – and I recorded a grass minimum of 3.3°C, indicating that there was a covering of hail for a time. Two further outbreaks of exceptionally violent thunderstorms and hail affected parts of the country in the next few days and lead us here into another diversion.

8.6   The Forecasting of Thunderstorms

This is not a success story. Many of the severe storms in this account were barely, if at all, forecast. There may have been a slight improvement over the years, but there are still many failures. The reasons are complex. At its simplest, we are dealing with one of nature's finest phenomena whose development often has disparate causes. It is often not possible to indicate more than that thundery outbreaks are likely and that detail is not possible (but that needs to be stated clearly, not fudged). There was, during my professional forecasting career, a school of thought that suggested that the word 'thunder' should only be used when we were certain, because it scared some folk and we did not want to do that unnecessarily. Very kind, but not very professional, it was an attitude that did not encourage proper attention to such events. More seriously, though, any chance of successful thunderstorm forecasting rests on a careful understanding of the weather patterns that encourage them. I was never taught this but slowly developed at least a modicum of expertise through close study (and my detailed records). If forecasters do not have that motivation, they will not easily pick up such skills. That has resulted in far too many serious errors over the years. The development of computer models of greater complexity and finer resolution has certainly reduced the error/omission count, but the programming of the models is by no means always up to dealing with this phenomenon.

On Sunday 5 June 1983, an exceptionally severe outbreak of thunderstorms affected south Dorset (Mortimore and Rowe, 1984). It was not forecast. I remember very clearly the (routine) conference that radio and television forecasters had with the chief forecaster at Bracknell on the previous afternoon. Jim Bacon, the day's television forecaster, expressed concern about the degree of instability in the upper air as shown in the tephigrams and said he did not trust the model's lack of forecast development. The chief did not share his concern. Jim persisted and was clearly unhappy at the concluding comments which left him unable to express more than a small risk of storms. The weather chart just prior to this outbreak is shown in Figure 8.7.

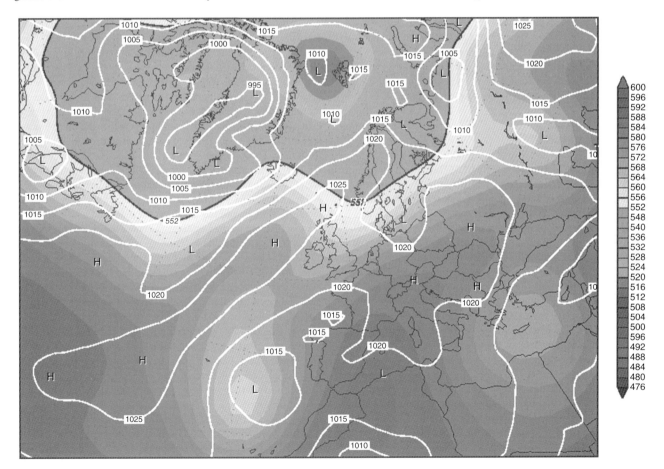

Figure 8.7   *Surface isobaric chart and 500 hPa chart for 0000 UTC on 5 June 1983. Courtesy of NOAA/NCEP reanalysis project.*

Later in the 1980s, I occasionally expressed concerns, or post-event criticisms, of a similar nature. I referred earlier to my discussions with Epping Forest District Council about flood risks from severe storms. I was able to pinpoint the weather patterns that seemed to bring the greatest risk and contacted them on a few occasions when I was off duty, though would not claim any great success beyond putting them on alert. Then on Saturday 22 August 1987, noting that official forecasts were not forecasting storms (after a correctly forecast overnight outbreak), I became very concerned that the pattern that afternoon was ripe for a severe event. I rang the duty engineer at lunchtime to update this concern. He said that he had been in touch with the London Weather Centre and been assured that the risk was minimal and they were going to take no action. That afternoon, a series of violent storms crossed the area, giving 45 mm of rain at Loughton and 77 mm at nearby Chigwell Row, with Abridge suffering serious flooding as the River Roding overflowed. There was much lightning damage too, and across Essex, it was one of the fire brigades' busiest days. The publisher's lawyers will not want me to repeat the legal phrase my notes have about that day. The day's forecasting was inexcusably poor. I never heard from the District Council again.

Two months later came the Great Storm of October 1987, where the forecasting failures were mirrored on a larger scale than those of the August event. A major change in forecasting practices and techniques ensued, and certainly, there has since then been a much greater awareness of the need for a constant vigil about severe weather. I am not entirely convinced that this extends sufficiently to thunderstorms. There has also of course been a growth of private-sector forecasting, much of it of high standard – but not all of it, and it is regrettably true that the more extreme 'cowboys' get a lot of publicity, leading to a tendency for the public to feel we call 'wolf' too often. For one reason or another, though, outbreaks of severe thunderstorms are still too often poorly forecast.

8.7 Back to the 1980s

Further heavy thunderstorms affected Loughton on the night of 22/23 June and the late afternoon of 6 July 1983. 1984 was a quieter year, with no notable summer thunderstorms, but the year did not start quietly with 2 days of thunder in January and a remarkable early morning outbreak on 8 February, the most severe winter thunderstorm I have observed. A cold front was racing south-eastwards across the British Isles, and thunderstorms came with it, a very active one affecting Loughton with fierce hail and a near gale between 4.15 and 4.45 a.m. There was an average of about two flashes a minute, and they were often nearly overhead. The individual cells must have been travelling at close to 50 mph, as is typical in winter storms which are more normally of the one flash-and-bang variety. On this occasion, new cells must have been generated continually for several minutes.

Although not noted at Loughton, a nearby oddity is worth recording here. On 10 March 1984, a weak northerly airflow gave mostly cloudy and occasionally rather damp weather over eastern Britain. Around 7 p.m., there was a solitary lightning strike, causing a power cut, near North Weald, about 8 miles north-east of Loughton.

In early October, the combination of a tropical storm remnant, a dig southwards of cold air in mid-Atlantic and warm seas resulted in a very vigorous depression that moved east–north-east into the Bay of Biscay on the 4th. During the 5th, it continued north-eastwards over France, and its occluded front pushed north-westwards into south-east England and East Anglia, where it came virtually to a halt for a few hours. Eventually, another depression formed at the triple point (i.e. the point where the cold front has just caught up with the warm front) over north Germany and moved north, absorbing the old tropical system within its circulation so that the occluded trough swung away eastwards in the evening. However, while it was over the south-east, it gave up to 8 hours of more or less continuous thunderstorms – from 9 a.m. to 5 p.m. at Loughton. The main storm band advanced from the east around 11 a.m. in the form of a squall line whose leading edge kept decaying. Thunder and lightning occurred fairly frequently with a steadily increasing intensity of rain. Large hail fell for a minute or two as the squall line finally came overhead, when it became very dark for a time. This leading edge then moved on towards London, while Loughton had steady rain and occasional thunder and lightning until, around 4 p.m., the trough edged back eastwards and a squall line advanced from the north–north-west. This brought a renewal of torrential rain and close thunder and lightning; 34 mm of rain fell at Loughton. Quite a day – I did not get much sleep between my two night shifts!

In 1985, after a near miss on 26 May when we were on the south-western edge of a storm that brought a major hailstorm to mid-Essex, we had just one notable storm, on 5 July. With the flooding of 4 June 1982 in mind, I noted that this storm arose out of a set of circumstances that have been found to be good predictors for damaging flash floods in the Epping Forest area. First, shallow low pressure moved north into England in the early hours, with much castellanus, but the potential instability was not released. During the day, there was further cooling aloft, and a stagnant mass of hot, humid air remained near the surface: the morning was sunny and the temperature at Loughton rose to 26°C. However, a patch of heavy rain appeared in the Bournemouth area around 10 a.m. and tracked north-east, apparently weakening for a time. Later, though, it grew into an active thunderstorm in the Bagshot area around 12.30 p.m., which continued north-eastwards with two further intensifications in the Wimbledon and Epping areas. The Wimbledon tennis championships were spectacularly flooded around 2 p.m.: a warning that a 'shower' was approaching allowed groundsmen to cover the courts just in time. It was here, also, that the first strong, gusty winds were reported. The storm came into earshot of Loughton around 2.15 p.m., and there was soon fairly frequent thunder and lightning and a very heavy downpour, preceded by squally winds. As the storm moved past and on to the Epping Forest ridge, it suddenly intensified further, and thunder and lightning became almost continuous with many earth strokes, while visibility was reduced to 500 m in intense rain. This very fierce storm then tracked across Epping and North Weald, causing considerable flooding – to a depth of 4 feet in places. A feature of this outbreak was that individual cells moved north–north-east – developing on the forward right side of the storm and decaying as they left its northern flank, so that the mean movement of the storm track was north-eastwards.

There was a 'close call' in a similar weather pattern on 14 July, when a sudden explosive development of a cell advancing from west–north-west into a south-westerly airflow produced a 'wall of water' with visibility reduced to 50 m. The torrent only lasted a couple of minutes, although 14 mm fell in 20 minutes. There was only a little thunder. On the following Saturday, 20th, there were eight separate thundery showers at Loughton in a cool, westerly airflow.

There was nothing of particular note in 1986, and the events of 22 August 1987 have already been recounted earlier. However, on 29 July 1987, 53 mm fell in 20 minutes in north Epping, with severe flooding there and in North Weald. Loughton had no rain at all, but a fair amount of thunder and lightning. Just to reaffirm that thunderstorm/flash-flooding forecasting is never straightforward, this event occurred as a showery trough followed a cold front in a north-westerly airflow – nothing like the 'normal' scenario for such events in the area (Figure 8.8).

It was September 1992 before I next encountered a heavy storm; there were several 3-rated storms in the intervening years, which were certainly worth watching but do not merit an account here – although it is noteworthy that thunder occurred on three of the first four days in February 1988, with a near miss on the 7th. February was rapidly losing its status as the quietest month of the year.

8.8 Thunderstorms of the 1990s

The early hours of 18 September 1992 brought Loughton's worst storm since the early hours of 1 June 1983 and was my first 1-rated storm since then. An area of low pressure was drifting north-eastwards into southern England, pushing a cold front ahead of it. After earlier isolated thundery activity on the front, the sferic count was almost nil during the evening, and the late evening television forecast dismissed the front (here we go again!). A rumble of thunder was heard to the south-west around 1 a.m., but sferics were still sparse as some thundery activity passed northwards to the west of us. There were three close discharges around 1.15 a.m. and shortly afterwards a brief spell of heavy rain. Then everything 'blew up'. A major storm area developed off the East Sussex coast and moved north from Kent to Norfolk, rather to the east of Loughton. There were simultaneously outbreaks of severe storms over some other parts of the country on the front. Loughton had a severe storm from 2 to 4 a.m. There was almost continuous medium-level thunder and lightning moving slowly from the south during the first hour – rather like a 'wall of light' gradually shifting northwards. Then, there were several violent earth strokes in quick succession just to the south-west and west from 3.10 to 3.25 a.m., after which it was back to the almost continuous medium-level display of light and sound. The last rumble was to the north-east at 4.30 a.m. There were short spells of torrential rain, but rain was never a major feature of the storm locally. The early morning satellite picture is reproduced in Figure 8.9.

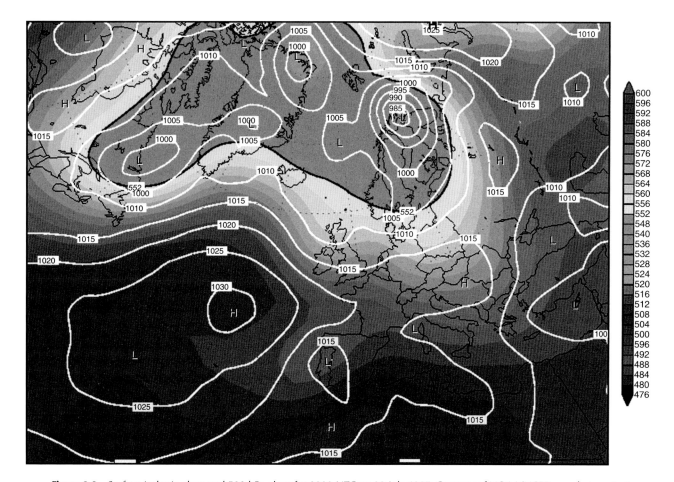

Figure 8.8 Surface isobaric chart and 500 hPa chart for 0000 UTC on 29 July 1987. Courtesy of NOAA/NCEP reanalysis project.

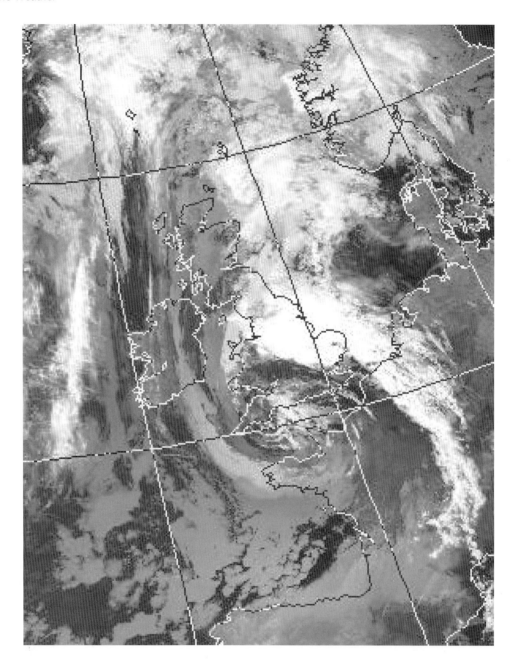

Figure 8.9   *Infrared satellite image for 0411 UTC on 18 September 1992. Courtesy of University of Dundee.*

Thunder appeared at fairly regular intervals during 1993, without ever hitting the heights at Loughton. However, North Weald suffered yet another extreme deluge on 10 June: 121 mm fell in 3 hours in the early evening, with 56 mm in 30 minutes. There was distant thunder at Loughton and a short shower, on a day of weak low pressure over hot, humid air which triggered fairly widespread thunderstorms.

An outbreak of severe thunderstorms on 24 June 1994 attracted more publicity than usual because it was linked to an 'epidemic' of serious asthma-like allergies. Sunny, warm weather in a brisk south-easterly airflow was replaced by low pressure and a cold front from the south-west (Figure 8.10). Widespread severe storms – presaged by extensive castellanus and the odd rumble of

thunder overnight – developed along the cold front in late afternoon. A band of, initially, cirrus approached Loughton from the south-west from 6 p.m. and slowly thickened. By 8 p.m., it was very humid, dark and airless as the wind fell calm. Asthma sufferers swamped Whipps Cross hospital outpatients, our nearest major hospital. I was among those feeling distinctly groggy but had no need of hospital treatment. I had been watching cricket at Ilford all day and soaking up the brisk breeze and all its pollen. My surmise was that all this pollen converged in the still conditions under the approaching storm. Some medical and meteorological experts produced papers linking the medical emergency to summer thunderstorms, but not a lot has been heard on this theme since then, and I suspect it was a special case. The storm affected

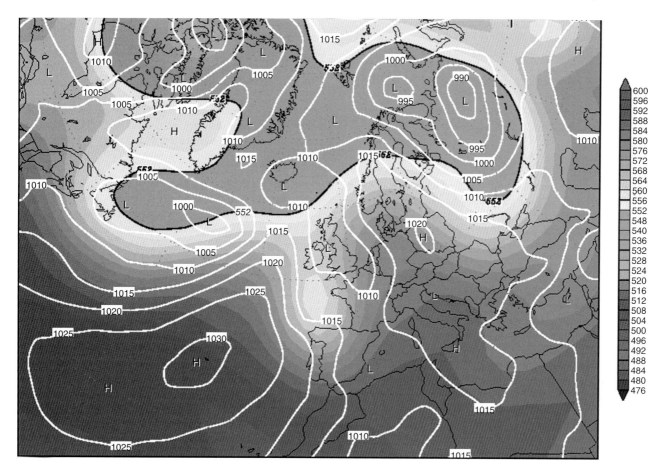

Figure 8.10 *Surface isobaric chart and 500 hPa chart for 0000 UTC on 25 June 1994. Courtesy of NOAA/NCEP reanalysis project.*

Loughton from 8.45 to 9.30 p.m.: very frequent and fierce flashes sprayed the ground to the south and east, mostly a few miles away so that the thunder was not particularly loud, nor the rain all that heavy. The lightning was phenomenal, though – very bright, with several lengthy flashes that repeated themselves along their path, as well as some bead lightning.

The next severe storm I witnessed was on a visit to friends in Cheshire and was observed from Gawsworth Hall, Macclesfield, on the afternoon of 24 July 1994, as a cold front moved eastwards into hot air. After some earlier weak thundery activity (as observed from Holmes Chapel), our hosts were keen on a country walk. I firmly scotched that idea, so they drove us to this historic house instead. Initially, around 3.30 p.m., there was a sudden very heavy downpour from a sullen sky – then the thunder and lightning began. During the next 90 minutes, several cells drifted northwards with fierce activity: frequent overhead thunder and lightning and torrential rain causing some flooding. The storm area slowly retreated to the north-east, and it was over by 7 p.m. Before leaving 1994, we may recall that during a conference that TORRO ran to mark the 70th anniversary of the Thunderstorm Census Organisation in London on 22 October, there was a brief thunderstorm – we were in a windowless room and did not see it!

No major storms were observed in 1995, or in 1996, when I noted that it was probably the quietest first half of a year in the British Isles since daily thunderstorm records began in 1931. The storm on 29 July 1996 merits a mention, though. My notes start by commenting that thundery showers over the country that day had bypassed us – 'seemingly, the month was to be frustrating to the end', an interesting remark in the light of the experience of recent years (at the time of writing), since there had been five days of thunder already, with a close earth flash on two of them. On this evening, at 6 p.m., anvil cirrus was in abundance with some darker cumulus, but nothing of note. At 6.08 p.m. came a clap of thunder and, almost instantaneously, the first large drops of rain. By 7.15 p.m., 40 mm had fallen – the second highest total for such a duration in 20 years. There were short spells of squally winds and local flooding. Radar confirmed that the storm developed just west of Loughton and decayed not far to the south-east: the anvil seen from central London was apparently most impressive, but so rapid was the decay that by 7.45 p.m., there was little evidence of a storm cloud from Loughton, with a lot of blue sky.

I had to wait until 3 July 1999 to next observe a heavy storm. A cold front was moving north-east across the country, and there were some thunderstorms associated with it. A particularly severe series of storms developed over Hampshire and Wiltshire around midnight and moved north-east, developing further across north London, the north and east Midlands and East Anglia during the early hours. From Loughton, there was almost continuous lightning from 2.15 to 3.30 a.m., including an overhead phase around 3 a.m. with 11 mm of rain in 5 minutes and a few powerful earth strokes. The storm system had cleared to the east by 4 a.m.

8.9   The Most Recent Thunderstorms: 2000–2013

My notes during my last few years at Loughton (to July 2004) are interspersed by comments about missing thunder that other places experienced. This was then compounded in May 2001: 'the week 9–16 May was perhaps the most frustrating for a thunderstorm observer in this record. It was the week of a booked holiday in the Isle of Man: no thunder was heard there, but it occurred on 5 days during the week at Loughton, including one of the worst storms on the entire record. There were no other storms in May'. It even hurts to transcribe that note, 12 years on! The background to the amazing storm on the night of 9/10 May was that a small cold pool (a region of anomalously cold air aloft) had slipped south-west through central England and south Wales on the 8th. Warm surface air then tried to push north on its eastern flank but was undercut by a persistent cool north-easterly airflow. An outbreak of extremely active thunderstorms began in the eastern English Channel, moving into Kent and East Sussex by 10 p.m. and transferring (with further development) across the centre of England and dying out off north Wales around 8 a.m. Loughton was hit between midnight and 1 a.m., with 19.6 mm of rain and virtually continuous thunder and lightning passing overhead. It was reported as a 'fantastic' storm to watch, all the more so because it was so unexpected and followed a grey, chilly day.

On 5 August 2002, a cold pool was centred over the Thames Estuary with a slack trough extending all the way westwards from a weak low pressure area over Russia. After a cloudy morning with a spell of rain, the sky darkened to the north-east around 1 p.m. A rumble of thunder was heard around 1.22 p.m. During the next 10 minutes or so, the darker cloud edged slightly closer on its way southwards. As it did so, the countryside was steadily obliterated from view by rain. The rain cloud extended round to the north but was initially weaker in that direction. However, rapid development occurred right through this band around 1.35 p.m.; rain began at my location, and as it did so, there were two loud cracks of thunder to the east and south-east. The sky then became jet black just to the north, the rain became very heavy, and there was an earth stroke less than a mile away out of this cloud. The next 15 minutes brought torrential rain (10 mm) and a short spell of a Force 5 northerly wind, and visibility reduced to 500 m, while several lightning strokes sprayed the immediate neighbourhood. As the overhead cell retreated to the south after 1.50 p.m., its thunder died out, but a new cell passed a mile or two to the west with another six earth strokes and huge claps of thunder.

The last storm of any note that I observed from Loughton was on 19 July 2003. A cold front was drifting east into hot air and had become inactive although there was extensive medium and high cloud. A patch of rain developed over Cherbourg just after 3 p.m. and reached West Sussex shortly after 6 p.m., and the first sferics were noted just off that coast around 7.15 p.m. Aggressive castellanus (an unusual feature of this storm being an absence

of cumulonimbus despite it developing well before sunset) appeared to the west of Loughton by 7 p.m., with darker cloud soon apparent to the south-west. This cloud turned thundery around 7.40 p.m., and a storm tracked north–north-east a few miles to the west. Things really came to life after 9 p.m., with a series of storms passing north–north-east a few miles to the east giving thunder and lightning about five times a minute during the next 30 minutes, and frequently, if more distantly, until about 10.30 p.m. There were many flashes of considerable length and many intricate patterns around the medium-level cloud. About 600 flashes were viewed in all.

It is worth recording the minor thundery shower on the hottest day so far recorded in the United Kingdom (at the time of writing), 10 August 2003. Most places had a cloudless sky that day, but a small patch of cloud from the south-west developed as it crossed London, and a couple of short, sharp showers (of hot rain!) and a spell of blustery winds affected Loughton – followed by some rumbles of thunder to the east around 6 p.m.

From mid-July 2004 onwards, my 'base' has been Claygate (Esher, Surrey). My first full month, there, August 2004, was a lively one for thundery activity – especially at Loughton! The 'best' storm I observed during it was actually at Sandiacre (5 miles west of Nottingham) on 18 August, when the village caught the edge of a fierce storm that moved north-east from Leicester to Nottingham between 4 and 5 p.m.; thunder was almost continuous (Figure 8.11). Large hail was reported along its track, in a showery south-westerly airflow.

9 September 2005 was interesting. On a hot day, cumulus built rapidly over Surrey in the early afternoon, and the North Downs became the trigger for a series of torrential downpours over a 5 hour period. Claygate was almost at the source of developments. The sky darkened here after 2 p.m. (after a virtually cloudless morning). A few rumbles of thunder were heard to the north-east between 2.45 and 3 p.m., from a shower that probably began over Epsom Downs. The afternoon showers drifted north–north-east, and this one went on to give flooding over Wimbledon and, especially, around Hammersmith an hour or so later; several underground railway stations were closed for a time. Meanwhile, a torrential downpour broke over Claygate shortly after 3 p.m. and lasted for about 15 minutes before drifting away; another shower passed to the west. Both (but, within earshot, perhaps more especially the western one) turned thundery some 6 miles to the north, with rumbling thunder from about 3.45 to 4.15 p.m. This storm complex then caused serious flooding to road, rail and property in Kingston, Richmond and Roehampton (where it was reported that some trees were blown down). Around 4.30 p.m., there were one or two rumbles of thunder to the south-east, followed by a torrential downpour from another developing cumulonimbus overhead; no thunder was heard locally from this, but it moved on to further flood south-west London. Sporadic thundery downpours continued on similar tracks until 7 p.m., and 27.6 mm was recorded in total.

I only had to wait another day for the first 'big one' at Claygate. London became the focus again for thunderstorms and torrential rain in the late afternoon and evening of the 10th, with several further reports of flooding across the capital. The storm zone lay across north Kent and north Surrey in the evening, with intense thunderstorms and torrential rain causing flooding around Greenwich and Dartford, and in west Surrey. Claygate was dry

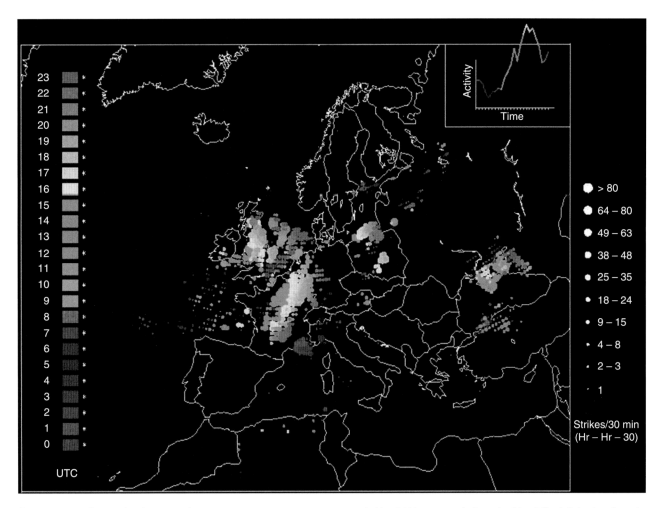

Figure 8.11 *Lightning distribution and intensity on 18 August 2004, as compiled by © Wetterzentrale from the Met Office's lightning detection network. (See insert for colour representation of the figure.)*

until 8 p.m., though there were spells of distant thunder from 4.20 p.m. From 8 to 9.15 p.m., a fairly static, heavy thunderstorm sat over the village and its immediate neighbourhood, giving 39.9 mm of rain. Close examination of the rainfall trace suggests 25 mm fell between 8.25 and 8.35 p.m., and comparison with the remarkable Loughton event of 4 June 1982 suggests this exceeds the peak rainfall rate on that occasion (when 69 mm fell in 70 minutes). It was, therefore, possibly the most intense short-period rainfall event I have observed. It was also the first time that in excess of 25.4 mm (1.00 in) of rain had been personally observed on two consecutive days (all sites) since September 1968 at Leigh-on-Sea (39.9 mm on 14th, 81.5 mm on 15th) – in a somewhat similar weather pattern. Visibility was reduced to 300 m for a time. Thunder and lightning occurred every minute or so; although some more distant lightning was also observed, most of it during this period was within 2–3 miles, and some of it was overhead. A fierce earth stroke was observed about half a mile to the west at 8.15 p.m., with a few others a little further away to the north-west and north. The rain eased around 9.15 p.m. (it had eased about 8.45 p.m., only to rapidly intensify again a few minutes later with renewed ferocity of lightning); there were still two or three flashes within a mile or two just after this – then it ceased

completely. Figure 8.12 shows the chart midway between these 2 days of storms.

Much as was noted for several summer months at Loughton, July 2006 looked like being a frustrating month when we never quite got under a worthwhile storm at Claygate. A weak lunch-time storm on the 27th reinforced this feeling. The afternoon was then hot and sunny with shallow cumulus and a cirrus overhang from French storms. Around 4.30 p.m., cumulus to the west began to grow, and as this edged round to the north-west by 5 p.m., it was clearly only a matter of time before thunder broke out; the first flash was seen about 5 miles away at 5.05 p.m. The next 30 minutes or so brought flashes and bangs at 2–3-minute intervals as different cells grew and moved on a track from about 200° around to the west and north-west; still rain did not seem imminent. However, by 5.40 p.m., the flashes and bangs were getting nearer, with some fierce earth flashes to the west. New cells were now beginning to appear to the south-west – and ragged cumulus overhead brought the first shower of large drops of rain. Rain set in consistently from 5.55 p.m.; between 6.05 and 6.35 p.m., it was torrential, giving 30 mm in 30 minutes – swept in on a fresh westerly wind – with visibility reduced to 800 m at times. The temperature fell from 30°C (at 4 p.m.) to 17°C. A further 10 mm fell

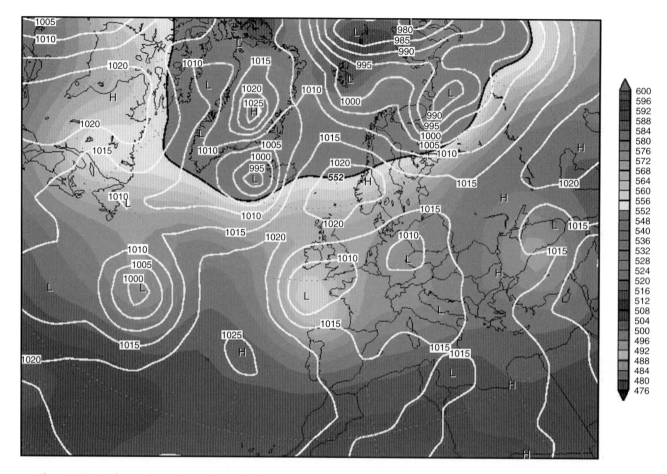

Figure 8.12   *Surface isobaric chart and 500 hPa for 0000 UTC on 10 September 2005. Courtesy of NOAA/NCEP reanalysis project.*

between 6.35 and 7.15 p.m., with the total from the storm 44.1 mm. Overall frequency of the thunder and lightning was about one per minute, with many powerful flashes and bangs within a mile, especially from 6.15 to 6.45 p.m. After 7.15 p.m. the blue sky, which had perhaps always been to the south-east had the rain allowed it to be seen, became more apparent to the south, and the storm zone slowly retreated whence it came, to the north-west and north – with several more big flashes and hefty bangs until it finally went quiet around 8 p.m. and the anvil receded.

The next few months were fairly active: April 2007 was the first thunder-free month for 11 months. On 9 July 2007, I was playing croquet at Surbiton as a thunderstorm approached from the west–north-west towards 4 p.m. Successive cells got steadily closer, but no rain fell. I advised my opponent that we should seek shelter from the lightning, and shortly after we did so – though the rain had started by now – a house was struck by lightning just a few yards away. Very heavy rain and hail fell for a time, and thunder and lightning occurred every minute or so, with several earth strokes in the vicinity. Nevertheless, youngsters carried on playing tennis not far from the croquet lawns throughout the deluge! Later, as I returned to Claygate, which, 4 miles to the south-west, had not had any rain from the Surbiton storm, a new cell passed a mile or so to the south from 6 to 6.30 p.m., with torrential rain, squally wind and hail (7 mm diameter). On 28 July, a cold front drifting north-east into a warm, slack airflow gave us a lively

storm in the late evening, and on 31 August, following a rare sunny, warm day, came Claygate's third notable thunderstorm of the summer – in a summer not at all notable for these countrywide (there were none over Epping Forest). A fairly narrow zone of active thunderstorms, originating near the Pyrenees the previous evening, moved directly north over the village. Rumbles of thunder were heard to the south-east from 8.05 a.m. – rather eerily at first as they came through thick fog (50–100 m visibility) and stratus; one or two flickers of lightning were discerned. A flash 2 miles south at 8.20 a.m. led into a lively storm passing overhead during the next 15 minutes, with torrential rain for a time, largely dispersing the fog. The lightning was impressive, including two or three earth flashes within a mile. Spells of rain, heavy at times, continued until 10.15 a.m., and there was another thunderstorm in the afternoon.

There was a rather odd development on 27 June 2009. Thunderstorms broke out within a broad convergence zone from London to Wales; they were severe with large hail in parts of the London area (and elsewhere). Claygate lay to the south of the convergence zone, and there was little sign of significant convection until around 4 p.m., when a few large drops of rain fell and several of the cumuli around the sky grew into cumulonimbi (glaciation (i.e. clouds composed of ice crystals rather than the water content of developing cumulus and cumulonimbus) being more obvious than any particularly dark bases). At 4.38 p.m., there was a rumble of thunder to the north-east; radar showed a small shower over Surbiton. From then until 5.30 p.m., there were a few more rumbles, and the odd flash, over the northern quadrant of the sky. The southern sky was innocuous, but the cloud that had earlier given a few spots had moved only a little but turned into a cumulonimbus (again without a dark base). From this, rain, which soon became torrential, fell from 4.50 to 5.10 p.m. Towards the end, as intensity eased, hail fell, and from about 5.10 to 5.20 p.m., there was a heavy hail shower – or, more accurately, an ice fall – from the very edge of the cloud as it eased away to the north. The hailstones measured up to 10 mm in diameter and were painful (but not in sufficient quantity to lie more than sparsely). It was surprising that all this development never led to overhead thunder (though, again, there was never a dark base). Radar could just

pick the deluge out – separated from the London storm (so that the Wimbledon tennis championships continued unhindered) – and, supported by visual observation, indicated the shower died out within a few minutes of leaving Claygate, where it had given 18 mm and localised flooding. On 16 July 2009, as a deepening depression moved north-east across south-west England, two cold fronts crossed south-eastern England in the evening. They gave a thundery evening at Claygate, leading into a very active storm (which may have been largely a new development) which passed overhead from 10.15 to 10.45 p.m. There were four or five flashes a minute, with a few big bangs and heavy/very heavy rain.

No heavy thunderstorms affected us in the following four and a half years to the end of 2013: in parts of the south, the hiatus was even longer. I only recorded thunder on 8 days in 2010, my lowest

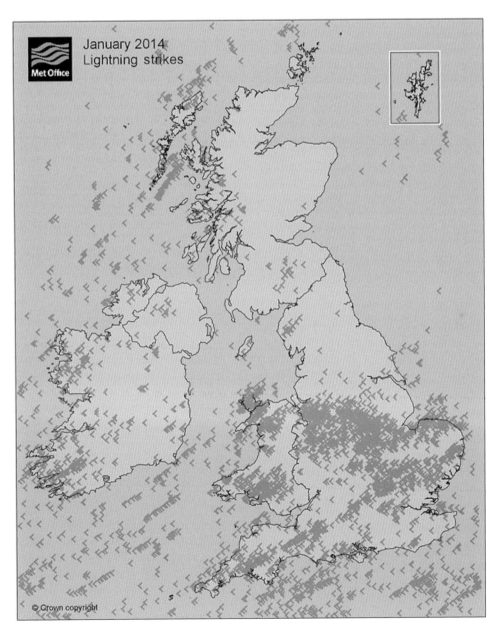

Figure 8.13 *Lightning distribution over the United Kingdom in January 2014, as detected by the Met Office's lightning detection network. Met Office, UK. © Crown copyright. (See insert for colour representation of the figure.)*

annual total anywhere, and none of these were given a better rating than 5. There was just one 3-rated storm, on 28 June, in 2011, when all the others merited only 5–7. 2012 was rather better, with 6 days of thunder in April and an interesting, and unforecast, storm on 19 August: on the boundary between hot air over south-east Britain and very hot air over France, an area of unstable medium-level cloud moved north from France, and a storm developed on the western edge of this area, off Cherbourg, in the late morning. There was then rapid development and movement of a storm system across West Sussex, west Surrey, London, west Essex, west Suffolk, leaving north Norfolk in the early afternoon. It crossed Claygate from 12.25 to 1.10 p.m. We also had quite a lively storm just before daybreak on Christmas morning. There were several close discharges, and 23 mm fell from 4 a.m. to midday with 7.8 mm in the storm, the most intense rainfall of 2012. In 2013, no storm merited better than a 4 rating.

This account was intended to finish at the end of 2013. However, January 2014 brought perhaps the most remarkable statistic of my thunderstorm-reporting career. Thunder was heard on 7 days at Claygate, a figure that would be remarkable for a summer month, and must be unprecedented locally for January. None of the storms were particularly active, but most of them included quite close discharges. Figure 8.13 shows how active this record-breaking wet month was for lightning across the United Kingdom, but another oddity was that, in the south-east, it was only north-west Surrey that caught so much thunder. It was heard on 6 days at Morden, about 10 km to the north-east of Claygate (those six were the same as at Claygate, where the seventh day was an isolated flash and bang quite close to the west of us). Only 1 or 2 days with thunder were reported over much of Berkshire and none at all over areas of Kent. Claygate had 32 days of thunder in 2014.

8.10   Concluding Remarks

This has been a personal account of over 60 years of thunderstorm observing. I have found it an absorbing hobby. I make no claims as to the importance of my records, which have only ever been kept for my own interest – some of which I hope I have conveyed in the preceding pages. Times change, and I am doubtful that many would follow the same hobby today. Certainly, it would be approached differently now with the wealth of information that is available on the internet, as will be readily appreciated by typing relevant storm-based words into a search engine: the thunderstorm enthusiast still has plenty to enjoy.

References

Jackson, M.C. (1977) Mesoscale and small-scale motions as revealed by hourly rainfall maps of an outstanding rainfall event: 14–16 September 1968. *Weather*, **32**, pp.2–17.

Mortimore, K.O. and Rowe, M.W. (1984) The hailstorm of 5 June 1983 along the English south coast. *J. Meteorol.*, **9**, pp. 331–336.

Pedgley, D.E. (1962) *A meso-synoptic analysis of the thunderstorms on 28 August 1958.* Geophysical Memoirs, no. 106. Meteorological Office, H.M. Stationery Office, London.

Pike, W.S. and Webb, J.D.C. (2004) The classic overnight thunderstorms of 22–23 and 23–24 June 1960 in the British Isles. *J. Meteorol.*, **29**, pp. 277–291.

Prichard, R. (1970) A Cheshire thunderstorm. *Weather*, **25**:11, pp. 495–499.

Prichard, R. (1992) Remembering 1968. *Weather*, **47**:12, pp. 473–480.

Prichard, R. (2012) The remarkably thundery month of June 1982. *Weather*, **67**:6, pp. 146–149.

9

Severe Hailstorms in the United Kingdom and Ireland: A Climatological Survey with Recent and Historical Case Studies

Jonathan D.C. Webb[1] and Derek M. Elsom[1,2]

[1] *Thunderstorm Division, Tornado and Storm Research Organisation (TORRO)*, Oxford, UK*
[2] *Faculty of Humanities and Social Sciences, Oxford Brookes University, Oxford, UK*

9.1 Introduction

9.1.1 Establishment of a Tornado and Storm Research Organisation Research Database of Hail Events

Data collection from current sources, now predominantly digitised communications, as well as extensive historical searches of meteorological journals (e.g. *British Rainfall*, 1861–1968), earlier scientific journals (e.g. *Philosophical Transactions of the Royal Society*), *Gentleman's Magazine*, *The Times* newspaper, regional newspapers and historical documents, has enabled the construction of a database containing more than 2500 hailstorm events in Britain and Ireland which extends back to the first known damaging hailstorm at Wellesbourne (Warwickshire) in 1141 AD. Requests for information have been made via press appeals and, more recently, the Internet. Initial public and media descriptive reports, usually involving comparisons with familiar objects like marbles and golf balls, are liable to slightly overestimate hail size as noted by Changnon (1971). However, with recent events, follow-up correspondence has resulted in clarification; for example, some initial reports of golf ball-sized hail[1] have been subsequently found to refer to hail in the 31–40 mm diameter range. This verification has been supported by damage surveys, photographic evidence and analysis of high-resolution radar. Annual summaries of hailstorm events have also been published in the *Int.*

Journal of Meteorology *since 1984, as part of the Tornado and Storm Research Organisation (TORRO) annual review. Details of all the most severe (H5+) events in the database are included in the Appendix A, www.wiley.com/go/doe/extremeweather.*

TORRO's wider research structure includes a forecasting division which focuses, within the scope of its research, on identifying, and thus predicting, the synoptic and meteorological conditions which give rise to damaging hailstorms. The recent destructive hailstorm of 28 June 2012 (peak intensity H6) occurred on a day for which a forecast of severe convective weather had been issued by TORRO (Knightley, 2013). A similar watch was issued for 1 July 2015 when severe thunderstorms across northern England deposited hail over 50 mm diameter. The main analyses that follow focus on Great Britain (i.e. England, Scotland and Wales). However, reference is also made to the more extreme events recorded in Ireland and also the Channel Islands. Comparisons with the results of some other hail climatology studies from continental Europe are also included to place the hail hazard in these islands in context.

9.2 Assessing the Intensity of Hail Falls

9.2.1 Hailstorm Intensity Scale

In 1986, the TORRO International Hailstorm Intensity (H) Scale was first introduced (Webb *et al.*, 1986), and a slightly revised and more compact scale has been in use since 2005 (Sioutas *et al.*, 2005, 2009; Webb *et al.*, 2009). The characteristic damage in Britain and Ireland associated with each increment (from H0 to H10) is listed in Table 9.1 but may need to be slightly modified

---

*http://www.torro.org.uk/

[1] The standard golf ball diameter (since 1990) is 43 mm.

Extreme Weather: Forty Years of the Tornado and Storm Research Organisation (TORRO), First Edition. Edited by Robert K. Doe.
© 2016 John Wiley & Sons, Ltd. Published 2016 by John Wiley & Sons, Ltd.

Table 9.1   *TORRO International Hailstorm Intensity (H) Scale.*

Intensity Category	Description	Typical Hail Diameter (mm)*	Probable Kinetic Energy, J-m²	Typical Damage Impacts
H0	Hard hail	5	0–20	No damage
H1	Potentially damaging	5–**15**	>20	Slight general damage to plants and crops
H2	Significant	10–**20**	>100	Significant damage to fruit, crops and vegetation
H3	Severe	20–**30**	>300	Severe damage to fruit and crops, damage to glass and plastic structures, paint and wood scored
H4	Severe	25–**40**	>500	Widespread glass damage, vehicle bodywork damage
H5	Destructive	30–**50**	>800	Wholesale destruction of glass, damage to tiled roofs, significant risk of injuries
H6	Destructive	40–**60**		Bodywork of grounded aircraft dented, brick walls pitted
H7	Destructive	50–**75**		Severe roof damage, risk of serious injuries
H8	Destructive	60–**90**		(Severest recorded in the British Isles) Severe damage to aircraft bodywork
H9	Super hailstorms	75–**100**		Extensive structural damage. Risk of severe or even fatal injuries to persons caught in the open
H10	Super hailstorms	>100		Extensive structural damage. Risk of severe or even fatal injuries to persons caught in the open

* Approximate range (typical maximum size in bold), since other factors (e.g. number and density of hailstones, hail fall speed and surface wind speeds) affect severity.

Table 9.2   *Hail sizes applicable to the TORRO Hailstorm Intensity Scale.*

Hail Size and Diameter (mm) in Relation to TORRO Hailstorm Intensity Scale

0	5–9	Pea
1	10–15	Mothball
2	16–20	Marble, grape
3	21–30	Walnut
4	31–40	Pigeon's egg > squash ball
5	41–50	Golf ball > pullet's egg
6	51–60	Hen's egg
7	61–75	Tennis ball > cricket ball
8	76–90	Large orange > soft ball
9	91–100	Grapefruit
10	>100	Melon

The size code is the maximum reported size code accepted as consistent with other reports and evidence.

for other countries to reflect differences in building materials and types. Table 9.2 reflects the hail size and diameter in relation to the Hailstorm Intensity Scale. Since the 17th century, only one British hailstorm (that of 15 May 1697 in Hertfordshire – see the following case study) is assessed to have attained an intensity of H8 on the H scale, while seven others have reached a definite H7. The H8 Hertfordshire event of 1697 is assessed as the same intensity as the great Munich storm of 11 July 1984 in Germany which resulted in some very serious injuries as well as widespread damage to buildings, vehicles and commercial aircraft.

9.2.2   Kinetic Energy

Use of simple hail gauges has been pioneered by individual observers in Great Britain, for example, by Meaden (1976) who discussed various designs and the potential for recording hail frequencies and (through calibration) diameters. However, on account of the relatively local and narrow swathed nature of hailstorms, a dense network of hail pads or full-time observers is required to provide useful climatological data. This has been achieved in southwest France and

has enabled a climatological survey, focussing on extreme hail days, to be published (Berthet *et al.*, 2013). A similar network of hail pads has been running in northern Greece (Sioutas *et al.*, 2005). The use of radar reflectivities to estimate hail size and kinetic energy (Waldvogel *et al.*, 1979) has been applied to several case studies using empirical relationships, for example, by Hohl *et al.* (2002a) and Schuster *et al.* (2006). Hail pad kinetic energy and corresponding radar data were compared by Schmidt *et al.* (1992), and there was good agreement in respect of the overall extent of hail swathes. However, one limitation of radar data is that high kinetic energies can result from either the presence of very large hailstones or a very high density of smaller stones (Hohl *et al.*, 2002a).

9.2.3   Hailstone Size and Damage

Maximum hailstone size, the primary basis of the TORRO classification, is the most important parameter relating to structural damage, especially towards the more severe end of the scale. Leigh (2007) found that hail around 40 mm diameter was the minimum size for damage to slate or pottery roof tiles, whereas the equivalent threshold for concrete tiles was hail of 45–50 mm diameter. Yeo *et al.* (1999) and Leigh (2007), in an analysis of the impact of the great 1999 Sydney hailstorm in Australia, noted that damage to car bodywork was mostly associated with hail over 30 mm diameter. Similar observations were made, with reference to central European hailstorms, by Hohl *et al.* (2002b) who also noted, however, that the enhanced impact of hailstones with strong horizontal winds could produce visible damage to vehicles with hailstones over 20 mm across. The aforementioned findings in respect of roof and vehicle damage have been strongly supported by evidence from TORRO's investigations and published case studies. Average hailstone diameter is the best guide in applying the hail intensity scale since larger hailstones are often oblate.

9.2.4   Other Factors Affecting Damage

The TORRO scale provides the flexibility to increase the rating by an increment on the basis of additional information, such as the ground wind speed, the quantity of hail on the ground and, if available, the recorded kinetic energy, and this assessment would be confirmed by the nature of the damage the hail caused. A recent example was a severe squall accompanied by horizontally blown 25–30-mm-diameter hailstones east of Hastings, Sussex, on 15 July 2007, assessed as H4–H5 (Webb, 2009); later, it was confirmed that some larger stones of up to 35 mm diameter

Figure 9.1 *Hailstones which fell at Elham, Kent, 15 July 2007.* © Peter Gay 2007.

were observed (Figure 9.1). Koontz (1991) noted that a 35-kt horizontal wind increased the impact energy of 50-mm-diameter hailstones by 30%.

Figure 9.2 plots the maximum hailstone size ranges among the significant damaging (H2+ intensity) hailstorms which occurred between 1930 and 2009; it refers to the 526 cases (out of a total of 684 events) where there are specific references to hail size. Assessment of the remaining events has been based on the recorded damage.

9.3 Annual Frequency of Hail

9.3.1 All Significant, Damaging Hailstorms

An analysis of all storms with a minimum intensity of H2 (sufficient to cause significant damage to fruit and crops) indicates general consistency in the reporting of these relatively severe events over the past 80 years (Figure 9.3). A separate analysis of hailstorms which reached the higher, severe intensity of H3 or more (sufficient to, at least, crack glass) indicated that the reporting has remained fairly consistent since the 1870s (Webb *et al.*, 2001).

The analysis in Figure 9.3 also indicates a relatively high frequency of H2+ events in the 1930s and since the mid-1980s. The high frequency in the 1930s may reflect that this was the period of greatest participation by the network of national thunderstorm observers in the Thunderstorm Census Organisation (TCO) (Bower, 1947), a UK organisation founded in 1924 which was subsequently incorporated into TORRO in the early 1980s. More recently, the TORRO membership scheme, coupled with the rapid expansion of electronic resources like the TORRO members' Internet forum, digital photography and social media, has greatly assisted the speed of data collection and confirmation.

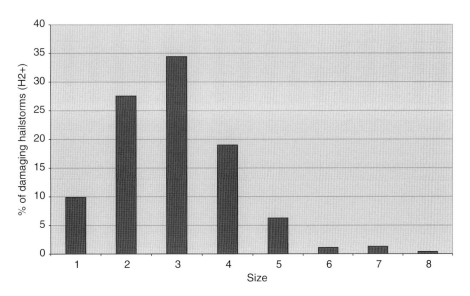

Figure 9.2 *Percentage of damaging hailstorms associated with specific hailstone maximum diameter ranges (mm), 1930–2009.*

Size code	1	2	3	4	5	6	7	8
Diameter	10–15	16–20	21–30	31–40	41–50	51–60	61–75	>75

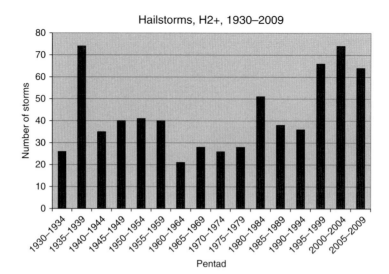

Figure 9.3   *Number of significant, damaging hailstorms (H2+ intensity) in Great Britain by pentads, 1930–2009.*

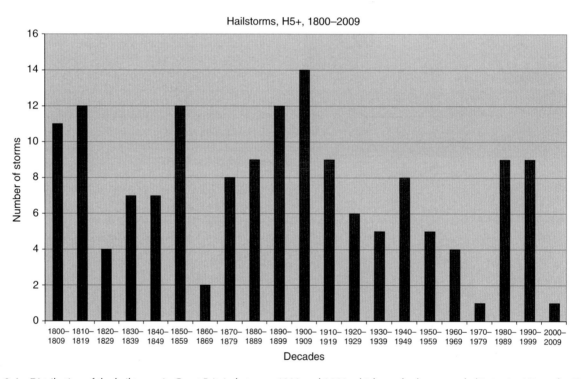

Figure 9.4   *Distribution of the hailstorms in Great Britain between 1800 and 2009 which reached or exceeded intensity H5 on the TORRO international scale, by decades.*

9.3.2   Frequency of Extreme, Destructive Hailstorms

Figure 9.4 shows the decadal distribution of the severest storms (the 6.6% reaching H5 intensity or more) since 1800. Historical consistency in the reporting of meteorological phenomena is highest for the most extreme events, although before 1800 under-reporting is likely on account of factors such as the relatively small number of newspapers in circulation (Clark, 2004).

Extreme events provide relatively small samples from which a cautious interpretation of trends is appropriate. Nevertheless, this distribution of extreme hailstorms shows some evidence of a downward trend, from a peak in the 1900s, through to the 1970s, with a reversion (at least temporarily) to a higher number of events in the last two decades of the 20th century.[2] Appendix A, www.

[2] However, there have been only three further events since 2000.

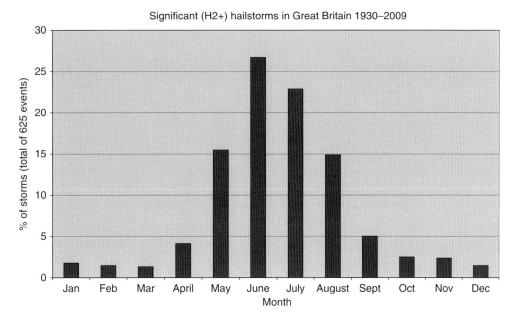

Significant (H2+) hailstorms in Great Britain 1930–2009

Figure 9.5 *Monthly analysis of significant (H2+ intensity) hailstorms, 1930–2009.*

wiley.com/go/doe/extremeweather and Table 9.6 do indicate that there has been a tendency for the most intense events to cluster together in several decades, such as 1935–1938, 1945–1947, 1958–1959, 1967–1968, 1983–1985 and 1994–1996.

9.3.3 Comparisons with Continental Europe

Sioutas *et al.* (2009) identified major hail zones in both the Northern and Southern Hemispheres including a European zone from the northern Iberian Peninsula across central Europe through to the northern Balkans. Webb *et al.* (2009) included comparisons between the incidence of damaging hail in Great Britain with the occurrence in some continental European countries evident from published studies by Tuovinen *et al.* (2007 and 2009) and Saltikoff *et al.* (2010) for Finland, Sioutas *et al.* (2009) for Greece and Dessens (1986) for south-west France.

9.4 Seasonal Occurrence of Hail

9.4.1 General Seasonal Incidence of Hail and Damaging Hailstorms

Using stations in the UK[3] synoptic and climatological network, Grant (1995) analysed data for 24 years (1972–1995). Although the occurrence of all hail (i.e. pieces of solid ice 5 mm diameter or more) showed a March peak, the incidences of hail 20 mm diameter or more were most common in June (28% of cases) and July (15%). This is confirmed by an analysis of TORRO's database for 1930–2009 which identified 620 hailstorms that reached a significant, damaging intensity of H2 or more. The early summer experiences the highest proportion of such storms, with 50% occurring in June (27%) and July (Figure 9.5). An analysis

[3] That is Great Britain and Northern Ireland.

for the slightly higher H3 intensity threshold (Figure 9.6) indicates a similar early summer peak.

9.4.2 Storms of H5 Intensity or More

Figure 9.7 presents an analysis of hailstorms of intensity H5 or more (specifically, for this analysis, H4–H5 or more) which have occurred in Great Britain from 1800 to 2009. As in a previous study (Webb, 1988), it indicates that the peak monthly incidence of these more severe storms is slightly later than for all hailstorms of intensity H2 or more, with July accounting for 41% of these events followed by June (21%); 93% occurred between May and August. A separate analysis of the 50 most intense British hailstorms for the past 350 years, that is, those that reached an even higher threshold of H5–H6 intensity or more (Webb *et al.*, 2001), also highlighted the high proportion of such events in July (40%).

Even during the winter months, severe hailstorms are occasionally reported near windward coasts. Figure 9.8 shows 35-mm-diameter hailstones which fell in the Carmarthen area of south Wales on 13 November 1974 (Smith, 1976); very unstable returning maritime polar air, with embedded troughs, was arriving around a very deep depression south-west of Ireland. During the night of 7–8 January 1998, the intense hail core of a severe thunderstorm, associated with a mesoscale waving coastal front, intermittently made landfall as it tracked east–north-east along the south coast from near Weymouth, Dorset, to Hastings in East Sussex. Hail up to 38 mm diameter accompanied the associated tornado event at Selsey, West Sussex (Pike, 1998).

9.4.3 Comparison with the Incidence of Thunderstorms

The seasonal distribution of all damaging hail events correlates closely with the incidence of thunderstorms in central and eastern England (Figure 9.9a, b). Grant (1995) identified 35 days between

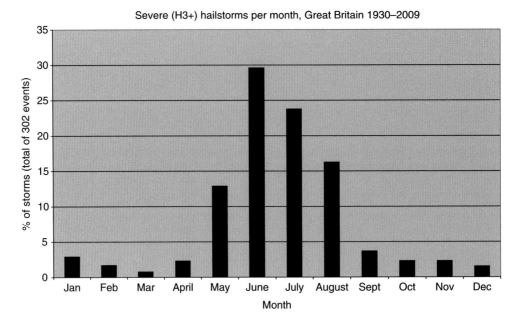

Figure 9.6   *Severe hailstorms (H3+) by month 1930–2009.*

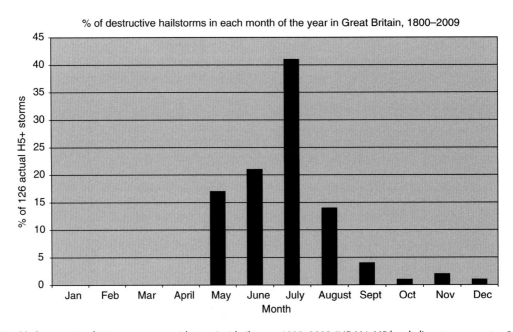

Figure 9.7   *Monthly frequencies of H5 or more severe (destructive) hailstorms 1800–2009 (NB H4–H5 borderline storms count as 0.5 of an event).*

1972 and 1994 when thunder was reported at over 30% of all stations in the UK synoptic and climatological network. The distributions of these especially thundery days were: 4 in May, 13 in June, 12 in July, 5 in August and 1 in September.

9.5   Geographical Distribution

9.5.1   Storms of H2 Intensity or More

The geographical distribution of hailstorms of H2 intensity or more is presented for the period 1930 to 2009 in England and Wales (Figure 9.10). Unit area analysis is applied using each

county (ceremonial/lieutenancy county boundaries for England and Wales are employed in this paper) and standardised as the number of recorded hailstorms per 1000 km[2] per 100 years. Some population bias is evident in the distribution with, for example, a relatively high frequency of storms in Greater Manchester (population density is 1945 people per km[2])[4] and an apparently low incidence in mid-Wales (where the population density is 25 per km[2] in Powys). In Scotland (Figure 9.11), only some eastern areas show a moderate incidence of damaging

[4] Source: Statistical Office.

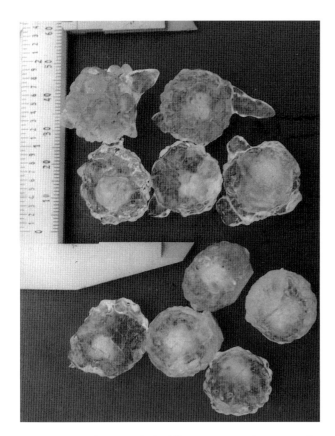

Figure 9.8 *Hailstones which fell at Towy Castle, Carmarthen (Dyfed), 13 November 1974. © D. Chris Smith (1974).*

hailstorms over the 1930–2009 period, these including Lothian and Fife and towards the Moray Firth. More recently, significant hail fell in the Lothian and Dundee areas on three occasions during the cyclonic Scottish summer of 2011 (Webb and Blackshaw, 2012).

9.5.2   Geographical Distribution of Storms of H5 Intensity or More

Figure 9.12 maps the incidence of the severest (H4–H5+) storms over England and Wales over the extended period 1800–2009, standardised as above. It indicates a conspicuous maximum occurrence towards the south-east Midlands and central East Anglia. However, all counties of England and Wales have been affected by at least one hailstorm of H5 or greater intensity since 1800. Hailstorms of this severity have been very rarely reported in Scotland or Ireland. Only one hailstorm *above* H5 intensity is known for Scotland, on 24 July 1818; this event in the Orkney Islands reached a remarkable intensity of H7 (see the following case study). The only other destructive (H5) hailstorms in Scotland since 1800 have occurred in the south-east regions (see Appendix A, www.wiley.com/go/doe/extremeweather).

The current database for Ireland includes only three such H5+ events. However, an H4 intensity hailstorm which accompanied a tornado across Dublin on 18 April 1850 (Lloyd and Hogan, 1850) ranks as one of the severest April hail events anywhere in these islands. No H5 storms are on record for the small land area of the

Channel Islands. However, there are three occasions during the last 30 years when hail of 25 mm diameter has been recorded at either Jersey or Guernsey Airports: on 5 June 1983 (Randon, 1983), on 29 June 1986 and during a typical fast-moving winter polar maritime thunderstorm on 26 January 1990.

9.5.3   Point Frequencies

It should be noted that while Figures 9.10, 9.11 and 9.12 give a general indication of the geographical distribution of hail events, the calculation of actual 'spot' frequencies is only possible where there is a network of hail pads. Moreover, meteorological phenomena are always erratic in their actual distributions. The Northampton area experienced two H5+ events in 1900 and another in 1935 (*see* Appendix A, www.wiley.com/go/doe/extremeweather). There is also a positive correlation between severity of individual hailstorms and the size of the area affected; the historic hailstorm of 1843 (Webb and Elsom, 1994) affected about 900 km² of Bedfordshire and Cambridgeshire.

9.5.4   European Comparisons

From hail pad data, Fraile *et al.* (2003) estimated that the point (1 km²) frequency of a storm with hailstones over 30 mm diameter and/or TKE (Total Kinetic Energy) above 500 J-m² (i.e. about H4+) was around 22 years in the most hail-prone departments of south-west France, north of the Pyrenees. This does confirm that severe hail is a relatively small feature of the climate of Britain and Ireland compared to parts of continental Europe. Dessens (1986) also discussed the distribution of damaging hailstorms in the Aquitaine region of France. The severity threshold used (hail 15 mm diameter or more or crop damage) was similar to the TORRO H2 level, and a map of frequencies per 900 km² squares indicated 1–2 severe hail falls per year across six departments of central Aquitaine. For northern Greece, Sioutas *et al.* (2009) analysed the distribution of damaging hail falls through (i) a hail pad network covering 2400 km² and (ii) insurance claims data for crop damage over a region of 6500 km². The hail pad data indicated 32 events of H2 intensity or more over a 9-year period: 3–4 per year (1.5 per 1000 km² per year). Figures 9.10, 9.11 and 9.12 provide little evidence of topography enhancing frequencies. However, this aspect needs more study in relation to storms in Britain.

9.6   Hailstorm Characteristics

9.6.1   Hail Swathes

Data for 1930–2009 indicated that only a minority of all damaging hailstorms, from H2 intensity upwards, affected a significant area. Of the 620 such events in this 80-year period, 380 involved reports, albeit sometimes numerous, from a single 5-km² location. However, there is a strong correlation between increasing severity and longer storm tracks. Between 1800 and 2009, path lengths are known for 123 of the 160 hailstorms in Great Britain of intensity H4–H5 upwards. Of these 123 events, 66% cases had swathe lengths of 20 km or more. Since 1800, all but three of the 42 hailstorms of intensity H5–H6 or more are known to have had swathe lengths of 20 km or more.

(a)

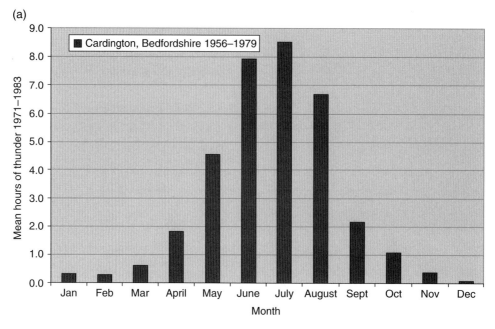

(b)

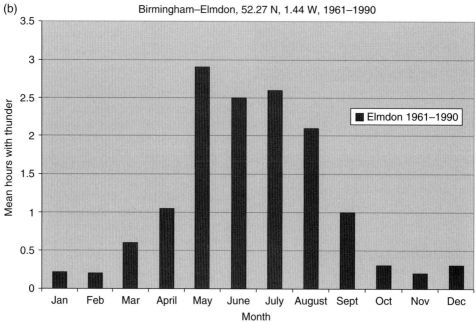

Figure 9.9 *(a) Monthly mean number of hours with thunder at Cardington (Bedfordshire), 1956–1979. Based on information supplied by the Met Office, UK (Marchant, 1990). (b) Monthly mean number of hours with thunder at Birmingham (Elmdon), 1961–1990. Based on information supplied by the Met Office, UK.*

In Ireland, at least one storm of exceptional areal extent has been observed: on 25/26 July 1985 (see also Chapter 7). Although the intensity was H3–H4 (30-mm-diameter hailstones were measured at Trassey climatological station in Co Down), severe hail damage to crops occurred from Co Kilkenny, north–north-eastwards across the east of the country to Co Armagh, a swathe length of about 200 km (Mortimore, 1986; Betts, 2003).

Since 1800, the 13 storms in Great Britain assessed as (i) H4–H5 intensity or more and (ii) with a known swathe length of 100 km or more have had an average maximum width[5] of 12 km, emphasising the magnitude of these events. Historical claims of hail swathes

exceeding a width of 20 km are probably attributable to two or more separate storms following parallel, even overlapping tracks such as with the Essex hailstorms of 26 May 1985 (Elsom and Webb, 1993; see also the following case study). Since 1974, TORRO has investigated some extreme hail events in detail (Table 9.3) by appealing for more observations of hail and undertaking local site surveys. Earlier UK hailstorms investigated in detail with published results include the Horsham storm of 5 September 1958 (Ludlam and Macklin, 1960; Meteorological Office, 1963).

[5] This breadth is evident for at least 10 km of the swathe length.

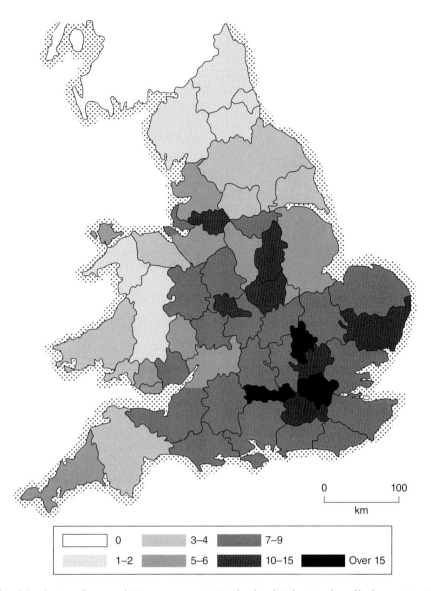

Figure 9.10 *Geographical distribution of storms of H2 or more severity, England and Wales. Number of hailstorms (per 1000 km² per 100 years) reaching or exceeding H2 intensity, for counties of England and Wales, 1930–2009 (legend above). (See insert for colour representation of the figure.)*

9.6.2 Radar and Hail Swathe Identification

The investigation of more recent events has had the advantages of the availability of precipitation radar displays and satellite imagery in addition to more widespread photographic evidence, though the latter needs careful verification. The 55-dBZ precipitation reflectivity threshold (100 mm/hour on the standard Marshall–Palmer conversion to rainfall rate but 127 mm/hour on the convective-specific conversion formula) has been widely used, internationally, as an indicator of the presence of significant hail. It has been specifically used in the identification of hail swathes, for example, by Schiesser (1990) in respect of crop damage and Schuster *et al.* (2006) and Hohl *et al.* (2002a) in respect of building damage, and these authors also extended the study to relate the radar reflectivity to associated hail kinetic energy. The relationship of reflectivity to particular hailstone sizes has been investigated, for example, by Auer (1972), by Smyth *et al.* (1999)

who closely studied the character of the falling precipitation to distinguish hail from rain and by Hardaker and Auer (1995) whose study combined radar with a cloud top height parameter. While ground observations have been the basis of swathe lengths in Tables 9.3 and 9.4, these reports have been supplemented by reference to radar.

9.6.3 Results of Hail Swathe Analyses

An analysis of all 123 storms of H4–H5 intensity or more since 1800 with known swathes (in Great Britain) indicates that the most common tracks were from SW to NE (30%) and SSW to NNE (21%), with 89% of storms moving from quadrants between SSE and WSW. The 16 most extreme storms regarding swathe length, that is, those exceeding 100 km (all but two of which were at least H5 intensity), have shown an especially strong tendency for a SW–NE track (Figure 9.13). This correlates with the

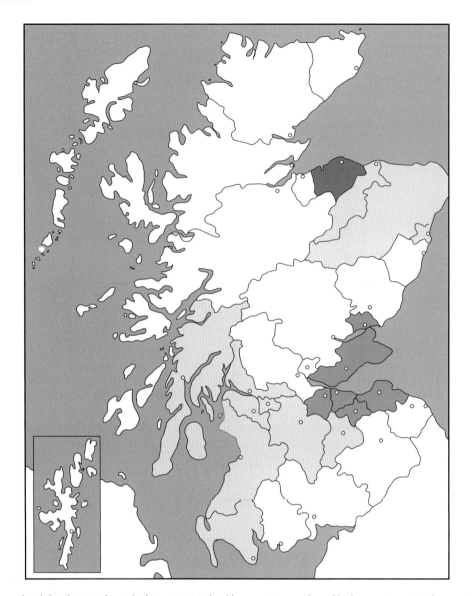

Figure 9.11 *Geographical distribution of H2+ hailstorms in Scotland by counties, number of hailstorms (per 1000 km² per 100 years) reaching or exceeding H2 intensity, for lieutenancy areas of Scotland, 1930–2009 (white = 0–1, light blue = 2, mid blue = ≥3, dark blue = 5). (See insert for colour representation of the figure.)*

most common mid-level (500–700 hPa) wind flow and also the tendency for supercell storms to propagate somewhat to the right of the general steering wind flow. Likewise, the track of the exceptional Ireland hailstorm of July 1985 was from south–south-west and just to the right of the southerly 500-hPa steering wind.

9.7   Synoptic Weather Types and Hailstorms

9.7.1   Specific Synoptic Background to Hailstorms

Prior to 1871, a reconstructed synopsis is available for a few extreme events that have been the subject for case studies. Surface and upper air data are available from the NCEP and ECMWF archives of reanalysis charts, from 1871 to 1957, respectively

(NCEP, Wetterzentrale, ECMWF, 2014). Synoptic information has been available (1871 onwards) in the numerous case studies published in meteorological journals, the latter including *British Rainfall*, the *Meteorological Magazine*, *Weather* and the *Int. Journal of Meteorology*, also from the *Daily Weather Report* (DWR) of the UK Met Office (to 1980) and the latter's successor, the *Daily Weather Summary*.

Upper air observations are increasingly referred to in case studies since the 1930s. Even prior to the availability of upper air soundings, an indication of mid-level wind flow can be gained from the confirmed direction and speed of movement of storms, considering that the cumulonimbus storm steering wind tends to be at about 500 hPa in situations favourable for severe convective storms. The favoured hailstorm tracks in the British Isles, from between south-east and south-west, are consistent with warm air

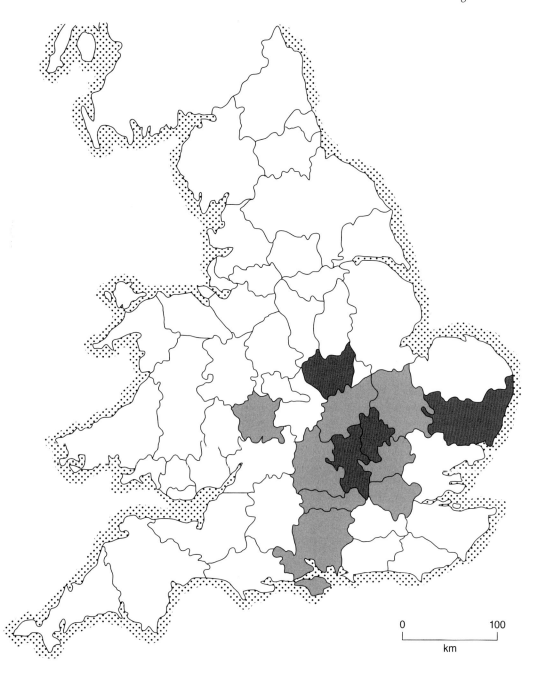

Figure 9.12 *Geographical distribution of destructive (H5+) hailstorms over England and Wales, 1800–2009. Number of hailstorms (per 1000 km² per 100 years) reaching or exceeding H5 intensity, for counties of England and Wales, 1800–2009 (white = 0–1, light shade = 2, darker shade = ≥3).*

advection on the forward edge of an upper trough approaching from the west, conditions associated with the severe thunderstorm Spanish Plume situations (Lewis and Gray, 2010). Such scenarios are evident in many of the case studies presented later in this chapter and are associated with large values of convective available potential energy (CAPE) (see also Chapter 12). CAPE is a measure of the energy available for convection and thus impacts the potential vertical updraught speed within a cumulonimbus. On an upper air ascent (e.g. see Chapter 3), CAPE is represented by the area between the environmental temperature profile and the 'curve' of a rising air parcel, that is, the area in which a rising parcel remains warmer than its environment and upwardly buoyant.

Case studies have confirmed that strong vertical wind shear, especially between ground level and 6 km, has been a consistent feature of the most extreme UK hailstorms. Strong shear is associated with supercell thunderstorms which have a persistent rotating updraught (mesocyclone). Not only is the long-lasting

Table 9.3 *Summary of observations of hail during four very severe hail events across England.*

Date of Event	Hail Swathe Length km	Start and Finish Locations	Number of Observations of Hail	Max Confirmed Hail Size mm	References
5 June 1983	200	50.62°N, 2.51°W to 50.77°N, 0.29°E	80	50+	Mortimore and Rowe (1984)
26 May 1985	75	51.57°N, 0.14°E to 52.12°N, 0.84°E	47	60–70	Elsom and Webb (1993)
*7 June 1996/swathe 1	80	51.93°N, 0.10°E to 52.43°N, 0.64°E	21	50	Webb and Pike (1998a)
7 June 1996/swathe 2	18	52.17°N, 0.13°W to 52.25°N, 0.09°W	12	50	Webb and Pike (1998a)
7 June 1996/swathe 4	165	50.54°N, 2.45°W to 51.85°N, 1.10°W	39	50	Webb and Pike (1998a)
7 June 1996/swathe 7	180	50.97°N, 1.73°W to 52.19°N, 0.02°E	40	50+	Webb and Pike (1998a)
28 June 2012	120	52.41°N, 1.64°W to 53.03°N, 0.25°W	42	80–90	Clark and Webb (2013)

* Eight hail swathes (with max hail size over 15 mm diameter) in total; of the other four, two produced max hail over 20 mm diameter.

Table 9.4 *Swathe directions for hail swathes over 100 km long (all intensity H5 or more), 1800–2013.*

	Date of Event	Hail Swathe Began (County and Location)	Hail Swathe Direction
1	9 August 1843	Gloucester – 51.92°N, 1.72°W	SW → NE
2	24 June 1897	Buckingham – 51.51°N, 0.59°W	WSW → ENE
3	2 August 1906	Buckingham – 52.07°N, 0.67°W	SW → NE
4	27 May 1913	Bedford – 52.23°N, 0.47°W	WNW → ESE
5	4 July 1915	Somerset – 51.24°N, 3.00°W	SW → NE
6	4 September 1935	Gwent – 51.58°N, 2.99°W	SW → NE
7	5 September 1958	W Sussex – 50.78°N, 0.90°W	SW → NE
8	9 July 1959*	Hampshire – 50.97°N, 1.73°W	SW → NE
9	2 July 1968	G Manchester – 53.40°N, 2.06°W	SW → NE
10	14 July 1975	West Midlands – 52.45°N, 1.94°W	SSW → NNE
11	5 June 1983	Dorset – 50.62°N, 2.51°W	W → E
12	7 June 1983	Shropshire – 52.93°N, 3.01°W	SSW → NNE
13	7 June 1996 (1)	Dorset – 50.54°N, 2.45°W	SSW → NNE →
14	7 June 1996 (2)	Hampshire – 50.97°N, 1.73°W	SW → NE
15	7/8 January 1998†	Dorset – 50.97°N, 1.73°W	WSW → ENE
16	28 June 2012	West Midlands – 52.41°N, 1.64°W	SW → NE

* Technically, there were two hail swathes from this storm, with a brief phase during which only rain reached the ground (Browning and Ludlam, 1960).
† Intermittently passed over the English Channel as well as onshore.

nature of such storms especially favourable for hail growth, but the mesocyclone gives the updraughts much greater momentum than the CAPE alone would produce (Weisman and Klemp, 1984). Such storms are also the most likely to produce strong downburst squalls and associated wind-driven hail which can result in more severe damage than the hail size alone would indicate. On 24 July 1994, a hailstorm in south Lincolnshire caused H6 intensity damage although the largest hailstones recorded were about 40 mm diameter. Vertical wind shear was strong; the Aughton ascent registered a surface wind of 110°/4 kt veering to 225°/25 kt at 700 hPa.

There have been 13 occasions during the past 60 years when hailstones 50 mm diameter or more have been confirmed in a storm in the United Kingdom (Table 9.5). In all these cases, strong wind shear was present and indeed was remarked upon in published case studies, a list of which is included in Webb *et al.* (2009). In 10 of these 13 cases, there was a hail swathe over 50 km long, and in each instance, the trajectory veered slightly to the right of the mid-level steering wind. Both these features are consistent with supercell storms. Favourable conditions for supercell development can provide updraughts to support significant hail even when CAPE values are relatively low. Westbrook and Clark (2013) described a recent such tornadic storm which deposited intense hail at least 15 mm diameter across Oxfordshire on 7 May 2012. Such shear is also important in generating the occasional severe winter hailstorms in the United Kingdom such as the south Devon storm of 13 December 1978 (Owens, 1980).

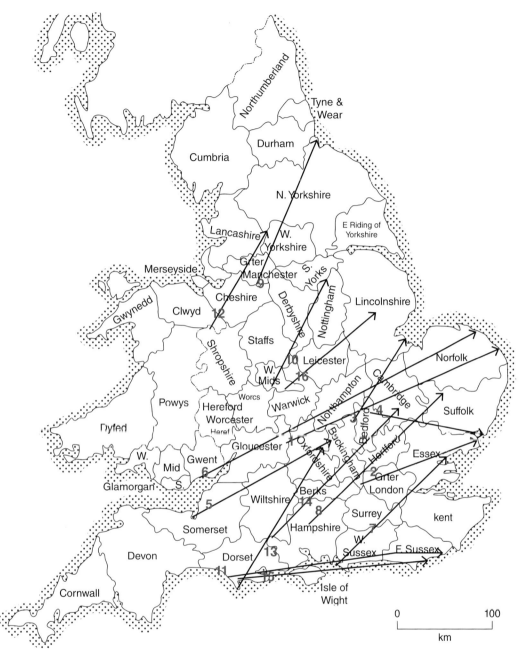

Figure 9.13   *Swathe tracks of 16 hailstorms which exceeded H5 intensity and had path lengths of over 100 km, 1800 to 2013. See also Tables 9.4 and 9.5 for dates.*

9.8   Hour-of-Day Distribution

While diurnal occurrence analysis of the 156 severest hailstorms on record since 1800 (Figure 9.14) indicated, not unexpectedly, a strong preference for occurrence near or just after the mean time of peak daily temperatures, it is also evident that some such events have occurred independently of surface heating overland (as the trigger for surface-based convection). It may be noted that all five events recorded between 0000 and 0600 GMT occurred between July and September although there is a recent case of 30-mm-diameter hail falling in at least three separate areas of England during the early hours of 1 May 2005 (Figure 9.15). The results may be compared to the diurnal record of thunder observed over a 60-year period in north-east Bristol, England (see Chapter 7). Dessens (1986) and Gaiotti *et al.* (2003), in south-west France and Northern Italy, respectively, noted that hailstorms peaked around 1600 GMT, but that the proportion of overnight events increased later in the season (i.e. July–September).

Table 9.5 *Occasions when hail with an average diameter of 50 mm or more has been confirmed as falling in the United Kingdom, 1950 to 2013.*

Date	Max Hail Diameter (mm)	Hail Swathe* Length (km)	Hail Swathe Trajectory	Surface Winds (Dir/Speed)	500 hPa Winds[†] (Dir/Speed)
5 September 1958	70–80	100	SW → NE	120°/05 kt	210°/29 kt
9 July 1959	50	220	SW → NE	030°/08 kt	210°/44 kt
13 July 1967	75	90	SW → NE	060°/06 kt	210°/41 kt
1 July 1968	68	90	SSW → NNE	040°/11 kt	180°/57 kt
14 July 1975	50–55	105	SSW → NNE	160°/2 kt	200°/37 kt
5 June 1983	50+	200	W → E	030°/15 kt	235°/38 kt
7 June 1983	50–75	130	SSW → NNE	150°/09 kt	200°/27 kt
26 May 1985	60–70	75	SSW → NNE	160°/05 kt	190°/36 kt
7 June 1996	50+	180	SW → NE	060°/07 kt	210°/51 kt
17 May 1997	50–70	10+	SSE → NNW	050°/07 kt	165°/35 kt
28 June 2012	50–90	120	SW → NE	160°/07 kt	215°/42 kt

* For hail over 10 mm diameter.
† Based on nearest radiosonde station.

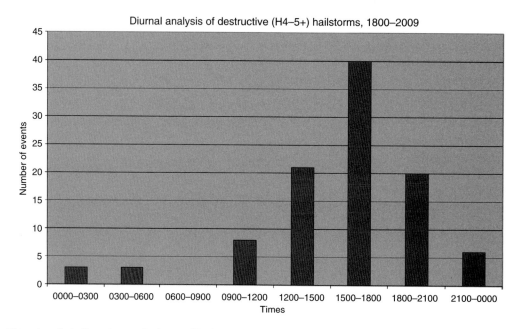

Figure 9.14 *Diurnal analysis (based on peak phase) of hailstorms of H4–H5 intensity or more, 1800–2009, based on 98 (out of 120) events for which times are known.*

Figure 9.15 *Hailstones which fell in north-east Bristol, England, early morning on 1 May 2005. © Bryan A. Williams (2005).*

9.9   Summary of TORRO's Overall Findings

More than 2500 British and Irish hailstorms are recorded in TORRO's database of which over 1300 are of hailstorm intensity H2 or more, including 160 British storms since 1800 that reached intensity H4–H5 or more on the TORRO H scale. The latter, more intense storms are shown to be mostly restricted to the period from May and September, with a well-defined peak during July, typically following a track from the S, SSW or SW to the N, NNE or NE with a swathe length of usually 20 km or more (exceeding 200 km in exceptional cases) and a swathe width sometimes reaching 10–16 km. TORRO's research is ongoing, so further historical events may come to light in the future, and, likewise, additional information may result in reassessments of storms currently in the database.

We now turn attention to case studies of 20 of the most severe hailstorm events in the TORRO databank. Several additional severe hailstorms are referred to in Chapter 7, including the great west

London hailstorm of 3 August 1879 and the south-east Scotland hailstorm of 29 August 1930.

9.10   Twenty of the Most Severe Hailstorms

9.10.1   1687: The Alvanley Storm

On 28 June 1687, Alvanley, Cheshire, was hit by a thunderstorm with hailstones up to 230 mm circumference (70 mm diameter). The fierce winds overturned a windmill, while the hail broke all windows facing south-west. Both slate and thatch roofs were torn and uncovered, and bark on trees was ripped as if by a knife. The storm is described in the Palatine Notebook (Anon, 1883) and in a published letter (Anon, 1687).

9.10.2   1697: Remarkable Late Spring Storms

A series of devastating hailstorms occurred over a period of 4 weeks. On 10 May, a severe hailstorm moved north–north-east across a 95-km swathe from St Asaph in north-east Wales, across the Liverpool area to Fleetwood, and on to Blackburn in Lancashire. The hailstones breached thatched roofs and snapped boughs off trees. The hailstones averaged 5 ounces (140 g), were up to 230 mm circumference (70 mm diameter) and were described as 'goose egg' size. The largest were reported to be 8–12 ounces (225–340 g). In Dyserth, Flintshire (Clwyd), lambs, poultry and a dog were killed by the hailstones, while people sustained severe head and body injuries. All windows were broken on the windward side. In Ormskirk, windows and roof tiles were broken. Although the city of Liverpool just missed the storm, it became extremely dark under the shelf cloud.

Besides the severity, this event was also significant for the identification of a major hailstorm swathe. Previous severe British hail events, like the Alvanley storm (see preceding text), had been described in some detail but at a local level. In contrast, accounts of the two extreme storms on 10 and 15 May 1697 pick up on the identifiable swathe of hail reports and damage extending along a track many kilometres long and attributable to a single storm system (Royal Society, 1697a, b).

On 15 May came a storm which, for sheer severity, ranks as the most extreme British hailstorm of the past 350 years (Tord, 1697). At Offley, near Hitchin in Hertfordshire, the hailstones were measured at 343 mm circumference with some anecdotal reports indicating 445 mm. These correspond to measurements of about 110 and 140 mm diameter, respectively, and the pieces of ice were described as 'some oval, some round, some flat'. At Offley, 7000 quarries of glass were broken at a house, the ground was torn up, and great oak trees were split. Tiles and windows of houses were all shattered to pieces. At least one human fatality was attributed to the hail, a young shepherd, while another person caught in the open was severely bruised but evidently recovered. The storm swathe was at least 25 km long, extending north-eastwards to Potton (Bedfordshire). At Hitchin, the hailstones were around 60 mm diameter and were reported to be piled 5 ft (150 cm) deep. At Ickleford, the hail broke most of the windows and tiles of a manor house. At Potton, 'two new houses were entirely levelled with the ground', and on the south-west side of the town, the rye crop was both burnt by lightning and beaten down by the hail, an interesting parallel to what happened in west Berkshire on 7 June 1996 (Webb and Pike, 1998a). Robert Tailor, an apothecary at

Hitchin, wrote, 'Thunder, lightning and heavy showers had occurred from 9 a.m. to 2 p.m., then the great storm approached from the southwest…the wind being east and blew hard'. This is an early observation associating strong wind shear with severe thunderstorms!

On 6 June, another round of devastating hailstorms was reported, this time on the England/Wales border. At Westhide, Herefordshire, many hailstones were more than 70 mm across, while at Pontypool, Gwent, they were 65 mm diameter.

9.10.3   1763: The Great Kent Storm of 19 August

This devastating hailstorm and squall cut a 65-km swathe right across Kent, 7–8 km wide in places, with the peak intensity between Tonbridge and Maidstone. At Yalden, eight barns were blown down and other buildings unroofed. At Tonbridge, the sky darkened so much it appeared 'like an almost total eclipse of the sun'. This preceded the violent thunderstorm from 1200 to 1240 h[6] when the town was hit by flash flooding and hailstones about 40 mm diameter. At Wateringbury and Nettlestead, the hailstones reached 80 mm diameter and accumulated 90–120 cm deep. Windows and tiles were broken to pieces, walls battered, and 'bark very much wounded and torn off'. At Barming, east of Maidstone, the 'oyster-shaped' hailstones measured around 70 mm diameter, and some 2000 acres of hops were destroyed.

9.10.4   1808: The Great Somerset Hailstorm of 15 July

This event was outstanding, not only for the record size of the hailstones but also for careful scientific data collection and review by a local teacher, A. Crocker, master of Frome School. In the introduction to his account of this historic storm, published in the *Monthly Magazine and British Register* (Crocker 1808), he highlighted (i) the value of careful investigation of storm events based on verified first-hand observations and (ii) the importance of getting such investigations published for the benefit of future generations.

At Buddon House, near Wincanton, 'hail descended with such velocity as to entirely strip or loosen bark, not one tree escaped'. Windows were broken with violence, and oak doors deeply indented. At Templecombe and Batcombe, the largest hailstones measured 105–109 mm diameter with many reported as weighing over 220 g. At Castle Cary, the pieces of ice were around 90 mm diameter and compared to split nutmegs. At Mells, hailstones were up to 70 mm diameter, 'wherever the hailstones struck a tree or branch, the bark was torn off' (Clark, 2004). At Newby House, Mells, all west-facing windows were destroyed, and many Cornish slates and pantiles broken.

Fortuitously, the main storm swathe passed between the cities of Bath and Bristol, affecting just the eastern outskirts of Bristol (as it was then) though the city experienced the violent thunderstorm with hail the size of small marbles. However, extensive damage was done in villages east and north of Bristol, such as Keynsham, Newton, Kelston, Preston and Corston. Brislington, Frenchay and Almondsbury were also affected. Here, most exposed windows were broken, small branches broken off, and fruit crops destroyed. The storm crossed the Severn, and in the

---

[6] Local time.

Monmouth area it was reported that 'boughs of trees were cut asunder'. Once again, the wind shear with a severe hailstorm was unintentionally described. In Bath, the 'tempest arose in the southwest, spread to northwest and died away in the northeast, opposed for nearly all its continuum by a strong breeze' (*Bath Chronicle*). There were two hail swathes, each up to 10 km wide, the main one at least 95 km long from north Dorset to Monmouthshire and a secondary swathe, some 15 km long, centred near Glastonbury. The latter event produced hailstones around 45 mm diameter at Ashgate and over 50 mm across at High Ham.

Later in the evening, violent thunderstorms occurred further north-east in Gloucestershire and Worcestershire with further large hail, mostly 13–25 mm diameter but sufficient to break glass in such places as Tetbury (Glos) and Broadway (Worcs).

9.10.5   1818: Stronsay, Orkney, 24 July

This was the farthest north that hailstones over 50 mm diameter have ever been recorded in Britain. This event reached an intensity of H7, creating a 32-km-long damage swathe from Stronsay to North Ronaldsay in the Orkney Islands, the extreme northern part of Scotland. The following vivid account of the storm's approach (between 12 noon and 1 p.m. local time) was given by an eyewitness (Neill, 1823), and an accompanying map of the hail swathe was plotted:

> A very dense jet-black cloud rose from the sea five miles south and approached steadily in a direct line towards the centre of the island. As it approached, it grew larger and the sky grew darker, with vivid flashes of lightning and tremendous peals of thunder. One flash of lightning was much brighter than the rest and it appeared that the cloud was cleft asunder so that you could see the Orkney mainland through the gap, and it appeared to strike the Stronsay Firth like a solid body smashing into the sea. The wind increased dramatically, the sea became very rough, and darkness like that of night, threatened to come on. Very large hailstones began to fall and one, the size of a goose egg, crashed through the glass of one of the windows and struck the floor violently. The wind increased almost to hurricane force, and instead of hailstones, pieces of ice of all shapes and sizes came crashing down, breaking all the glass in our south facing windows. Lumps of ice were seen striking the water, making it appear as if it was boiling, and lightning in the form of balls or masses of fire was striking the sea with such ferocity, that the sea was dashed up as high as the masts of ships.

He added that 'the hailstones had fallen not only from a great height, but, owing to the strength of the wind, at a very considerable angle'. Mr Taylor was obliged to flee from one room to another in his house, in order to avoid the fragments of glass, which were driven to the farther side of the apartment. In his bedroom, the wash-hand basin, although standing at some distance from the window, was shivered in pieces by a hailstone.

The meeting house and manse of the Reverend Mr Taylor were also in the direct path of the storm. In both dwellings, the south windows were wholly shattered, not only the glass but the wooden astragals (framework) being broken. He observed by his watch the duration of the violent wind and heavy hail and states it to have been little more than 8 minutes.

The *Caledonian Mercury* reported intense thunderstorm activity which also affected parts of mainland Scotland on this day. Lightning struck and damaged Dumbarton Castle, and a carriage was struck near Inveruglas in the Trossachs, while a violent shower of large hailstones was reported on Cairn o' Mount, Aberdeenshire; in Montrose, much loud thunder was heard from this latter storm as it passed to the north-west of the town.

9.10.6   1843: The Great Hailstorm of 9 August

In terms of both severity and the area affected, this is believed to be Britain's severest recorded hailstorm. There was 12–18 hours of continuous thunderstorms across the Midlands on this historic day and lightning, and flash flood damage was also widespread (Webb and Elsom, 1994). The general storm track was south-west to north-east, but the overall track of the great hailstorm was west–south-west to east–north-east, and the hail swathe was 15 km wide in places, for example, across north Oxfordshire. On several occasions, this principal hailstorm temporarily veered to a more easterly track than the other storms, reflecting a propagation to the right similar to that noted during recent British supercell hailstorms such as that on 28 June 2012 (Clark and Webb, 2013).

The severity of the damage indicated that this 1843 storm reached an intensity of H7 on the TORRO international hailstorm scale. The wholesale devastation of many cereal and fruit crops along the swathe prompted a series of appeal funds to be launched and also led to the founding of the General Hail Storm Insurance Society which later merged with the Norwich Union. Two other very damaging hailstorms affected Berkshire[7] and Kent.

An anticyclone drifted eastwards into northern France on the 7th introducing a humid, rather cloudy south-westerly airstream. This anticyclone declined and moved away eastwards on the 8th as a cold front edged slowly south-eastwards into the Irish Sea; the stagnant air over England became increasingly warm and oppressive. Meanwhile, a shallow, thundery depression developed over Brittany, and the cold front subsequently became almost stationary from the Bristol Channel to the Humber by the 9th (Figure 9.16). As pressure rose to the west of Ireland, a much cooler, brisk north-easterly airflow extended across Wales and northern England. During the following 24 hours, the depression over Brittany transferred north-eastwards to the southern North Sea with the cold front continuing very slowly and erratically south-eastwards. Both the movement of the main hailstorm (at about 20 kt per hour) and the previous synoptic weather type suggest that there was a fresh south-westerly airflow at 500 hPa. It is interesting that the reconstructed synoptic situation resembles that for the 'Wokingham storm' of 9 July 1959 (Figure 9.24). The light southerly winds on the warm side of the cold front would have provided an ideal inflow of very humid surface air supplying additional energy to the developing storms.

Thunderstorms broke out in the Bristol Channel area between 0200 and 0400 GMT on 9 August 1843. Exceptionally severe thunderstorms affected the south-west Midlands between 0400 and 1000 GMT. The turnpike road from Gloucester to Painswick was flooded for 3 km, and the mills at Upton St Leonards were inundated. Seven houses in Worcester were struck by lightning, and a 'fireball' was reported to have entered

---

[7] As per pre-1974 county borders.

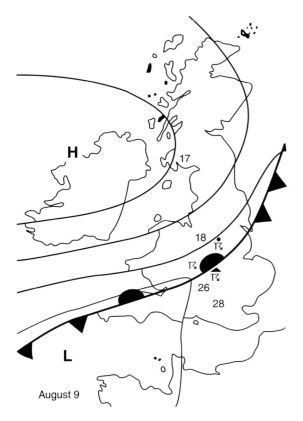

Figure 9.16 *Synoptic chart for 9 August 1843, temperatures in °C. First published by Webb and Elsom (1994), based on information provided by John Kington.*

a house at Gaydon, Warwickshire, via the chimney, causing considerable damage. There were a few isolated reports of hail during this period, notably one of 'hailstones 25 mm thick' at Evesham, Worcestershire. Between 0500 and 0700 GMT, thunderstorms spread north-east across Leicestershire, Nottinghamshire, Rutland, Lincolnshire and East Yorkshire. These counties experienced the most prolonged thunderstorm activity that remarkable day, lasting for over 12 hours and including many serious lightning incidents.

The main hail swathe (Figure 9.17) was 255 km long and was 15–16 km wide in places! This great hailstorm developed over the northern Cotswolds around noon. At Stow-on-the-Wold, a severe thunderstorm occurred between 1200 and 1300 GMT, accompanied by 'large jagged lumps of ice which damaged windows and crops in neighbouring villages' (*Gloucester Journal*). A swathe about 10 km wide, extending across north Oxfordshire, experienced hailstones with a maximum diameter exceeding 30 mm. A letter in *Jackson's Oxford Journal* described the storm over Chipping Norton:

> About 1 o'clock … a tremendous storm never witnessed by the oldest inhabitant; the vivid lightning showed itself in the glare of midday; the artillery of the clouds discharged their awful cannonade and the hail began to descend with pitiless fury. Who can describe the smashing of windows in the town, the rattling of the broken panes as they fell among various articles, or were driven across various rooms.

The *Banbury Guardian* reported that about 50,000 panes of glass were broken in Chipping Norton during the 20-minute onslaught. Large quantities of slates were also broken by the hailstones, many of which were described as measuring 20–23 cm in circumference (60–75 mm diameter). Apples were split in two with 'one half left hanging on the tree'. At nearby Heythrop, the gardener's house required reroofing following damage to the slates. The *Memoir of a Hailstorm* by Rev John Jordan of Enstone provided a very detailed, scientific account and survey, recording such impacts as severe damage to Welsh and Stonesfield slate roofs:

> All my young fruit trees were so seriously injured by long pieces of bark stripped off them … they must be cut down to produce new shoots. Several houses having been slated with a slate which is not so hard as the Stonesfield slates, the roofs were entirely pounded to pieces, so as no longer to afford the least protection. Those of my own house, which were some of the best in the neighbourhood, were many of them broken. The slater has assured me that he could not, with a stone as large as his fist, strike a blow, by throwing it, of sufficient power to break the slates as the hailstones did.

He measured and weighed a sample of eight hailstones, all spheroidal, at an average of 47 mm diameter and 57 g. The largest in the vicinity was 60–65 mm diameter (200 mm circumference). Similar-sized hailstones were measured at Tracey Farm, Great Tew, but with the additional hazard of 'a complete hurricane'; 'all windows facing the storm were smashed in'. At nearby Sandford St Martin, 'windows were smashed in, leads and all, and a cartload of slates knocked off a roof'.

The storm was also devastating in the vicinity of Biggleswade, Bedfordshire, where hailstones the size of walnuts were accompanied by violent squalls. All houses in the town that faced north, north-west or north-east had broken windows, and the windows of the church were reported to have been totally demolished.

At Cambridge Observatory, James Glaisher reported the continuous rumbling of thunder to the west and north-west from 2 p.m. (1400 GMT):

> From 4 to 4.45 p.m., the storm approached rapidly in an almost due east direction passing rather northward; some large drops of rain fell in the interval and the flashes of lightning became very vivid and of a brilliant purple colour. At 4.45 p.m. the hailstorm began and for 20 minutes continued. Many hailstones measured an inch (25 mm) in diameter; some were even larger. They fell as closely as the drops of rain from a waterspout and with their weight and a brisk northeast wind caused immense destruction. The hail fell upon the earth where temperature was considerably higher and then a mist or almost steam arose.

The *Cambridge Advertiser and Free Press* and the *Cambridge Independent Press* gave a very comprehensive account of the storm damage. Hailstones the size of hen's eggs, 6 in. (about 15 cm) in circumference, were reported to have been picked up in the Trumpington Road district. Many buildings in the town, including the Town Hall and St Clement's church, sustained the almost total destruction of north-facing glass; 800 ft² (70 m²) of glass was broken at the County Courts, about 3000 panes of glass were destroyed at the Botanic Gardens, and nearly 5000 window panes were smashed at the University of Pitt Press. Every college

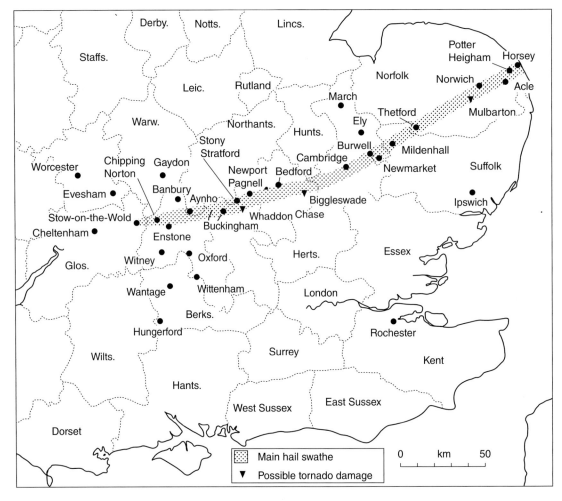

Figure 9.17 *The great hailstorm; course of the main hail swathe of 9 August 1843. First published by Webb and Elsom (1994).*

reported some broken glass, and some leaden window frame-work was battered in; even iron was reported bent. There was also much destruction of exposed slates and chimney pots in the town. The storm was also very destructive through the Cambridgeshire villages of Grantchester, Barton, Cherry Hinton, Bottisham and Quy. Fields of cereal crops were completely threshed. Almost all north-facing windows in Quy were broken, and the adjacent road to Anglesey Abbey was knee-deep in hail-stones. At houses in Grantchester, 'the leadwork of window casements was broken out'.

The great hailstorm tracked steadily north-east from Thetford, where 'the hail and the hurricane broke almost every window that faced the onslaught', to reach Norwich between 1900 and 1930 GMT and subsequently Ludham, Potter Heigham and Horsey near the coast. The effects of the storm were devastating between Old Buckenham and just east of Norwich. Hailstones more than 50 mm in diameter were accompanied by violent squalls. At Markshall, trees were 'absolutely barked', and at Mulbarton, hedges were 'completely peeled' and trees left leaf-less and stripped of bark. Some houses in the villages of East Carleton, Bixley, Mulbarton and Wreningham had every window broken; indeed, the *Norwich Mercury* reported that 'from

Swardeston Dog to Mulbarton scarcely a pane was left entire'. Nearly all the windows of St Edmund's yarn factory in Norwich were broken. There was much destruction of greenhouses and the tops of gas lamps in the city. The flooding in Norwich was so severe that the stalls in the marketplace were swept away.

9.10.7   1893: Northern England and Southern Scotland on 8 July

The week of the wedding of the Duke of York (the future George V) in 1893 was meteorologically eventful. Violent thunderstorms with hail had already affected northern England on 2 July when an historic cloudburst occurred in the Cheviot Hills (Clark, 2005). Then, on 8 July, a hailstorm and violent squall broke 200,000 window panes of glass and damaged tiled roofs in Richmond, Yorkshire. During the storm, hailstones 50 mm diameter were collected and photographed by professional photographer H. J. Metcalfe (Rollo, 1893), making this probably the first British hailstorm with photographic evidence (120 years later, such an event would have generated hundreds of digital images from cameras and mobile phones, many uploaded to social media!). Another Yorkshire town, Harrogate, was simultaneously affected

Figure 9.18 *Large/damaging hail reports on 8 July 1893. Δ = hail ≥15 mm diameter, ▲ = hail ≥40 mm diameter. Data plotted by author. Outline map reproduced from Ordnance Survey map data by permission of Ordnance Survey. © Crown copyright, 2013.*

by a separate storm of almost equal intensity with hailstones 40 mm diameter.

Pressure was low west of Ireland and high over Scandinavia with surface east to south-east winds across central and northern Britain. At 500 hPa, a major upper trough had moved into Ireland. In a classic 'plume', temperatures exceeded 30°C on the 7th and 8th (33.3°C at Cambridge Observatory), and ahead of a sharp surface trough, temperatures rose quickly on the morning of the 8th to reach 28.3°C in Sheffield. Plots of the Dumfries, Richmond and Harrogate hail swathes all indicate the storms tracked from south–south-east to north–north-west (Figure 9.18).

Ten significant hailstorms occurred on this day. A violent squall and hail affected the Nith Valley, including Dumfries and Closeburn, around noon. Hailstones were measured at 40 mm diameter (round with a snowy core) with other irregular pieces of ice measuring 45 mm by 22 mm. At Maybole, Ayrshire, hailstones 25 mm across broke glass throughout the town. Further south, a violent thunderstorm squall with hail affected the Peterborough area, also towards noon, while at Binbrook and other locations in Lincolnshire, walnut-sized hailstones broke greenhouse glass.

Following the hailstorms, William Marriott of the Royal Meteorological Society visited the Harrogate and Richmond areas, the former just 2 days afterwards. He confirmed the initial reports of the storms. In Richmond, the storm commenced about 1420 GMT accompanied with a strong wind. Hailstones and glass were hurled to the opposite sides of rooms and even 8 m down a passage. A tile over 19 mm thick was broken with a 90-mm-diameter fracture. All windows facing east and south were broken. The largest hailstones were around 60 mm diameter and weighed 100 g. In Harrogate, 100,000 panes of glass were broken within a 5-mile (8 km) radius (J. Farrah, the R. Met. Soc. observer). Most windows facing south-east were broken as well as skylights and conservatory glass. Nearly all panes of glass on second floors were broken, and trees were blown down (Marriott, 1894).

Elsewhere, at Kirkby Stephen, Cumbria, hailstones 40 mm across were reported. Violent hailstorms affected the lower valleys of the Ouse and Wharfe rivers in the later afternoon, the hail typically described as plum to marble size. At Bramham, near Wetherby, hail up to pigeon's egg size was heaped 60 cm deep.

9.10.8   1897: The Diamond Jubilee Storm of 24 June

Devastating thunderstorms hit south-east England, East Anglia and the far south-east Midlands on 24 June 1897, causing havoc in the run-up to Queen Victoria's Diamond Jubilee celebrations. Worst affected were Essex and Bedfordshire where violent squalls and hailstorms caused immense damage to property and crops. The event was sometimes referred to as 'the Essex tornado', though the wind damage appeared attributable to a violent squall. The NCEP 500-hPa reanalysis and surface chart for 0000 GMT on 24 June 1897 showed a classic Spanish Plume scenario (quite similar to the severe storms of 24 June 1994 described by Young (1995)) with a bulge of very warm air over western Europe ahead of a sharpening upper trough approaching Ireland and Biscay. A brisk south-westerly airflow is evident at 500 hPa with a shallow surface depression over north-west France.

The surface analysis for 0800 GMT on 24 June 1897 (Figure 9.19) shows shallow low centres across eastern England. The 0800 GMT observations indicated very humid air across the south and east of England; temperatures were already 21–22°C with dew points of 16–18°C. A fine morning saw screen temperatures rise rapidly, attaining over 27°C widely with 32°C at Brixton, London.[8]

9.10.8.1   Thunderstorms Erupt between 1200 and 1400 UTC

Queen Victoria spent a 'comparatively quiet day' at Windsor Castle. However, it was remarked (*Daily News*, 25 June 1897) that the Lords of the Admiralty arrived at Windsor railway station at about 1315 GMT to the accompaniment of flashes of lightning and peals of thunder; these pyrotechnics sent garden diners scurrying indoors! *British Rainfall* described the violent storm which burst over Slough from 1315 to 1415 GMT when lightning struck and fired two houses; this storm also produced the first report of large hail. Further reports of hail came from Wealdstone, Edgware

and Finchley between 1330 and 1400 GMT; at Wealdstone, hailstones weighed over 1 ounce (c. 38 mm diameter). At Wembley Park, a thunderstorm squall swept away a large marquee where young people were sheltering, and several were injured but not seriously.

A dry squall swept across other parts of west and north London. In West Kensington, the sudden squall at 1330 GMT was accompanied by a temperature fall of over 8°C. At Ealing, the squall at 1350 GMT raised 'an extraordinary volume and density of dust' and also did appreciable damage to the Jubilee decorations. At Camden Square, two sudden strong squalls occurred between 1401 and 1409 GMT, and observer George Symons remarked that 'in NW and N a thick darkness covered everything'.

9.10.8.2   The Great Hail and Windstorm across Essex

At Chelmsford, a *British Rainfall* observer measured hailstones averaging 35 mm across (50 mm in longest diameter). At Ingatestone, one measured 45 mm diameter and another weighed 3.5 ounces indicating maximum diameters of around 60 mm.[9] At Theydon Bois, one averaged 57 mm, while at Kelvedon, one stone weighed 2.5 ounces (just over 50 mm diameter).

The hail swathe (with hail 15 mm diameter or more) extended 115 km from Slough[10] (Buckinghamshire) to Brightlingsea near Colchester. During the storm's severest phase in mid-Essex, the hail swathe was up to 15 km wide. Severe hail (20 mm diameter plus) fell on a continuous swathe at least 70 km long from Epping Forest to just south of Colchester. Reliably reported times of the squall and hail indicate the storm progressed east–north-east at around 30 mph (25–30 kt).

At Gaynes Park, Epping, the Jubilee dinner was 'washed from the tables', and villagers deserted the park in alarm as hailstones fell fast and large. At Danbury, a rushing wind 'approached in whirls and eddies'. Thirty trees were blown down in Blackmore, while at Ingatestone, 40 (many large) trees were uprooted at a farm and adjacent park. In Mill Green, Mountnessing, Stock and Writtle, there were incidents of uprooted trees, roofs and chimneys blown down and wagons overturned. A sheet of iron was carried 30 ft and a windmill wrecked. At Mill Green, a tiled roof was entirely demolished on the windward side. Umbrellas were riddled by the hail! Not only were windows entirely shivered on the windward side, but blinds were shredded by the hail onslaught (*see* Essex Field Club, 1897; Symons, 1897).

9.10.8.3   Bedfordshire Storm

A 22-km-long swathe, south-west to north-east, experienced extensive damage from hail (Herts Field Club, 1897). In the village of Henlow, nearly all windows were broken by hail, many trees were blown down, and the roofs of several cottages were stripped. At Langford, hailstones 'resembling rough pieces of ice with sharp edges' severely bruised and cut bark on fruit trees, broke glass and killed rooks and poultry.

---

[8] Stevenson screen but probably a slightly sheltered site like Camden Square.

[9] Assuming a hard spheroidal hailstone with a typical density of 0.9 g/cm³.
[10] The rainfall observer at Maidenhead reported marble-sized hail, but without times, so it is assumed this occurred in the early evening storm which severely affected nearby Marlow.

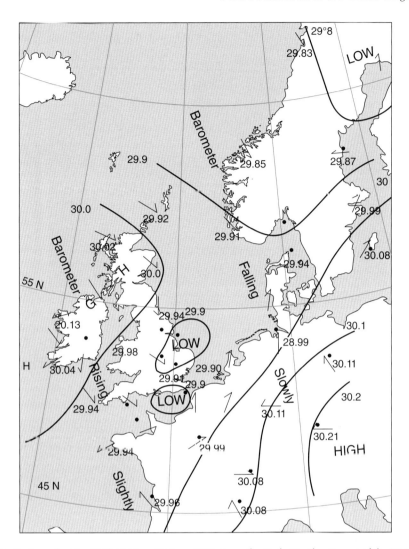

Figure 9.19 *Surface chart for 0600 GMT on 24 June 1897. From the Daily Weather Report of the Met Office, UK.*

9.10.8.4 The Thames Valley Storm

This intense storm struck a small area around Marlow about 1808 GMT, but there was severe damage from wind, intense rainfall and hail. At Marlow Mills, 36 mm of rain fell in 20 minutes. Streets were flooded, trees were blown down in all directions, portions of roofs were blown off, roof tiles were carried some 18 m, and two pinnacles of the church were brought down. Windows were broken by hail and many boats on the Thames were overturned.

9.10.9 1915: 4 July

This was one of the longest-track hailstorms on record, extending at least 170 km from north Somerset to Buckinghamshire, and, assuming this overland swathe was related to severe hail which also fell in north Devon, for 250 km. A trailing frontal zone extended from Portugal across Biscay and southern England to the North Sea (USWB[11]). A shallow depression was situated over Biscay and moved north-east (Figure 9.20). Temperatures peaked

[11] US Weather Bureau Northern Hemisphere charts.

at 29 °C in south-east England. The broadscale situation resembled the chart for 9 August 1843!

This is one of the earliest hailstorms for which TORRO received an account in response to a press appeal. This eyewitness described the storm at East Down, near Combe Martin in north Devon. This 'started off with flat pieces of ice, then hailstones as big as hens eggs'. Large branches were broken off trees which were also left with permanent marks on the bark, all roof slates were broken and windows smashed, while many poultry were killed at Bowden Farm. At nearby Honeywell Farm, 'one hailstone made a clean hole right through of one of the thick red pantiles' (W. Pile, personal communication).

The main storm, possibly linked to the north Devon event, can be tracked over a 170-km-long swathe from Burnham-on-Sea, Somerset, to Buckinghamshire (Figure 9.21). At Shipham, Somerset, the hailstones were up to 140 mm circumference (45 mm diameter), while at Eastville, Bristol, the hailstones were up to 38 mm diameter. Extensive damage was reported in the Chew Valley where many trees were snapped or uprooted by the windstorm and hail smashed or riddled most exposed

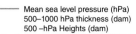

Mean sea level pressure (hPa)
500–1000 hPa thickness (dam)
500 –hPa Heights (dam)

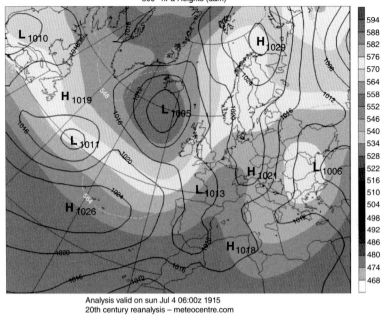

Analysis valid on sun Jul 4 06:00z 1915
20th century reanalysis – meteocentre.com

Figure 9.20 *NCEP reanalysis chart, 4 July 1915, 0600 GMT. Courtesy of NOAA ESRL/PSD reanalysis project (see Compo et al., 2011), Meteocentre. (See insert for colour representation of the figure.)*

Figure 9.21 *Locations of reports of large hail on 4 July 1915. Δ = hail ≥15 mm diameter, ▲ = hail ≥40 mm diameter. Data plotted by author. Outline map reproduced from Ordnance Survey map data by permission of Ordnance Survey. © Crown copyright, 2013.*

windows. The *North Wiltshire Herald* described the storm damage across the southern Cotswolds which occurred between 1500 and 1600 GMT. At Poole Keynes, 'lead window frames were twisted and battered into all sorts of fantastic shapes and windows were pierced like with bullets'. In the Malmesbury area, plaster was knocked off the front wall of a house, metal roofing was pierced, and tree bark was split.

Further north-east, the Radcliffe Observatory, Oxford, recorded a 'violent' thunderstorm with hail from 1700 to 1800 GMT. At Benson Observatory, 19 km south-east, J. S. Dines observed marked wind shear between 1225 and 1730 GMT with a surface north-easterly wind of 15 mph, while 'the wind at moderate height, as shown by cumulus and strato-cumulus clouds (estimated 4000 ft), blew from south'. Between 1700 and 1730 GMT, he observed a near-instantaneous change of surface wind through south to west and pressure changes and cloud observations to the north-west indicative of a supercell and an associated mesocyclone. Thunder was heard to the south-west and west at 1800 GMT. Other reports in the vicinity of the hailstorm describe the light easterly surface wind changing to a violent westerly squall. There were also two secondary hail swathes, one in the Exe Valley and the other from Monmouthshire (Gwent) to Malvern, near where the hail squall damaged hops in the Frome Valley.

9.10.10   *1935: 22 September (Sometimes Referred to as the 'Great Northamptonshire Hailstorm')*

This storm produced the longest UK hail swathe on record at 335 km, traversing England from the Severn Estuary to the North Sea! On 21 September 1935, an anticyclone was situated over Germany, and increasingly warm and humid air was being advected northwards over Britain behind a diffuse warm front; temperatures reached 24°C in Jersey and 26°C in Paris. *British Rainfall 1935* stated that 'conditions very favourable for the development of thunderstorms were present, namely a lapse rate exceeding the saturated adiabatic from a height of about 6000 ft to great heights and an adequate supply of moisture in the lower layers'. A shallow depression (1013 hPA) was located over Biscay; this subsequently moved north-eastwards to reach the Bristol Channel by 0100 GMT on the 22nd (*see* Webb *et al.*, 1994), accompanied by a complex series of fronts. By 0700 GMT, the depression had deepened to 1001 hPa and was situated across the Humber. The main cold front then lay across East Anglia, while the axis of the second cold front (or backbent occlusion) was lying north/south across the Midlands. A small inversion was reported at about 3500 ft which may have been crucial in 'keeping the lid' on convection until it was explosively triggered by general uplift associated with the approach of the developing depression and associated upper trough. Radiosonde data for the evening of 21 September 1935 also indicated very substantial wind shear from the surface to at least 2 km. At Manston (Kent), the surface wind was east by south 10 mph; at 7000 ft, this had veered to SW/W 25 mph (22 kt). The track of the hailstorm can be closely identified with the medium-level 'steering wind'.

British Thunderstorms 1935 described how four thunderstorm 'squall lines' crossed England during the night of 21/22 September 1935. The principal squall line, associated with the main cold front, was responsible for the record 335-km-long hailstorm swathe. Vigorous thunderstorms reached the Bristol

Channel area soon after midnight. A severe thunderstorm affected the Newport area between 0115 and 0225 GMT, accompanied by marble-sized hailstones; 52 mm of rain was recorded there overnight. *The South Wales Argus* described the effects of the storm in the Wye Valley near Chepstow. At Langstone, hailstones punched holes 'as large as apples' in nursery glass, while at a farm near Tintern, 29 windows were broken. Violent winds uprooted trees at St Mellons, near Chepstow, while a roadside hedge between St Arrans and St Briavels collapsed under the weight of 120-cm (4-ft) hail drifts. *British Rainfall 1935* described the havoc that 38–50-mm-diameter hailstones caused at Itton Court where hundreds of panes of greenhouse glass were smashed.

The hailstorm temporarily decreased in severity after crossing the River Severn, but serious damage recommenced towards the Gloucestershire/Oxfordshire border with greenhouses smashed at Notgrove and Cold Aston. The Turkdean to Notgrove road was blocked by huge hail drifts. From the Wychwood area (west Oxfordshire) north-eastwards to north Huntingdonshire and Cambridgeshire, the storm raged with greatest severity. The hail swathe was often around 16 km wide; for instance, in Oxfordshire, damage extended from Great Rollright in the north-west to Steeple Aston and Upper Heyford on the southern flank of the storm. At Hook Norton, windows were smashed with such violence that the golf ball-sized hailstones struck on opposite sides of rooms (*Banbury Guardian*). Trees were uprooted by the violent winds. A *British Rainfall* observer reported that the west–north-west gale, and hailstones 38 mm by 19 mm, 'smashed hundreds of panes of glass in 2 minutes' at the village of Sibford Ferris where 30 mm of rain fell in 25 minutes.

A broad swathe across the length of Northamptonshire was swept by the storm. Scattered reports of damage were received from the rural south of the county, for example, much horticultural glass was broken at Litchborough. The more urban area of central Northants, including Northampton, Wellingborough, Rushden and Higham Ferrers, witnessed the full force of the hailstorm. From Newport Pagnell, just across the border in Buckinghamshire, R. H. Primavesi, writing in the *Journal of the Northamptonshire Natural History Society*, described the approaching storm:

'At 0345 GMT a structureless grey-green haze which was alive with flickering light, by 0415 GMT there appeared to be three layers of storm cloud, the central one being bombarded by lightning flashes from above and below; the effect was really awe-inspiring, thunder was a continuous, menacing rumble'.

The *Wellingborough News* and *Northampton Advertiser* reported that at Irthlingborough:

> …lights appeared in almost every house and many people hurriedly got up to see if any damage had been done. Many windows were smashed and residents had a busy time dealing with the water which came through into the houses. The homely cup of tea soothed many ruffled nerves when the storm had passed!

An eyewitness account received in response to TORRO'S press appeal confirms this account:

> The storm at Irthlingborough was preceded by continuous lightning for about an hour. The entire household deemed it wise to go downstairs. In two of our garden frames there was not one

unbroken pane. The hailstones, averaging golf ball size, struck the brick wall of the house as though it was being bombarded with large pebbles. We had a garden shed with a corrugated iron roof; for years afterwards it contained small dents caused by the hail. Several slates on an adjoining building had holes pierced through them while ripening tomatoes in the garden were literally torn to shreds. (T.E Saxby)

The largest hailstones were typically the size of hen's eggs: for instance, a stone 6 in. (150 mm) in circumference and weighing 4 ounces (consistent measurements for a spheroidal hailstone, i.e. 45–50 mm diameter) was picked up at Weston Favell, Northampton (Pollard, 1936). Bilham (1936) reported that holes up to 100 mm diameter were punctured in an asbestos cement roof. One well-documented hailstone which fell at Rushden was the size of a tennis ball with four distinct spheres; another there measured 75 mm by 57 mm.

Inevitably, there was severe and widespread destruction of horticultural glass; for instance, a letter received by TORRO recalls that in parts of the Rushden area, not a single pane was left in greenhouses. In Wilby, all but 24 panes of a 2500-pane glasshouse were broken. Massive damage to glasshouses occurred in Northampton, and there was a huge subsequent boom in the glazier's trade! There were numerous incidences of other spectacular damage. Thatched roofs were breached in Duston. In Rushden, a hailstone 'scored a direct hit' on a porcelain insulator which was split in two, bringing down a wireless aerial. At Wilby, a large hailstone burst through 6-mm-thick glass and landed on a piano, breaking several keys! In Wellingborough, hailstones broke through the glass roof of a tannery and damaged leather skins in a drying room. In Bythorn, broken glass was scattered about bedrooms.

Some of the largest hailstones, observed as over 70 mm diameter, fell as the storm squall crossed north-west Huntingdonshire

(now Cambs), especially among a cluster of villages just south of Peterborough, including Holme, Glatton and Norman Cross (Figure 9.22). The largest pieces of ice weighed at least 4 ounces (113 g) and exceeded 50 mm diameter. Nearly every house in Holme had broken windows and slates. Glass globes at the petrol pumps were smashed. The *Peterborough Advertiser* reported that slate roofs were smashed to pieces at Stilton and Conington Castle. A drawing of hailstones over 100 mm diameter at Holme on 22 September 1935, sent by Holme schoolmaster J.W. Ingham, was published in the local press. Further east, slates and chimney pots littered some streets in Chatteris following the storm. At Pymore, many slates and windows were broken with consequent flooding of upstairs rooms. The hailstorm became somewhat less severe while crossing Norfolk, although at Litcham and Billingford, hailstones 25–30 mm diameter caused considerable damage to greenhouses, glass lights, etc.

9.10.11 1958: The Horsham Hailstorm of 5 September

Often referred to as 'the Horsham storm', this was almost certainly the first UK hailstorm to be tracked by radar (Ludlam and Macklin, 1960). An old rippling front appeared on the 0600 GMT *DWR* chart, from approximately Bournemouth–Bristol–Aberystwyth–Dublin; this had been 'dropped' from the corresponding 1200 GMT chart. A depression was slow moving well to the west–south-west of the British Isles, while pressure was high over eastern Europe. The general surface wind flow was south-easterly. However, by 1800 GMT, a shallow surface low was evident near the Normandy coast, ahead of which low-level winds backed east–north-east (Figure 9.23). On the 500-hPa contour chart, a cold vortex was evident west of Ireland. The

Figure 9.22 *Giant hailstones covering the ground at Glatton, near Peterborough, on 22 September 1935. Supplied to TORRO, c. 1993, original photographer unknown.*

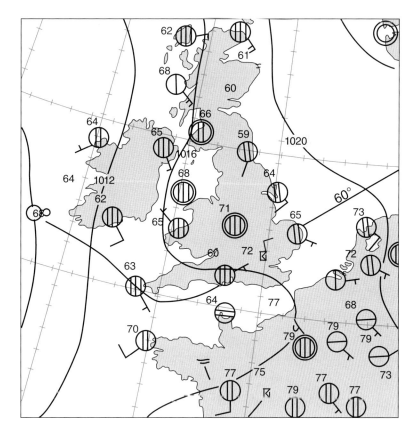

Figure 9.23   Chart for 5 September 1958, 1800 GMT. From the Daily Weather Report of the Met Office, UK.

Larkhill radiosonde indicated that winds veered with height to south–south-westerly over central south England; they were south–south-west 25–35 kt at mid levels and 60–70 kt near the tropopause.

A precursor to the 'Horsham' event was a sharp thunderstorm with hail up to 25 mm diameter which affected the west of the Isle of Wight in the early afternoon. The main severe hail swathe (represented by reports of hail 20 mm diameter or more) extended about 100 km, from near Chichester (1600 GMT) to the Thames Estuary around Dartford (1850 GMT). Hail, albeit smaller in size, continued further north-east across Essex and was still intense enough to damage crops around Chelmsford. Observations of hail with a maximum size of 10 mm diameter or more extended along a total swathe length of 180 km.

This hailstorm was at its peak in the Horsham area of Sussex, where hailstones of 70–80 mm diameter were measured. One was weighed in at 190 g (6¾ ounces), the largest confirmed weight of a hailstone in Britain, although larger hailstones almost certainly fell in this and other hailstorms such as May 1697 and July 1808. An airliner approaching Gatwick airport was severely dented in flight. Passengers on a train ran into intense darkness in the Horsham area and were startled by a '3 in. (75 mm) thick' hailstone smashing through the carriage window! A woman passenger was cut in the face, but worried passengers were advised it was safer to stay on board rather than alight at Rudgwick station. Many roofing tiles and windows were broken violently, tree bark was split,

and the ground was pitted 5 cm deep (Ludlam and Macklin, 1960). Fruit orchards were devastated. At Hills Green near Kirdford, a rainfall observer described how 50 acres of apple orchards was devastated by hail during the developing stage of the storm (1645–1730 GMT) with 'half the apple crop left on the ground and the rest cut and scarred'. Bark was also cut and split. A TORRO appeal respondent, recalling the storm at Billinghurst, described how 'there was not a fragment of glass left in any window of the house apart from a south facing roof window'. Another remarkable feature was the width of the hail swathe in West Sussex: 10–15 km, comparable with the severest hailstorms on record for Great Britain like August 1843. A further aspect of the storm, indicative of a supercell, was that it veered somewhat to the right of the mid-level steering wind, that is, propagating north-east, and this subsequently set it on course for merger with another severe thunderstorm tracking on a more north–north-easterly course from the Brighton area (see Chapter 7 for additional details of this storm system).

9.10.12   *1959: 9–11 July* (Including the 'Wokingham Storm')

The Wokingham storm of 9 July 1959 was the first hailstorm which prompted a detailed study with radar data, and this was complemented by an extensive ground truth survey (Browning and Ludlam, 1960; 1962). The study contributed much to the understanding of severe convective storms. The hail swathe was

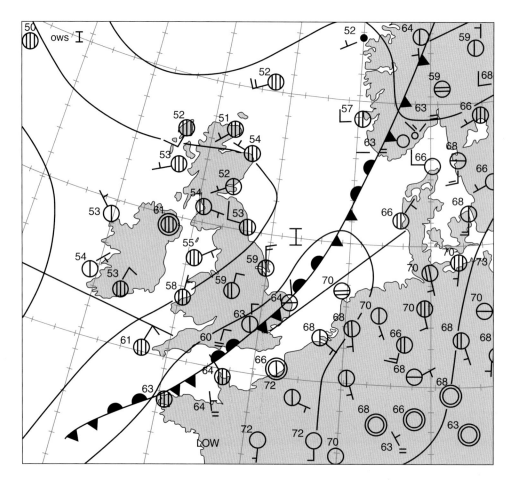

Figure 9.24 *Surface chart for 9 July 1959, 0600 GMT. From the Daily Weather Report of the Met Office, UK. (cf. chart for 9 August 1843 in Figure 9.16).*

around 220 km[12] long, and the largest hailstones measured were about 50 mm diameter around Wokingham, Berkshire, and 38 mm diameter at Saffron Walden, Essex (see also Figure 3.6 in chapter 3).

The 8th was very hot and sunny over most of England and Wales with over 15 hours of sunshine and temperatures peaking at 32.8°C at Mildenhall. A cold front, which lay from around Pembroke to Edinburgh at 1200 GMT, was (then!) a weak feature with no more than a trace of rainfall recorded at any DWR stations to 2100 GMT. However, the front was by now showing the first 'ripples' as it slowed to a standstill between 1800 and 0000 GMT on 8th. At 0600 GMT on 9th (Figure 9.24), the DWR shows the front lying from the Isle of Wight to Norfolk, but if anything it had edged back west at 1200 GMT. Temperatures were still in the 'hot' category in the extreme south-east on the 9th (maximum 30°C at Hastings), while it was exceptionally hot over continental northern Europe with 37°C recorded in central France and north Germany. At 500 hPa level (Figure 9.25), the winds were south-westerly ahead of a pronounced trough west of the British Isles (210°/44 knots in the Crawley 0000 GMT observation which showed 65 kt at 300 hPa).

[12] However, with a short break where hail apparently did not reach the ground.

The main storm was an elevated supercell which developed rapidly off the Dorset coast between 0830 and 0900 GMT, crossing central southern England to reach East Anglia by 1400 GMT (Browning and Ludlam, 1960). Some intense rainfalls were recorded as the storm moved north-east on the 9th: 23 mm fell in 15 minutes at Silchester (1108–1123 GMT) and 15 mm in 9 minutes at Maidenhead (1133–1142 GMT). Hail caused extensive damage. At Wokingham, Winnersh and Swallowfield, car windscreens were shattered, shop windows broken, tiles torn off roofs, and blinds shredded. At Arborfield nurseries, 3000 panes were broken in the glasshouses, while outdoor crops 'looked as though they had been viciously slashed off below the buds by a scythe'; 15,000 outdoor chrysanthemums were destroyed. A local farm also suffered severe damage to oat and barley crops. Hailstones were described as hen's egg size. Moreover, before broken glass at Arborfield could be repaired, the greenhouses were flooded during the further thunderstorms overnight 10th–11th (see following text).

At Saffron Walden, Essex, the morning of the 9th was described as warm and very muggy with a light south-east wind (J.C. Banks, personal communication). At about 1300 GMT, it became overcast with the western sky a 'browney black'; the wind increased to gale force from the south-west (blowing leaves and dust

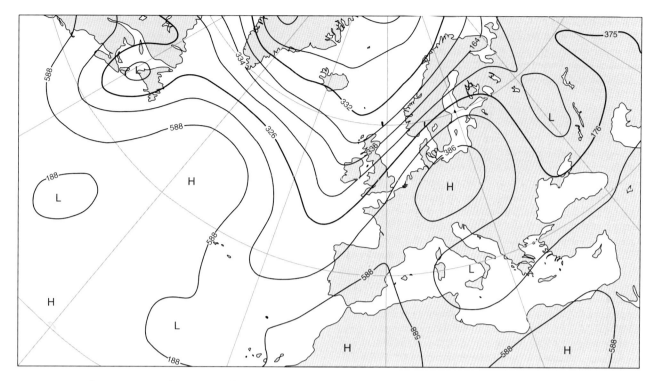

Figure 9.25 *500-hPa contour chart for 9 July 1959 at 0600 GMT. Courtesy of ECMWF-ERA Interim Analysis.*

around) and then large, albeit scattered chunks of ice measuring 38 mm across began to fall. There was an amusing incident when a man chased a schoolboy who he thought had thrown an object at him! These large stones were observed to be fairly soft and 'snowball'-like, many breaking on impact. However, after about 2 minutes, the hail fall changed to an intense fall of slightly smaller but hard, marble-sized hailstones, accompanied by overhead thunder, and it was these stones that inflicted much of the damage. The hail turned to heavy rain before skies brightened from the west, and the wind veered to the north-west accompanied by a sharp fall in temperature.

Fifty panes of glass were broken at The Bull Inn, Langley, while at Brent Pelham, window and greenhouse glass was broken. Severe damage to crops (especially barley, but also sugar beet and potatoes) occurred in several villages including Wendens Ambo, Newport, Quendon, Horseheath, Ashdon, Radwinter, Littlebury and Duddenhoe End. Further north, walnut-sized hail was reported at Bury St Edmunds where the canvas roof of a van was riddled.

The front came to life again across the south and east of England overnight 10th–11th, this time with a widespread episode of thunderstorms. Ripples moved north–north-east along the frontal zone with small low pressure centres developing. At Hope House, Hindolveston, Norfolk, where a total of 88 mm of rain was recorded, 83 mm fell in 4 hours and 15 minutes, and 63.5 mm was estimated to have fallen in only 20 minutes, equalling the duration record set at Sidcup 10 months previously. This cloudburst at Hindolveston was accompanied by hailstones 13–25 mm diameter which completely covered the ground. Thetford recorded 53 mm of rain in an hour (0330–0430 GMT)

accompanied by hailstones up to 25 mm diameter which damaged local greenhouses.

9.10.13   1967: The Wiltshire Hailstorm of 13 July

A shallow depression over northern France drifted north into southern England. Surface winds were light east–north-east, veering to south-west at mid levels and reaching 50 kt at 300 hPa, ahead of an upper trough west of the British Isles. Initial thunderstorm development occurred over north Dorset around 1700 GMT. A 90-km-long hail swathe extended from the Mendips in Somerset, across north-west Wiltshire to south Gloucestershire. Hailstones up to 75 mm across were observed and photographed in north Wiltshire (Hardman, 1968; Meaden, 1984). There was extensive destruction of glasshouses and windows and pitting of car bodywork.

In Holt, damage to one market garden was so great that 20 tonnes of glass shards was picked up over the following 3 weeks together with broken timber frames. In the north of Bradford-on-Avon, many houses sustained broken roof tiles and car roofs looked as if they had been battered all over with a large hammer. All the glass in a market garden was broken; '90% of the hailstones were at least golf ball size and many were tennis ball size' and rebounded from the lawn to heights of 3 m or more.

9.10.14   1968: The 'Dust Fall' Storms of 1–2 July

This was a classic Spanish Plume episode. At the surface, there was a slow-moving frontal zone across western Britain, while east to south-east winds were advecting very hot surface air into south-east England (33°C was recorded in the London area). At 500 hPa, there were strong south–south-west winds ahead of a deep upper trough (Figure 9.26). A shallow surface depression over north-west France on the 1st drifted north into central England on the 2nd, subsequently deepening as it continued northwards.

A succession of severe thunderstorms moved north–north-east across western and northern regions of England and Wales with multiple severe hail events (Figure 9.27), including at least one supercell producing hail 50–75 mm diameter. A TCO (Thunderstorm Census Organisation) observer at Bicton Heath near Shrewsbury, having observed seven separate storms during the 1st and the early morning of the 2nd, commented on his reports, 'I give up!' Nine hail events reached the severe (H3+) category with four swathes of hail damage over 50 km long (Table 9.6). A daily rainfall of 91 mm was recorded at Welshpool on the 1st, while St Athan (Glamorgan) recorded 108 mm for the 1st–2nd. West Baldwin Reservoir, Isle of Man, recorded 185 mm over 48 hours.

On the morning of the 1st, west Somerset and south Wales were battered by hailstones over 50 mm diameter between 0930 and 1100 GMT. At Exebridge and Dulverton (Somerset), hailstones measured at least 50 mm diameter; cars were dented and greenhouses smashed. At Minehead (Somerset), hailstones 50–65 mm diameter were measured by the TCO observer between 0945 and 1000 GMT; this followed another storm cell earlier which deposited 13-mm-diameter hail. There was extensive damage to greenhouses in the town. On the opposite side of the Bristol Channel, at Cardiff Airport (Glamorgan), elliptical hailstones which fell at 1026 GMT were measured at 75 mm by 60 mm (*Station Daily Register*). Photographs of the hailstones were published in Stevenson (1969, p 131) and Macklin *et al.* (1970, p. 472, Plate XII). The local *Barry and District News* reported

that an aircraft hangar roof was penetrated and the elevator in the tail of a Dakota plane was damaged. Much glass was broken at local nurseries. At Pontypridd (Glamorgan), cars were dented and windows broken (G. Williams, personal communication). The Pontypridd observer also reported that the hailstones broke roof slates, demolished greenhouses and broke many windows in Treforest and Cilfynydd. In Treharris, further up the Taff Vale, hail caused considerable damage to greenhouses.

The next severe hail event crossed Cornwall between 1150 and 1230 GMT (*St Mawgan Met Office Daily Register*). At St Dennis and St Stephen (near St Austell) in Cornwall, windows were broken by hailstones described as the size of billiard balls, up to 50 mm across. Later, this storm clipped the Hartland area of north-west Devon. In a first pulse (between 1305 and 1340 GMT), hailstones up to 31 mm diameter were observed at RAF Hartland Point (*Hartland Coastguard Station – Daily Register*). About an hour later (just after 1400 GMT), hail 18 mm diameter fell at Hartland Coastguard. The *Hartland Times* reported that glass in many greenhouses in Hartland village was shattered and the roofs of parked cars were dented. The report said, 'a piece of ice 4 in. long was found in one garden'. The Hartland Coastguard noted that thunderstorms commenced at 1255 GMT and thunder was heard during 14 hours between then and noon on 2 July.

On the evening of the 1st, in the Burnley and Colne Valley area of Lancashire, hailstones, described as like 'golf balls', smashed greenhouses, windows and skylights and dented cars. A photograph of the hailstones was published in the *Lancashire Evening Telegraph*. Hail also damaged gardens in the Liverpool area.

Another wave of violent thunderstorms spread north into Devon late in the evening of the 1st (Figure 9.28). Thunder began at 2200 GMT at Slapton, 2240 GMT at Plymouth and 2300 GMT at Exeter. The Slapton area was hit by hailstones 50 mm across or more during the thunderstorm from 2200 GMT

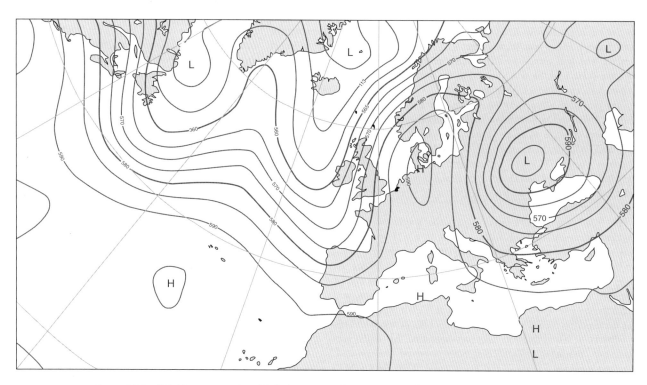

Figure 9.26 *500-hPa contour chart 2 July 1968, 0000 GMT. Courtesy of ECMWF-ERA Interim Analysis.*

Figure 9.27 *Large hail reports, 1–2 July 1968. Δ = hail ≥15 mm diameter, ▲ = hail ≥40 mm diameter. Data plotted by author. Outline map reproduced from Ordnance Survey map data by permission of Ordnance Survey. © Crown copyright, 2013.*

Table 9.6 *Multiple hail events, 1800 to date.*

Date (s) of Episode	Number of Severe (H3+) Hail Events	Number of Significant (H2+) Events
20–21 June 1936	12	14
1 August 1846	11	12
1–2 July 1968	9	9
8 July 1893	9	10
27 June 1866	8	9
1–2 June 1889	8	10
11–12 June 1900	8	9
15 May 1833	7	7
5 July 1852	7	9
17–18 July 1926	7	7
7 June 1983	7	9
9 August 1843	6	7
21 June 1851	6	6
12 August 1938	6	8
2–3 July 1946	6	6
7 June 1996	6	8
19 June 2005	6	7
29–30 May 1944	5	8

Monthly distribution: 2 May, 8 June, 5 July and 3 August.

to midnight, and there was enormous destruction of glass. Slapton climatological station reported a heavy thunderstorm with hail 20 mm diameter or more from 2330 to 0000 GMT. The local press reported that the village 'looked as if every cricket ball in creation had been hurled at its windows!' The largest hailstones were reported to weigh 113 g. Many glass-houses were reduced to wooden frames, walls whitewashed and pitted, roof tiles broken, and corrugated iron pierced (A. Barr, personal communication to M. Rowe). Similar damage was reported from Blackawton, Stokenham, Chillington and Torcross. At Wadstray House, the hail or 'spears of ice' broke greenhouse glass and indented the ground. Although Exeter and Chivenor Met stations missed the destructive hail, both stations recorded dramatic thunderstorm weather. At Chivenor, the aircraft runway was struck and damaged by lightning, and the Met Office window was blown open by a 50-kt squall. At Exeter, a roaring noise was reported from 0002 to 0005 GMT, and a 'violent noise' to west–north-west around 0006 GMT was noted as possible hail. Much spectacular lightning activity was observed overnight and into the morning of the 2nd, with a 'ball of lightning close to runway' observed at 1030 GMT,

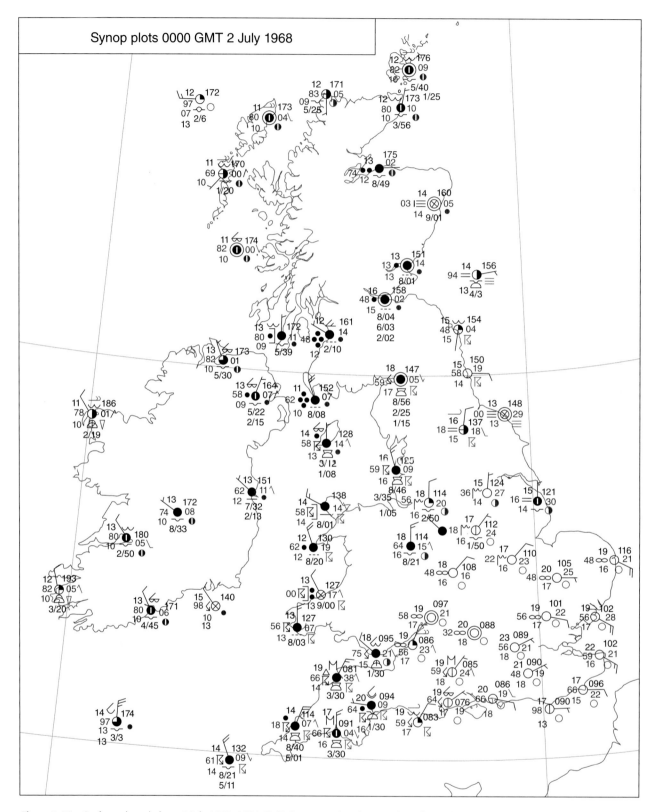

Figure 9.28   *Surface plotted chart, 2 July 1968, 0000 GMT. Courtesy of Paul Brown based on information supplied by the Met Office, UK.*

accompanied by an 'explosion-like gunfire'; pea-sized hail followed at 1040 GMT.

Between 0400 and 0500 GMT on the 2nd, Shropshire was hit by severe hail. At Craven Arms and in the Telford/Wellington area, golf ball-sized hailstones totally demolished greenhouses, dented cars and broke windows. At Newport, a large number of car windscreens were smashed. Slightly later on the morning of the 2nd, a severe hailstorm moved from Cheshire and Manchester north-eastwards across West and North Yorkshire to Teesside, with large hail along a 125-km swathe. In Huddersfield and Liversedge, windows were smashed. Great damage to glasshouses occurred in Harrogate, Knaresborough and Nidderdale. At Yeadon, near Leeds, hailstones were piled 45 cm deep (Hodgson, J). At Leeming, 36 mm of rain was recorded in 9 minutes, while the contemporary TCO headquarters in Huddersfield recorded 28 mm in 20 minutes accompanied by conical hailstones 38 mm by 32 mm.

9.10.15   1983: South Coast Hailstorms of 5 June

An anticyclone was building strongly across Scotland. Surface winds across southern England were moderate to fresh north-easterly, while at 500 hPa, a pronounced trough was moving south-east across England and Wales (see also Figure 8.7). In the early hours, a severe hailstorm affected the western side of the Channel Island of Jersey with hail 25–50 mm diameter (Randon, 1983). Then, from late morning, a series of classic 'elevated' supercell thunderstorms (Figure 9.29) developed on the north (cold) side of a frontal zone over the English Channel in the same manner as 3 August 1879 and 9 July 1959, albeit with the storm steering wind flow oriented more west to east. These storms tracked along the south coast of England, and the consolidated hail swathe, mostly attributable to the first supercell, was over 200 km long from Weymouth

to Eastbourne and 10–12 km wide in places (Hill, 1984; Rowe and Mortimore, 1984; Meaden, 1995).

South coast resorts have rarely featured in severe hailstorms, but on this occasion, the Bournemouth area experienced large hail on four separate occasions (Figure 9.30). These four storms all passed overhead of Poole Harbour and Bournemouth, at least two depositing hail over 30 mm diameter. There was widespread damage to glasshouses, car and caravan bodywork. In Poole harbour, an unfortunate wind surfer ran into problems as he was repeatedly hit by large hailstones which were 'plonking' into the water. Several people in Poole received painful bruises from the hailstones. Winfrith (AERE) climatological station in Dorset reported hailstones ranging from 25 to 75 mm diameter. At Church Knowle, the vinyl roof of a mobile home was left 'leaking like a sieve'. Another eyewitness was cycling home with companions to Stoborough from church in Wareham that Sunday and encountered the storm with 'darkness at noon' followed by huge lumps of ice the size, texture and appearance of fist-made snowballs; a view of forked lightning striking Creech Barrow (just south of Stoborough) added to the drama (P. Withers, personal communication to M. Rowe).

9.10.16   1983: Violent Hailstorms in North-West England on 7 June

A series of severe supercell thunderstorms produced several hail swathes south–south-west to north–north-east across north Wales and north-west England. The anticyclone over Scotland moved away eastwards into Europe. A trough moved north-eastwards ahead of a cold front which crossed the country early on the 8th. There was a light south-easterly wind flow at low levels, while at 500 hPa, there was a strong south–south-westerly flow ahead of a deep upper trough. Nine separate hailstorms have been identified. One hail swathe was 130 km long and overlapped with another 90 km long to give a damage swathe up to 15 km wide. Hailstones 50–75 mm diameter were measured in north Powys, north Cheshire, Greater Manchester and Lancashire. There was extensive damage to glasshouses, windows, roof tiles and cars. Several photographs of hail 50–60 mm diameter were published by Dent and Monk (1984). In the Met Office climatological station network, St Harmon (Powys) recorded hailstones 50–75 mm diameter between 1700 and 1800 GMT, and Oswestry

Figure 9.29   *NOAA visible satellite image, 5 June 1983, 1426 GMT.* © *University of Dundee.*

Figure 9.30   *Hailstones at Christchurch, Dorset, on 5 June 1983. Photograph by Allen White.*

(Shropshire) reported hail 25–50 mm diameter. At Gateacre, Liverpool, thunderstorms persisted for over 5 hours with golf ball-sized hailstones falling for several minutes around 2100 GMT (K. Ledson; COL 1983).

TORRO received numerous eyewitness accounts of these storms:

Caersws, Powys: Greenhouses damaged by hail.

Ash Farm, Sale (Manchester): Hailstones 50 mm diameter or slightly more. Glass broken in windows and greenhouses.

Partington, Greater Manchester: Hail about the size of golf balls. Gutters were pierced all along a road as if someone had gone along the row of houses with a shotgun. Windows and tiles smashed and greenhouses destroyed.

Near Manchester Airport: Hailstones 50 mm by 38 mm with sharp edges. Car bodywork covered in deep dents on bonnet, roof and boot.

Swinton, Greater Manchester: Hailstones averaged 25 mm diameter. Perspex and carport roofing wrecked and reinforced glass roof cracked.

Urmston, Manchester: Hail about the size of golf balls. Widespread damage to greenhouses, windows, doors and car bodywork.

Barnoldswick, Lancashire: Size ranged from 19 mm diameter to golf ball size. Tiles damaged and car bodywork seriously dented.

9.10.17   1985: The Essex 'Dunmow' Hailstorm of 26 May

A trailing cold front across the English Channel returned north as a warm front overnight 25th–26th, while an Atlantic depression, approaching a developing upper ridge over central Europe, slowed down south-west of the British Isles at around 45°N, 20°W. As this depression drifted slowly north, its fronts crossed Ireland and Scotland. The 500-hPa chart for 1200 GMT on 26 May shows a deep upper trough over the west of Ireland.

Thunderstorms affected many areas of England during the early morning and afternoon on Sunday, 26 May 1985. Although the storms caused localised flooding and lightning damage, the outstanding feature of the afternoon thunderstorms was the destructive hailstorm which cut a swathe of severe damage across rural areas of north-west Essex. An appeal for information concerning the effects of this storm was issued soon afterwards by David Brooks, a weather forecaster for Anglia Television, as well as by TORRO in local newspapers. This produced an overwhelming response from the public in the form of typed or handwritten letters (these being pre-Internet days!), containing detailed accounts of the storm, the size of the hailstones and the damage caused. Local weather observations were provided by members of the Anglia Television Weather Correspondence Network, the TCO, the Climatological Observers Link (COL) and TORRO, in addition to the meteorological observations available from the Meteorological Office.

Two severe hail swathes, which eventually merged, tracked across Essex and into Suffolk (Figure 9.31). Extensive damage was done to west-facing windows and tiled roofs. Largest hailstones were measured at 40–65 mm diameter around the villages of Good Easter, High Easter, Barnston, Pleshey and North End (Elsom and Webb, 1993) (Figure 9.32a–c). Hailstones exceeding 30 mm diameter (Figure 9.31) fell across an area approximately 25 km long by 7 km wide, while the overall 'damaging' hail swathe (hail 10 mm diameter or more) extended for 75 km with a swathe width of up to 15 km during the severest phase. The secondary hail swathe resulted from another storm which followed about 20 minutes later several miles to the east and which later merged as the first system slowed. The second storm deposited hail over 30 mm diameter in the Chelmsford area but was overall (fortunately, as the track crossed a more populated area) less severe than the 'Dunmow storm'.

Nearly all the glasshouses along the 25-km-long swathe of severest hail damage lost panes. Reports from High Easter, North End, Pleshey and Lindsell confirm that some lost every single pane of glass. At North End, a man had a narrow escape when a greenhouse was wrecked close to where he had been sunbathing a few minutes earlier. In Great Dunmow, hailstones not only smashed greenhouses but also, in one observed case, split plant pots in two. Few premises in the villages of Good Easter and High Easter escaped without broken window panes. At a farmhouse near Stebbing, every window on the windward side was broken. A plastic garage roof in Little Dunmow was so punctured with holes that broken sections fell to the ground. Damage to tiled roofs was reported from places as far apart as Pleshey and Lindsell. At a house in High Easter, 15 slates on the roof were holed, and the guttering was totally destroyed; the house had been re-slated only 2 years before. In Good Easter, the severe hail squall ripped 150 slates off one roof. The early Victorian stained glass window at the western end of Pleshey church was 'holed and splintered in no less than 19 places' – an appeal fund was launched subsequently to enable it to be restored. Several stained glass windows at Good Easter village church were also holed. In North End, tiled roofs were severely damaged, and even a corrugated iron roof was dented (T. Watson, personal communication). Motor vehicles and caravans were pitted, some so severely that they were reduced to insurance write-offs. Cars were typically stippled on the roof, bonnet and boot lid, while there were numerous instances of windscreens being smashed. Some spectacular effects were noted as the hail hit relatively warm water such as one pond near Little Dunmow which was observed to have resembled a 'boiling cauldron'. The dramatic fall in temperature which accompanied the storm was evident at Stansted Airport where the temperature fell from 22°C at 1400 GMT to 12°C 1 hour later. Fortunately, with regard to potential human injuries, the hailstorm crossed a predominantly rural area where the absence of urban noise also gave better warning of the approaching storm. One man's arm was bruised while opening a car window, but injuries were much fewer than during the Greater Manchester hailstorm of 7 June 1983.

Forecaster Nigel Bolton, then based at Wattisham Met Office, observed the aftermath of the storm in north-west Essex. He visited a house just outside Felstead which lost eight roof tiles, a windscreen wiper off a car and a cold frame (smashed together with the wooden frame). The householder suffered painful injuries when he was hit on the hand and then on the back of the head by hailstones. The stones here were about 40 mm across and almost perfectly round. Between High Easter and Pleshey, he came across hailstones that were still, over 4 hours after the storm,

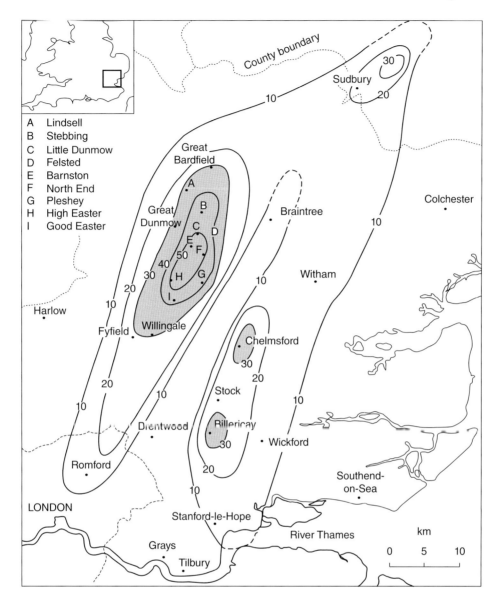

Figure 9.31 *Distribution of maximum hailstone size in diameters (mm) produced by two hailstorms affecting Essex during the afternoon of 26 May 1985. First published in Elsom and Webb (1993).*

around 50 mm diameter; these were more oblate spheres and must have originally been considerably bigger. Indeed, he added, 'as far as I am aware, the largest hailstone measured was just west of Pleshey at 70 mm'. In this area, potato crops were reduced to mush, and the largest stones had made holes in people's lawns. Every single pane of glass in most greenhouses had been shattered, and 'many people who had left their cars outside had bodywork damaged, looking like someone had gone around with a hammer'. Figure 9.32a and b shows typical examples of the spherical and oblate hailstones up to approximately 50 and 65 mm diameter, respectively.

According to a National Farmers' Union (NFU) survey, at least 60 farms (several press reports cited 80 or even 100) were affected, with about 40 farms reporting over 85% of their entire crop destroyed. Most farms were not insured, and a hail disaster fund was launched by the Essex branch of the NFLT. The proceeds

were divided between farmers who had been in business less than 3 years, farmed under 100 acres, and who were among those losing over 85 per cent of their crop. Interest-free loans were arranged for some other farmers to carry them over to the next harvest.

9.10.18 1996: The Storms of 7 June

Eight damaging hail swathes affected areas between Dorset and Norfolk following one of the hottest early June days on record with a temperature of 33°C recorded in London (Appendix A, www.wiley.com/go/doe/extremeweather). Two swathes were over 150 km long (Webb and Pike, 1998a, b). A large storm area rapidly developed from the Chiltern Hills across East Anglia in the early evening. The second zone of severe storms (Figure 9.33) formed just ahead of the cold front, tracking from Wessex to the

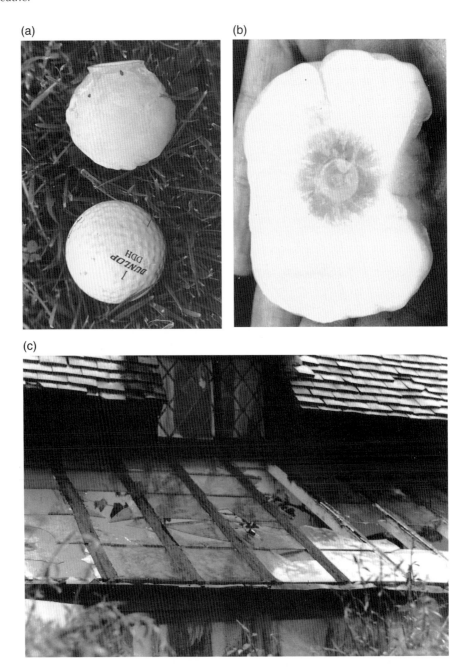

Figure 9.32 *(a–c) Hailstorm at Barnston and North End, Essex, on 26 May 1985. (a) Spherical hailstone c. 50 mm diameter. Courtesy of Braintree and Witham Times. (b) Oblate hailstone c. 65 mm across. © Richard W. Hammond (1985). (c) Damage to glass conservatory and tiled roof. © Terry K. Watson (1985).*

south-east Midlands. They were fed by an inflow of warm air on the northern flank of a frontal wave depression and its extension in the form of an elongated heat low. Markedly, convergent surface winds were evident over Oxfordshire just before the storms commenced.

During the early evening storm at Bourn (Cambridgeshire), hailstones averaged the size of a golf ball, and the largest observed was 63–75 mm diameter. One farmhouse sustained 26 broken roof tiles, 24 broken sash window panes and 21 smashed greenhouse glass sheets. An entire barn roof had to be replaced. Almost

simultaneously, another hail swathe a few miles further east deposited 38–50-mm-diameter hailstones in south-east Cambridge and the nearby villages of Great and Little Shelford (Figure 9.34). Likewise, two separate hail swathes affected Dorset: one which passed through Sherborne (Braunholtz, 1996) produced 50–63-mm-diameter hail at Oborne, while the other swathe extending north-east from Portland was associated with hailstones up to 50 mm across at Crossways near Weymouth. Later in the evening, hailstones broke 70 roof tiles on a pair of semi-detached bungalows in Luton. At South Fawley, Berkshire,

Figure 9.33 *Infrared satellite image, 7 June 1996, 1856 GMT. ©*
Crown Copyright The Met Office.

Figure 9.34 *Hailstones, Little Shelford, Cambs, 7 June 1996. © Lucy*
Jenkins.

both hail and lightning caused severe damage to crops. Windows
were cracked by hail at East Lockinge, near Wantage, where
74 mm of rain was recorded; radar indicated that 72 mm of this
fell in 2 hours.

The intense hail dramatically cooled the ground, causing steam
fog to form. At Owermoigne, near Dorchester, 40-mm-diameter
hailstones pierced corrugated iron and broke much glass at
Owermoigne nursery. In the aftermath of the storm, visibility was
reduced to just 20 m in the thick fog which lasted for 30 minutes.
The storms were also memorable for the intense lightning activity
with observers in Berkshire, Oxfordshire and Buckinghamshire

noting 20 or more discharges per minute during peak storm periods (Webb and Pike, 1998b). Many buildings were struck, and in
the Newbury area, 100,000 homes lost power after lightning damage; further north, lightning also cut power to 20,000 homes in
south Lincolnshire.

9.10.19   1997: The Severe Storms of FA Cup Final on Saturday, 17 May

During the early afternoon of Saturday, 17 May 1997, intense
thunderstorms broke out over, broadly, central southern England.
These were associated with some destructive hailstorms as they
moved northwards into the south-east Midlands (Webb 2013).
Other areas of England, south Wales and south-east Scotland
experienced numerous incidents of lightning damage and flash
flooding during a series of thunderstorms from the evening of 16
May to the morning of 18 May 1997.

9.10.19.1   Synoptic Background

The 500-hPa synoptic chart for the morning of 17 May indicated
that a sharp upper trough was situated across Biscay. A surface
warm front, the leading edge of a plume of very humid air, had
pushed slowly northwards into England and Wales on the 16th,
accompanied by vivid thunderstorms in places, especially from
Hampshire to the Thames Valley in the evening. Where low cloud
and mist cleared relatively early, surface temperatures within the
stagnant air mass associated with a shallow area of low pressure
shot up to 26.6°C at Heathrow and 25.5°C at Gatwick.

The Larkhill ascent at 0600 UTC indicated deep instability
once surface temperatures rose above about 20°C. Surface air was
moister just to the north. At Benson, a maximum temperature of
21.1°C (dew point 17°C) was recorded at 1250 GMT, 15 minutes
before the first thunder was heard to the south. The 1200 GMT
radiosonde data from Herstmonceux indicated that the 5–10-kt
surface wind 'sheared' to SSE 35 kt at 500 hPa and to southerly
55 kt at 250 hPa.

By 1200 GMT, a rippling cold front was edging slowly into the
south-west approaches, while there was evidence of a diffuse
'inner' cold front lying across England. Surface winds had veered
to south–south-east at Gatwick and Winchester by 1400 GMT and
from Heathrow to Salisbury by 1500 GMT, apparently in association with a sea breeze front. Meanwhile, light north-easterlies persisted across the South Midlands (e.g. at Brize Norton, Malvern,
Elmdon and Bedford), providing the additional trigger of converging surface winds across central southern England. The combination of very humid surface air, a 'capping' stable layer and very
unstable air aloft provided the background for explosive cumulonimbus eruption. Satellite pictures (Figure 9.35) at 1600–1700 GMT
indicated that some cumulonimbus tops were 'overshooting' the
34,000-ft-high tropopause by a considerable margin.

9.10.19.2   North Hampshire/West Berkshire Downs

The first radar echoes almost exactly coincided with the earliest
ground reports of a thunderstorm developing just to the east of
Andover at 1255–1300 GMT. This storm deposited hailstones
18 mm diameter in the villages of Longparish and Hurstbourne
Priors. Further cells developed in the same area between Andover
and Walbury Hill and tracked north–north-west, while fresh cells

developed near Newbury and Basingstoke around 1330 GMT and drifted north into Oxfordshire. Hail 12–20 mm diameter was observed at several places from Hartslock to Wheatley between 1430 and 1600 GMT.

Yet another intense cumulonimbus cell (probably a 'daughter cell' of the above) developed near Sonning Common, at the southernmost edge of the Chilterns north of Reading, around 1415 GMT, and radar pictures indicated that this was the origin of the severe hailstorm in the Thame area where a tornado of T2 intensity struck the village of Towersey. The swathe of hail over 30 mm diameter extended from the south-east edge of Thame to several kilometres distance north of the Oxfordshire/

Buckinghamshire border (an area at least 10 km long and 5–7 km wide). The severest damage places this storm at H5–H6 intensity on the TORRO hailstorm scale. A selection of eyewitness descriptions is presented herewith:

Thame, Parliament Road (Oxon): around 1515 GMT. Some hailstones were spherical, golf ball size, up to 45 mm across; others were irregular, for example, 40 mm by 20 mm. Greenhouse glass was broken, and PVC roofing was holed. Several other eyewitness accounts from Thame indicated that these largest hailstones predominantly fell on the eastern side of the town. Figure 9.36 shows typical hailstones of up to 40 mm diameter.

North Weston (Oxon): 1530 GMT, 'very large ice cubes'. Panes broken in greenhouse, cars dented, at least one person suffered head bruises.

Long Crendon, Bicester Road (Bucks): 1530 GMT, hailstones up to 50 mm diameter. Clay roof tiles chipped, caravan pockmarked, car bodywork dented, many greenhouse panes/garden frames broken.

Long Crendon, The Square: 1530 GMT, hail about 50 mm across. Many exposed window panes broken.

Chearsley (Bucks): 1530 GMT – the largest hailstones were described as 'slightly bulgy discs over 50 mm long'. The Church belfry windows were broken. Car bodywork was dented. Pantiles and clay tiles were broken, as were some west- and south-facing house windows.

Haddenham (Bucks): 1530 GMT, pieces of ice up to 70 mm diameter, some spheroidal, but many irregular with sharp/pointed edges. PVC roofing was holed.

Ashendon (Bucks): Hailstones up to 63 mm diameter. Greenhouse glass was destroyed, windows were broken, plastic guttering cracked, and cars severely dented.

A Beavers Cubs and Scouts camp at Long Crendon was thrown into turmoil as tents were blown over and the children and adults dived for cover. The camp organiser was bruised on her back by a hailstone. Similar chaos hit a wedding at Chearsley when a marquee collapsed under the weight of hailstones. Another major cumulonimbus development occurred between Slough and Amersham in mid-afternoon. This was the origin of the severe

Figure 9.35 *Infrared satellite image, 17 May 1997, showing three huge supercell thunderstorms, 1703 GMT.* © University of Dundee.

Figure 9.36 *Hailstones which fell in Queen's Road, Thame, Oxfordshire, on 17 May 1997.* © David Colbourne.

storm which affected the area between Milton Keynes and Bedford. Highlights of the storm included:

Woburn Sands: 86.4 mm of rain was measured (at an official Met Office site), nearly all falling between 1630 and 1800 GMT. Hailstones up to golf ball size caused considerable denting of car and car bodywork (indicative of H4 severity). Floodwaters up to 1.5 m deep burst open the door of some high-street stores. Sand was washed down into the town from neighbouring woods.

Woburn: Spheroidal hailstones 25–38 mm diameter. A car roof was damaged. Another first-hand report described two bursts of hail, the first (just after 1630 GMT) of hail 10 mm diameter and then, after a lull, the second event with hailstones between 'golf ball and chickens' egg size. Plants were ripped to shreds, much greenhouse glass broken and a caravan severely dented (Anon, 1998).

Aspley Guise: Greenhouses were damaged by hail and cars were dented.

Brogborough: Hail 'the size of 50 pence pieces' dented car bodywork.

Lidlington: Hailstones up to golf ball size, confirmed by photographs (*Lidlington Advertiser*, 1997). There was damage to greenhouses, car windscreens and car bodywork.

Wootton: A tornado of T3 intensity carried a caravan 30 yards over a shed (Meaden, 1998). There were also reports of 'golf ball'-sized hailstones falling in the village.

Figure 9.35 highlights the anvils sheared ahead of these two (probable) supercell thunderstorms in the winds of over 50 kt near the tropopause.

9.10.19.3 North Somerset and Bristol

At 1530 GMT, a patch of light showers appeared as radar echoes north-west of Yeovil (P. Stevenson, personal communication), and within 15 minutes, this cell was displaying intense echoes. This was the origin of the severe thunderstorm which affected the Bristol area.

At *Midsomer Norton* (Somerset), 56 mm rain was recorded, flooding 'hundreds of properties'. At *Warmley* (just east of Bristol), hailstones up to 30 mm diameter were measured. Leaves were stripped off trees, covering parked vehicles (A.P. Head, personal communication). At Thornbury (south Glos), a funnel cloud was photographed at about 1700 GMT (McPhillimy, 1997).

9.10.20 2012: The Destructive English Midlands Hailstorm of 28 June

This event produced some of the largest hailstones, by diameter, recorded in Britain (Table 9.3); however, most of the largest hailstones were very oblate. Surface charts show a double-structured cold front moving north-east across England and Wales with prefrontal troughing (see also Chapters 3 and 14). At 500 hPa, there was a major upper trough west of the British Isles. However, the overall mid-level flow remained progressive, with south-westerly winds of 40–50 kt; the situation was consistent with the classic Spanish Plume. This major event is also referred to in chapters 3 and 12 where the storm mode and forecasting aspects are discussed.

Very large hail (20 mm diameter plus) fell in association with a single, severe 'supercell' thunderstorm, along a 110-km-long swathe from Coventry, West Midlands, to Sleaford,

Figure 9.37 *Hailstones 45–75 mm diameter at Burbage, Leics, 28 June 2012. © John Purt.*

Lincolnshire (see map of swathe in Chapter 3, Figure 3.24). Hail 50 mm or more in diameter fell along much of this swathe, with hail 75 mm (3 in.) in diameter or more falling in the Hinckley/Burbage area (Figures 9.37 and 9.38) and (more locally) just west and north of Melton Mowbray (Figures 9.39 and 9.40). Radar returns indicate the swathe of hail 10 mm diameter or more extended for around 120 km. The hail resulted in extensive damage to crops, vehicles, glasshouses and roofing materials; the severest damage rated H6 on the TORRO scale. As an example of the impacts during the severest phase of the hailstorm of 28 June 2012, the following is a first-hand description of damage at one property in Burbage, Leics (M. Gaisford, personal communication) (Figure 9.38):

About 20 tiles broken (Marley concrete type), about 15 dents in car, uPVC guttering bracket broken, two holes in uPVC guttering, 10 leaded panes broken in front (north side) of house, most of greenhouse roof broken, blue engineering brick wall coping chipped, old cast iron downpipe into soft water tank dislodged, lots of leaves and twigs from oak and magnolia trees broken off.

It was the most severe hail event across Leicestershire since June–July 1900 (see Appendix A, www.wiley.com/go/doe/extremeweather). Other aspects of this storm are discussed in Chapter 3, and a more detailed account of this extreme event has been published by Clark and Webb (2013). At least two other more localised large hail events occurred on this day, in the Brecon area of east Wales and at Telford in Shropshire.

9.11 Concluding Remarks

TORRO has been directly involved in detailed investigations of most of the very severe (H5+) hailstorms which have occurred in the past 40 years, for example, in published studies by Mortimore and Rowe (1984), Elsom and Webb (1993), Webb and Pike (1998a, b) and Clark and Webb (2013). This research has been concurrent with the ongoing development of an historical hailstorm database. The latter includes sufficient data on storms of H5 intensity or more (typically representing hail over 40 mm

Figure 9.38 *Huge hailstones covering the ground, 28 June 2012, Burbage, Leics. © Mike Gaisford.*

Figure 9.39 *Hailstone which fell at Brooksby Hall, Leicestershire, 28 June 2012. © Megan Cooper.*

Figure 9.40 *Hailstone around 50 mm diameter, Melton Mowbray, Leicestershire, 28 June 2012. © Linda Manners.*

diameter) to present a realistic climatology of the most hazardous storms also based on over 200 years of data. The databank of all significant (H2 or more) storms is sufficiently consistent over a period of 80 years for seasonal and geographical analyses to be presented with some confidence.

A reliable, long-period database of significant hail events not only enables a realistic hail climatology to be presented but also has the potential to assist in predicting future trends, in the context of wider climatic change. A recent study, which included TORRO data and involvement, projected changes in the incidence of severe hailstorms in the United Kingdom during the rest of the 21st century (Sanderson *et al.*, 2014). The future prospect of higher-resolution models incorporating wind shear and other parameters (in addition to CAPE) is of particular importance.

Acknowledgements

The authors acknowledge the importance of the information, including the results of site damage surveys, on hailstorms supplied by the network of TORRO members, observers and contacts, especially throughout the British Isles. Special thanks are due to Neil Lonie (Dundee University) for satellite data and due to staff at the National Meteorological Library and Archive at Exeter for assistance in locating publications and observational data.

References

Anon (1687) *A true relation of the great thunder, lightning, rain, great wind and prodigious hail that happened at Alvanley in the parish of Frodsham, Cheshire, on Sunday the 19th day of June 1687.* D. Mallet, London.

Anon (1883). The Great storm at Alvanley, near Frodsham, A.D. 1687. *Palatine Notebook*, **3**, 31 July 1883, Manchester.

Anon (1998) The events of 17th May 1997. *Bedfordshire Naturalist, Bedford*, **52**, 12–16.

Auer, A.H. (1972) Distribution of hail and graupel with size. *Mon. Weather Rev.*, **100**, 325–328.

Berthet, C., Wesolek, E., Dessens, J., Sanchez, J.L. (2013) Extreme hail day climatology in southwestern France. *Atmos. Res.*, **123**, 139–150.

Betts, N.L.(2003) Analysis of an anomalously severe thunderstorm system over Northern Ireland. *Atmos. Res.*, **67–68**, 23–34.

Bilham, E.G. (1936) The great Northamptonshire hailstorm of 22nd September 1935, in: *British Rainfall*, pp. 281–283.

Bower, S.M. (1947) *British thunderstorms 1934–1937* TCO, Huddersfield.

Braunholtz, T. (1996) The Sherborne hailstorm and tornado of 7 June 1996. *J. Meteorol.*, **21**, 382–384.

Browning K.A., Ludlam, F.H. (1960) *Radar analysis of a hailstorm*. Tech note 5, Department of Meteorology, Imperial College, London.

Browning, K.A., Ludlam, F.H. (1962) Airflow in convective storms. *Q. J. R. Meteorol. Soc.*, **88**, 117–135

Changnon, S.A. (1971) Hailfall characteristics related to crop damage. *J. Appl. Meteorol.* **10**, 270–274.

Clark, C. (2004) The heatwave over England and the great hailstorm in Somerset, July 1808. *Weather*, **59**, 172–176.

Clark, C. (2005) The cloudburst of 2 July 1893 over the Cheviot Hills, England. *Weather*, **60**, 92–97.

Clark, M.R., Webb, J.D.C. (2013) A severe hailstorm across the English Midlands on 28 June 2012. *Weather*, **68**, 284–291.

(COL 1983). Climatological Observers Link (COL) Bulletin June 1983.

Compo, G.P., J.S. Whitaker, P.D. Sardeshmukh, N. Matsui, R.J. Allan, X. Yin, B.E. Gleason, R.S. Vose, G. Rutledge, P. Bessemoulin, S. Brönnimann, M. Brunet, R.I. Crouthamel, A.N. Grant, P.Y. Groisman, P.D. Jones, M. Kruk, A.C. Kruger, G.J. Marshall, M. Maugeri, H.Y. Mok, Ø. Nordli, T.F. Ross, R.M. Trigo, X.L. Wang, S.D. Woodruff and S.J. Worley (2011) The twentieth century reanalysis project. *Q. J. R. Meteorol. Soc.*, **137**, 1–28.

Crocker, A. (1808) Account of the tremendous thunderstorm which fell in Somersetshire on 15 July 1808. *Monthly Magazine and British Register*, London pp 302–308.

Dent, L., Monk, G.A. (1984) Large hail in north-west England on 7 June 1983. *Meteorol. Mag.*, **113**, 249–264.

Dessens, J. (1986) Hail in south western France. 1. Hailfall characteristics and hailstorm environment. *J. Clim. Appl. Meteorol.*, **25**, 35–47.

ECMWF (2014) 40 year archive of re-analysis charts. Available from: http://apps.ecmwf.int/datasets/ (accessed 29 April 2015).

Elsom, D.M., Webb, J.D.C. (1993) Destructive hailstorms in Essex on 26 May 1985. *Weather*, **48**, 166–173.

Essex Field Club (1897) The great storm of midsummer day 1897. *Essex Naturalist*, **1897**, 113–129.

Fraile, R., Berthet, C., Sanchez, J.L. (2003) Return periods of severe hailfalls computed from hailpad data. *Atmos. Res.*, **67–68**, 189–202.

Gaiotti, D., Nordio, S. Stel, F., (2003) The climatology of hail in the plain of Friuli Venezia Giulia. *Atmos. Res.*, **67–68**, 247–259.

Grant, K. (1995) Thunderstorm-days and hail-days in the United Kingdom, 1972–1991. *J. Meteorol.*, **20**, 281–297.

Hardaker, P.J., Auer, A.H. (1995) The separation of rain and hail using simple polarization radar and IR clod top temperature. *Meteorol. Appl.*, **1**, 201–204.

Hardman, M.E.(1968) The Wiltshire hailstorm of 13 July 1967. *Weather*, **23**, 404–414.

Hertfordshire Natural History Society and Field Club (1897) *Transactions*. Gurney & Jackson, London.

Hill, F. (1984) The development of hailstorms along the south coast on 5 June 1983. *Meteorol. Mag.*, **113**, 345–363.

Hohl, R., Schiesser, H.H., Aller, D. (2002a) Hailfall: the relationship between radar-derived hail kinetic energy and hail damage to buildings. *Atmos. Res.*, **63**, 177–207.

Hohl, R., Schiesser, H.H., Knepper, I. (2002b) The use of weather radars to estimate hail damage to automobiles: an exploratory study in Switzerland. *Atmos. Res.*, **61**, 215–238.

Knightley, P. (2013) Severe weather Forecasts in the British Isles 2012. *Int. J. Meteorol.*, **38**, 172–175.

Koontz, J.D. (1991) The effects of hail on residential roofing products? *Proceedings of the Third International Symposium in Roofing technology*, April 1991, Montreal, NRCA/NIST, 1991.

Leigh, R. (2007) Hailstorm — one of the costliest natural disasters. *Coastal Cities Natural Disasters Conference*, 20–21 February 2007, Sydney, NSW.

Lewis, M.W., Gray, S.L. (2010) Categorisation of synoptic environments associated with Mesoscale Convective Systems over the UK. *Atmos. Res.*, **97**, 194–213.

Lidlington Advertiser (1997) Village hit by freak storm. Available from: http://www.lidlington.org/news/aut97/index.html (accessed 8 March 2014).

Lloyd, H., Hogan, W.(1850) On the storm which visited Dublin on the 18th April, 1850. *Proc. R. Ir. Acad.*, **4** (515–520), 520–522.

Ludlam, F.H., Macklin, W.C. (1960) The Horsham hailstorm of 5 September 1958. *Meteorol. Mag.*, **89**, 245–251.

Macklin, W.C., Merlivat, L., Stevenson, C.M. (1970) The analysis of a hailstone. *Q. J. R. Meteorol. Soc.*, **96**, 472–486.

Marriott, W. (1894) Thunder and hailstorms over England and the south of Scotland – 8th July 1893. *Q. J. R. Meteorol. Soc.*, **20**, 31–39.

Marchant, H. (1990) Thunder at Cardington, Bedfordshire. *J. Meteorol.*, **15**, 43–49.

McPhillimy, H. (1997) Funnel cloud north of Bristol, 17 May 1997. *J. Meteorol.*, **22**, 347.

Meaden, G.T. (1976) Practical hail-gauges for climatological stations. *J. Meteorol.*, **1**, 313–319.

Meaden, G.T. (1984) Trowbridge-Melksham tornado and severe local storm of 13 July 1967: an addendum to M. E. Hardman's 'The Wiltshire hailstorm'. *J. Meteorol.*, **9**, 288–290.

Meaden, G.T. (1995) The fall of ice-coated coke, clinker, coal, grit and dust from the hailstorm-cumulonimbus of 5 June 1983 over Poole, Bournemouth and neighbouring regions. *J. Meteorol.*, **20**, 367–380.

Meaden, G.T. (1998) TORRO Tornado division report: May 1997. *J. Meteorol.*, **23**, 142–143.

Meteorological Office (1963) *British Rainfall 1958*. HMSO, London, Part 3, pp. 15–21.

Mortimore, K.O. (1986) The severe Irish thunderstorms of 25/26 July 1985. *J. Meteorol.*, **11**, 299–306.

Mortimore, K.O., Rowe, M.W. (1984) The hailstorm of 5 June 1983 along the English South Coast. *J. Meteorol.*, **9**, 331–336.

Neill, P. (1823) Notice respecting a remarkable shower of hail which fell in Orkney on the 24th of July 1818. *Trans. R. Soc. Edinb.*, **9** (1) January 1823, 187–199.

Owens, R. (1980) Severe winter hailstorm in south Devon. *Weather*, **35**, 188–199.

Pike, W.S. (1998) The overnight tornadoes of 7–8 January 1998 and a coastal front. *Weather*, **53**, 244–258.

Pollard, A.E. (1936) Large hailstones at Northampton. *Meteorol. Mag.*, **71**, 15–16.

Randon, D. (1983) Severe storms in Jersey, 31 May and 5 June 1983. *J Meteorol.*, **8**, 320–322.

Rollo, R. (1893) *On hail*. Edward Stanford, London.

Royal Society (1697a) A letter from Mr Halley at Chester, giving an account of an extraordinary hail in these parts, on the 29th April last. *Philos. Trans, 1695–1697*, **19**, 570–572.

Royal Society (1697b) Part of another letter, dated May 1, giving a larger account of the same hail-storm. *Philos. Trans, 1695–1697*, **19**, 572–576.

Saltikoff, E., Tuovinen, J.P., Kuitunen, T., Hohti, H., Kotro, J. (2010) A climatological comparison of radar and ground observations of hail in Finland. *J. Appl. Meteorol. Climatol.*, **49** (1), 101–114.

Sanderson, M.G., Hand, W.H., Groenemeijer, P., Boorman, P.M., Webb, J.D.C., McColl, L.J. (2014) Projected changes in hailstorms during the 21st century over the UK. *Int. J. Climatol.*, **35**, 15–24.

Schiesser, H.H. (1990) Hailfall: the relationship between radar measurements and crop damage. *Atmos. Res.*, **25**, 559–582.

Schmidt, W., Sciesser, H., Waldvogel, A. (1992) The kinetic energy of hailfall Part 1V Patterns of hailpad and radar data. *J. Appl. Meteorol.*, **31** (10), 1165–1178.

Schuster, S.S., Blong, R.J., McAneny, H.J. (2006) Relationship between radar-derived kinetic energy and damage to insured buildings in Eastern Australia. *Atmos. Res.*, **81**, 215–235.

Sioutas, M.V., Meaden, G.T., Webb, J.D.C. (2005) Hail frequency and intensity in Northern Greece. *Presentation 5th EMS Annual Meeting –7th ECAM 2005*, 12–16 September 2005, Utrecht, The Netherlands.

Sioutas, M., Meaden, G.T, Webb, J.D.C. (2009) Hail frequency, distribution and intensity in Northern Greece. *Atmos. Res.*, **93**, 526–533.

Smith, D.C. (1976) 35 mm diameter hailstones from a late-autumn storm [Carmarthen, 13 November 1974]. *J. Meteorol.*, **2** (1976–77), 152.

Smyth, T.J, Blackman, T.M, Illingworth, A.J, (1999) Observations of oblate hail using dual polarization radar and implications for hail-detection schemes. *Q. J. R. Meteorol. Soc.*, **125**, 993–1016.

Stevenson, C.M. (1969) The dust fall and severe storms of 1 July 1968. *Weather*, **24**, 126–132.

Symons (1897) The storm of 24 June 1897. *Symons Meteorol. Mag.*, **1897**, 84–91.

Tord, I. (1697) *A full and true relation of the most terrible and dreadful tempest of thunder & lightning, rain and hail that ever yet was seen or heard in England*. J. Wilkins, London.

Tuovinen, J.P., Punkka, A.J., Teittinen, J., Hohti, H. (2007) A climatology of large hail in Finland (1930–2006). *4ᵗʰ European Conference on Severe Storms*, 10–14 September 2007, Trieste, Italy. Available from http://www.essl.org/ECSS/2007/abs/07-Climatology/tuovinen-1177098934.pdf (accessed 29 April 2015).

Tuovinen, J.P., Punkka, J.R., Hohti, J.H., Schultz, D.M. (2009) Climatology of severe hail in Finland: 1930–2006. *Mon. Weather Rev.*, **137**, 2238–2249.

U.S Weather Bureau (1915) Historical frontal charts – Historical weather maps. Northern hemisphere. Sea level. Daily synoptic series (1899–1938). Available in National Meteorological Library and Archive, Washington, DC.

Waldvogel, A., Federer, B., Schmidt, W. (1979) The kinetic energy of hailfalls: part 2, radar and hailpads. *J. Appl. Meteorol.*, **17**, 1680–1693.

Webb, J.D.C. (1988) Hailstorms and intense local rainfalls in the British Isles. *J. Meteorol.*, **13**, 166–182.

Webb, J.D.C. (2009) A note on the severe hailstorm in Sussex and Kent, UK – 15 July 2007. *Int. J. Meteorol.*, **34**, 229–232.

Webb, J.D.C. (2013) Severe thunderstorms and large hail on 17 May 1997. *Int. J. Meteorol.*, **38**, 77–88.

Webb, J.D.C., Blackshaw, J.K. (2012) Notable Scottish thunderstorms in summer 2011. *Weather*, **67**, 199–203.

Webb, J.D.C., Elsom, D.M. (1994) The great hailstorm of August 1843. *Weather*, **49**, 266–273.

Webb, J.D.C., Pike, W.S. (1998a) Severe thunderstorms and hail in England on 7 June 1996. *J. Meteorol.*, **23**, 192–203.

Webb, J.D.C., Pike, W.S. (1998b) Thunderstorms and hail on 7 June 1996: an early season 'Spanish plume' event. *Weather*, **53**, 234–241.

Webb, J.D.C., Elsom, D.M., Meaden, G.T. (1986) The TORRO hailstorm intensity scale. *J. Meteorol.*, **11**, 337–339.

Webb, J.D.C., Rowe, M.W., Elsom, D.M. (1994) The frequency and spatial features of severely damaging British hailstorms: an outstanding 20th century case study monitored by the T.C.O. on 22 September 1935. Fifth TORRO Conference, Imperial College, London, Tornadoes and Storms 3. Part 2. *J. Meteorol.*, **19**, 335–345.

Webb, J.D.C., Elsom, D.M., Reynolds, D.J. (2001) Climatology of severe hailstorms in Great Britain. *Atmos. Res.*, **56**, 293–310.

Webb, J.D.C., Elsom, D.M., Meaden, G.T. (2009) Severe hailstorms in Britain and Ireland, a climatological survey and hazard assessment. *Atmos. Res.*, **93**, 587–616.

Weisman M.L., Klemp J.B. (1984) The structure and classification of numerically simulated convective storms in directionally varying wind shears. *Mon. Weather Rev.*, **112**, 2479–2498.

Westbrook, C., Clark, M. (2013) Observations of a tornadic supercell over Oxfordshire using a pair of Doppler radars. *Weather*, **68**, 128–134.

Wetterzentrale. NCEP re-analysis charts. Available from: http://www.wetterzentrale.de/topkarten/fsrea2eur.html (accessed 8 April 2015).

Yeo, S., Leigh, R., Kuhne, I. (1999) The April 1999 Sydney hailstorm. Natural Hazards Research Centre. *Natural Hazards Q*, 1999, **5**, 2.

Young, M. (1995) Severe thunderstorms over southeast England on 24 June 1994; a forecasting perspective. *Weather*, **50**, 250–256.

10

Lightning Impacts in the United Kingdom and Ireland

Derek M. Elsom[1,2] and Jonathan D.C. Webb[2]

[1]*Faculty of Humanities and Social Sciences, Oxford Brookes University, Oxford, UK*
[2]*Tornado and Storm Research Organisation (TORRO)*, Oxford, UK,*

10.1 Lightning as a Weather Hazard

Lightning is a significant weather hazard in the United Kingdom and Ireland as it is in many countries of the world (Lane, 1948; Viemeister, 1961; Uman, 1986, 2008; Rakov and Uman, 2003; Smith 2008, Bouquegneau and Rakov, 2010, Elsom, 2015a). It is a transient high-voltage electrical discharge or spark produced by a thunderstorm (Figure 10.1). Each brilliant flash lasts only milliseconds. Lightning superheats a column of air, many kilometres in length but less than 5 cm in diameter, to around 30,000°C.

Lightning is the response to turbulent airflows within a thunderstorm causing positive and negative electrical charges to separate. Typically, tiny ice crystals and splinters acquire positive charges on striking large ice pellets (graupel) and are swept upwards within the thunderstorm. The heavier graupel and water droplets, carrying negative charges, descend to the lower parts of the cloud. The negatively charged thunderstorm base repels electrons on the ground away from the vicinity such that the ground beneath the storm base becomes positively charged. Eventually, the electrical separation between the regions of strong electrical charges becomes so great that lightning is triggered. Lightning may be initiated between the cloud and the ground (cloud-to-ground lightning), wholly within the cloud (intra-cloud lightning), between nearby clouds (inter-cloud lightning) and between the cloud and the surrounding air (cloud-to-air lightning). Intra-cloud lightning is the most frequent form of lightning and is commonly called sheet lightning. It is seen as a diffuse brightening of the cloud since the lightning channel is obscured by the cloud. Cloud-to-ground lightning is the intensely bright, branching form of lightning with which most people associate the lightning

weather hazard. Around 20–25% of all lightning is of this type, and in the United Kingdom and Ireland, there are, on average, 200,000–300,000 cloud-to-ground lightning strikes each year.

Currently, lightning is responsible for many injuries to people each year and, in some years, deaths too. In the course of a year, hundreds of buildings and their contents may be damaged as a result of the explosive force of lightning striking a roof or wall, electrical currents surging through electrical circuits within a building (and the electrical and electronic equipment connected to them), and, on occasions, fires ignited by the excessive heat generated in the wiring and connected appliances. Lightning striking industrial premises may ignite flammable and hazardous material causing conflagrations and releasing toxic pollutants which threaten nearby residents who have to be evacuated. Woods, grass, heaths, moors and peat, and the wildlife of those ecosystems, may be destroyed if lightning-initiated fires are not tackled quickly. Decades of investment in the growing of large-scale commercial forests may be lost from lightning-initiated fires. Lightning causes costly disruption to electricity power distribution systems to homes and businesses (Figure 10.2) as well as direct damage to energy production through strikes to power stations, wind turbines and solar panel arrays. Production from oil drilling platforms in the North Sea may be disrupted if lightning strikes the infrastructure or the helicopters transporting personnel. Communication systems and all forms of transport may be disrupted by lightning strikes. National and international commercial flights may have to be suspended when lightning poses a threat to refuelling of aircraft at airports. Flight delays may occur when lightning strikes commercial aircraft in flight although such strikes are relatively uncommon. Although aircraft are designed to withstand lightning strikes, an aircraft struck by lightning is usually given a detailed check as a safety precaution after landing.

*http://www.torro.org.uk/

Extreme Weather: Forty Years of the Tornado and Storm Research Organisation (TORRO), First Edition. Edited by Robert K. Doe.
© 2016 John Wiley & Sons, Ltd. Published 2016 by John Wiley & Sons, Ltd.

(a)

(b)

Figure 10.1 *(a) lightning seen from Charmouth, Dorset, on 22 July 2013. © Matthew Clark; (b) lightning seen from Hayling Island, May 2011. © Jane Burridge.*

10.2 Historical Research into Lightning

Since 1974, the Tornado and Storm Research Organisation (TORRO) and, since 1924, the Thunderstorm Census Organisation (TCO, which TORRO took over responsibility for in 1982) have undertaken a wide range of data collection, analyses and research concerning lightning incidents in the United Kingdom and Ireland. The TCO was particularly active in the 1930s. By 1937, there were 3077 voluntary observers and organisations contributing lightning and thunderstorm observations. Many of these observations were recorded on preprinted postcards and sent to Morris Bower who coordinated the network. *Annual Summaries* were published. Initially, these formed the basis of two volumes

(1931–1933 and 1934–1937). Subsequently, these lengthy reports were replaced by a four-page *Annual Summary* that was to continue in this format until 1980 (Mortimore, 1990). They provided important early insights into the spatial and seasonal distribution of thunderstorms in the United Kingdom and Ireland. Although the emphasis of TCO observers was on the timing and duration of thunder and lightning, they were encouraged to include notes on the impact (local damage, injuries, deaths) caused by lightning and other thunderstorm-related phenomena such as intense rainfall, strong winds, large hail and tornadoes.

An example of the effectiveness of the TCO was its ability to document over 250 damaging lightning incidents caused by severe thunderstorms centred on Devon on 4–5 August 1938

Figure 10.2 *Lightning above power lines, Cardiff. © Chris Cameron-Wilton. (See insert for colour representation of the figure.)*

(see also Chapter 7). One of the projects that the TCO undertook was a 4-year survey of the incidents when a tree was struck by lightning in the United Kingdom and Ireland (Bower, 1936). There were 164 tree incidents recorded, and the species struck most frequently were oak (37%), elm (20%), ash (16%), poplar (8%) and Scots pine (6%). In some incidents, the lightning left no evidence, while in others it left scorched marks, mostly straight but a few spiral. The bark along these strips was ejected explosively because the sap was heated intensely and then expanded and vapourised. The bark became potentially lethal shards threatening anyone standing under the tree, let alone the other danger they faced if the lightning side-flashed to their bodies as the electrical current passed down the tree trunk towards the ground. Some trees were set alight by the lightning. The survey revealed that birch, holly, hornbeam and horse chestnut were not known to have been struck, and there was only one known incident involving beech. One of the conclusions from this limited survey was that smooth-barked trees were not so frequently struck as rough-barked ones although it was noted that smooth-barked trees, especially when wet, may not show evidence of a lightning strike. It was recognised that a much more comprehensive and longer survey was needed in the future, taking into account the total population of each tree species, in order to reach any firm conclusions. The study stressed the danger of anyone standing near a tree during a thunderstorm by highlighting that 'at least 4500 insured animals' are killed annually (in the 1930s) while sheltering under solitary trees in the United Kingdom and Ireland.

The frequency and impacts of lightning received much attention by the TCO, particularly in the 1930s and 1940s, but unfortunately very few documents relating to these investigations have survived. Other organisations such as the Electrical Research Association (1948–1974) and the Climatological Observers Link (1970–present) complemented the thunderstorm data collection, analyses and research of the TCO (1924–1980) and TORRO (1974–present).

10.3   Research into Lightning Impacts

Since TORRO was formed in 1974, and *The Journal of Meteorology* launched in 1975, TORRO has provided systematic documentation of lightning incidents affecting people, animals, buildings, property, trees, motor vehicles and aircraft. The collection of this information, which cannot be measured using scientific instruments, is vital in order to improve our understanding of the nature and impact of this weather hazard in the United Kingdom and Ireland. This type of data collection relies on a network of volunteers and researchers to explore lightning incidents they have observed, collect details from media reports of other incidents and conduct library research of historical events mentioned in local newspapers, weather diaries and other sources.

Tuffnell (2011a, b) provides an example of the insights into the lightning hazard to be gained from historical research of newspapers. He investigated reports of thunderstorms throughout northern England from 1750 to 1799. The start of this period was the time when Benjamin Franklin (1706–1790) conducted electrical experiments to confirm that lightning was 'electric fluid' and which resulted in his invention of the lightning rod (conductor). Tuffnell (2011b) highlights that even with such advances being made in understanding the nature of lightning, many people at the time were not fully aware of the risks it posed and the actions they could take to reduce the risk. In this context, the *Leeds Intelligencer* offered an early attempt at public education about lightning safety on 16 September 1783:

> The cause of so many horses and sheep being lately killed by lightning was owing to their seeking shelter under trees, which undoubtedly are conductors to the lightning. This is a matter that should be thoroughly known, in order to warn people who happened to be near trees at the time of thunder how dangerous their situation is.

During the period from 1750 to 1799, a time when the national population was relatively low compared with today, more than 90 people were reported by newspapers to have been killed by lightning in northern England. Farmers and farm labourers were often lightning victims as it was not unusual for them to carry on working during a thunderstorm. Newspapers from this period contained numerous reports of horses, cattle, sheep and dogs being killed by lightning too. For most of this period, lightning rods were absent from buildings so tall ones, notably churches, often featured in reports of lightning damage along with domestic buildings. Damage to the masts of wooden sailing ships around the coast of northern England was also reported as it was not until 1820 that British inventor William Snow Harris (1791–1867) developed a permanent, effective lightning protection system for ships (Cannell, 2011).

10.4   Annual Number of Lightning Incidents Causing Injuries and Deaths

The incidence of lightning strikes to people in recent decades has been documented and analysed since TORRO initiated a new detailed National Lightning Incidents Database in 1994 (Elsom, 1994). Preliminary findings were discussed by Elsom (1996) and Elsom (2001). Elsom and Webb (2014) extended the database retrospectively to 1988 to provide a 25-year database (1988–2012) for the United Kingdom and Ireland for analysis. Data collection continues today. The details of individual incidents cited in this paper are taken from the UK and Ireland Lightning Incidents Database.

Elsom and Webb (2014) summarised the results from the 25-year UK Incidents Database (1988–2012) which showed there were 445 incidents in which one or more persons are known to have experienced an electrical shock. The annual average of 18 incidents a year is likely to be an underestimate of the actual number of incidents that took place. Additional incidents likely occurred where a person experienced only a minor electrical shock from lightning causing no apparent injury. Such incidents are unlikely to have been recorded as only family and friends may have been told of it. Of the 445 total incidents known for the 25-year period, slightly more than half (52%) were outdoors, in the open air, and nearly half (47%) of those experiencing an electrical shock due to lightning were indoors. The remaining 1% of incident locations comprised four incidents in a motor vehicle and two in an aircraft, a glider.

More than one-quarter (28%) of the incidents resulted in more than one person experiencing an electrical shock from lightning. This meant there was a total of 722 people struck by lightning for the 25-year period, that is, an average of 29 people per annum.

The locations of each incident where people were struck outdoors and those incidents resulting in a fatality are summarised in Table 10.1. Many different activities were being undertaken when death occurred including leisure pursuits (e.g. camping, climbing, fishing, golf, team sports, walking and water sports) and working outdoors (e.g. farming, construction and fishing).

10.5   Lightning Injuries

Injuries outdoors from lightning in the United Kingdom from 1988 to 2012 were in most cases more serious than those occurring indoors (Table 10.2). Moreover, all of the deaths for this period occurred outdoors. This analysis excludes people who died because of fires, explosions and roof collapse caused by lightning but had not experienced an electrical shock.

In recent decades, during which knowledge of cardiopulmonary resuscitation (CPR) among the public has become more widespread and paramedics have defibrillators to deploy if needed, resuscitation of some lightning victims has been reported. For example, from 1988 to 2012, 17 people were resuscitated after being struck by lightning and who may have died otherwise.

Serious injuries from lightning include full-thickness burns, unconsciousness, severe fractures and spinal injuries. Minor injuries may range from superficial or partial-thickness burns, bruising of arms or legs, temporary damage to the eardrums or eyes, temporary weakness, numbness or pain in the shoulder, arm or legs to minor tingling of arms or legs (Andrews *et al.*, 1992; Callagan, 1999; Cooper *et al.*, 2007; Elsom, 2015a).

The lightning injuries reported for an incident usually focus on the acute or immediate injuries. However, some injuries may continue to affect a survivor's life for months, years and even a lifetime. For example, a 16-year-old farm worker was struck by lightning as he moved a herd of cattle on a Dorset farm on 7 June 1996. Although he was resuscitated after a cardiac arrest, he suffered 30% burns to his body which required skin grafts, was left paralysed and experienced 85% brain injuries requiring 24-hour care every day.

Some injuries caused by lightning may not develop for several months or even a year afterwards. A cataract may develop following a lightning strike because of the sensitivity of the eyes to the electrical currents and intense heating. For example, a man was killed while walking in open grassland on the cliff top of Flamborough Head, East Yorkshire, on 11 May 1997. His girlfriend was holding his hand at the time. She was thrown to the

Table 10.1 *Locations of all outdoor (in the open air) lightning incidents in the United Kingdom, 1988–2012 (percentages are rounded to nearest whole number).*

Type of Location	% of Total Outdoor Incidents	% of Total Incidents with a Fatality
Sheltering under a tree	16	20
Mountain, hill, moor or cliff top	11	22
Low-lying farm land or countryside	5	7
Golf course	13	7
Sports ground or recreation park	20	20
Urban setting (*excluding above*)	20	9
Near, in or on water	11	13
Others, for example, airfield	4	2

Table 10.2 *Degree of medical impact of lightning on people outdoors and indoors in the United Kingdom, 1988–2012 (based on the 'worst' recorded medical impact known for each incident).*

Effects	Outdoors (% of Incidents)	Indoors (% of Incidents)
Death	21	0
Resuscitated	7	0
Serious	16	2
Minor	56	98

ground, her clothes scorched, and she suffered burns to her face, legs and wrist, the latter where the electrical current heated the metal of her watch strap to a very high temperature. Her shoes were blown off and burned badly, and all her toes were burned. One of her toes later required amputation. Four days after the incident, she reported discomfort to her right eye, and after 2 months, a cataract developed requiring surgery (Cazabon and Dabbs, 2000).

10.6   Electrical Routes by Which Lightning Causes Injuries

The routes by which people may experience an electrical current from lightning are illustrated in Figure 10.3 (Golde and Lee, 1976; Rakov and Uman, 2003, Cooper *et al.*, 2007; Uman, 2008; Elsom and Webb, 2014; Elsom, 2015a):

a.   Direct strike
b.   Ground current (step voltage effect)
c.   Side flash (splash)
d.   Contact voltage (touch potential)
e.   Ground surface arcing
f.   Upward streamers
g.   Subsequent discharge from an insulated object, for example, metal roof

In addition to the above ways in which lightning delivers an electrical shock to a person, lightning may affect people indirectly. For example, a lightning strike close to a 75-year-old motorist on 13 September 1994 at Grimstone, Dorset, caused him to lose control, possibly because of suffering a heart attack, and he was killed when the car overturned. However, there was no evidence that he had suffered an electrical shock. In July

Figure 10.3   *Routes by which the electrical current from lightning reaches a person. A: Direct strike, B: Ground current (step voltage effect), C: Side flash (splash), D: Contact voltage, E: Ground surface arcing, F: Upward steamer, G: Subsequent flash from an insulated object. © Derek Elsom/Clare Elsom.*

1974, an 11-year-old schoolgirl died from a fractured skull when lightning struck an oak tree at Basildon, Essex, in England, and vapourised the sap which then expanded explosively sending bark flying from the tree as lethal shards. When lightning struck a tree at Great Torrington, Devon, on 30 April 2011, splinters of branches and trunk were blasted up to 45 m away, and only a splintered stump remained. Fortunately, nobody was in the vicinity when it happened. If lightning strikes rock, it may shatter and the fragments thrown into the air as lethal shrapnel which has the potential to cause injury to anyone close by. On 7 June 2007, lightning struck a single-lane asphalt road at Kirkside, Aberdeenshire, leaving a 50-cm-wide trench with 'chunks of tar bigger than cricket balls' being thrown over 100 m away. A witness later examined the road and reported a 'hot, burning smell; it was like newly laid tar with steam rising from it'. Although the fragments had damaged a sign, nobody was close enough to be hurt.

People may be injured by falling roof tiles and masonry when lightning strikes a house. Fires initiated by lightning may lead to the occupants suffering burns and the effects of smoke inhalation. Occupants who experience damage to their homes may need to be taken to hospital for treatment of trauma shock and/or suspected heart attacks. Lightning igniting flammable liquids and materials stored on industrial premises may cause conflagrations and explosions that result in many casualties. A man was killed by a fire when lightning struck a fuel container at Avonmouth Refinery, near Bristol, in August 1994. Farm animals may stampede if lightning and thunder frightens them, and this may result in injury to anyone nearby should they be knocked over and trampled. Horse riders may be thrown if their mounts are startled, as happened at Chigwell, Essex, on 15 May 2012 when lightning struck 5 m away from a group of disabled riding pupils. Fortunately, none were seriously hurt.

10.7   Lightning Strikes to Groups of People

Over half of all fatal incidents (56%) in the United Kingdom from 1988 to 2012 – all outdoor incidents – took place with one or more persons nearby also experiencing an electrical shock. Fortunately, with the exception of the incident on 22 September 1999 when two women died while sheltering under a tall tree in Hyde Park, London, all other people survived in such incidents. However, many of them were knocked to the ground and suffered burns and other injuries which required treatment in hospital.

When a group of people are affected by lightning, the magnitude of the electrical current they experience may vary greatly as may the route by which the current is delivered to them. This may be from one or more of the routes illustrated in Figure 10.3 as happened on 2 September 1995. A football match at Aylesford, Kent, involving 10-year-old children had to be stopped when lightning and heavy rain began. Some players and spectators took shelter in a nearby building, but others sought shelter under a nearby tall tree. Suddenly, lightning struck the tree and side flashed to a man holding an umbrella. The man, 3 other adults and 13 boys received electrical shocks as electrical currents passed down the umbrella shaft to the man and between him and those nearby. The ground current spreading out from the man

also affected the group. The group was thrown to the ground. Many had burns and experienced numbness in their legs, making it difficult to stand or walk initially. Damage to eyes was common too. Although 17 people were taken to hospital, six soon recovered and were discharged. The others needed further treatment. Eight were discharged after 12–24 hours, but the remaining three, who had suffered cardiopulmonary arrest and been resuscitated, were kept in the hospital. An 8-year-old boy had to be resuscitated twice after his cardiorespiratory arrest. He also suffered 18% full-thickness burns to his back and buttocks and remained in hospital for 2 weeks. Many of the patients had tiny, circular, full-thickness burns on the sides of the soles of the feet and tips of the toes. This 'tip-toe lightning sign' was caused by the electrical current when it arced to the ground from their skin. This group highlighted that some effects of a lightning strike may last weeks and months. Three of the adults suffered lower limb paraesthesia (intense tingling and numbness) lasting from 1 to 12 weeks, while some individuals variously experienced amnesia, confusion, depression, mood swings, panic attacks and even psychotic behaviour (Webb *et al.*, 1996; Fahmy *et al.*, 1999; Elsom and Webb, 2014).

Other large groups struck by lightning included 11 teenagers sheltering under a metal playground slide at Walsall Park, West Midlands, on 10 June 1993. They received electrical shocks from lightning, with two receiving minor burns but none needing hospital treatment. On 23 July 1996, 14 Italian students, all aged 14–15 years, were struck by lightning while walking on a bank of the River Thames at Richmond Lock in south-west London. One girl was seriously hurt and placed in intensive care with burns to her stomach and legs. Some experienced minor burns to their hands. Others were treated for minor burns, pains in their arms, bruising and trauma shock. At Great Barr Leisure Centre, Birmingham, as many as 19 people were thrown to the ground or suffered minor burns, eye injuries, aches and bruises during a lightning incident on 10 August 2003. The worst affected was a 40-year-old woman who suffered severe extensive burns and had to be resuscitated after a cardiac arrest.

One of the largest groups of people to have been struck by lightning in the United Kingdom took place at Ascot race course in Berkshire on 14 July 1955 when around 50 people were taken to local hospitals when lightning struck a refreshment tent, opposite the Royal Enclosure. Witnesses described seeing the lightning run along the metal fencing before affecting many people. Twelve people were rendered unconscious for more than a few moments. A 28-year-old pregnant woman died at the scene, and a 51-year-old man died next day in hospital. The immediate effects of the lightning for those admitted to hospital included burns, dysphasia (problems with communicating), temporary paralyses of their limbs, paraesthesia, tremors, temporary deafness, numbness and headache (Arden *et al.*, 1956; Elsom and Webb, 2015; Shaw and York-Moore, 1957). Effects of the lightning strike were noted in some people a year later including depression, bouts of dizziness, intense headaches, mood changes, partial deafness, persistent paraesthesia and the development of cataracts. On the same day as the Ascot incident, five other people died from lightning strikes, and there were 15 deaths in total for 1955 in England and Wales. Fortunately, no year since then has reached this total.

10.8 Locations to Avoid during Thunderstorms

Table 10.1 indicates that the injuries and deaths from lightning in the United Kingdom from 1988 to 2012 occurred in all types of outdoor locations. Nearly two-thirds of all deaths happened in three locations: near or under a tree (20%), on a mountain, hilltop, moor or cliff top (22%) or in a sports or recreation ground park or race course (20%).

Despite well-known advice not to seek shelter under a tall tree during a thunderstorm, many people continue to do so. They seem to worry more about getting wet from the heavy rainfall than the possibility of being struck by lightning. On 13 June 2006, a 40-year-old woman was killed while sheltering under a tree. Paramedics arrived within about 10 minutes, but the unconscious woman, who was burned from her neck to her elbow, had suffered irreversible internal organ failure and brain damage. She died 3 days later in hospital. On 27 June 1999 at Small Heath Park in Birmingham, a thunderstorm interrupted six teenagers playing cricket so they took shelter under a nearby tree. Lightning struck the tree, and one boy suffered serious burns and cardiac arrest before being resuscitated, but he died in hospital 3 days later. The others suffered minor injuries but survived.

Mountain and hill locations are dangerous places to be when thunderstorms develop as there is no immediate shelter. This highlights the importance of planning walking and climbing trips with reference to weather forecasts. Moreover, if someone is struck, then it may take considerable time to get medical assistance and, in the most serious cases, to be transferred to hospital. This was especially true in earlier decades. On 25 and 26 May 2010, there were three separate lightning incidents in the mountains of Cumbria, but helicopters were used to transfer the patients to hospital. This meant the patients, who all survived, received medical attention more quickly than if they had had to be carried down the mountain and then transported by ambulance to the hospital.

Although deaths and serious injuries have occurred on golf courses from lightning, around two-thirds of such incidents result in only minor injuries. In the past 15 years or so, professional golf courses have been installing warning sirens (klaxons) to tell golfers when lightning threatens and that they should cease playing and head for the relative safety of the clubhouse. Unfortunately, some do not heed the warning. On 7 June 1998, one golfer ignored the warning siren to clear the course at Bathgate, West Lothian, and putted out the hole before opening his umbrella. His umbrella was then struck by lightning causing him to drop it. He sought shelter under a nearby tree where he was struck again, this time by a side flash from the tree. Fortunately, he suffered only a sore arm. On 14 May 2003, lightning struck a 50-year-old golfer twice at Peterborough. It struck via his umbrella on the 14th hole, and he experienced an electrical shock down his arm but was otherwise unhurt. About half an hour later, on the 17th hole, he was struck again with the umbrella flying out of his hand. He said it was a lot worse the second time, like needles passing down his arm from his shoulder, against which the umbrella was leaning. He carried on and finished the round, an action which, given some of the serious injuries that lightning has caused to golfers, was both reckless and, if a warning siren had sounded, against the rule book.

Sometimes, golfers recognise it is unsafe to continue after seeing lightning in the distance even though the siren has not yet sounded. This happened at Newbury, Berkshire, on 28 June 2005, when two golfers discontinued play and headed across the fairway towards the safety of the golf club. As they walked, one was struck on his shoulder by lightning and died despite attempts at resuscitation. The other was knocked to the ground but was unhurt. It emerged that a nearby golf course, unlike their golf course, had sounded their siren 1 hour before the fatal lightning incident. The resulting coroner's inquest highlighted the need for all golf courses to implement improved monitoring of the lightning risk and timely sounding of the warning siren in the future.

Sports and recreation grounds and parks are locations where children have been struck by lightning because of their frequent use of such facilities. On 2 June 1999, a group of six children were huddled under an umbrella in a playing field in York when they were struck. They were all temporarily blinded, and all reported being thrown violently to the ground and feeling numb. A 13-year-old girl holding the umbrella suffered burns to a hand and foot. The others were not burned but reported pains in their ears and legs. At Chalgrove, Oxfordshire, on 1 August 1998, lightning struck a boy on a recreation ground while he was running for shelter from a developing thunderstorm. He suffered cardiac arrest but, despite receiving prompt treatment from paramedics, died 4 hours later in hospital. Two other boys, running either side of him, were knocked unconscious temporarily and suffered minor burns, dizziness, headache and lower limb weakness.

Being on or near a river, lake or sea when lightning threatens increases the risks of being struck because a person may be the tallest object in the vicinity. On 6 August 2012, a kayaker was found washed up on the beach at Bradwell-on-Sea, Essex. He had been struck by lightning at sea the previous day, his clothing was burned, and he had probably died from cardiac arrest. On 9 August 1999, lightning killed a man while he was fishing at the edge of a dyke at Reydon, Suffolk. His teenage sons were sat either side of him, and they felt only a tingling sensation. He was conscious initially after being struck and signalled his sons to get help, but paramedics were unable to resuscitate him when they arrived.

10.9 Lightning Incidents Affecting People Indoors

Table 10.3 indicates the indoor situations in the United Kingdom between 1988 and 2012 when the occupants experienced an electrical shock from lightning. Lightning may strike a person inside a building directly through an open window or door. People may be tempted to stand at a window or at an open door to observe a spectacular lightning display. Unfortunately, this location was responsible for 15% of indoor occasions from 1988 to 2012 in the United Kingdom when people experienced an electrical shock from lightning (Table 10.3).

More often, the reason why occupants experience an electrical shock from lightning happens when lightning strikes the building and the electrical current surges along television aerial cables and electrical circuits (Table 10.3). It may cross over to the plumbing circuit of metal pipes connected to radiators, sinks, showers and baths. An occupant touching electrical and plumbing circuits and

Table 10.3 *Locations of all indoor lightning incidents in the United Kingdom, 1988–2012 (percentages rounded to nearest whole number).*

Location Near or Touching …	% of Total Indoor Incidents
Corded telephone	26
Electrical circuit and appliances, including wired computers	19
Window or external door	15
Plumbing circuit	15
Other or unspecified	25

equipment may experience an electrical shock (contact voltage route) or a side flash if they are near to them when the current jumps to their body which is a better electrical conductor. Around midnight on 26 July 2006 at Stanford-in-the-Vale, Oxfordshire, a woman was observing a thunderstorm from her open bedroom window when she experienced a pain in her elbow and knee and was thrown backwards into the room. When lightning had struck her house, her knee had been touching a radiator, and she thought her elbow may have been in contact with the metal window latch too.

Not surprisingly, being near or touching computer, telephone or electrical equipment features in a significant number of indoor incidents. As most are fitted with some form of protection against electrical current surges (surge arrestors), the electrical shock experienced is usually much smaller than that experienced from a direct strike. Increasingly, laptops, tablets and telephones are wireless which may reduce the risk posed to occupants as these are not connected to the electric circuits along which the electrical current generated by the lightning may be passing.

The commonest indoor situation of those affected by lightning in the United Kingdom for the period 1988–2012 (Table 10.3) was when the occupant was using a corded telephone along which

lightning had sent an electrical current. The user may experience a sudden electrical and acoustic shock ('loud bang or click') which may result in a ruptured eardrum, tinnitus (ringing in the ears) and minor burns. The electrical current may cause their muscles to spasm violently throwing them across the room and resulting in bruises, fractures and cuts. They may even be knocked unconscious for a short time. The closer the handset is to the ear, the more serious the injuries.

On 5 July 2006 at Anfield, Liverpool, lightning struck the house of a 38-year-old woman while she was using the telephone. She was thrown across the room and suffered burns and a swollen elbow. She recollected, 'I saw an electrical spark come out of the telephone … it seemed to disappear but when I gripped the receiver it literally came out of my elbow … I felt a sudden blinding headache'. Although she did not appear to suffer any lasting serious effects, in the worst cases, balance problems, headaches, sleep disturbance and problems with the eyes and ears may last for months or even years. Cataracts may develop many months or even a year later. A 9-year-old boy received an electrical shock while using a corded telephone on 8 June 1992 at Doncaster, South Yorkshire, when lightning struck his home. He lost consciousness for a few seconds and suffered superficial burns on his right cheek and forehead. After 4 months, he began suffering blurred vision, and by 7 months, a cataract was diagnosed in his right eye which required surgery (Dinakaran *et al.*, 1998).

10.10   The Frequency with which Lightning Strikes a Person

Using annual counts of cloud-to-ground lightning strikes, Table 10.4 highlights that, for the 14-year period, 1989–2002, a lightning incident in which a building, structure, tree, animal or

Table 10.4 *Annual number of cloud-to-ground lightning strikes and lightning strike incidents (buildings, structures, trees, animals and people) in the United Kingdom and Ireland, 1989–2002.*

Year	Number of Cloud-to-Ground Lightning Strikes	Total Known Lightning Incidents	Number of Incidents When People Were Struck	Number of Incidents Causing Death*
1989	163,069	130	20	4
1990	127,737	88	7	1
1991	168,110	84	10	2
1992	279,644	246	24	4
1993	201,399	158	23	3
1994	548,966	257	49	5
1995	347,412	255	26	1
1996	194,205	114	17	6
1997	293,342	157	17	1
1998	200,212	106	22	3
1999	281,892	316	28	4
2000	309,837	161	15	0
2001	402,773	226	23	0
2002	207,132	166	10	1
Total	3,725,730	2,464	291	35
Annual average	**266,124**	**172**	**21**	**2.5**
Incidents to number of strikes	–	**1:1500** (rounded to nearest 100)	**1:13,000** (rounded to nearest 1,000)	**1:107,000** (rounded to nearest 1,000)

Annual lightning counts were provided by EA Technology for a similar area (based on the coordinate system of −200 km east, 1300 km north and 700 km east, −100 km north).

*All annual incidents were single deaths except for 1999 when two people died in one incident.

person in the United Kingdom and Ireland was struck occurred once every 1500 cloud-to-ground lightning strikes. The frequency with which a person was struck was once every 13,000 cloud-to-ground strikes, and a fatality resulted once every 107,000 strikes. These figures update the analysis undertaken by Elsom (2001) who used a shorter time period (1993–1999), but the results are similar.

A difference in the lightning risk may be evident between the earlier and later years listed in Table 10.4. For the initial 7-year period (1989–1995), the occurrence of a fatal incident was once every 92,000 strikes, but for the later period (1996–2002), the risk lessened to once every 126,000.

10.11   Fewer Deaths from Lightning Over Time

The average number of deaths due to lightning in the United Kingdom and Ireland has decreased markedly in the past century and a half (Lawson, 1889; Golde and Lee, 1976; Baker, 1984; Elsom, 1993, 2001, 2015b; Elsom and Webb, 2014). Detailed information on lightning fatalities is available for England and Wales since 1852. In the latter half of the 19th century (1852–1899) in England and Wales, there were 19 deaths per annum, on average. The worst single year was 1872 when 46 people were killed. During the first half of the 20th century (1900–1949), lightning fatalities had fallen to 13 deaths per annum, on average, and by the second half of the 20th century (1950–1999), it was five deaths each year on average. Decreases in England and Wales have occurred despite the population more than tripling since the mid-19th century (Elsom, 1993, 2001, 2015b). Marked long-term decreases are evident for Scotland, Northern Ireland and Ireland too although the data records are not as long.

During the first decade of the 21st century (2000–2009) in England and Wales, the mean annual death total had fallen to one death, on average (Table 10.5). There were even some years in this decade (2000, 2001, 2007, 2008) when no deaths were recorded in England and Wales.

There are many reasons for the marked reductions in lightning fatalities, and it is not due to a significant decrease in thunderstorm frequency. One of the most important is that there are fewer people working in outdoor occupations that exposed them to the thunderstorm risk. Farms and farm labourers featured in many deaths in the 19th century in the United Kingdom and Ireland, but numbers employed in agriculture have fallen greatly since then. There is greater awareness among farmers of the need to stop work and seek shelter in a substantial building or enclosed vehicle (tractors have had to have metal cabins since the 1970s). These provide protection by providing a safe passage of the lightning to ground on the outside of the building or vehicle.

Those employed in the outdoor construction industry and power line repairers now enjoy more stringent health and safety regulations to ensure they are not placed at risk when thunderstorms are forecast. Today, the majority of people work and live in towns and cities where they are protected by substantial buildings, some of which are fitted with lightning protection systems to ensure lightning is earthed safely to the ground. If lightning does lead to an electrical current entering the building or electrical surges being generated from nearby lightning strikes, surge protectors are fitted to telephones and computer system to minimise or even eliminate the electrical current to which a user may be exposed when lightning strikes.

In the 19th century, most people had to rely on walking, riding a horse or sitting in an open carriage when travelling, and this exposed them to being struck by lightning. By the 20th century, in contrast, most people were travelling in enclosed motor vehicles, trains and even underground railways (e.g. London Underground). Today, travelling even longer distances by aircraft is a relatively safe way form of transport given all the research and testing of aircraft to ensure that a lightning strike has minimal, if any, effect on an aircraft and its passengers and crew.

Today, the people facing the greatest risk from lightning tend to be those participating in outdoor leisure activities, such as mountain climbing and fell walking, and in sports such as golf, fishing and water sports. School and public education have made most people better informed about the risks of lightning than they once

Table 10.5   *Mean annual total of lightning deaths in England and Wales in each decade since the 1850s.*

Decade	Mean Annual Number	Highest Annual Total (and Year)	Half-Century Mean Annual Number	% of Male Fatalities
1852–1859	20.5	45 (1852)	19	84
1860–1869	13.3	26 (1861)		
1870–1879	22.6	46 (1872)		
1880–1889	17.6	33 (1884)		
1890–1899	20.8	43 (1895)		
1900–1909	13.2	28 (1900)	13	89
1910–1919	17.6	31 (1914)		
1920–1929	9.3	17 (1925)		
1930–1939	12.3	24 (1939)		
1940–1949	10.0	15 (1946)		
1950–1959	10.0	15 (1955)	5	82
1960–1969	4.3	9 (1960)		
1970–1979	3.8	11 (1970)		
1980–1989	4.2	14 (1982)		
1990–1999	2.6	5 (1994)		
2000–2009	0.8	2 (2005 and 2006)	n/a	n/a

Updated from Elsom (2001).

were. Organisations, such as schools and sports clubs, are more willing to take responsibility for their members and reschedule or discontinue outdoor activities when thunderstorms are forecast. One of the reasons for the greater willingness and ability of organisations and individuals to take such action is because weather forecasts are readily available and more accurate than they once were. Warning systems have been installed at golf courses to alert golfers when to discontinue play immediately and to seek shelter in the clubhouse. Applications are now available for mobile smartphones which give the distance of the nearest lightning and may even tell the user to 'seek shelter now' if lightning is too close.

Medical improvements have resulted in fewer deaths in recent decades. The cause of immediate death from lightning is cardiopulmonary arrest rather than the serious burns that may be experienced (Cooper *et al.*, 2007). Since the 1960s, an increasing proportion of the population now know how to administer CPR. Better communication systems mean that ambulances and paramedics can be quickly summoned for emergency treatment to be administered, including using a defibrillator to restart the heart. The widespread use of smartphones means that medical assistance can be summoned promptly even from remote locations. Transfer to hospital of lightning victims is much speedier today – helicopters may be used for remote locations such as mountains – so a casualty can benefit in a short time from effective and specialist treatment. Nurses and doctors are now better trained regarding how a lightning victim needs to be treated and cared for.

10.12 Lightning Strikes to Animals

Many incidents in which lightning injured or killed animals have been recorded in TORRO's Lightning Incidents Database for the United Kingdom and Ireland. They include farm animals (cows,

goats, horses, ponies and sheep), stud farm horses, rare zoo animals, family pets and birds. Fish farm stock may be killed too by a lightning strike.

Four-legged farm animals feature relatively frequently in lightning fatalities for two reasons. First, they tend to crowd together for shelter from the thunderstorm under a tree, against a metal fence or around a metal water trough or feeding rack. This makes them vulnerable to side flashes from the tree or metal object as well as the transfer of electrical currents between the animals because they are in contact with each other (contact voltage) or near to each other (side flash). Second, the ground current (stepped voltage effect) gives rise to higher electrical currents generated through their bodies because their legs are wide apart (Kessler, 1993). The greater the distance between points of contact of a body to the ground within the field of the ground current, the greater the electrical charge generated. In contrast to a four-legged animal, a person's legs are less far apart so they are subjected to less electrical current caused by the stepped voltage effect (see lightning route B in Figure 10.3). Moreover, the current passes through vital organs in the case of animals, including the heart, lung and liver, whereas this is not the situation with people (Gomes, 2012). The ground current may be felt up to 10 or even 30 m from the lightning strike which may be a tree, post, open ground or even another person or animal.

The largest known number of animal fatalities caused by a lightning strike in the United Kingdom and Ireland occurred at Campshead Farm in Lanarkshire, Scotland, when lightning killed 320 sheep and a farmer's son in June 1748 (Finlay, 1969). During 19 thunderstorm days between 4 July and 21 September 1884, livestock deaths totalled 44 horses, 181 cattle, 239 sheep, 3 pigs and 2 dogs (Scott, 1985). TORRO's records contain many recent incidents resulting in the deaths of a large number of animals (Table 10.6, Figure 10.4).

Table 10.6 *Examples of groups of animals being killed by lightning in the United Kingdom since 1994 (details extracted from TORRO Lightning Incidents Database).*

Date	Location	Details
25 August 2012	Shepway, Kent	9 cows killed while sheltering under a tree (which disintegrated)
14 June 2009	Pencaitland, East Lothian, Scotland	16 bullocks killed while sheltering under a tree
2 July 2006	Hennock, Devon	91 sheep killed while sheltering under a tree. On the same day, lightning killed four cattle at Yealmpton in the same county
28 June 2005	Pepperstock, Luton, Bedfordshire	12 Holstein–Friesian pedigree cattle killed while sheltering in waterlogged ground under a tree
18 August 2004	Newmill, Penzance	15 rare heifers (Friesian-cross-Limousins) killed
10 August 2003	Trevor, near Llangollen, north Wales	15 cows in calf and the stock bull killed while sheltering under a tree. They were worth more than £40,000 and represented half of a herd of pedigree d'Aquitaines
29 February 2000	Yell, Shetland, Scotland	14 sheep killed
25 August 1999	Musehill Farm, near Tiverton, Devon	10 cows killed and 4 injured (1 lame and 3 with eye damage) while sheltering under a tree. The loss represented one-quarter of the farmer's dairy herd
27 May 1999	Chagford, Devon	13 cows killed
4 January 1999	Ayrshire, Scotland	24 sheep and a bull killed
23 October 1998	Shenton, Leicestershire	13 Holstein–Friesian calves killed while sheltering under an oak tree
18 August 1994	Thetford, Norfolk	22 pregnant Friesian cows, worth £60,000, killed as they crowded around a metal feeding rack when lightning struck it
24 June 1994	Holmesfield, Derbyshire	32 cows were killed – one-third of the farm's milking herd – while sheltering under a tree (Figure 10.4)

Figure 10.4   *Cows (32) killed by lightning at Holmesfield, Derbyshire, on 24 June 1994. Courtesy the* Derbyshire Times.

Some modern farm practices have helped reduce the threat of animal deaths. Animals housed in barns are now better protected by well-grounded metal structures, rather than wooden ones, which means that when lightning strikes, the electrical current passes over the outside of the building before earthing to the ground. Tall isolated trees in a pasture pose a potential danger to animals crowding around it, should it be struck by lightning. To reduce the possibility of side flashes from the tree, Gomes (2012) suggests fitting a metal wire ring (cross section about 8 mm²) around the tree at about 3 m above the ground. Connected to this ring at equal spacing should be 3–4 wires, with similar cross sections and about 10 m in length, extending downwards to the base of the tree. Tie rings should be added around the tree, at about 1-m intervals, connected to the downward wires. At the base of the tree, the downward wires should be buried at about 0.5 m below ground level and extended radially away for about 6–7 m. To reduce the step voltage effect, the ground should be covered with a 10–20-cm layer of gravel or any other earth material that has extremely high resistivity. The greater the area of coverage with gravel, the better the safety of the animals. An alternative to installing this arrangement would be to fence off the tree with wooden and non-conductive fencing to keep the animals away it. This would eliminate the potential danger of a lightning strike to the tree affecting the animals but would prevent them benefitting from shade on days when thunderstorms do not threaten.

10.13   Lightning Impacts on Aircraft and Motor Vehicles

TORRO receives reports of lightning striking an aircraft several times each year in the United Kingdom and Ireland but rarely does any damage occur as all aircraft are designed to withstand being struck. As a precaution, after landing, the aircraft may be given a rigorous inspection before continuing with its flight. A small burn or pitting of the outer skin of the fuselage may be the only evidence of where lightning struck.

The year 2007 provides an illustration of the types of aircraft incidents in which lightning struck. On 19 March, 100 km east of the Shetlands over the North Sea, lightning struck an oil industry Scotia Super Puma helicopter with 15 people on board, but it returned safely to its base. On 2 July, up to four aircraft leaving Dublin Airport were struck by lightning with at least one having to return promptly to be checked out. On 15 August, a Belfast-bound passenger aircraft had to be diverted to Manchester Airport after lightning struck its nose cone soon after taking off from Birmingham. It landed safely, and it was reported that the nose cone had been scorched and two rivets removed but that this had posed no danger to passengers or crew. However, there are rare occasions when serious damage may occur. For example, when lightning struck an oil company's helicopter north east of Aberdeen on 19 January 1995, the helicopter was forced to ditch in the North Sea, but all the crew and passengers were rescued.

Gliders are designed to withstand lightning strikes too, but when lightning struck a glider at Dunstable, Bedfordshire, on 17 April 1999, it fused the control rods, caused the right wing to fall away and stripped some of the glass fibre skin from the fuselage. It seems the intense heat from the lightning strike caused air and/or moisture trapped within the glass fibre to expand rapidly tearing it apart. The pilot received burns to the back of his head and neck and was briefly unconscious. He and his trainee suffered damaged eardrums. However, both parachuted from around 800 m and landed safely.

Enclosed vehicles are relatively safe locations to be during thunderstorms. If lightning strikes the vehicle, it passes over the bodywork and eventually arcs to the ground from a tyre which may display a small burn mark. Occupants do not experience an electrical shock unless they are touching the metal sides or window to cause some of the current to be conducted inside. On rare occasions, there are reports of the vehicle's electrics being damaged and the engine catching fire. On 23 January 2001, lightning struck a moving police vehicle between Hindon and East Knoyle, Wiltshire. The police officer reported 'a loud massive thunderous noise and a great big blue flash went right through the car … arcing across the bodywork … and through my head. I felt a tingling up my arms, back and shoulders … was shocked and stunned … and slowed down as the car jolted'. His police radio was put out of action.

10.14   Increasing Awareness of the Lightning Risk

One of TORRO's aims has been to raise awareness of the risks posed by extreme weather, including lightning, in the United Kingdom and Ireland. It has provided lightning safety advice for use by the media and organisations charged with public safety advice such as The Royal Society for the Prevention of Accidents (2007). There continues to be an urgent need for individuals and groups of people to be aware of the lightning risk and ensure they take thunderstorm forecasts into account when planning outdoor activities. This is becoming even more important given the increasing number of people participating in outdoor leisure activities. Sports clubs should have a lightning safety policy and plan to ensure that those responsible for the safety of participants know they should delay the start or stop an activity if thunderstorms are developing.

Lightning incidents in TORRO's National Lightning Incidents Database highlight that although awareness of the lightning threat is generally greater than in previous decades, there continue to be occasions when people fail to recognise the magnitude of the risk. They carry on regardless with their team sport, golfing round, fishing, playing in a park or walking on the hills. Alternatively, they do interrupt their activity during a thunderstorm, although sometimes because of the heavy rain accompanying it rather than the threat of lightning, but then participants seek shelter under a nearby tall tree instead of a large, substantive building or enclosed motor vehicle. Similarly, many people indoors continue to use a corded telephone even though thunderstorms are in the vicinity. This suggests that although there is much information and advice already available about the lightning risk, more educational and public awareness programmes are needed.

In addition to the threat to personal safety, lightning in the United Kingdom and Ireland continues to pose a risk of damage to buildings and their contents, disruption of electrical supplies and transport and communication systems, deaths and injuries to farm animals and damage to grassland, heathland and forests. The seriousness of these potential impacts highlights the importance of TORRO continuing its systematic collection and analyses of lightning incidents in the United Kingdom and Ireland as well as highlighting the need for increased investment in research into practical ways to minimise these impacts.

Acknowledgements

Thanks are extended to TORRO members, researchers and the public who have provided details of lightning incidents in the United Kingdom and Ireland to the authors and TORRO. Many individuals have given much time to investigating lightning incidents they have observed, collecting details from media reports of other incidents and conducting library research of historical events mentioned in local newspapers, weather diaries and other sources.

References

Andrews, C.J., Cooper, M.A., Darveniza, D. and Mackerras, D. (1992) *Lightning Injuries: Electrical, Medical and Legal Aspects*. Boca Raton, FL: CRC Press.

Arden, G.P., Harrison, S.H., Lister, M.D. and Maudsley, R.H. (1956) Lightning accident at Ascot. *Br. Med. J.*, **1**, 1450–1453.

Baker, T. (1984) Lightning deaths in Great Britain and Ireland. *Weather*, **39**, 232–234.

Bower, S.M. (1936) *British Thunderstorms Containing Summer Thunderstorms: Fourth Annual Report 1934*. Huddersfield: Thunderstorm Census Organisation, pp. 36–44.

Bouquegneau, C. and Rakov, V. (2010) *How Dangerous Is Lightning?* New York: Dover Publications.

Callagan, J. (1999) Medical aspects of lightning injuries. *J. Meteorol.*, **24** (242), 280–284.

Cannell, H. (2011) *Lightning Strikes: How Ships are Protected from Lightning*. Brighton: Book Guild Publishing.

Cazabon, S. and Dabbs, T.R. (2000) Lightning-induced cataract. *Eye*, **14**, 903.

Cooper, M.A., Andrews, C.J. and Holle, R.L. (2007) Lightning injuries, In P.S. Auerbach (ed.) *Wilderness Medicine*, Fifth Edition, Philadelphia: Elsevier, pp. 67–108.

Dinakaran, S., Desai, S.P. and Elsom, D.M. (1998) Telephone-mediated lightning injury causing cataract. *Injury*, **29**, 645–646.

Elsom, D.M. (1993) Deaths caused by lightning in England and Wales, 1852–1990. *Weather*, **48**, 83–90.

Elsom, D.M. (1994) Injuries and deaths caused by lightning in the United Kingdom: a new national database. *J. Meteorol.*, **19** (193), 322–327.

Elsom, D.M. (1996) Surviving being struck by lightning: a preliminary assessment of the risk of lightning injuries and death in the British Isles. *J. Meteorol.*, **21** (209), 197–206.

Elsom, D.M. (2001) Deaths and injuries caused by lightning in the United Kingdom: analyses of two databases. *Atmos. Res.*, **56**, 325–334.

Elsom, D.M. (2015a) *Lightning: Nature and Culture*. London: Reaktion Books.

Elsom, D.M. (2015b) Striking reduction in the annual number of lightning fatalities in the United Kingdom since the 1850s. *Weather*, **70**, in press.

Elsom, D.M. and Webb, J.D.C. (2014) Deaths and injuries from lightning in the United Kingdom, 1988–2012. *Weather*, **69**, 221–226.

Elsom, D.M. and Webb, J.D.C. (2015) Lightning tragedy at the Royal Ascot Racecourse, Berkshire, 14 July 1955. *Int. J. Meteorology*, **40** (390), 48–56.

Fahmy, F.S., Brinsden, M.D., Smith, J. and Frame, J.D. (1999) Lightning: the multisystem group injuries. *J. Trauma Injury Infect. Crit. Care*, **46**, 937–940.

Finlay, R. (1969) *Touring Scotland: The Lowlands*. Henley-on-Thames: Foulis, p. 82.

Golde, R.H. and Lee, M.D. (1976) Death by lightning. *Proc. Inst. Electrical Engineers*, **123**, 1163–1180.

Gomes, C. (2012) Lightning safety in animals. *Int. J. Biometeorol.*, **56**, 1011–1023.

Kessler, E. (1993) Twelve cattle killed by lightning. *Weather*, **48**, 178–181.

Lane, F.W. (1948) *The Elements Rage*. London: Country Life, pp. 76–96.

Lawson, R. (1889) On the deaths caused by lightning in England and Wales from 1852 to 1880. *Q. J. R. Met. Soc.*, **15**, 140–146.

Mortimore, K.O. (1990) Thunderstorm climatological research in Great Britain and Ireland: a progress report and aims for future study. *Weather*, **45**, 21–27.

Rakov, V.A. and Uman, M.A. (2003) *Lightning: Physics and Effects*. New York: Cambridge University Press.

Royal Society for the Prevention of Accidents (2007) Lightning at leisure. www.rospa.com/leisuresafety. Accessed 6 April 2014.

Scott, J.L. (1985) Lightning deaths in Great Britain and Ireland. *Weather*, **40**, 28.

Shaw, D. and York-Moore, M.E. (1957) Neuropsychiatric sequelae of lightning stroke. *Brit. Med. J.*, **2** (5054), 1152–1155.

Smith, C.B. (2008) *Lightning: Fire from the Sky*. Newport Beach, CA: Dockside Sailing Press.

Tuffnell, L. (2011a) Thunderstorms in Northern England, 1750–1799: Part 1: Thunderstorm characteristics. *Int. J. Meteorol.*, **36** (363), 219–227.

Tuffnell, L. (2011b) Thunderstorms in Northern England, 1750–1799: Part 2: Thunderstorm impacts. *Int. J. Meteorol.*, **36** (365), 291–300.

Uman, M.A. (1986) *All About Lightning*. New York: Dover Publications.

Uman, M.A. (2008) *The Art and Science of Lightning Protection*. Cambridge/New York: Cambridge University Press.

Viemeister, P.E. (1961) *The Lightning Book*. New York: Doubleday.

Webb, J., Srinivasan, J., Fahmy, F. and Frame, J.D. (1996) Unusual skin injury from lightning. *Lancet*, **347**, 321.

11

Ball Lightning Research in the United Kingdom

Mark Stenhoff[1] and Adrian James[2]

[1] Ball Lightning Division (1985–1992), Tornado and Storm Research Organisation (TORRO)*,
London, UK,
[2] Ball Lightning Division (1991–2000), Tornado and Storm Research Organisation (TORRO)*,
London, UK,

11.1 Introduction

Of the meteorological phenomena discussed in this book, ball lightning is probably the most controversial and possibly the least well understood. It has been a phenomenon of scientific interest in the United Kingdom since the foundation of the Royal Society of London in 1660, yet even in the modern era, its very existence has occasionally been called into question. Arago, the pioneer of the scientific study of ball lightning, described the phenomenon as 'one of the most inexplicable problems of physics today'.[1] While the study of ball lightning is not at the forefront of modern physics research, it is nevertheless true that research proposing various new or adapted hypotheses or models to explain the nature of ball lightning continues to be published frequently in respected scientific journals. Theories and laboratory experiments designed to explain the nature of ball lightning often relate to developments in contemporary mainstream physics. The complexity of many of these theories is beyond the scope of this chapter which is essentially a historical review.[2] Case histories and ball lightning theories are presented here with limited evaluation, and theories are merely outlined. Readers wishing to evaluate case histories or ball lightning theories in more detail are referred to the *References* at the end of the chapter.

---

* http://www.torro.org.uk/

[1] 'Les éclairs en boule…me paraissent aujourd'hui un des phénomènes les plus inexplicables de la physique' (Arago, 1854b).

[2] For a more detailed and technical survey of theories and experiments covering the period until the end of the 20th century, the reader is referred to Stenhoff (1999).

11.2 Definitions

11.2.1 Lightning

Lightning 'is a very long electric spark, "very long" meaning greater than about 1 km' (Uman, 2008), formed between charge centres of opposite polarity. The most common type of lightning occurs within a cloud, but the most familiar kind of lightning is that which occurs between a cloud and the ground, and it is this that is most closely related to reports of ball lightning (other than events reported within aircraft). Less often, lightning can be a discharge from cloud to air or from one cloud to another. A complete lightning discharge is called a flash, and this lasts typically only a few tenths of a second. The diameter of the cloud-to-ground lightning channel has been estimated at about 2.5 cm. Lightning is generally associated with thunderstorms, but rarely lightning can strike from a clear, blue sky (Uman, 1986).

Benjamin Franklin, Thomas-François d'Alibard and others established the electrical nature of lightning in the mid-18th century. However, studies of electricity were then in their infancy, and electricity did not really become a subject for systematic study until more than 50 years later. Franklin correctly believed that current electricity was a flow phenomenon and identified positive and negative charges, but he and his followers sometimes hinted that 'electric fluid' had properties distinct from those of ordinary matter.[3] A qualitative and then a quantitative understanding of electricity gradually evolved from the work of Michael Faraday, Georg Ohm and others in the early 19th century. It was not until the closing years of the 19th century that science became aware of the existence of negatively charged particles

---

[3] Dibner (1977), pp. 23–49.

Extreme Weather: Forty Years of the Tornado and Storm Research Organisation (TORRO), First Edition. Edited by Robert K. Doe.
© 2016 John Wiley & Sons, Ltd. Published 2016 by John Wiley & Sons, Ltd.

called electrons that can carry electric currents in wires. The associated positively charged particles called protons that balance the charge of most atoms thus making them electrically neutral were not discovered until 1920.[4]

Because of the devastation wreaked by lightning, early investigators believed lightning to be composed of an unusual state of highly combustible matter then described as 'fulminating matter'.[5] Fulminating matter featured in some hypotheses concerning the nature of ball lightning as recently as the 20th century.

11.2.2 Ball Lightning

In the absence of a consensus on the underlying mechanism responsible for ball lightning, it is necessary to define ball lightning more loosely based on eyewitnesses' descriptions. Ball lightning is reported to have the following characteristics. It is generally associated with thunderstorms. It is luminous and roughly spherical, with a modal diameter of 20–50 cm and a lifetime of several seconds. It moves independently through the air, often in a horizontal direction (Stenhoff, 1999). Although many reports claiming to describe ball lightning conform only partially to this description and others diverge considerably from it, it is clear that ball lightning is very different from lightning itself – in its form, dimensions and reported duration. This indicates that ball lightning is not a form of lightning as defined previously.

Ball lightning is described as a transient and often inconspicuous phenomenon, rarely (if ever) recorded in photographs or on film or videotape or by other instruments. In only a minority of reports are damage or traces attributed to ball lightning; hence, it mostly commonly interacts minimally with the environment in which it is observed. Where damage is reported, this can very often be attributed to ordinary lightning. We must therefore depend heavily on the perception and memory of those who happen to see it, both of which are fallible. Nevertheless, there is a surprising degree of consistency in many reported characteristics of ball lightning.

Especially during the 18th and 19th centuries (but less often in the 20th), ball lightning was more often described in British and UK sources as 'globular' or 'globe' lightning. In early sources, ball lightning might also ambiguously be described as a fireball or, even more vaguely, as a 'thunderbolt'.

11.3 What Ball Lightning Is Not

Stenhoff (1999), Chapter 3, pp. 39–54, provides an overview of phenomena that can be mistaken for ball lightning.

In the experience of the authors, deliberate hoaxes account for no more than a handful of reports of ball lightning,[6] nor have the authors found any evidence that ball lightning witnesses have experienced abnormal psychological phenomena such as hallucinations. Normal psychological factors can, however,

contribute to misidentification of well-understood phenomena as ball lightning. Factors might include the effect of anxiety in exaggerating the recollection of duration of events during thunderstorms.[7]

Another perceptual phenomenon potentially relevant to ball lightning is auto-kinesis: the illusion of movement of a stationary luminous source seen in conditions of low ambient lighting, for example, stars or planets observed with the naked eye at night. Although debate about the exact mechanism continues (Levy, 1972), many psychologists believe that this phenomenon is caused by natural, minute movements made by the eye and disregarded by the brain when a source is reconciled with its background yet discernible without a visible background.

It is notoriously difficult to estimate the true size of an unknown object seen in the open air; hence, these data are unreliable for ball lightning seen against the sky. As a general rule, reports of ball lightning observed within enclosed spaces are considerably more reliable and yield far more data for scientific analysis than those of ball lightning seen in the open air. Reports of ball lightning seen in closed spaces yield far more reliable estimates of distance and hence size. Many phenomena that might be misinterpreted as ball lightning in the open air, for example, astronomical phenomena (especially when observed under unusual atmospheric conditions),[8] unusual atmospheric phenomena,[9] insect swarms, streetlights, firework displays or distress flares, vehicle headlights, landing and other lights on aircraft and balloons such as weather balloons, are not observed within enclosed spaces. Cloud-to-ground lightning can initiate combustion or vaporisation of objects on the ground or chemical reactions or perhaps initiate the burning of methane in marshy regions, again producing luminous regions such as *ignis fatuus*. Useful reviews of such phenomena are to be found in Anon. (1968), Condon (1969), Minnaert (1993), Stenhoff (1999) and Maunder (2007).

Other phenomena naturally associated with thunderstorm electricity might also be misidentified as ball lightning. Unfamiliar electrical discharge phenomena such as electric arcs or St Elmo's fire are sometimes observed during thunderstorms. St Elmo's fire, also known as St Elmo's light,[10] corona discharge, glow discharge or brush discharge, is a luminous phenomenon sometimes observed when the electric field is increased around earthed objects (especially sharply pointed ones), for example, ships' masts (Figure 11.1), electric pylons, the radomes or wingtips of aircraft and the heads and hands of humans. Whereas St Elmo's fire maintains contact with an electrical conductor, ball lightning is described as moving freely through the air.

Under some circumstances, discharge phenomena can also be observed in the open or within enclosed spaces. Electrical cables, TV antennae and metal pipes penetrate modern buildings, and

---

[4] Electrons and protons are found in all neutral atoms.

[5] We now know that lightning is composed of plasma, by far the most abundant state of matter in the universe, although occurring naturally only in the case of storm-related phenomena near the surface of the Earth.

[6] Many of those reporting the phenomenon choose anonymity rather than seeking publicity.

[7] Stetson *et al.* (2007) ask, 'Does time really slow down during a frightening event?', concluding that time slowing is a function of recollection rather than perception, perhaps through richer encoding of a salient event.

[8] For example, fireballs, bolides, meteors, the sun, the moon, planets or stars and occasionally comets, orbiting artificial satellites or satellite re-entries.

[9] For example, parhelia or *ignis fatuus*. Lightning within clouds (intra-cloud lightning) can also give the impression of ball lightning.

[10] The spelling was 'St Helmo' in some early sources. It was also occasionally named 'St Helen's fire'.

Figure 11.1 *St Elmo's fire on masts of ship at sea Hartwig, (1875).*

St Elmo's fire might be observed in their vicinity. If the potential of such a conductor reaches a sufficiently high voltage, it is possible for a much more dramatic arc discharge or long spark to be formed inside a house.[11] This might in turn trigger combustion or vaporisation of nearby materials.

Combustion or chemical reactions can take place at the point of impact of lightning on an object on or near the ground, thus producing a region of enhanced persisting brightness. It has been argued that the intensity of a nearby lightning flash or an arc discharge can be so great that it generates a positive after-image.[12] After-images move with movement of the eye. Arago discussed this phenomenon as a possible explanation of ball lightning, and it continues to be proposed occasionally in scientific journals. Further discussion appears in the following. Combustion, chemical reactions or after-images might explain luminosity that persists beyond the very short duration of a discharge of conventional lightning.

A more exotic perceptual illusion than the after-image is that of 'magnetic phosphenes' ('magnetophosphenes'). Ordinary lightning is associated with very high currents that produce intense and rapidly changing magnetic fields. It has been suggested that these can induce electrical activity in the retina that produces the illusion of luminous regions in the periphery of the visual field (Peer and Kendl, 2010).

Investigators of ball lightning should eliminate all reports that can plausibly be explained by these or other physical or psychological processes.[13] In evaluating reports of damage or traces allegedly left by ball lightning, it is also essential to question whether the damage might have been caused by ordinary lightning. Ordinary lightning can cause a range of spectacular and sometimes bizarre effects including electrical, thermal and mechanical damage to buildings, aircraft, trees and other structures and electric shock and burns to and electrocution of humans and other animals.[14] A much more detailed analysis of possible confusion between the effects of ball and ordinary lightning is presented by Stenhoff (1999).[15]

11.4 Ambiguity: Ball Lightning, Fireballs, Meteors and Meteorites

The published works of Arago and Flammarion in the 19th century did much to define and identify ball lightning, but the notion that lightning may assume globular form long preceded their research. James Thomson thus described a summer thunderstorm in his poem 'The Seasons'.

> *Down comes a deluge of sonorous hail,*
> *Or prone-descending rain. Wide-rent, the clouds*
> *Pour a whole flood, and yet, its flame unquenched,*
> *Th'unconquerable lightning struggles through,*
> *Ragged and fierce, or in red whirling balls,*
> *And fires the mountains with redoubled rage.*
>
> (James Thomson, 1802)

Thomson's poem seems to evoke ball lightning, but in the late 17th and 18th centuries, the terms 'fireball' and 'ball of fire' were used to describe a number of different phenomena. Under the page heading 'Balls of Lightning', a letter in *The Gentleman's Magazine* of 1752 relates that upon the top and two arms of a church cross at Plausac in France, whenever any thunderstorm passed over or near it, there settled three balls of fire and that from time immemorial it had been a tradition among the local people that when these balls appeared, no danger was to be apprehended from lightning (Anon, 1752). St Elmo's fire, not lightning, is recognisable in this description, and the confidence of the locals in their security from lightning strikes was certainly misplaced!

Many poorly understood natural phenomena were discussed at the Royal Society's early meetings and published in its journals. There were many references to fireballs (or 'balls of fire'), some associated with lightning and others with meteor showers. The phenomena thus described included 'lightning bolts' or 'thunderbolts', St Elmo's fire, meteors and meteorites and what we would now call ball lightning (in modern terminology, fireball describes

---

[11] This is via a mechanism called the 'side flash'. See Stenhoff (1999).

[12] The more familiar negative after-image observed after looking at a source of light has the complementary colour to that of the light source. Very intense sources, such as a lightning flash or other electrical arc, can produce positive after-images whose colour is the same as that of the stimulus.

[13] See, for example, Cooray and Cooray (2008) and Egorov and Stepanov (2008).

[14] Literature on lightning protection discusses many of the effects of lightning, for example, Uman (2008) and Golde (1973, 1977). Some of the capricious and apparently arcane effects of lightning are discussed in Friedman (2008) and Bouquegneau and Rakov (2010).

[15] Stenhoff (1999), Chapters 4–8, pp. 55–128.

an exceptionally bright meteor). Meteorites have been collected for thousands of years, but it was not until the early 19th century that their extraterrestrial nature was accepted. Until then, Western scientists believed them to be an atmospheric phenomenon like lightning.[16] However, meteors are usually easy to distinguish from thunderstorm-related phenomena in the published accounts.

An early fellow of the Royal Society (FRS), the naturalist Robert Plot, wrote of a

> …meteor…of a globular figure, seen Nov. 22, Anno 1672, about 12 or one at night, not in motion but stationary, against the west door of Wednesbury Church….which shone so bright, that it gave light at half a mile's distance.

Plot, to his credit, recognised this 'meteor' as St Elmo's fire and compared it with instances in classical authors such as Seneca, where 'meteors….slide down and rest upon fit subjects, such as the masts of ships at sea, the spears and engines of soldiers at land' (Plot, 1686).

The Royal Society published an account of a phenomenon, possibly ball lightning, observed from a ship near the Lizard in Cornwall on 4 November 1749. The title of the report, *An Account of an Extraordinary Fireball Bursting at Sea*, exemplifies the ambiguous use of terminology. A Mr Chalmers observed:

> …a large ball of blue fire rolling on the surface of the water, at about three miles distance from us: we immediately lowered our topsails….but it came down upon us so fast, that before we could raise the main tack, we observed the ball to rise almost perpendicular, and not above forty or fifty yards from the main chains; it went off with an explosion as if hundreds of cannon had been fired at one time….We had no thunder nor lightning, before nor after the explosion.
>
> (Chalmers, 1749)

It is more difficult to evaluate reports, which are very numerous from about the middle of the 18th century, of a fireball or ball of fire falling upon a structure or a living being. A typical instance is a report from 1770 when 'there happened at Chester, a most dreadful storm of hail and rain, attended by lightning and thunder, at which time a ball of fire happened to fall on the spire of Trinity church, which was so much damaged, that it must be taken down and rebuilt' (Anon., 1770). The characteristics of this 'ball of fire' are not stated, and it cannot be identified as ball lightning. The damage inflicted on the church is consistent with a high-energy cloud-to-ground lightning flash, whereas ball lightning is probably a low-energy form of lightning which would not cause such severe structural damage.[17] It is clear that there was much scope for confusion between these various phenomena, and as a number of newspaper reports confirm, this persisted until at least the mid-19th century.

11.5   Early Beliefs about Lightning and Ball Lightning

It has been argued that 'early observations of ball lightning provide very limited historical material'.[18] Our understanding of lightning has changed greatly with time, but it would be surprising if we were unable to deduce the occurrence of ball lightning from accounts of thunderstorms that predate the development of atmospheric science. Nevertheless, it is hard to discern sightings of the phenomenon in early literary sources. A solitary reference in the Anglo-Saxon Chronicle to 'immense flashes of lightning, and fiery dragons….flying in the air' is to be associated with the opaque realm of marvels and portents, signs and apparitions. In an age when natural phenomena were inexplicable, only mythology provided images for people to give account of extraordinary sights, and until Benjamin Franklin demonstrated the identity of lightning with electricity in the middle of the 18th century, Lucretius and Aristotle remained recognised authorities on the subject of thunder and lightning.

Before being recognised as an electrical phenomenon, lightning was associated with a multitude of myths and superstitions (Schonland, 1964; Prinz, 1977). All ancient civilisations incorporated lightning and thunder within their religious beliefs (Rakov and Uman, 2003). Many cultures worshipped and feared gods (notably the Norse god Thor) who expressed their wrath by hurling 'thunderbolts' towards the earth.[19] Prinz (1977) wrote: 'In many ethnological representations from prehistoric times lightning is depicted… as a stone falling from heaven or a stone axe hurled from the skies. Its destructive force, which becomes manifest in nature by split trees, broken rock and dead animals, is compared with the effects of a Stone Age tool. …Magic powers have always been attributed to the thunderstone…'.[20] Vestiges of this myth continue into the modern era. As recently as 1946, one author wrote: 'The idea that when there is a stroke of lightning something solid descends from sky to earth is quite erroneous, yet it is still widely held'.[21]

The foundation of the Royal Society,[22] which received its charter from Charles II in 1660, marked the beginning in England of systematic science and the adoption of the 'scientific method' and was attended by a new interest in meteorological phenomena of all kinds. The natural philosophers of the 17th and 18th centuries were dedicated to assembling data for the advancement of knowledge; many were also antiquaries, by habit researchers and recorders, with an empirical habit of mind unconstrained by theory. They were mostly men of a sceptical temper, in deliberate

---

[16] The extraterrestrial nature of meteors was not established until triangulation measurements in 1798. The extraterrestrial origin of meteorites was not confirmed until the work of Ernst Chladni and Jean-Baptiste Biot between 1794 and 1803, although scepticism persisted within and beyond the scientific community for some time afterwards.

[17] Stenhoff (1999), p. 195.

[18] Singer (1971), p. 6.

[19] Gold (1952) wrote, 'In ancient and classical times, the word translated thunderbolt was usually used for ordinary lightning discharge "begotten from thick clouds piled on high". A reference to Abelard's letter to a friend – "the high summits call the thunderbolt" – implies that also in medieval times the thunderbolt was ordinary lightning. But nowadays the word thunderbolt usually implies an unusual phenomenon such as ball lightning'.

[20] See also Goodrum (2008).

[21] See Qakley (1946).

[22] Full title: 'The Royal Society of London for Improving Natural Knowledge'.

contradistinction to what they perceived as the superstition and credulousness of former times.[23] Before this, much folklore and many irrational beliefs were attached to what we now call natural phenomena.

11.6 Early Reports of Ball Lightning

The earliest record in the United Kingdom of what may have been a ball lightning event comes from Little Sodbury in Gloucestershire:

In the yeere 1556, in less than two months dyed Maurice Walsh, together with seven of his children, occasioned by a fiery sulphureous globe rolling in at the parlour door at dinner time, which struck one dead at table, and occasioned the death of the rest. It made its passage through a window on the other side of the room.

(Atkyns, 1712)

11.6.1 Ball Lightning over Land

There are many early accounts of fireballs, balls of fire or thunderbolts falling from the sky. When education and literacy were largely the privilege of the wealthy, it would take some time for scientific ideas to be assimilated into general knowledge. A far greater proportion of the population worked on the land, so there was no shortage of potential observers of meteorological events, even if their interpretation of these events was often somewhat fanciful. Given the preconceived notion that many people had about lightning, it should come as no surprise if ordinary cloud-to-ground lightning was described in this way, observers imagining the lightning channel to be the path followed by the envisaged falling phenomenon.

A spectacular thunderstorm in London on 7 August 1794 produced 'in Great Windmill-street, two balls of fire (which) fell within 10 minutes of each other, the direction of which extended towards the south, of prodigious length, but without much injury' (Anon., 1794). At Worcester on 16 July 1750, 'balls of fire issued from the lightning, one fell in or very near one of our principal streets' (Anon., 1750). On the same day in Cambridge, 'several fire balls were seen in the air', while at Dorking a number of persons are said to have seen 'large balls of fire, which, as they fell upon the houses or ground, divided into innumerable directions' (Anon. 1750). A ball of fire 'from the clouds' fell among horses in camp near Salisbury and bounded along the ground on 6 August 1757 (Anon., 1757). A tradesman at Mixbury, Northamptonshire, on 3 July 1724 noted 'a sort of fire-ball, as large as a man's head, to burst in four pieces near the Church', and the rector of Aynho 'heard the hiss of a ball of fire, almost as big as the moon', which flew over his garden from SE to NW (Wasse, 1725).

The earliest account in England of a ball of fire falling in the open during a thunderstorm dates from July 1665, when there broke 'a most terrible clap of thunder just over Norwich that shaked the whole city, and a ball of fire fell upon Dr. Browne's house, and did some little damage... and so ran (in the sight of divers people) down the walk in the market-place, but (God be praised) did not hurt any person' (Anon., 1665).

A thunderstorm began near Wakefield on 1 March 1774 at half past six in the evening. The witness, a Mr Nicholson, whom Joseph Priestley commends in the published account as 'a good electrician', saw 'a flame of light, dancing on each ear of the horse that I rode, and several others much brighter on the end of my stick'. After 20 minutes,

...the storm abated, and the clouds divided, leaving the northern region very clear: except that, about ten degrees high, there was a thick cloud, which seemed to throw out large and exceedingly beautiful streams of light, resembling as Aurora Borealis, towards another cloud that was passing over it; and, every now and then, there appeared to fall from it such meteors as are called falling stars. These appearances continued till I came to Wakefield, but no thunder was heard.

About nine o'clock a large ball of fire passed under the zenith, towards the south-east part of the horizon. I have been informed, that a light was observed on the weather-cock of Wakefield spire....all the time that the storm continued.

(Priestley, 1774)

A number of discrete phenomena are recorded in this account and not all are demonstrably electrical in nature. The lights seen on the horse's ears and the weathercock of Wakefield spire were St Elmo's fire. The 'large and beautiful streams of light' that appeared to be cast from the higher-level to a lower-level cloud resist ready interpretation. Perhaps, they too were a form of 'glow discharge', although the thundercloud had by then moved away; or else, it was indeed the aurora borealis that Mr Nicholson observed, and the shooting stars were just what they seemed. The fireball that passed over the sky long after the end of the thundery activity may have been a meteor and unrelated to the preceding thunderstorm. Winter thunderstorms are usually fast moving and of short duration. It is possible that the witness has conflated astronomical with meteorological occurrences.

Between 1 and 2 p.m. on 3 July 1830, during a violent thunderstorm,

...the inhabitants of Middle Scotland-yard [in Westminster in London] were much terrified and alarmed at hearing a loud and tremendous explosion in the direction of Messrs. Dalgleish and Taylor's coal-wharf. A number of persons soon repaired to the spot, when it was discovered that the electric fluid had struck the mast of a coal barge... moored in the river [Thames], a short distance from the wharf. Messrs. Dalgleish and Taylor's clerk was sitting in the counting-house at the time, when he observed a ball of fire strike the mast of the vessel and explode with a terrific report, which shook the building to its very foundation, and at the same instant he saw the fragments of the mast hurled with great force in all directions, one of which struck violently against the counting-house. A lad who was standing near the mast, had a most miraculous escape, having been thrown a distance of six or seven paces by the shock; he was completely stunned, but

---

[23] The pedant Thomas Rudder, for instance, in revising Atkyns' *History of Gloucestershire* in 1779, obviously had his doubts about the 'fiery sulphureous globe' that supposedly accounted for Maurice Walsh and his family, as he omits all mention of it in his paragraph on the incident, commenting only that 'the lightning entered at the parlour door, and forced its way out of the window on the opposite side of the room' (Rudder, 1779).

escaped without the least injury; three men were also knocked down, and had a similar fortunate escape. Mr. Mallet, the owner of the barge, had only quitted her a few minutes prior to the awful visitation.[24]

11.6.2　Ball Lightning over Rivers and the Sea

From the end of the 17th century until well into the 20th century, Britain's Royal Navy was the most powerful navy in the world. The heyday of the wooden sailing ship, or clipper, was during the 18th and early 19th centuries when they were used extensively for trading. In the 40 years from 1793 to 1832, more than 250 wooden ships of the British navy were damaged by lightning. Masts were shattered, cargo was destroyed, one in eight ships was set on fire, and more than 200 mariners were killed or severely disabled (Anderson, 1879). Physician Sir William Snow Harris FRS (1791–1867) was also an electrical researcher whose enthusiasm for the phenomena of atmospheric electricity and invention of a new lightning conductor to protect ships earned him the nickname 'thunder-and-lightning Harris'. His research was summarised in his book *On the Nature of Thunderstorms* (Harris, 1842) and, inter alia, refers to globular lightning from a nautical perspective (further discussion follows).

The Rev. E. Budge reported to a meeting in Truro of the Royal Institution of Cornwall on 10 November 1843 that he had corresponded with Harris about an observation made of a 'meteorological appearance' recorded by people of integrity (Anon., 1843). On 2 August 1843, an American ship with a cargo of timber was lying in the middle of Helford Harbour just opposite Bosaban in Cornwall. At about 9 p.m. there was misty rain and a fresh wind, and crew stationed on deck were alarmed to see a strange looking light rapidly approaching them over the water from a short distance up the river in the same direction as the prevailing wind. It soon passed by the vessel at a distance of about 200 yards. At its closest approach, it was similar in both size and appearance to a burning tar barrel, although at one point it seemed to divide into two parts. After passing the ship, it continued in the same direction towards the open sea, remaining visible for several minutes. When on the distant horizon, it appeared somewhat like the light of a steamer. Its estimated speed was at least 30 miles per hour, and while it was in sight, it covered a distance of at least 5 miles. The weather during the previous day was very unsettled. In the early morning there had been heavy rain, and although the afternoon was rather windy with showers, there was no lightning or thunder. Harris considered that this observation could be explained as a brush discharge. Budge noted the many similarities to the event recorded by Chalmers (1749).[25]

There are many reports describing fireballs falling onto ships. On 14 February 1809, the vessel 'Warren Hastings', which had set sail from Portsmouth a few days previously, was struck by lightning on three occasions in a very short time. Each time, the lightning carried itself towards the masts of the ship in the form of a globe of fire (Arago, 1837). About 3 p.m. the western sky was very overcast and a storm appeared to be approaching. Several sailors were sent aloft

…to strike the top gallant masts as speedily as possible, but while lowering them the wind blew tremendously, and the rain fell in torrents, accompanied by heavy claps of thunder. In the midst of the confusion, occasioned by the storm, three balls of fire at short intervals were emitted from the heavens: one of them fell into the main-topmast cross-trees, killed a man on the spot, and set the main mast on fire, which continued in blaze for about five minutes, and then went out. The seamen both aloft and below were almost petrified with fear. At the first moment of returning recollection, a few of the hands ran up the shrouds to bring down their dead companion, when the second ball struck one of them, and he fell, as if shot by a cannon, upon the guard-iron in the top, from which he bounced off, into the cross-jack braces. Finding that he still survived, he was relieved from his perilous situation, and brought upon deck with his arms much shattered and burnt. This poor fellow was expected to undergo an immediate operation, as the only means of saving his life. The third ball came in contact with a Chinese, killed him, and wounded the main-mast in several places; the force of the air, from the velocity of the ball, knocked down… the Chief Mate, who fell below, but was not much hurt. For some time after the storm subsided, a nauseous, sulphrous smell continued on board the ship.[26]

On 26 May the same year,

…a foreign ship, lying in the gallions below Woolwich [on the River Thames in London] had her top and main-mast struck by a thunderbolt, which shivered[27] them to pieces, killed one man, and wounded another.[28]

Balls of fire are sometimes reported to have fallen over the sea. A thunderstorm at sea has been described in 1789 in which the flashes of lightning were 'of three kinds, one such a sudden flash as we frequently see, another which went zigzag along the cloud, and the third, which appeared to be a ball of fire, descending to the water' ('Agricola', 1789). On 17 September 1780, on a thundery morning at Eastbourne, 'a horrid black cloud appeared, out of which Mr. Adair saw several balls of fire drop into the sea successively' (Brereton, 1781). A flash of lightning described by witnesses as 'a meteor' immediately followed, injuring Mr Adair and killing two of his servants.

William Borlase composed two descriptions of remarkable thunderstorms in Cornwall. His memoir of a storm at Ludgvan described the lightning as, 'Sometimes….pointed as a dart; in some places edg'd as a scythe, now but one thin sheet or stream, then two or three, and then one again. Now it fell as several separate balls of fire; but upon the house as a large gush, or torrent' (Borlase, 1753).

The account is impressionistic, and the reference to 'balls of fire' may be as fanciful as the rest of the content. More particular is the description by Borlase of a 'fire-ball, about 5 in. in diameter, somewhat sharp, and pointed in the fore-part', which after a clap of thunder fell upon a boat in the River Tamar containing a number of fishermen, killing one outright, on 2 August 1757 (Borlase, 1758). There are events on record in which the agency of ball lightning may be suspected, but whose nature must remain a subject for speculation:

---

[24] *London Standard*, Monday 5 July 1830.

[25] It is interesting to note that this article refers to the event under the subheading 'Meteors', confirming continued ambiguity in the use of terminology.

[26] *Northampton Mercury*, 25 February 1809.

[27] Shivered = splintered or shattered.

[28] *Northampton Mercury*, 27 May 1809.

From time to time the west coast of Wales seems to have been the scene of mysterious lights. In the fifteenth century, and again on a larger scale in the sixteenth, considerable alarm was created by fires that 'rose out of the sea'. Writing in January 1694, the rector of Dolgelly stated that sixteen ricks of hay and two barns had been burned by a 'kindled exhalation which was often seen to come from the sea'…..We have a statement from Towyn that within the last few weeks 'lights of various colours have frequently been seen moving over the estuary of the Dysynni river and out at sea. They are generally in a northerly direction, but sometimes they hug the shore, and move at high velocity towards Aberdovey, and suddenly disappear'.

(Coleman, 1878)

There are similarities in this account with the aforementioned report by Chalmers (1749). In none of these instances is ball lightning evidently implicated, nor is it apparent that electrical phenomena are involved. The blue colour of the ball of fire off the Cornish coast may suggest an affiliation with the *ignis fatuus*, or will-'o-the wisp; we can only conjecture.

11.6.3 Ball Lightning Associated with Churches

Accounts of fireballs falling upon churches, or infiltrating their interiors, are very numerous. The earliest of these concerns an incident at Wells Cathedral in 1596, when

…there entred in at the west window of the Church a darke and unproportionable thing of the bigness of a footballe, and went along the wall on the pulpit side, and sodainly it seemed to breake,

but with no lesse sound and terror, than if an hundred double canons had been discharged at once, and therewithal came a most violent storm and tempest of lightening and thunder, as if the Church had bene full of fire.

(Stowe and Howes, 1631)

The chronicler implies that this visitation was attributable to the careless omission of a prayer by the priest when he began a discourse of spirits and their properties! To 17th-century worshippers, such a horrifying apparition had demonic origins. At Widecombe church in Devon on 21 October 1638, 'some said they saw a great fiery ball come in at the window and pass through the Church'. A certain 'Master Lyde with many in the Church did see…as it were a great ball of fire, and the most terrible lightening come in at the window'. The church appears to have been struck by a tornado, which killed one person. Luminous phenomena associated with tornadoes may assume the aspect of fiery globes; Rowe speculated that the 'two long bright flares of light' seen in the East Anglian tornado of 1646 may be a garbled reference to ball lightning (Rowe, 1980) (Figure 11.2).

Several people were talking in the church porch at Sampford Courtenay in Devonshire on 7 October 1711 when 'of a sudden, a great fireball fell in between them, and threw some one way, some another, but no-one was hurted'. Meanwhile the ringers 'looking out of the belfry into the church, saw four fire-balls more, a little bigger than a man's fist, which of a sudden broke to pieces; so that the church was full of fire and smoak' (Chamberlayne, 1712). During a thunderstorm at South Molton on 6 June 1751, 'a fireball, as they call'd it', passed between people standing by the

Figure 11.2 Woodcut illustrating the event at Widecombe, Anon (1638).

Figure 11.3   *Globe of fire descending into a room. Hartwig (1875, p. 267).*

south window of the church, while 'another ball, to appearance', rolled towards the west end and entered the belfry where it reportedly broke a very large stone into pieces; thence, it ascended the steeple and melted the iron wire of the chimes and clock into grains before entering the bell chamber and throwing a large bell off the brass it hung upon, besides forcing the brass out of the beam and frame of the bell. Hence, the ball seems to defy gravity and cause successive episodes of structural damage without dissipating its energy. All the damage, however, can be ascribed to a cloud-to-ground discharge. Perhaps the onlookers received mistaken impressions of what they saw. The published account leaves open this possibility (Palmer, 1752). Many other instances of balls of fire falling upon churches to split steeples and pinnacles are on record, but in all cases of damage to masonry, we may suspect the agency of linear lightning. On 23 July 1763, 'a fireball burst with a terrible explosion' in a gallery of All Saints in Hertford (Anon., 1763), and in a letter dated 13 June 1783, the poet William Cowper wrote from Olney that 'yesterday morning at 7 o'clock two fireballs burst either on the steeple or close to it. William Andrews saw them meet at that point… the noise of the explosion surpassed all noises I ever heard' (Cowper, 1804).

11.6.4   Ball Lightning within Houses

From the foregoing evidence, it is plain that not all the events described as fireballs in 18th-century thunderstorms necessarily denote ball lightning. The energy required to cast down heavy masonry belongs to cloud-to-ground discharges. In some cases, the onlooker may be seeing an apparent fireball formed by combustion at the point of contact of a linear lightning strike. The descriptive terminology is liable to be indistinct. On 2 August 1750 in Derby,

'lightning, or, some say, a fireball made its way down an chimney into the house of James Charlesworth, and split a deal box all to shivers' (Glover, 1829). On 20 July 1783, 'a blaze of lightning came down the chimney of a house at Olney, leaving its inhabitant speechless' (Anon., 1783). During a violent thunderstorm with very heavy rain on 3 September 1800 at about 2 p.m.,

> …what is vulgarly called a thunder-bolt entered the chimney of a house in Nottingham; the upper room was occupied by three men weaving stockings; they were driven from their looms and considerably bruised; the whole house is so shattered, the chimney in particular, that it must come down.[29]

A blaze of lightning to one person may seem a fireball to another; the difference in perception is not wide. The conditions under which these observations were made cannot be recreated. Modern houses have narrow chimneys or none, and all churches are provided with lightning conductors. When domestic heating was obtained by burning coal, the insides of chimneys were coated with soot, and this was probably the cause of the 'fireballs' emitted from chimneys struck by lightning, often described as having a smell like that of sulphur (Figure 11.3).

Our understanding of electricity is far different from Benjamin Franklin's notion of electricity as fluid distinct from ordinary matter. For this reason, lightning strikes in the second half of the 18th century were described in Franklinesque terms; they are seen as 'stream of electrical fire', 'a body of fire' and 'electric fluid' that 'falls', impacting upon solid matter. It is quite conceivable that some reports of 'fireballs' derive from a stylised way of

---

[29] *Hull Packet*, 9 September 1800.

looking at lightning. Such stylised perceptions are not uncommon. Linear lightning, for instance, has often been depicted in art as jagged or angular in form, which it is not, but the error has persisted. Artists have imitated others' misconceptions rather than representing what they actually see.

There is a detailed account of a thunderstorm at Steeple Ashton in Wiltshire in 1772. A woman in the village is reported to have seen 'a large quantity of lightning come out of a cloud'. From the extensive damage to the vicarage house on which the lightning 'is supposed to have fallen', it may be assumed that the building was struck by linear lightning. In the north parlour of the house, however, two men

> '...were conversing about a loud clap of thunder that had just happened, (when) they saw on a sudden a ball of fire between them....They described it to have been the size of a sixpenny loaf, and surrounded with a dark smoke; that it burst with an exceeding loud noise, like the firing of many cannons at once.... What is remarkable, Mr. Pitcairn remembers very well to have seen the ball of fire in the room for a short time, a second or two, after he had found himself struck with the lightning....' The author adds that 'a painter of Trowbridge during the storm, observed a ball of fire vibrate forward and backward in the air over some part of Steeple Ashton, and at last dart down perpendicularly, which in all probability was the ball of fire Mr. Wainhouse and Mr. Pitcairn saw in the north parlour of the vicarage house'.
>
> (King, 1773–1774)

W. Snow Harris, reviewing the evidence for ball lightning in the incident at Steeple Ashton, considered that it 'was evidently a species of brush or glow discharge, preceding the stroke of lightning which damaged the house'.[30]

11.6.5 Ball Lightning as a Precursor to Cloud-to-Ground Lightning

An instance of ball lightning as a precursor to a cloud-to-ground discharge can be deduced from a report of 27 July 1785, when at Worthington,

> ...a ball of fire, apparently the size of a goose's egg, passed between the scythe and the legs of a man mowing a field. It split a willow tree at a hundred yards distance, and shivered it in a hundred pieces. The man received no hurt. (Anon., 1785)

A luminous entity so small in size could not possess enough energy to destroy a fully grown willow. The appearance of the ball must have been followed by a cloud-to-ground discharge which struck the tree.

11.7 Interpreting Early Reports

Some of the phenomena described in these early accounts seem conformable with our contemporary understanding of ball lightning. It is best to be cautious. The terms 'fireball' and 'ball of fire' were widely and loosely applied. The available evidence suggests,

however, that observations of luminous globes occurring during thunderstorms have been made for as long as an interest in meteorology has existed in Britain.

11.7.1 1833: Early British Opinion about the Nature of Ball Lightning by Luke Howard FRS (1772–1864)

The earliest British scientist to publish comment on ball lightning was probably chemist and meteorologist Luke Howard FRS, sometimes called 'the father of meteorology', responsible for the system of cloud classification used to this day. In 1833, Howard remarked: 'What are called Fire-balls, in the accounts commonly given of thunder-storms, I believe are merely the body of electric matter, moving so slowly that its form may be distinctly seen in the passage; which is not commonly the case'.[31] Some of the case histories given here are reported in Howard's work.

11.7.2 1837–1859: *Sur le Tonnerre* and Other Works by François Arago (1786–1853)

Frenchman François Arago (1786–1853), like Benjamin Franklin both scientist and statesman, served for a short period as prime minister of France. Arago was an accomplished academic, contributing to a wide range of fields in physics, astronomy and mathematics. His work on ball lightning (Fr: 'éclairs en boule' or 'foudre en boule'), part of a broader study of thunderstorm phenomena, served as the stimulus for scientific research about the phenomenon.

Arago was an occasional visitor to scientific meetings in Britain, and his scientific research was familiar to many British scientists by 1815.[32] He was a foreign member of the Royal Society of London and was awarded its Copley Medal in 1825. He was also an honorary Fellow of the Royal Society of Edinburgh. Arago's first publication about thunderstorms *Sur le Tonnerre* (*About Thunder*) was published in 1837.[33] He wrote,

> Have these globes of fire or fireballs ... really existed? Was not the spherical form attributed to them an optical illusion? Would not a flash of [conventional linear] lightning, if we suppose it to be cylindrical, if its direction were exactly towards an observer, appear to him to be circular or at least globular?
>
> This objection would have weight if the spheroidal form had only ever been seen by those who, being situated exactly in the path of the lightning, should have been struck by it. But an observer placed outside the path of the lightning, viewing it transversely, and seeing it strike a neighbouring or distant house, could not attribute to it the form of a globe, unless it really were globular. These positional circumstances were almost always applicable in the following examples. The objection does not therefore merit further attention.[34]

---

[30] See Harris (1842), p. 39.

[31] Howard (1833), p. 218.
[32] For example, Babbage and Herschel (1815).
[33] Arago (1837), pp. 249–266.
[34] Translated from Arago (1837), pp. 258–259.

He then provided brief details of just over 20 possible ball lightning events, half of which were from the United Kingdom.[35] These reports were presented without critical evaluation, that is, without any attempt to explain them by means of already-understood phenomena such as St Elmo's fire or the effects of conventional lightning. Alas, this set a precedent for many subsequent scientific surveys that took 'ball lightning' reports at face value.

An abbreviated English translation of *Sur le Tonnerre* was published in Edinburgh in 1839.[36] Between 1842 and 1843, Arago's work on thunder and lightning, including ball lightning, was mentioned in British local newspapers.[37] The terminology used in the English translation is somewhat archaic, referring, without disambiguation, to 'thunderbolts', and speculates that lightning is composed of combustible 'fulminating material', and it is clear that Arago was working with Franklin's model of the 'electric fluid' and attempts to explain the high speed of propagation of conventional lightning in the context of this model. Arago wrote,[38]

> The lightning which appears in the form of balls, or fireballs, of which we have collected so many examples … and which are so extraordinary, first of all by the slowness and uncertainty of their motions, and then by the extent of the devastation they occasion in bursting, appear to me … among the most inexplicable phenomena in physics. These balls, or globes of fire, appear to be agglomerations of ponderable substances strongly impregnated with the very essence of the thunderbolt. How, then, are these conglomerations formed? in what regions are they produced? whence do they obtain the substances which compose them? what is their nature? and why are they sometimes suspended for a long period, only that they may precipitate themselves with the greater rapidity? &c. &c. To all these inquiries, science remains mute, and can make no reply.

Having posed these questions (many of which have yet to be answered 175 years later), he speculated briefly on a possible mechanism, touching on the notion of a chemical reaction between atmospheric gases to form nitric acid.

11.7.3   1838: Comments on Ball Lightning by Michael Faraday FRS (1791–1867)

Michael Faraday FRS (1791–1867), the great experimental physicist, studied electrical discharges as part of his extensive research into electricity. In 1838, he expressed scepticism, not of the existence of ball lightning, but rather of the idea that ball lightning was an electrical discharge:

> Electrical discharges in the atmosphere in the form of balls of fire have occasionally been described. Such phenomena appear to me to be incompatible with all that we know of electricity and its modes of discharge. As time is an element in the effect, … it is

possible perhaps that an electric discharge might really pass as a ball from place to place; but as everything shows that its velocity must be almost infinite, and the time of its duration exceedingly small, it is impossible that the eye should perceive it as anything else than a line of light. That phenomenon of balls of fire may appear in the atmosphere, I do not mean to deny; but that they have anything to do with the discharge of ordinary electricity, or are at all related to lightning or atmospheric electricity, is much more than doubtful.

(Faraday, 1838)

11.7.4   1842: 'On the Nature of Thunderstorms' by Sir William Snow Harris FRS (1791–1867)

The research of Sir William Snow Harris FRS (1791–1867) on lightning and lightning protection and his book on the subject (Harris, 1842) have already been mentioned in the context of lightning protection for ships. Harris agreed with Arago and Faraday about the reality of ball lightning but differed with their views on its nature. He appreciated their distinction from the properties of an 'electric spark' like ordinary lightning. He concluded that all ball lightning cases were due to a brush or glow discharge, in other words St Elmo's fire, often as a precursor to a flash of ordinary lightning.[39] He wrote:

> A great deal has been said relative to these appearances, and some doubts have been entertained of their real existence as mere balls of electrical light. Nevertheless the evidence of the existence of a form of disruptive discharge, faithfully conveying to the observer such an impression, is beyond question.

He then referred to the 'curious instance' reported by Chalmers (1749), to which we refer earlier.

He remarked:

> The great number of accounts of such appearances, and the remarkable coincidences in them all, extending as they do through nearly a whole century, and consequently given by observers in no way connected with each other, leave not the least doubt of their existence. … It is by no means easy to explain these appearances applicable to the ordinary electric spark: the amazing rapidity with which this proceeds, and the momentary duration of the light, renders it almost a matter of impossibility that the discharge should appear under the form of a ball of fire; it would be a transient line of sight: we must look, therefore, to some other source for an explanation of these appearances.

> Now it is not improbable, that in many cases in which distinct balls of fire of sensible duration have been perceived, the appearance has resulted from the species of brush or glow discharge already described, and which may often precede the main shock. …In short, it is not difficult to conceive, that before a discharge of the whole system takes place, that is to say, before the constrained condition of the dielectric particles of air intermediate between the clouds and the earth becomes as it were overturned, the particles nearest one of the terminating planes, or other bodies situate on them, may begin to discharge upon the succeeding particles, and make an effort to restore the neutral condition of the system by a gradual process….

---

[35] 7 October 1711, Sampford Courtenay, Devon; 3 July 1725, Aynho, Northamptonshire; 16 July 1750, Dorking, Surrey; 1752, Ludgvan, Cornwall; 20 June 1772, Steeple Ashton, Wiltshire; 1 March 1774, Wakefield, Yorkshire; September 1780, Eastbourne, Sussex; 14 February 1809, Ship 'Warren Hastings' which set sail from Portsmouth, Hampshire; 1809, Newcastle upon Tyne, Tyne and Wear; April 1814, Cheltenham, Gloucestershire (Arago, 1837, pp. 259–265).

[36] Arago (1839).

[37] For example, *Sussex Advertiser*, 30 August 1842; *Leicestershire Mercury*, 3 September 1842; *Reading Mercury*, 11 February 1843.

[38] Arago (1839), pp. 84–85.

[39] Harris (1842), pp. 36–40.

If therefore we conceive the discharging particles to have progressive motion from any cause, then we shall immediately obtain such a result as that obtained by Mr Chalmers … in which a large ball of blue fire was observed rolling on the surface of the water, towards the ship from *to windward*. This was evidently a sort of glow discharge, or St. Helmo's [St Elmo's] fire, produced by some of the polarized atmospheric particles yielding up their electricity to the surface of the water, much in the same way, [as in another observation], the stationary cloud did on the land.

He then referred to a number of historic ball lightning reports included in Arago's 1838 essay and in *Philosophical Transactions*, concluding thus

It is therefore highly probable, that all these appearances so decidedly marked as concentrated balls of fire, are produced by a St. Helmo's [St Elmo's] fire in a given point or points of the charged system, previously to the more general and rapid union of the electrical forces; whilst the greater number of discharges described as globular lightning are … probably nothing more than a vivid and dense electrical spark, in the act of breaking through the air, – which, coming suddenly on the eye, and again vanishing in an extremely small portion of time, has been designated a ball of light.[40]

The book by Charles Tomlinson FRS (1808–1897), published 6 years after that of his older friend Harris, also discussed globular lightning,[41] referring to several reports from Arago (1839) and other authors. Tomlinson outlined Arago's classification of lightning which recognised globular lightning as a distinct form and shared Harris's opinion that globular lightning was a form of brush discharge.

11.7.5 1854–1868: English Translations of French Works

Arago's *Oeuvres Complètes* (*Complete Works*) were published the year following his death (Arago, 1854a). Arago's *Meteorological Essays* were published in English translation in London the next year (Arago, 1855). In London in 1868, a popular book by W. de Fonvielle was published in English translation under the title *Thunder and Lightning* (de Fonvielle, 1868). There was a section on globular lightning (de Fonvielle, 1868, pp. 32–39), several times described as 'meteors'. De Fonvielle claimed that 'sometimes globular lightning has been seen to explode in the presence of five or six hundred spectators assembled in a church or in a theatre!' He reinforced the evidence for the phenomenon by reference to the '150 authentic cases of globular lightning collected by Dr Sestier in his work' (Sestier, 1866). De Fonvielle referred to the matter of which ball lightning is allegedly composed as 'fulgurating substance' and goes on to ask rhetorically, 'What, in reality, is this fulgurating matter of which Arago speaks so frequently in his *Notice Sur le Tonnerre*? We will not undertake, any more than he has done, to define it. Who has ever explained what life is? Nevertheless physiologists are always speaking of it'.

11.7.6 The Late 19th Century: Ball Lightning, Spiritualism and Parapsychology

Before we consider the attitudes of scientists to ball lightning in the late 19th century, it is important to remark on something that, in retrospect, seems incongruous in the development of Western scientific thought during this era. At this time, many very reputable British scientists were very impressed by the claims made for spiritualism[42] and psychic phenomena. In English-speaking countries, spiritualism reached its peak in the period from 1840 to 1920, claiming more than 8 million adherents in the United States and Europe. A prominent supporter of spiritualism was Sir William Crookes FRS (1832–1919), the chemist, physicist and meteorologist who first referred to a 'fourth state of matter' that later became known as 'plasma' and was the phase of matter of which lightning – and perhaps ball lightning – is composed. Other proponents of spiritualism and psychic phenomena included several other FRS: eminent physicists Sir Oliver Lodge, Lord Rayleigh (president of the Royal Society, 1905–1908), and Balfour Stewart (Director of the Kew Observatory from 1859 to 1871)[43] and electrician and engineer Cromwell Fleetwood Varley (1828–1883), who also carried out experimental research related to ball lightning (Varley, 1870).

Camille Flammarion (1842–1925) was a French astronomer who entered the Paris Observatory in 1858, founding the Juvisy Observatory in 1883. He was also a great populariser of science, authoring books on astronomy and ballooning, and also psychic phenomena, in which he was a believer. In 1874, he published his book *L'Atmosphère* (Flammarion, 1874), journalistic and florid in style, which incorporated a collection of 50 ball lightning reports. The book appeared in English translation in 1905 (Flammarion, 1905). The latter is apparently the earliest English language source where the terms 'globular lightning' and 'ball lightning' were used interchangeably.

Several years after the publication of Flammarion's original book, Professor Balfour Stewart, professor of physics at the Owens College, Manchester, made remarks about globular lightning at the first general meeting of the Society for Psychical Research in London, of which he was then vice president, in London on 17 July 1882 (Stewart, 1882–1883). He used what he perceived to be the diminished scepticism of the scientific community towards globular lightning to suggest that the same sea change might eventually happen with psychic phenomena. In an article based on this paper in *The Spectator* (29 June 1882), he described globular lightning as

…a phenomenon that has been frequently observed by trustworthy observers, but that until very recently has hardly been accepted at all as anything that could possibly have occurred … It was said[44] in objection to all the evidence with reference to globular lightning, that is to say, a thunderbolt travelling at a slow rate, and afterwards exploding and giving rise to lightnings of the ordinary kind, that what occurs is an electrical discharge, and that all electric discharges must necessarily take place in a moment of time inappreciably small. Of late years, however, some physicists have suggested that this globular lightning, instead of being an ordinary

---

[40] Harris (1842), pp. 36–40.
[41] Tomlinson (1848), pp. 90–95.

[42] Spiritualism is a religion that claims that spirits of the dead can communicate with the living.
[43] See the following.
[44] Here, he presumably refers to Faraday, op.cit.

electric discharge, is really a sort of travelling Leyden jar, and I believe one foreign observer[45] has shewn in some experiments that something analogous to that on a small scale may be artificially produced. I think that I am entitled to say that a change of tone has consequently taken place among physicists with regard to the evidence for globular lightning. The evidence, of course, remains as before. A little additional evidence accumulates now and then, but the great bulk remains as it was. The fact that we are now able to explain this phenomenon without overthrowing entirely our received views on electricity, has certainly enabled people to accept that evidence that they would not have accepted before. Thus we see that the reason why this evidence was not accepted before, was because the hypothesis with regard to electric discharges was insufficient. We imagined that there could not be anything but an ordinary electric discharge; we did not imagine the possibility of what may be called a travelling Leyden Jar.

Much more recently, in 1937, the Fourth Baron Rayleigh (Robert John Strutt) referred to ball lightning in his presidential address to the Society for Psychical Research. He remarked that instances of ball lightning are 'scarcely understood any better than before. The reason of failure is no doubt that we cannot command them at pleasure so as to devise and carry out experimental tests. That is the same kind of difficulty as we meet with in connexion with our own problem [psychic phenomena]. Some people think it is in itself a ground for incredulity: but this is a train of thought which I have never been able to follow' (Strutt, 1937).

The association of globular or ball lightning with spiritualism and psychic phenomena, subjects soon afterwards discounted by the majority of the scientific community after the exposure of a number of cases of fraud, probably did little to improve the reputation of ball lightning studies. The limited scientific debate of ball lightning in British journals in the subsequent four decades might be partly a consequence of this. Perhaps more significantly, during this period, there was also a great deal of progress in basic physics, and this probably displaced 'fringe' phenomena, of which ball lightning was considered an example.

11.7.7   1921: The Meteorological Office 'Ball Lightning Enquiry'

The Meteorological Office received a report of ball lightning at St John's Wood, London, seen during a severe thunderstorm at 2 a.m. on 26 June 1921. As a result, the Met Office made an appeal via national newspapers for further eyewitness accounts of this or other ball lightning events. They were rewarded by reports from 115 more correspondents, which are now on file at the National Archives. Of these, 65 (57%) seemed to describe ball lightning (Clarke, 2012). The results of the enquiry were published by geophysicist and astronomer Harold Jeffreys[46] of the Meteorological Office in the *Meteorological Magazine* in September 1921 (Jeffreys, 1921). He reported that there was a broad diversity of phenomena other than ball lightning represented within the sample and that among the reports that were probably of ball lightning, there was an unexpectedly wide variation in fundamental reported

characteristics such as colours, shapes (although most were spheroidal) or estimated duration or size (Clarke, 2012).

11.7.8   1923: Survey of Ball Lightning Reports

The study of ball lightning was rendered more respectable 2 years later by the sober analysis by German physicist Walther Brand entitled *Der Kugelblitz* as this was based on a much more substantial collection of reported observations (Brand, 1923). However, this book did not become available in English translation until 1971 (Brand, 1971), so it is difficult to be sure of its influence on research in Britain at the time of its original publication.

11.7.9   1870–1934: Speculations on the Nature of Ball Lightning

A variety of models were proposed to explain ball lightning in the period from 1870 to 1910. A large number of papers attempted to relate ball lightning to experimental studies of direct current electrical discharges, as pioneered by Faraday. Alternative theories were also proposed, such as the impregnation of cosmic dust with combustible gas as it passed through protuberances from the sun and chemical reactions involving ozone. The decade from 1924 to 1934 saw the revival of discussions of 'fulminating matter' and the suggestion that ball lightning could be powered by thermonuclear reactions initiated by lightning. Many of these models now seem quaint and archaic. Other ball lightning models have survived the test of time and have been further developed more recently. Stenhoff (1999) provides a comprehensive list of references.

11.7.10   1936: Does Ball Lightning Exist?

Humphreys in the United States, initially of the opinion that the number and high quality of ball lightning reports indicated that it could not be an optical illusion (Humphreys, 1929), collected and critically analysed 280 reports and eventually drew the conclusion that none of them was certainly ball lightning if defined as a leisurely moving and approximately spherical body, luminous by virtue of its electrical state or condition. He dismissed all but two or three of the reports as due to a few causes: 'brilliant flash at the point struck; persistence of vision; broken discharge path, giving separate flashes; meteorites; will-o'-the wisp [*ignis fatuus*]; falling molten metal; lightning seen end-on; brush discharge' (Humphreys, 1936).

11.7.11   1936–1937: The 'Tub-of-Water' Event and the Estimated Energy Content of Ball Lightning

A very frequently cited event attributed to ball lightning took place in Dorstone, Herefordshire, England, and was reported in the *Daily Mail* on 5 November 1936 under the heading 'A Thunderstorm Mystery'. The primary source was a letter from a correspondent, Mr W. Morris, who reported that

> …during a thunderstorm I saw a large, red hot ball come down from the sky. It struck our house, cut the telephone wires, burnt the window frame, and then buried itself in a tub of water which was underneath. The water boiled for some time afterwards, but when it was cool enough for me to search I could find nothing in it.

---

[45] This appears to refer to the work of de Tessan (1859a, 1859b).

[46] Later to become Sir Harold Jeffreys and to be elected FRS and president of the Royal Astronomical Society.

The newspaper published a response from the then Astronomer Royal Sir Harold Spencer Jones FRS:

> It would seem that your correspondent saw a very rare phenomenon, known as globular or ball lightning. This is much the rarest of all forms of lightning and has never been satisfactorily explained.
>
> The lightning in this form appears as a luminous ball of fire, anything from a few inches to two or three feet in diameter. It usually remains visible for several seconds, but exceptionally it can be seen for one or two minutes; it generally moves slowly and ultimately bursts with a loud report like a bomb.
>
> It is not, of course, a 'thunderbolt,' and your correspondent naturally found no material object in the tub.
>
> *Daily Mail* (London), 5 Nov 1936

The following year, Prof B. L. Goodlet presented a paper on lightning to the Institution of Electrical Engineers (Goodlet, 1937). Goodlet's paper included a section on ball and bead lightning. He said that there was considerable evidence that ball lightning really did occur and referred to the outcome of Brand's survey (1923). In the discussion that followed, there was a communicated comment by Sir Charles V. Boys, who had been actively involved in experimental lightning research, which, inter alia, referred to the *Daily Mail* report.[47] Boys wrote[48]:

> The importance of this observation is that for the first time, so far as I know, a measure of the energy in the red hot ball has been roughly found. I have ascertained that the ball appeared to be the size of a large orange, and after 20 minutes the water was too hot for Mr Morris to put his hand into it. The amount of water was about 4 gallons [approximately 18 litres].

Goodlet used the information from Boys to estimate the energy of the ball. He assumed that all the water in the rain butt was raised to 100°C and that 1.8 kg of water was vaporised and arrived at a very substantial figure of roughly 4–11 MJ.[49] Barry used these data to estimate that the energy density of the ball was between 7.3 kJ cm^{-3} and 19 kJ cm^{-3}.[50] Barry also referred to calculations by various other authors that yield somewhat different figures.[51] It should be noted, however, that there is no published evidence that Mr Morris indicated that any water had been vaporised.

The source of data has been taken by many authors at face value. Unfortunately, there is no evidence that data was obtained from the eyewitness other than by correspondence between him and Boys. It seems unlikely that Boys (81 years old in 1936), carried out a field investigation, and there is no evidence that anyone else did so. Goodlet simply worked

from the information provided by Boys. The apparent endorsement to the newspaper report provided by Sir Harold Spencer Jones seems to be nothing more than a comment on Mr Morris's letter.

In any case, the average energy available from a typical lightning flash is about 200 MJ, more than sufficient to evaporate the quantity of water described previously.[52] There may be a more mundane explanation to explain the estimated high energy. Impure water is a poor electrical insulator. Perhaps, then, the current of a lightning channel travelled to earth via the water in the water butt, and the observer simply saw the bright light of the nearby flash, its perceived duration being extended, perhaps, by a positive after-image.[53]

Lightning expert Professor Martin Uman (pers. comm., 1997) suggested an alternative scheme in which lightning caused an arc from nearby power lines, thus enabling the mains electricity supply to provide a continuing current to maintain the arc, hence vaporising the water. From the observer's viewpoint, the outcome would be very similar. Unfortunately, in the absence of a field study, we do not know whether there were power lines in the vicinity. Nor do we know whether the tub was composed of metal, a good electrical conductor, or of wood, a poor insulator, or other material.

The high energy content and energy density calculated from this observation were extremely influential in motivating and shaping the development of ball lightning theory throughout the world. Many models subsequently developed were designed to account for these high figures and were developed with the apparent promise that understanding the mechanisms associated with ball lightning might enable the development of exciting new energy sources or that some radically new aspect of physics might be required to explain the phenomenon. The positive outcome of this case is that it stimulated scientific interest in ball lightning. The negative consequence is that it set a precedent in encouraging many scientists to invest a great deal of effort in developing exotic theories to account for the supposed high energy of ball lightning in this and similar cases where ordinary lightning was the more probable source of energy while failing to look critically at reports by witnesses of what may have been ball lightning to determine the characteristics that should be modelled by theories.

11.7.12 1937–1957: Quantum-Mechanical and Nuclear Hypotheses for Ball Lightning

Internationally, debate concerning the nature of ball lightning continued. Theodore Neugebauer published a quantum-mechanical model (Neugebauer, 1937). W. I. Arabadji proposed a nuclear theory where cosmic rays were focused into small regions by thunderstorm electric fields (Arabadji, 1956). A. Dauvillier proposed a different nuclear scenario in which ball lightning was composed of radioactive carbon-14 created by the action of thermal neutrons, supposedly liberated in lightning, on atmospheric nitrogen (Dauvillier, 1957).

---

[47] The date for the newspaper report is given by Boys as 3 October 1936. Uman Barry and Singer give the date 5 November 1936 (Uman, 1969, 1984, p. 247), (Barry, 1980, p. 259), (Singer, 1971, p. 157). Barry holds a photocopy of the newspaper article confirming the latter date (pers. commun.), reproduced as a facsimile by Stenhoff (1999), p. 7.
[48] Goodlet (1937), pp. 32–33.
[49] Barry (1980), p. 55.
[50] Barry (1980), p. 48.
[51] Barry (1980), pp. 48–49, 66.

[52] Golde (1977), p. 185.
[53] Lightning frequently discharges via bodies of open water, but the amount of water evaporated would not be measurable.

11.7.13   1955–1972: Plasma Hypotheses for Ball Lightning

By the end of the 1950s, the greatest emphasis was on plasma models. Russian Nobel Prize winner in physics Pyotr Kapitza (1894–1984) wrote a very influential paper outlining a model in which ball lightning was composed of plasma generated by an electromagnetic standing wave (Kapitza, 1955). This model was designed to account for the supposed exceptionally high energy density of ball lightning, and he thus argued that it required the electric field of the thunderstorm as an external energy source.

Another Russian physicist Vitalii D. Shafranov, a prominent pioneer in theoretical plasma physics and controlled fusion research, studied the possibility of stable plasma configurations in the form of toroids and spheres. He called these 'plasmoids' (Shafranov, 1957). Several experimental physicists confirmed that plasmoids could be generated. This led to the notion that ball lightning might consist of a plasmoid with some kind of internal structure like a vortex ring and that this would enable the energy of the ball to be contained and its structure to be preserved. In the United Kingdom, vortex plasmoid models were developed by Scottish electrical engineer Charles Bruce of the Electrical Research Association and Eric Wooding of Royal Holloway College, University of London (Bruce, 1963a; Bruce, 1963b; Bruce, 1964; Wooding, 1963). Both proposed formation mechanisms for a ball lightning plasmoid. Bruce's was based on the escape of plasma from a normal lightning channel, while Wooding's suggested the formation of a plasma vortex ring when normal lightning struck or penetrated a solid surface. He supported his theory with the analogy of the ablation of a solid surface by a high-powered laser pulse (Wooding, 1972). Both these mechanisms used the fluid model of a plasma described in Shafranov's analyses.

There are difficulties with plasmoid models, the simplest being that if the plasmoid is hot, convection (upward vertical motion) would be expected, yet this is described in very few ball lightning reports. Barry argues that the mass density of ball lightning should be close to that of the ambient air because otherwise it could not remain stationary, move horizontally or fall from a high altitude.[54] A further objection is that hot gas or plasma would rapidly expand and diffuse into the surrounding air. Plasma vortex models of ball lightning suggested that the ball was confined against expansion or diffusion by strong magnetic fields generated by electric currents within the vortex.

Models such as Kapitza's could explain the absence of convection because the ball would be confined at a particular point in the standing wave pattern. A development of earlier direct current discharge models was published by Finkelstein and Rubenstein (1964). This suggested that ball lightning was formed by a dielectric inhomogeneity in the d.c. electric field under a thundercloud. Other authors produced a mathematical version of a similar theory (Uman and Helstrom, 1966), but Uman later acknowledged that such models could not explain the formation of ball lightning inside Faraday cages, for example, the fuselage of a metal aircraft (Uman, 1968).

11.7.14   1964: *The Flight of Thunderbolts* by Sir Basil Schonland FRS

Sir Basil Schonland FRS (1896–1972), born in South Africa but who also lived and was partly educated in England, was noted for his research on lightning and for the development of radar during the Second World War. He was director of the Atomic Energy Research Establishment at Harwell from 1958 to 1960 and director of the research group of the newly created UK Atomic Energy Research Establishment from 1960 to 1961. His book on the physics and folklore of lightning, *The Flight of Thunderbolts*, briefly made reference to ball lightning.[55] Schonland argued, as others, that frequent reports that ball lightning coincided with or was seen just after a ground flash of lightning meant that 'most of the reported cases are extremely likely to have been caused by an optical illusion', that is, an after-image. He continued: 'Moreover, anyone who is close to a flash of lightning is hardly in a position to give a reliable account of what happened, and stories of fireballs passing down chimneys and bouncing round the house with a sulphurous smell must be treated with considerable reserve'. He then went on to state, incorrectly, that 'no professional observers of the weather, such as meteorologists, have ever seen a fireball…'. In fact, James Durward, a former deputy director of the UK Meteorological Office had reported observing ball lightning on two occasions: once in Scotland in the open air and, on a subsequent occasion, inside an aircraft in summer 1938 (Gold, 1952).

Schonland was, however, more persuaded by a number of ball lightning reports from the Austrian Vorarlberg. Although he dismissed some of these accounts as descriptions of St Elmo's fire, he identified five cases where there were multiple witnesses (thus making an after-image less probable). In three of these, the ball was described as rolling down a fissure or gorge and was thus somehow related to the flow of a mountain stream. In each of these reports, the witnesses rejected the possibility of St Elmo's fire or an after-image. Schonland considered whether the special topography of the region was relevant and suggested that there might be an association with 'giant' lightning flashes that completely discharge the upper, positively charged part of a thundercloud and that would be more prevalent in mountainous regions. He also suggested that when the ground was dry, a lightning discharge would follow a path of least electrical resistance, for example, a stream, and produce large bubbles of gas such as methane or hydrogen–oxygen mixtures liberated from electrolysis of water whose combustion might produce the effects observed.

11.7.15   1969: Eminent UK Scientists Report Ball Lightning in Aircraft

Roger Jennison (1922–2006), an eminent radio astronomer who had worked at Jodrell Bank and Manchester University and who in 1965 became professor of electronics at the University of Kent at Canterbury, provided a detailed account of a personal experience of such an event (Jennison, 1969). He was the only passenger, seated near the front of an all-metal commercial aircraft flying over the US East Coast during a thunderstorm on 19 March 1963 at 0005 EST. There was much turbulence and the aircraft was apparently struck by lightning. Some seconds later, a perfectly

---

[54] Barry (1980), pp. 45–46.

[55] Schonland (1964), pp. 54–47.

symmetrical glowing sphere of diameter 20–24 cm emerged from the cockpit and travelled at constant height and speed, about 75 cm above the floor at about 1–2 m s⁻¹ relative to the aircraft in a straight line down the central aisle, passing him at a distance of only about 50 cm. The blue-white sphere had no structure and appeared almost opaque, although it appeared limb darkened. He sensed no heat. He estimated the optical power as 5–10 W. It was also witnessed by a panic-stricken member of cabin crew as it vanished into the toilet compartment at the rear of the aircraft (Jennison, 1972).

There are many examples of ball lightning seen within aircraft, and these pose a particular challenge for theory. Ball lightning models based on fluids (e.g. plasmoid and chemical models) could only explain ball lightning within aircraft if the aircraft were penetrated in some way. In fact, most lightning strikes to aircraft produce only damage to the outer surface – the metal skin is not punctured. Because of the Faraday cage effect, lightning models based on electric fields do not readily explain ball lightning seen within a metal aircraft, although aircraft are imperfect Faraday cages.[56]

Roger Jennison was not the only UK scientist to provide a first-hand account of a ball lightning observation. Others have included FRS Dr L. Harrison Matthews (scientific director of the Zoological Society of London from 1951 to 1966)[57] and radio-astrophysicist Sir Martin Ryle, who was awarded the 1974 Nobel Prize in Physics and who was Astronomer Royal (1972–1982) (Davies, 1987). Other eminent scientists who have reported seeing ball lightning have included Dr Eric Dunford,[58] director of space science at the Rutherford Appleton Laboratory until his retirement in 1998 and, as already mentioned, James Durward, deputy director of the Meteorological Office (Gold, 1952).

11.7.16 1969: *The Taming of the Thunderbolts* by C. Maxwell Cade and Delphine Davis

1969 saw the publication of the first UK book dedicated entirely to ball lightning. *The Taming of the Thunderbolts: The Science and Superstition of Ball Lightning* by C. Maxwell Cade and Delphine Davis was largely aimed at a popular readership but included some scientific content. The book opened with a chapter entitled 'Thunderbolts in Legend and Literature', followed by a discussion of the controversy about the existence of ball lightning and overviews of St Elmo's fire and conventional lightning. A selection of case histories was followed by an analysis of the characteristics of ball lightning and a review of several theories. There were then more speculative discussions linking ball lightning to unidentified flying object (UFO) reports and other 'fringe' phenomena. The book concluded with speculations about ball lightning theories leading to the development of new energy sources such as nuclear fusion (Cade and Davis, 1969).

11.7.17 1971: Ball Lightning: An Optical Illusion?

In 1971, a flurry of correspondence was stimulated by a communication to *Nature* by Canadian astrophysicist Edward Argyle from the Dominion Radio Astrophysical Observatory in

Columbia. Argyle revived the suggestion of Humphreys (1936) and Sir Basil Schonland FRS (1964) that ball lightning was an optical illusion. Argyle suggested that the intensity of an ordinary lightning flash was sufficient to stimulate a positive after-image that might be interpreted by an observer as ball lightning (Argyle, 1971).

A subsequent issue of *Nature* carried serious criticisms of Argyle's hypothesis from, perhaps unsurprisingly, ball lightning witness Professor Roger Jennison and also from Dr Neil Charman, from the Ophthalmic Optics Department at the University of Manchester Institute of Science and Technology (UMIST), and Dr Paul Davies of the Institute of Theoretical Astronomy in Cambridge (Jennison, 1971; Davies, 1971; Charman, 1971a). Charman also wrote a paper about perceptual aspects of ball lightning reports the same year (Charman, 1971b) and went on to publish a brief review of ball lightning in *New Scientist* in 1972 (Charman, 1972) and a more substantial review in *Physics Reports* in 1979 (Charman, 1979). As an expert in the field of perception and vision, Charman was critical of the positive after-image hypothesis, rejecting it as a general explanation for ball lightning reports because, in contrast with most reported ball lightning, (i) positive after-images decay steadily in apparent brightness with time, (ii) after-images do not maintain constancy of colour, (iii) an after-image that has disappeared may be recovered by blinking and (iv) the apparent size of an after-image depends on the distance of the background against which it is viewed. The hypothesis also struggles to explain consistent multiple-witness reports of a single event or reports of odour or noise.[59]

11.7.18 1971: Micrometeorites of Antimatter?

The same issue of *Nature* carried a paper by Dr David Ashby of the UKAEA Culham Laboratory and Dr Colin Whitehead of the UKAEA Atomic Energy Research Establishment entitled 'Is ball lightning caused by antimatter meteorites?' (Ashby and Whitehead, 1971). This 'speculative hypothesis' was proposed to account for certain properties of ball lightning, such as penetration into buildings and aircraft, and its estimated energy of 0.1–1 MJ and depended on the postulate that antimatter and matter can coexist because of a barrier between solid matter and antimatter. A tiny micrometeorite of antimatter (mass about 5×10^{-13} kg) could annihilate with ordinary matter to produce 0.1 MJ of energy. The luminosity of ball lightning would be caused by ionisation of the surrounding air produced by the annihilation.

The authors further proposed that the micrometeorites would become negatively charged by the emission of positrons and through secondary emission caused by recoiling fragments. In the electric field of a thunderstorm, they would experience an electrostatic force towards the ground. On 7 February 1972, BBC television broadcast an episode in its *Horizon* science series called 'The Day It Rained Periwinkles'. This focused on unexplained phenomena including ball lightning and included interviews with Roger Jennison and David Ashby.

[56] Stenhoff (1999), pp. 109–122.
[57] Ibid., p. 170.
[58] Ibid., pp. 168–169.

[59] Charman (1979), p. 281.

11.7.19 1972–1995: Crew's Ionised Jet-Stream Hypothesis for Ball Lightning (and Some UFOs)

British engineer Eric W. Crew (c. 1924–2011), a former colleague of C. E. R. Bruce, published a series of papers and other communications about ball lightning in various sources including *New Scientist* and TORRO's *Journal of Meteorology* between 1972 and 1995 (Crew, 1972).[60] Like Cade and Davis, Crew suggested that ball lightning might be the cause of some UFO reports. He proposed a model for ball lightning and some UFOs based on jets of ionised and neutral gas ejected from lightning channels. He proposed that if a stream of ionised air ejected in this way approached an earthed object, a charge of opposite polarity would be induced in the object. Provided the jet stream had sufficient velocity and charge, a steady, luminous discharge might take place near the object. Alternatively, two oppositely directed charged streams of air might produce a vortex phenomenon as described previously (Crew, 1980).

11.7.20 1974: TORRO and the *Journal of Meteorology*

In 1974, Dr G. Terence Meaden formed TORRO – the Tornado and Storm Research Organisation. TORRO began to publish the *Journal of Meteorology* (later the *International Journal of Meteorology*) in 1975. From its first volume, the journal carried ball lightning reports and the occasional paper on ball lightning theory.

11.7.21 1976: A Close Encounter with a Fiery Ball Raises Questions of Ball Lightning Energy

In 1976, Mark Stenhoff of Royal Holloway College, University of London, published a report in *Nature* from a woman in Smethwick who appeared to have been struck by ball lightning the previous summer (Stenhoff, 1976). The report received worldwide media attention and Stenhoff was contacted by large numbers of people who also reported ball lightning. Dr Eric Wooding, also from Royal Holloway, carried out energy estimates that yielded far lower values than those obtained in 1937 from the 'tub-of-water' event (Wooding, 1976).

On 8 August 1975 from about 6 p.m., a violent thunderstorm was in progress in Smethwick in the West Midlands. The storm caused extensive damage. A lady was in the ground floor kitchen at the rear of her terraced house (Figure 11.4).

She had filled a kettle with water and turned towards the gas stove when she saw a luminous ball about 10 cm in diameter appear between and above the stove and a refrigerator, both of them next to the open door to the rear garden. Above the stove was a flue leading to an open chimney, sealed with a metal plate inside the kitchen. Above the refrigerator was an air vent (see Figure 11.5.) Outside the kitchen door was a 3 m conifer tree.

The ball had a bright blue-to-purple core that was surrounded by a 'flame-coloured' halo. The ball moved towards her and she felt a burning heat that made her 'glow all over' and she heard a sound 'something like a rattle'. Instinctively, she gave a scream

[60] See Stenhoff (1999), p. 284, for a full list of references.

Figure 11.4 The house in Smethwick.

and brushed the ball away from her with her left hand, when it exploded, having been in sight for only about 1 second. She could smell singeing and discovered that a hole had been melted in her polyester dress and nylon tights. Her legs were numbed and reddened and her left hand showed redness and swelling, although her skin was unbroken, and she found it necessary to force off her wedding ring under running cold water. Her ring was perhaps heated by an inductive process. The damage to her dress appeared to be entirely thermal (see Figures 11.6 and 11.7). Estimates by Wooding suggested that the total thermal energy of the ball was no more than approximately 3 kJ. This would only provide enough energy to heat and vaporise about 9 g of water from room temperature, in contrast to the considerably greater mass of water apparently vaporised by the ball in the 'tub-of-water' event.

11.7.22 Ball Lightning as Electromagnetic Radiation

Dr Geoff Endean (1938–2005) of the Department of Engineering Science at the University of Durham was active in the field of ball lightning research from 1976 until shortly before his untimely death. In 1976, he introduced a model in which ball lightning consisted of electromagnetic field energy trapped in a spherical vacuum that was separated from the surrounding air by an ionised sheath. In common with some earlier theorists, he envisaged two misaligned, oppositely directed components of a developing lightning channel with opposite charges that would set up a rapidly rotating dipole. He developed and refined this model over subsequent years (Endean, 1976). The technical aspects of this

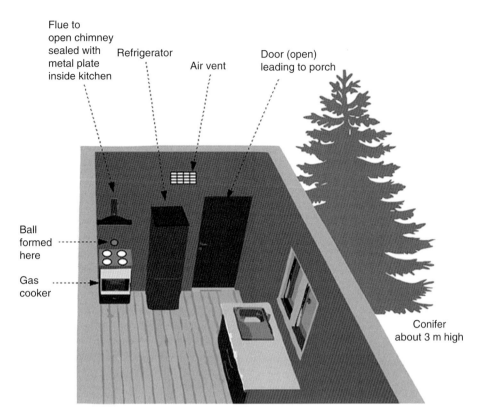

Figure 11.5   *Sketch of interior of kitchen in Smethwick (1975) case.*

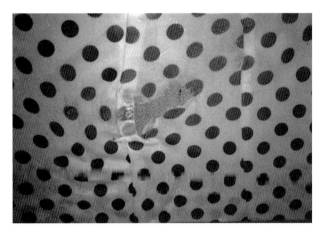

Figure 11.6   *Hole in dress fabric. Photograph: © Brian Tate.*

Figure 11.7   *Detail of damage to fabric. Photograph: © Brian Tate.*

development are beyond the scope of this text, but a summary endorsed by Endean was published by Stenhoff (1999).[61]

11.7.23   Reviews of Ball Lightning

In 1977, a two-volume set *Lightning*, edited by Dr R. H. Golde formerly of the Electrical Research Association and published in London, contained a chapter (in volume 1) about ball light-

ning by Dr Stanley Singer (Golde, 1977). This and the 1980 book *Ball Lightning and Bead Lightning* by Dr James Dale Barry from the United States made reference to the Smethwick case, and Barry carried out more detailed energy estimates, comparing these with other cases including the 'tub-of-water' event (Barry, 1980). In 1979, Dr Neil Charman of the Department of Ophthalmic Optics at the UMIST published a detailed review of ball lightning that made a significant contribution to this field of study. His paper discussed the observational characteristics of ball lightning and theories and experiments related to ball lightning (Charman, 1979).

[61] pp. 199–200; 211–212.

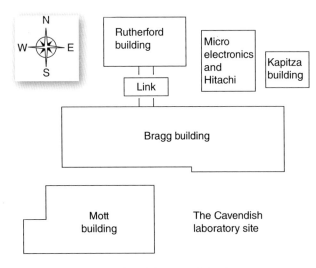

Figure 11.8   *Plan of the Cavendish Laboratory.*

11.7.24   Ball Lightning Reported at Cambridge University

Professor Sir Brian Pippard FRS (1920–2008) reported a particularly well-attested event observed at the Cavendish Laboratory of the University of Cambridge just after 4 p.m. during a violent thunderstorm on 3 August 1982 (Pippard, 1982). The storm was so severe that staff on the upper floor of the two-storey building considered going downstairs.[62] Lightning struck the laboratory and its vicinity several times, although there was no structural damage. Immediately following one flash near the Bragg Building, several observers reported that they saw at least one luminous ball. A physicist seated with his back to a window on the ground floor of the Mott building saw his room briefly illuminated as if by a very bright object moving rapidly westwards between the Bragg and Mott buildings (see Figure 11.8). It would be difficult to explain his observation as a positive after-image. A second observer on the first floor saw the space between the buildings 'filled with a luminous haze' at least to first-floor level, and he interpreted this as sheet lightning. However, on looking to the west, he observed a blue-white light of approximately the apparent size of the moon which appeared motionless about 10–15° above the horizon and was in sight for 3–4 seconds. Another observer in the same room may have seen the same phenomenon just before then, as she had the impression that it was receding while expanding from its original size which was about that of a grapefruit. The data are consistent with an approximate distance from the observers of 12 m. Three other people made a further report of a very bright, blue-white ball moving above the ground to the west. They also said that it subtended about the same angle as the moon. It was in sight for about 4–5 seconds before it vanished.

A closer encounter was experienced by an administrative assistant in a duplicating room on the ground floor. She was closing a window when she was startled by a noise that suggested the window had been knocked in. A bright, spinning, sparkling object of pyrotechnic appearance entered past her head, rebounded from a copying machine and departed as it had arrived. She said, 'It came

in through the window, spinning, rolling, throwing out all sorts of sparks like a Catherine wheel. I was terrified'.[63] The window was undamaged. Another person in the same room was convinced that something had entered the room.

A video recording of lightning was made during a thunderstorm at Ashford in Kent in 1989. On later examination, several frames of the recording showed an image that was thought to have been ball lightning (Jennison *et al.*, 1990). Analysis of the tape showed that the image was probably generated by a street lamp illuminating the stop plane of the recorder while the recorder was defocused (Bergstrom and Campbell, 1991).

11.7.25   1985–1999: The TORRO Ball Lightning Division

The TORRO Ball Lightning Division was formed in 1985, with Adrian James as archives director and Mark Stenhoff as scientific director. In April 1990, TORRO devoted volume 15, number 148, of the *Journal of Meteorology* entirely to 'Ball Lightning Studies' (Meaden, 1990a). A number of papers discussed ball lightning reports,[64] two examining the Ashford video recording.[65] Others proposed or developed ball lightning theories,[66] one of these[67] discussing a mechanism proposed for crop circles[68] that might relate to ball lightning. The special issue also contained a bibliography of ball lightning items published in the *Journal of Meteorology* from 1975 to 1990.[69]

TORRO's fourth conference, held in 1992 at Oxford Polytechnic,[70] was themed 'Ball Lightning' (Stenhoff, 1992). The first session focused on ball lightning reports. Dr Eric Wooding[71] considered the feasibility of recording ball lightning, concluding that it would not be economically or practically viable to set up an automated system dedicated to this purpose. Adrian James discussed fatalities attributed to ball lightning, tentatively concluding that ball lightning might occasionally cause injury comparable to that caused by conventional lightning. Steuart Campbell[72] contended on the basis of several cases that he had examined in depth that the evidence for ball lightning as an atmospheric electrical phenomenon was weak. Prof Roger Jennison[73] gave a critique of some observations and theories of ball lightning and of attempts to reproduce the phenomena experimentally. Mark Stenhoff considered physical evidence for the existence of ball lightning, indicating the limited value of purely anecdotal reports and thus the importance of reports where traces or damage were attributed to ball lightning. He concluded that ball lightning exists but rarely causes damage and that it contains relatively little energy. High-energy damage could generally be attributed to ordinary lightning.

---

[62] See 'Test of fire for Cambridge physicists', The Times, Monday 23 August 1982, p.2.

[63] Ibid.

[64] Meaden (1990b), Meskill (1990), Ohtsuki and Ofuruton (1990), Rowe and Meaden (1990), Rowe (1990) and Jennings (1990).

[65] Meaden (1990c) and Stenhoff (1990).

[66] Witalis (1990).

[67] Meaden (1990d).

[68] The present consensus among scientists seems to be that crop circles are hoaxes.

[69] Ball lightning bibliography from *The Journal of Meteorology* 1975–1990: issue numbers 1–147, in (Meaden, 1990a).

[70] Now Oxford Brookes University.

[71] Emeritus reader in experimental physics at Royal Holloway College, University of London.

[72] Science writer from Edinburgh, Scotland.

[73] Emeritus professor, the Electronics Laboratories, University of Kent, at Canterbury.

Figure 11.9 *TORRO Ball Lightning Conference 1992. Left to right: Dr Geert Dijkhuis, Mark Stenhoff, Emeritus Prof Roger Jennison, Dr Geoffrey Endean, Steuart Campbell, Dr Eric Wooding, Bob Prichard (Chair), Adrian James and Dr Xue-Heng Zheng. © Terence Meaden.*

The second session concerned ball lightning theory. Dr Geert Dijkhuis[74] discussed the statistics and structure of ball lightning. Dr V. Geoffrey Endean[75] presented a paper about recent developments in electromagnetic field energy models. Finally, Dr Xue-Heng Zheng[76] discussed how the maximum rate at which microwaves could be transformed into heat in plasmas could explain the long reported lifetimes of ball lightning.

11.7.26 1999: *Ball lightning: An Unexplained Phenomenon in Atmospheric Physics* by Mark Stenhoff

Mark Stenhoff's book *Ball Lightning: An Unexplained Phenomenon in Atmospheric Physics*, the second book entirely on ball lightning by a UK author, was published late in 1999 (Stenhoff, 1999) (Figure 11.9). Stenhoff had studied ball lightning since 1975. The book discussed the study of ball lightning, reviewed its history and outlined reported characteristics. It discussed phenomena that are or might be mistaken for ball lightning. A detailed evaluation followed of damage and traces attributed to ball lightning and relating these to the known effects of conventional lightning. Stenhoff concluded that most damage imputed to ball lightning was most likely caused by conventional lightning. An evaluation of many photographs and some videotapes and films alleged to be of ball lightning followed, with the conclusion that most could be otherwise explained. Stenhoff then discussed arguments for and against the existence of ball lightning, concluding that there was a core of cases that were difficult to attribute to other causes. The book concluded with a detailed discussion of ball lightning theories and experiments and the most comprehensive bibliography on the subject published so far, with over 2400 references.

11.7.27 2002: Royal Society Theme Issue on Ball Lightning

The Royal Society published a theme issue entitled 'Ball Lightning' in January 2002, compiled by Dr John Abrahamson.[77] This publication gave strong endorsement to the scientific validity of ball lightning studies (Abrahamson, 2002).

11.7.28 2006: Publication of Ministry of Defence: *Unidentified Aerial Phenomena in the UK Air Defence Region* (2000)

In 2006, as a result of an application made under the Freedom of Information Act, a formerly secret report commissioned by the United Kingdom's Ministry of Defence was released to the public. The 465-page report, completed in 2000, was entitled *Unidentified Aerial Phenomena in the UK Air Defence Region*.[78] The author was a contractor employed by the Defence Intelligence Staff (DIS) on a long-term contract, and the author's identity was withheld under the terms of the Data Protection Act 1998. The report was essentially a review of reports of UFOs received by the Ministry of Defence over many decades, and it concluded that while UFOs did not represent a significant hazard or the evidence provide information valuable to national security, among the reports were descriptions of some interesting, relatively rare natural phenomena that are not yet fully understood. The report stated:

> Considerable evidence exists to support the thesis that the events are almost certainly attributable to physical, electrical and magnetic phenomena in the atmosphere, mesosphere and ionosphere. They appear to originate due to more than one set of weather and electrically-charged conditions and are observed so infrequently as to make them unique to the majority of observers. There seems to be a strong possibility that at least some of the events may be

---

[74] Zeldenrust College and Convectron NV, Terneuzen, the Netherlands.
[75] School of Engineering and Computer Science, University of Durham.
[76] Department of Engineering, University of Cambridge.

[77] Although this was published in the United Kingdom, none of the authors was United Kingdom based.
[78] Scientific and Technical Memorandum – No. 55/2/00.

triggered by meteor re-entry, the meteors neither burning up completely nor impacting as meteorites, but forming buoyant plasmas. The conditions and method of formation of the electrically-charged plasmas and the scientific rationale for sustaining them for significant periods is incomplete or not fully understood.[79]

The report contained discussions of ball lightning[80] and plasma phenomena[81] and experimental generation of plasma formations by pulsed discharges.[82] It appears that the report, while extensive, was not subjected to peer review, presumably because of its secret status. Its discussion of plasma phenomena is highly speculative, and many of its hypotheses would not withstand the level of critical appraisal to which much of the literature on ball lightning has been subjected.

11.7.29   2000–2014: TORRO Ball Lightning Division

TORRO continues to contribute to academic research on ball lightning. TORRO's Dr Robert K. Doe published a review of the subject in the book *Forces of Nature and Cultural Responses* (Doe, 2013). TORRO's most significant contribution to ball lightning research continues to be its focus on collecting eyewitness reports of ball lightning. Some of the more interesting reports follow. They are from the annual TORRO Ball Lightning Division reports published in the *International Journal of Meteorology* by the current Ball Lightning Division director, Peter van Doorn.

11.8   A Selection of Ball Lightning Reports Recorded by TORRO 2000–2014

11.8.1   Ball Lightning inside Houses

Ball lightning continues to be reported inside houses, although rarely enters via a chimney. More usually it enters through open or closed windows. Interaction is frequently with electrical or electronic equipment such as televisions and computers.

11.8.1.1   *Ball Lightning Entering a House*

On 22 July 2003 at approximately 04:30 GMT, during a thunderstorm at St Peter in the south-west of Jersey in the Channel Islands, a small ball of light entered the bathroom of a house through an open window. The ball was composed of a mass of bright sparks clustered together into a spherical form. It landed on the floor near the bath and then disappeared silently. Although it passed close to the witness and made contact with the floor, there was no injury or damage (van Doorn, 2004).

11.8.1.2   *Ball Lightning Seen at Close Proximity within a House*

The incident occurred at a house near Ascot Racecourse in Berkshire at the peak of an unusually violent thunderstorm on 27 March 2001 at 23:17 BST. A globular object, about the size of a baseball, emitting a light like a blue flame, travelled in a straight line through the room past a seated woman and her cat. The globe was estimated to have been little more than 60 cm from the woman's face as it passed by, but neither she nor the cat suffered any injury. The object first appeared as a small glowing object surrounded by a halo of light, within which numerous blue sparks moved around the nucleus. The globe sparks seemed to attach themselves to the light-emitting core or nucleus. When it had expanded to the size of a baseball, the object crossed the room in a perfectly straight line towards the window, leaving a 'misty grey/taupe trail with sparkling silver glitter that floated and then disappeared' in its wake. It seems that the phenomenon entered and exited the room through window and door panels of double glazing, although a view of the assumed passage of the ball through the glass was obscured by wooden blinds (van Doorn, 2002).

11.8.1.3   *Two Reports in the Same Storm, Both inside Houses*

On 24 February 2005, ball lightning was seen in a house at L'Islet on Guernsey in the Channel Islands. It entered via a glass porch as a streak of light and passed through a double-glazed door and into the lounge where a couple was watching television. It then changed to a globular form and moved towards the television, stopping approximately 30 cm from the screen, perhaps 75 cm above the floor. After a fraction of a second, it retraced its path out of the house. It was blue and white in colour and 'absolutely crystal clear'.[83]

On the same date, 24 February 2005, and also in Guernsey, an elderly man in St Sampson's was about to retire to bed when there was the loudest clap of thunder he had ever heard. Almost immediately, a ball entered his lounge, passing unimpeded through the window glass and curtains, emitting a hissing sound as it travelled through the air. It directed itself at the 'dead centre' of the back of his television set. The ball was football sized and yellow white in colour. The TV circuitry was destroyed, but there was no visible damage to the rest of the set, and the witness was surprised to find that it was not even warm to the touch.[84]

11.8.1.4   *Multiple Ball Lightning within Houses*

During a thunderstorm in the afternoon of 9 November 2008 and immediately after a lightning discharge followed by thunder, a man and woman were sitting in the living room of their house in Llandyfaelog near the coastal town of Cydweli, Carmarthenshire, West Wales, when a 'very, very bright white' stationary globe of light about 38–45 cm in diameter appeared between them about 1 m above the floor in the centre of the room. Although the ball was visible for only a second or so, it was clearly observed. The woman said in an interview to BBC Wales News[85]:

---

[79] Ibid., executive summary, p. 7.

[80] Ibid., vol. 2 – information on associated natural and man-made phenomena: Chapter 2 (ball and bead lightning).

[81] Ibid.: vol. 2 – information on associated natural and man-made phenomena: Chapters 16 (sunspot, aurora and seismic correlations), 19 (charged dust aerosols), 21 (ionospheric plasma), 24 (sprites, elves and blue jets). Vol. 3 – Chapter 4 (UAP work in other countries, former Soviet Union, plasma research).

[82] Ibid., Annex A (Russian generation of plasma formations by pulsed discharges).

[83] (van Doorn, (2006), p. 21).

[84] ibid.

[85] Report including streaming audio of interview at news.bbc.co.uk/1/hi/wales/7719107.stm.

It was just so bizarre. The ball just appeared around about head height, it didn't move, it just appeared. The light was so bright that even though we had lights on in the room at the time, it cast shadows, and then it just disappeared with this huge bang…. The ball disappeared with this huge crack—the sound was definitely inside the house: it wasn't like the thunder outside, it was definitely within the room—then the power went off and then came back on 10 or 15 seconds later.

Simultaneously, a man, in a bedroom of the same house, saw 'an incredibly bright light' enter the room through the open doorway from the hall. He did not discern any shape within the light, perhaps because it was so vivid. Despite the proximity of the phenomena to the three witnesses, none suffered any physical injury during the event or at its explosive climax. The objects caused no damage to the house and left no physical traces or odours (van Doorn, 2009).[86]

A further event was reported to have occurred during the same thunderstorm, this in the living room of a house at Llanfynydd Village near Wrexham (about 24 km NNE of Cydweli), where there was also flooding. A witness seated near a west-facing window by a table, on which stood a telephone and a laptop computer (which was not online), saw a 'ball of light' suddenly appear about 30 cm above the table and between the computer and telephone. The object was about 11 cm in diameter, 'yellow to orange yellow' and visible for no more than 2 seconds. It remained stationary, produced no odour and was silent until its disappearance which was marked by a 'loud pop to a small bang' (van Doorn, 2009).

11.8.1.5   Ball Lightning inside a Bedroom

A man in Solihull, West Midlands, reported the following experience on 1 September 2005 at 01.18 BST.

> I had just switched off my bedside lamp and was ready for sleep. My eyes were closed; however, I was aware of a bright light in the room. I opened my eyes to see a bright ball of fuzzy bright blue with internal bright white "sparks" at the bottom of my bed, floating at a height of 1.4 m (approx.); the whole of the room was illuminated by the sphere. The ball gently floated over my head at a constant height and a constant gentle velocity, travelling just over 7 feet [2.1 m] in about 15 seconds and exited the room through a framed print and the wall just above my head. There was no discernible sound, no smell, no heat and no evidence of [residue or damage] on either the picture or the wall left by the phenomenon. Disconcertingly, the sphere definitely changed course: the sphere was initially at the bottom left of my bed (not above) but it altered direction to move directly over the bed, moving up the full length of my body and over my head before exiting through the wall. Additionally, the sphere appeared to "hover" briefly above my head before exiting through the wall.[87]

11.8.2   Experienced Observer

An experienced astronomer and senior meteorologist reported an unusual phenomenon to the Meteorological Office who passed details to the TORRO Ball Lightning Division. On 13 September

2004 at about 22:35 in Aldershot, Hampshire, he was suddenly blinded by a very intense, brilliant, white, spherical flash of light that appeared as a football-sized, spherical white light about 15–20 m across the road from him above the roof of a house. It was in sight for a split second.[88]

11.8.3   Earthquake Ball Lightning?

On 27 February 2008 just before 01:00 GMT, the biggest earthquake in the United Kingdom for nearly 25 years shook homes across England. The British Geological Survey stated that the earthquake registered 5.2 on the Richter scale and that its epicentre was near Market Rasen, Lincolnshire.[89] There were many reports across Lincolnshire of unexplained flashes of light or 'mysterious lightning'. However, a woman in Westgate, Louth, described a grapefruit-sized glowing sphere that entered the ground floor bedroom of her home exactly when the earthquake sounds passed through the town. She stated: 'This thing seemed to be coming across the room straight at me. I was very frightened'. The ball caused no damage and vanished abruptly as if switched off like an electric light.[90]

Other reports included a single 'lightning' flash during the quake, three 'lightning' discharges just after and also multiple flashes of light during the disturbance including some that illuminated a bedroom window in Westgate in the manner of car headlights (van Doorn, 2009).

Transient luminous phenomena have long been associated with earthquakes. 'Earthquake lightning' is poorly understood. A possible explanation is the electric field associated with seismic strain,[91] although some reports might be explained by flashovers in electrical power systems disrupted by earth movement.[92]

11.8.4   A Recent Ball Lightning Event Reported to TORRO

Around midnight on the night of 15–16 December 2013, ball lightning was reported at Habost Lochs, Isle of Lewis, by a witness who said that he got caught out in a ball lightning storm near Habost Church Hall in South Lochs:

> [It was] amazing to hear and observe it as I hadn't expected any lightning despite the squally conditions. The rain started lightly as I was going over the bridge into Kershader at about 10 p.m. but it became heavy within two minutes becoming horizontal and squally as I retreated to Habost church hall for shelter. I first heard the swish and clank sounds through the trees and, very suddenly, the street lights became more intense and died one after the other. Once back at the house, my wife mentioned that she had observed low intensity flashes from the front room which is 200 metres further West on the other side of the street and also in Habost … The storm resumed … about 11:52 p.m. and I noticed moving orange spheres on Loch Erisort which were much less bright than normal lightning. I observed one of them in detail moving from West to East over the Loch towards the Kinloch churches near Laxay where

---

[86] See 'Woman's terror over ball lightning', Wales Online, 11 November 2008.

[87] van Doorn (2006), p. 220.

[88] van Doorn (2005), p. 225.

[89] See 'Earthquake felt across much of U.K.'.

[90] See 'Mysterious "lightning" spotted as 'quake hits Louth', Louth Leader, 28 February 2008.

[91] The piezo-electric effect.

[92] Uman and Rakov (2003), pp. 667–668.

it became an orange flash this was noted on my laptop as 00:02 16/12/2013. I was relieved in the morning when my son confirmed that an acquaintance of his on the West Side had tweeted him about seeing ball lightning around the same time over there.

Weather conditions at the time were squally, with showers of hail, sleet and rain, most with thunder. Very large thunderclouds were widespread across the sky.[93]

11.9   2014: Ball Lightning in the UK Media

Continued interest of the UK media in ball lightning was confirmed by widespread reporting[94] of an accidental instrumented observation of ball lightning by scientists in China. They obtained two slitless spectrographs[95] of the ball which was formed after a cloud-to-ground lightning strike. The ball moved horizontally. The spectral analysis indicated that radiation from soil elements was present throughout the duration of the ball (Cen *et al.*, 2014). This provided some support to a model in which ball lightning is formed from soil particles (Abrahamson and Dinniss, 2000). The model is not successful in explaining the formation of ball lightning in enclosed spaces or in the absence of a cloud-to-ground lightning strike.

11.10   Concluding Remarks

The phenomenon of ball lightning has engaged the minds of some of the most eminent and brilliant scientists in the United Kingdom, many of them Fellows of the Royal Society. The opinions of these scientists span the entire range from complete acceptance of the reality of ball lightning to scepticism about its existence. Ball lightning has also been of interest to government organisations such as the Meteorological Office and the Ministry of Defence. The United Kingdom has also been the source of some of the most interesting eyewitness reports of ball lightning, and the phenomenon appears to have been described in many historic UK documents. Other phenomena are frequently mistaken for ball lightning and the effects of ordinary lightning attributed to ball lightning, and greater efforts must be made to evaluate reports thoroughly and critically. TORRO continues to play an active and significant role in collecting, evaluating and disseminating ball lightning reports.

Acknowledgments

The authors are most grateful for the kind assistance and advice provided by the following: Matthew Clark, observations scientist at the Met Office; Peter van Doorn, director of the TORRO Ball

Lightning Division (2000–); Dr Richard Noakes, senior lecturer in history, University of Exeter; Helen Rossington, senior meteorologist at MeteoGroup UK; and Tom Ruffles, communications officer, Society for Psychical Research. They would also like to record their gratitude for the work of TORRO staff, especially Mike Rowe and Terence Meaden, in investigating, collecting and publishing many ball lightning reports.

References

Abrahamson, J. (ed.) (2002) Ball lightning (theme issue). *Philosophical Transactions of the Royal Society A*, Volume **360**(1790), pp. 1–152.
Abrahamson, J. and Dinniss, J. (2000) Ball lightning caused by oxidation of nanoparticle networks from normal lightning strikes on soil. *Nature*, Volume **403**, p. 519.
Agricola (1789) The nature and effect of sea breezes, the trade winds, etc. *Gentleman's Magazine*, Volume **59**(2), p. 597.
Anderson, R. (1879) *Lightning Conductors – Their History, Nature and Mode of Application*. London: E. & F. N. Spohn.
Anon. (1638) *A True Relation of Those Sad and Lamentable Accidents, Which Happened in and About the Parish Church of Withycombe in the Dartmoores*. London: Robert Harford.
Anon. (1665) The Intelligencer and the Newes, number 55, p. 600, 28 June.
Anon. (1750) Historical chronicle. *Gentleman's Magazine*, Volume **20**, p. 330, July.
Anon. (1752) Extract of a letter from the curate of Plausac, near Clermont in France. *Gentleman's Magazine*, Volume **22**, p. 252.
Anon. (1757) Historical chronicle. *Gentleman's Magazine*, Volume **27**, p. 334, July.
Anon. (1763) Historical chronicle. *Gentleman's Magazine*, Volume **33**, p. 409, August.
Anon. (1770) Historical chronicle. *Gentleman's Magazine*, Volume **40**, p. 438, September.
Anon. (1783). Advices from the country. *Gentleman's Magazine*, Volume **53**, p. 708.
Anon. (1785) Effects of thunder-storms at home and abroad, in the course of the year. *Gentleman's Magazine*, Volume **55**(2), p. 1033.
Anon. (1794) Historical chronicle. *Gentleman's Magazine*, Volume **64**(2), p. 856
Anon. (17 November 1843) Royal Institution of Cornwall. The Cornwall Royal Gazette, Falmouth Packet and Plymouth Journal, Issue 4074.
Anon. (1968) *Aids to Identification of Flying Objects*. Washington, DC: US Government Printing Office.
Anon. (1921) Ball lightning. *Nottingham Evening Post*, Saturday 30 July 1921, p. 2.
Anon. (1982) Test of fire for Cambridge physicists. *The Times*, Monday 23 August 1982, p. 2.
Anon. (2008a) Woman's terror over ball lightning. *Wales Online*, 11 November 2008.
Anon. (2008b) Earthquake felt across much of U.K. *BBC News*, 27 February 2008.
Anon. (2008c) Mysterious 'lightning' spotted as 'quake hits Louth, *Louth Leader*, 28 February 2008.
Anon. (2013) Extremely rare 'ball lightning' over Isles. *Stornaway Gazette*, 17 December 2013.
Anon. (2014a) Ball lightning caught on camera by scientists for the first time. *Huffington Post*, 20 January 2014.
Anon. (2014b) Great balls of lightning! Bizarre glow that has eluded scientists for centuries is captured on video for the first time. *Daily Mail*, 20 January 2014.
Arabadji, W. I. (1956) K teorii yavlenii atmosfernogo elektrischestva. *Minskogo Gosudarstvennogo Pedagogicheskogo Instituta*, Volume **5**, p. 77.

---

[93] See 'Extremely rare "ball lightning" over Isles', Stornaway Gazette, 17 December 2013.
[94] For example, Michael Slezak (2014); 'Ball lightning caught on camera by scientists for the first time' and 'Great balls of lightning! Bizarre glow that has eluded scientists for centuries is captured on video for the first time', Daily Mail, 20 January 2014.
[95] See physics.aps.org/articles/v7/5 – retrieved 2 March 2014.

Arago, F. (1837) *Sur le Tonnerre: Annuaire pour l'An 1838 Présenté au Roi, par Le Bureau des Longitudes*. Paris: Bachelier, Imprimeur-Libraire du Bureau des Longitudes et de l'Ecole Polytechnique.

Arago, F. (1839) On thunder and lightning. *The Edinburgh New Philosophical Journal*, Volume **26**, pp. 81–143; 275–291.

Arago, F. (1854a) *Oeuvres Complètes*. Paris: Gide et J Baudry, Editeurs.

Arago, F. (1854b) *Oeuvres Complètes – Tome Quatrième: Le Tonnerre*. Paris: J Claye.

Arago, F. (1855) *Meteorological Essays*. London: Longman, Brown, Green and Longmans.

Argyle, E. (1971) Ball lightning as an optical illusion. *Nature*, Volume **230**, p. 179.

Ashby, D. E. T. F. and Whitehead, C. (1971) Is ball lightning caused by antimatter meteorites? *Nature*, Volume **230**, p. 180.

Atkyns, T. (1712) *The Ancient and Present State of Gloucestershire*. London: Robert Gosling.

Babbage, C. and Herschel, J. F. W. (1815) Account of the repetition of M. Arago's experiments on the magnetism manifested by various substances during the act of rotation. *Proceedings of the Royal Society of London*, Volume **2**, pp. 249–250.

Barry, J. D. (1980) *Ball Lightning and Bead Lightning: Extreme Forms of Atmospheric Electricity*. New York: Plenum Press.

Bergstrom, A. and Campbell, S. (1991) The Ashford 'ball lightning' video explained. *J. Meteorol.*, Volume **16**(160), pp. 185–190.

Borlase, W. (1753) An account of a storm of thunder and lightning, near Ludgvan in Cornwall. *Philosophical Transactions*, Volume **48**, pp. 86–93.

Borlase, W. (1758) *The Natural History of Cornwall*. Oxford: s.n.

Bouquegneau, C. and Rakov, V. (2010) *How Dangerous is Lightning?* New York: Dover Publications, Inc.

Brand, W. (1923) Der Kugelblitz. In: *Probleme der Kosmischen Physik*. Hamburg: H. Grand.

Brand, W. (1971) *Ball Lightning*. Washington, DC: National Aeronautics and Space Administration.

Brereton, O. (1781) Account of the violent storm of lightning at East-Bourn, in Sussex, Sept. 17, 1780. *Philosophical Transactions*, Volume **71**, pp. 42–45.

Bruce, C. E. R. (1963a) Ball lightning, stellar rotation and radio galaxies. *Engineer*, Volume **216**, p. 1047.

Bruce, C. E. R. (1963b) Ball lightning. *Journal of the Institution of Electrical Engineers*, Volume **9**, p. 357.

Bruce, C. E. R. (1964) Ball lightning. *Nature*, Volume **202**, p. 996.

Cade, C. M. and Davis, D. (1969) *The Taming of the Thunderbolts: The Science and Superstition of Ball Lightning*. London, New York, Toronto: Abelard-Schuman.

Cen, J., Yuan, P. and Xue, S., (2014) Observation of the optical and spectral characteristics of ball lightning. *Physical Review Letters*, Volume **112**(3), pp. 035001.

Chalmers, (1749) An account of an extraordinary fireball bursting at sea. *Philosophical Transactions*, Volume **46**, pp. 366–367.

Chamberlayne, J. (1712) A relation of the effects of a storm of thunder and lightning at Sampford-Courtney in Devonshire, on October 7th 1711. *Philosophical Transactions*, Volume **27**, pp. 528–529.

Charman, W. N. (1971a) After-images and ball lightning. *Nature*, Volume **230**, p. 576.

Charman, W. N. (1971b) Perceptual effects and reliability of ball lightning reports. *Journal of Atmospheric and Terrestrial Physics*, Volume **33**, p. 1973.

Charman, W. N. (1972) The enigma of ball lightning. *New Scientist*, Volume **56**, p. 632.

Charman, W. N. (1979) Ball lightning. *Physics Reports*, Volume **54**(4), p. 261.

Clarke, D. (2012) *The UFO Files*. 2nd ed. London: Bloomsbury Publishing Plc.

Coleman, E. H. (1878) Mysterious lights. *Notes and Queries*, Volume **5th series**(9), pp. 87–88.

Condon, E. U. (ed.) (1969) *Scientific Study of Unidentified Flying Objects*. New York: Bantam Books, pp. 559–761

Cooray, G. and Cooray, V. (2008) Could some ball lightning observations be optical hallucinations caused by epileptic seizures? *Open Atmospheric Science Journal* Volume **2**, pp. 101–105.

Cowper, W. (1804) Extract of a letter to J. Newton. *Philosophical Magazine*, Volume **19**, p. 246.

Crew, E. W. (1972) Ball lightning. *New Scientist*, Volume **56**, p. 764.

Crew, E. W. (1980) Meteorological flying objects. *Quarterly Journal of the Royal Astronomical Society*, Volume **21**, pp. 216–219.

Dauvillier, A. (1957) Foudre globulaire et réactions thermonucléaires. *Comptes Rendus Hebdomadaires des Séances de l'Académie des Sciences*, Volume **245**, p. 2155.

Davies, P. C. W. (1971) Ball lightning or spots before the eyes? *Nature*, Volume **230**, p. 576.

Davies, P. C. W. (1987) Great balls of fire. *New Scientist*, Volume **116**(1592/1593), p. 64.

Dibner, B. (1977) Benjamin Franklin. In: R. H. Golde, (ed.) *Lightning, Vol. 1: Physics of Lightning*. London, New York, San Francisco: Academic Press, pp. 23–49.

Doe, R. K. (2013) Ball Lightning: An Elusive Force of Nature. In: K. Pfeifer and N. Pfeifer (eds.) *Forces of Nature and Cultural Responses*. Dordrecht, Heidelberg, New York, London: Springer, pp. 7–26.

van Doorn, P. (2002) TORRO ball lightning division report for 2001. *J. Meteorol.*, Volume **27**(266), pp. 180–183.

van Doorn, P. (2004) Ball lightning summary for the British Isles 2003. *J. Meteorol.*, Volume **29**(290), p. 215.

van Doorn, P. (2005) Ball lightning summary for the British Isles 2004. *J. Meteorol.*, Volume **30**(300), p. 221.

van Doorn, P. (2006) Ball lightning summary for the British Isles 2005. *Int. J. Meteorol.*, Volume **31**(310), pp. 217–218.

van Doorn, P. (2009) Ball lightning division review for the British Isles 2008. *Int. J. Meteorol.*, Volume **34**(340), pp. 194–198.

Egorov, A. I. and Stepanov, S. I. (2008) Properties of short-lived ball lightning produced in the laboratory. *Technical Physics*, Volume **53**(6), pp. 688–692.

Endean, V. G. (1976) Ball lightning as electromagnetic radiation. *Nature*, Volume **263**, p. 753.

Faraday, M. (1838) Experimental researches in electricity. Thirteenth series. *Philosophical Transactions*, Volume **128**, p. 162.

Finkelstein, D. and Rubenstein, J. (1964) Ball lightning. *Physical Review*, Volume **135**, p. A390.

Flammarion, C. (1874) *The Atmosphere*. New York: Harper and Brothers.

Flammarion, C. (1905) *Thunder and Lightning*. London: Chatto and Windus.

de Fonvielle, W. (1868) *Thunder and Lightning*. London: Sampson Low, Son, and Marston.

Friedman, J. S. (2008) *Out of the Blue: A History of Lightning: Science, Superstition, and Amazing Stories of Survival*. New York: Random House Inc.

Glover, S. (1829) *The History of the County of Derby*. Derby: Henry Mozley.

Gold, E. (1952) Thunderbolts: the electrical phenomena of thunderstorms. *Nature*, Volume **169**, p. 561.

Golde, R. H. (1973) *Lightning Protection*. London: Edward Arnold.

Golde, R. H. (ed.) (1977) *Lightning. Volume 1 – Physics of Lightning*. London, New York, San Francisco: Academic Press.

Goodlet, B. L. (1937) Lightning. *Journal of the Institution of Electrical Engineers*, Volume **81**(487), pp. 1–56.

Goodrum, M. R. (2008) Questioning thunderstones and arrowheads: the problem of recognizing and interpreting stone artifacts in the seventeenth century. *Early Science of Medicine*, Volume **13**, pp. 482–508.

Harris, W. S. (1842) *On the Nature of Thunderstorms; and on the Means of Protecting Buildings and Shipping Against the Destructive Effects of Lightning*. London: John W. Parker.

Hartwig, G. L. (1875) *The Aerial World: A Popular Account of the Phenomena and Life of the Atmosphere*. New York: D. Appleton and Co.

Howard, L. (1833) *Meteorological Observations made in the Metropolis and Various Places Around it*. London: Harvey and Darton; J. and A. Arch; Longman and Co.; Hatchard and Sons; S. Highley; R. Hunter.

Humphreys, W. J. (1929) *Physics of the Air*. 2nd ed. New York: McGraw-Hill.

Humphreys, W. J. (1936) Ball lightning. *American Philosophical Society Proceedings*, Volume **76**, p. 613.

Jeffreys, H. (1921) Results of the ball lightning enquiry. *Meteorological Magazine*, Volume **56**, p. 208.

Jennings, P. R. (1990) Mysterious scorching of a curtain during a violent squall on 21 February 1990. Was ball lightning involved? *J. Meteorol.*, Volume **15**(148), pp. 162–164.

Jennison, R. C. (1969) Ball lightning. *Nature*, Volume **224**, p. 895.

Jennison, R. C. (1971) Ball lightning and after-images. *Nature*, Volume **30**, p. 576.

Jennison, R. C. (1972) Ball lightning. *Nature*, Volume **236**, p. 278.

Jennison, R. C., Lobeck, R. and Cahill, R. J. (1990) A video recording of ball lightning. *Weather*, **45**(4), pp. 151–152.

Kapitza, P. L. (1955) O priroda sharovoi molnii. *Doklady Akademii Nauk SSSR*, Volume **101**, p. 245.

King, E. (1773–1774) Account of the effects of lightning at Steeple Ashton and Holt, in the County of Wilts, on the 20th June, 1772. *Philosophical Transactions*, Volume **63**, pp. 231–240.

Levy, J. (1972) Autokinetic illusion: A systematic review of theories, measures, and independent variables. *Psychological Bulletin*, Volume **78**(6), pp. 457–474.

Maunder, M. (2007) *Lights in the Sky: Identifying and Understanding Astronomical and Meteorological Phenomena*. London: Springer-Verlag.

Meaden, G. T. (ed.) (1990a) Ball lightning studies. *J. Meteorol.*, Volume **15**(148), pp. 113–181.

Meaden, G. T. (1990b) Low overhead lightning in Aylesbury, England, followed by an incidence of ball lightning inside a house, 20 December 1989. *J. Meteorol.*, Volume **15**(148), pp. 143–149.

Meaden, G. T. (1990c) Preliminary analysis of the video recording of a rotating ball-of-light, 10 September 1989. *J. Meteorol.*, Volume **15**(148), pp. 128–140.

Meaden, G. T. (1990d) Latest news on the circles effect and its creator vortex. *J. Meteorol.*, Volume **15**(148), pp. 164–169.

Meskill, D. (1990) Horizontal lightning bolt and a 'ball-of-light' in a house, South Limerick, Ireland, 23 February 1990. *J. Meteorol.*, Volume **15**(148), pp. 153–154.

Minnaert, M. G. J. (1993) *Light and Color in the Outdoors*. New York, Berlin and Heidelberg: Springer-Verlag.

Neugebauer, T. (1937) Zu dem Problem des Kugelblitzes. *Zeitschrift für Physik*, Volume **106**(7, 8), p. 474.

Ohtsuki, Y. H. and Ofuruton, H. (1990) Observations of ball lightning in Japan. *J. Meteorol.*, Volume **15**(148), pp. 154–156.

Palmer, J. (1752) An account of the effects of lightning at South Molton in Devonshire. *Philosophical Transactions*, Volume **47**, pp. 330–333.

Peer, J. and Kendl, A. (2010) Transcranial stimulability of phosphenes by long lightning electromagnetic pulses. *Physics Letters A*, Volume **374**(29), pp. 2932–2935.

Pippard, B. (1982) Ball of fire? *Nature*, Volume **298**(5886), p. 702.

Plot, R. (1686) *The Natural History of Staffordshire*. Oxford: s.n.

Priestley, J. (1774) An account of a storm of lightning observed on the 1st of March 1774, near Wakefield, in Yorkshire, by Mr Nicholson. *Philosophical Transactions*, Volume **64**, pp. 350–352.

Prinz, H. (1977) Lightning in History. In: R. H. Golde (ed.) *Lightning: Vol. 1: Physics of Lightning*. London, New York, San Francisco: Academic Press.

Qakley, K. P. (1946) The world of science – thunderbolts. *Illustrated London News*, 26 October 1946, (5610), p. 474.

Rakov, V. A. and Uman, M. A. (2003) *Lightning: Physics and Effects*. Cambridge, UK: Cambridge University Press.

Rowe, M. (1980) The supposed East Anglian meteorite of 1646. *Journal of the British Astronomical Association*, Volume **90**, pp. 478–479.

Rowe, M. W. (1990) Twelve more reports of ball lightning observed indoors. *J. Meteorol.*, Volume **15**(148), pp. 159–162.

Rowe, M. W. and Meaden, G. T. (1990) A case of ball lightning inside a bedroom in Greater Manchester. *J. Meteorol.*, Volume **15**(148), pp. 157–158.

Rudder, S. (1779) *A New History of Gloucestershire*. Cirencester: S. Rudder, p. 676.

Schonland, B. (1964) *The Flight of Thunderbolts*. 2nd ed. Oxford: Clarendon Press.

Sestier, F. (1866) *De la Foudre, de ses Formes et de ses Effets*. Paris: Ballière et Fils.

Shafranov, V. D. (1957) On magnetohydronamical equilibrium configurations. *Soviet Physics – JETP*, Volume **6**, p. 545.

Singer, S. (1971) *The Nature of Ball Lightning*. New York: Plenum Press.

Slezak M. (2014) Natural ball lightning probed for the first time. *New Scientist*, 16 January 2014, Volume **221**, p. 17.

Stenhoff, M. (1976). Ball lightning. *Nature*, Volume **260**, p. 596.

Stenhoff, M. (1990) Comments on a video-tape of a possible ball lightning event at Willesborough, Ashford, Kent on 10 September 1989. *J. Meteorol.*, Volume **15**(148), pp. 141–143.

Stenhoff, M. (ed.) (1992) *Proceedings: Fourth TORRO Conference – Ball Lightning*. Richmond: TORRO.

Stenhoff, M. (1999) *Ball Lightning: An Unsolved Problem in Atmospheric Physics*. New York, Boston, Dordrecht, London and Moscow: Kluwer Academic/Plenum Publishers.

Stetson, C., Fiesta, M. P., Eagleman, D. M. (2007) Does time really slow down during a frightening event? *PLoS One*, Volume **2**(12), p. e 1295.

Stewart, B. (1882–1883) Note on thought-reading. *Proceedings of the Society for Psychical Research*, Volume **1**, pp. 36–43.

Stowe, J. and Howes, E. (1631) *Annales, or a General Chronicle of England*. London: s.n.

Strutt, R. J. (1937) Presidential address. *Proceedings of the Society for Psychical Research*, Volume **45**, pp. 1–18.

de Tessan, M. (1859a) Sur la foudre en boule. *Comptes Rendus Hebdomadaires des Séances de l'Académie des Sciences*, Volume **49**, p. 189.

de Tessan, M. (1859b) Sur la foudre en boule (summary). *Fortschrift für Physik*, Volume **15**, p. 62.

Thomson, J. (1802) *The seasons*. London: A. Strahan, p. 69.

Tomlinson, C. (1848) *The Thunder-Storm; or, An Account of the Nature, Properties, Dangers, and Uses of Lightning, in Various Parts of the World*. London: Society for Promoting Christian Knowledge.

Uman, M. A. (1968) Some comments on ball lightning. *Journal of Atmospheric and Terrestrial Physics*, Volume **30**, p. 1245.

Uman, M. A. (1969, 1984) *Lightning*. New York: Dover Publications, Inc.

Uman, M. A. (1986) *All About Lightning*. 2nd ed. New York: Dover Publications.

Uman, M. A. (2008) *The Art and Science of Lightning Protection*. New York: Cambridge University Press.

Uman, M. A. and Helstrom, C. W. (1966) A theory of ball lightning. *Journal of Geophysical Research*, Volume **71**, p. 1975.

Uman, M. A. and Rakov, V. A. (2003) *Lightning: Physics and Effects*. Cambridge U.K., New York, Port Melbourne, Madrid and Cape Town: Cambridge University Press.

Varley, F. (1870) Some experiments on the discharge of electricity through rarefied media and the atmosphere. *Proceedings of the Royal Society of London*, Volume **19**, pp. 236–244.

Vicars, J. (1643) Prodigies & Apparitions, or Englands warning piece being a seasonable description by lively figures & apt illustrations of many remarkable & prodigious fore-runners & apparent Predictions of Gods Wrath against England, if not timely prevented by true Repentance. Written by J[ohn] V[icars]., British Library, Early English Books, 1641-1700/479:22, T. Bates/R. Markland, London[?]

Wasse, J. (1725) Two letters on the effects of lightning, from the Reverend Mr Jos. Wasse, Rector of Aynho in Northamptonshire, to Dr Mead. *Philosophical Transactions*, Volume **33**, pp. 366–370.

Witalis, E. A. (1990) Ball lightning as a magnetized air plasma whirl structure. *J. Meteorol.*, Volume **15**(148), pp. 121–128.

Wooding, E. R. (1963) Ball lightning. *Nature*, Volume **199**, p. 272.

Wooding, E. R. (1972) Laser analogue to ball lightning. *Nature*, Volume **239**, p. 394.

Wooding, E. R. (1976) Ball lightning in Smethwick. *Nature*, Volume **262**, p. 279.

Part III
Extremes

12

Forecasting Severe Weather in the United Kingdom and Ireland

Paul Knightley[1,2]

[1] *MeteoGroup UK Ltd, London, UK*
[2] *Forecasting Division, Tornado and Storm Research Organisation (TORRO)*, London, UK*

12.1 Introduction

Tornadoes are the atmosphere's most violent windstorms, and it is therefore no surprise that attempts to forecast them have been made for many years. The roots of tornado forecasting can be traced back to the 19th century. The work of J.P. Finley[1] was the key contributor to these early efforts (Galway, 1985). In 1884, he issued experimental forecasts based on a checklist. However, although he had some success issuing forecasts while with the US Army Signal Corps, the word 'tornado' was banned in 1886, as the chief signal duty officer believed that they would do more harm than good (Corfidi, 1999; Grazulis, 2001). Finley was something of a visionary, though, as in 1881, he suggested that a special observer be stationed at Kansas City during May, June and July, in order to report on tornadic activity (Finley, 1881). In 1954, the National Weather Service (NWS) transferred a unit dedicated to issuing tornado forecasts to Kansas City.

As European settlers made the dash westwards into the US's interior, the violence of the thunderstorms and tornadoes which they must have experienced would likely have been much greater than any they had experienced previously. However, these early forays into prediction were of little overall use. It was not until the 1950s that tornado forecasting began to be taken seriously (Doswell et al., 1993). The most famous early attempt at tornado forecasting is probably that undertaken by the US Air Force (USAF) meteorologists Major Ernest Fawbush and Captain Robert Miller, who subsequently produced pioneering work on the subject. On 20 March 1948, a tornado ripped through Tinker Air Force Base, in Oklahoma, injuring several men and causing more than $10 million damage.[2]

On the day of this tornado, the forecast had been for gusty winds. An inquiry by the Air Force on the following day ruled that a tornado 'was not forecastable, given the present state of the art'. On 22nd March, and for the next 3 days, Fawbush and Miller made a highly concentrated effort to gather, analyse and document every piece of available meteorological information which had preceded the tornado. This included obtaining as much information as they could about the upper air. Furthermore, they scrutinised data from previous tornado outbreaks in an attempt to determine which parameters were present and whether there was some kind of definable 'pattern' to the weather beforehand.

The weather maps for 25th March showed patterns which were remarkably similar to those which they had noted in their research. By 1430 local time, a strong squall line was approaching the base from the south-west. The base general, after having been briefed earlier on the threat of severe storms, came again to the weather station and asked (Newton et al., 1978):

> …if the forecasters believed tornadoes were likely in the vicinity of the base…. When the general was informed that the probability was considered high enough to justify issuance of a warning, his only comment was: 'Do it!' At 1500 the first operational tornado forecast was issued, with a warning for Tinker Air Force Base valid from 1600 to 1800.

Remarkably, a tornado hit the base at around 1800 local time, causing $6 million damage.[3] Miller estimated that the odds of a

---

*http://www.torro.org.uk/

[1] J.P. Finley was a US meteorologist who set out to prove that tornadoes could be forecast, as with other weather phenomena. He set precedents in meteorological forecasting that are still valid today.

[2] This equates to approximately $97 million in 2014.
[3] This equates to approximately $58 million in 2014.

Extreme Weather: Forty Years of the Tornado and Storm Research Organisation (TORRO), First Edition. Edited by Robert K. Doe.
© 2016 John Wiley & Sons, Ltd. Published 2016 by John Wiley & Sons, Ltd.

tornado hitting the base within a few days of the initial tornado were around 20 million to 1. However, thanks to the warning, no one was injured. From that point on, the Air Force began to use this so-called Fawbush and Miller technique to issue tornado forecasts. Unsurprisingly, the media caught wind of this. Several subsequent forecasts were also deemed successful. To put this into perspective, it must be appreciated that it is extremely unusual to have successfully predicted a tornado for a specific site. Today, forecasters would not attempt to predict a tornado striking a particular location 3 hours in advance, and this is despite the huge advances in the science of weather forecasting. The original forecast relied heavily on chance.

12.2   Modern Forecasting

The success of Fawbush and Miller's prediction, and of the subsequent Air Force forecasts, made many people's minds up: tornado forecasts should be made publicly available, in order for people and businesses to take evasive and protective action. In 1952, a tornado forecasting centre was set up, which was the forerunner of today's US Storm Prediction Center. The tornado forecasting centre was rebranded the Severe Local Storms Center (SELS) in 1953 and moved to Kansas City, Missouri, in 1954.

Early forecasts, rather like Fawbush and Miller's, were largely empirically based, relying on pattern recognition. It had been noted that larger-scale severe thunderstorm and tornado outbreaks were typically associated with certain atmospheric environments. Examples of these include large areas of low pressure (extratropical cyclones), abundant low-level moisture and the proximity of these features to jet streams. However, it was readily apparent that no single set of features was present in each and every case. It is now recognised that subtle differences can mean the difference between a few isolated storms and a major severe thunderstorm or tornado outbreak. Thus, while a general forecast could be issued using pattern recognition, when certain large-scale features were present, it was not possible to predict the likely magnitude or precise timings of severe weather and tornado outbreaks with much certainty, given the information available to the forecasters.

Since the 1950s, numerical weather prediction has formed the backbone of forecasting techniques, and today's meteorologists have a plethora of information to deal with. In addition to this, the understanding of the environment in which severe thunderstorms and tornadoes develop has increased enormously, although we are still some way off understanding exactly how tornadoes form. Day-to-day weather forecasting is much more accurate than it was 15–20 years ago, thanks to advances in the understanding of the atmosphere along with hugely more powerful computers. A trend towards using physical principles over pattern recognition and empirical techniques has taken place. However, features of the temporal and spatial scale of a tornado cannot currently be resolved by operational numerical weather models, and this is unlikely to change in at least the near future. This is because the atmospheric environment immediately surrounding tornadoes and the processes by which they develop are not very well understood.

12.3   Severe Storm Forecasting in the United States

The United States, especially the central states, is particularly prone to severe thunderstorms and tornadoes. As such, it has the longest history of tornado forecasting and the most advanced forecast and warning service of such storms in the world. Based on this, its model has been used and developed elsewhere. The Storm Prediction Center (SPC), based in Norman, Oklahoma, is responsible for the issuance of severe weather forecasts for the 48 contiguous states of the United States (herein, 'severe weather' will be used to describe the occurrence of tornadoes and severe thunderstorms). A severe thunderstorm is defined by the NWS as one where at least one of the following occurs: hail greater than 25 mm diameter; wind gusts greater than 58 mph, or damaging winds; one or more tornadoes.

The risk of severe weather is forecast for up to eight days ahead, using what are known as 'convective outlooks'. When severe weather is expected, regions in which the risk is deemed to be significant are highlighted. For days 1, 2 and 3, the risk of severe weather is given in a probabilistic and categorical form. Categorically, the SPC uses 'marginal', 'slight', 'enhanced', 'moderate', and 'high' risk to describe the risk of severe thunderstorms. Each term has a probability associated with it, and full details of these can be found on the SPC website.[4] As an example, a greater than 30% probability of a tornado occurring within 25 miles of a point on day one would constitute a high risk. The outlooks are issued at various times through the day, with the shorter-term outlooks having the more frequent updates. These forecasts are designed to give people an indication of the likelihood of severe weather in their general area, as well as giving the local forecast offices of the NWS and commercial meteorologists the ability to read the rationale behind the outlooks.

When organised severe weather is deemed likely in the coming several hours, the SPC will issue a severe weather watch.[5] This will be either a severe thunderstorm watch or a tornado watch – the former when large hail and damaging thunderstorm winds are expected, while the latter when these threats are accompanied with the possibility of multiple tornadoes.[6] These watches typically cover a fairly large area, perhaps 20,000–40,000 square miles, and are sometimes valid for more than 6 hours. Severe weather watches are designed to inform members of the public that severe weather is possible in the next few hours. Rarely, the wording 'particularly dangerous situation' or PDS is used, most often for a tornado watch. This is used when long-lived strong and violent tornadoes are possible, although it may also be used occasionally for a severe thunderstorm watch for exceptionally intense and well organised convective windstorms. The decision as to whether to issue a PDS watch lies with the forecaster and is a subjective decision.

---

[4] www.spc.noaa.gov
[5] Almost any thunderstorm can, at the most vigorous phase of its life, produce severe weather. However, for an enhanced threat of severe weather, storms require some degree of organisation. See the section on 'the ingredients-based approach' for more explanation.
[6] http://www.spc.noaa.gov/faq/#2.1

There has been a huge campaign over the years to educate the public so that they can minimise risk to themselves, for example, by practising severe weather drills and fitting a 'safe room' or basement in which to seek refuge if severe weather approaches. The issuance of a watch is typically a 'heads-up' that people should be on the alert to rapidly changing weather conditions and be prepared to seek shelter, possibly at short notice, should a severe storm approach.

Although people should take it upon themselves to keep a 'weather eye' on the sky in these 'watch' situations, most people only become aware of impending severe weather when severe weather warnings are issued by local NWS offices. These are issued when severe weather has either been indicated by radar or confirmed by observation. Visual confirmation is often achieved by a network of 'storm spotters'. Volunteers, usually members of the public or emergency service personnel, receive specialised training in the art of visually identifying features of storms that might indicate severe weather is occurring or may occur imminently. Examples might include the visual confirmation of low-level rotation within a thunderstorm or the existence of a wall cloud or funnel cloud. When a severe thunderstorm or tornado watch is issued, there is a call to arms of storm spotters. They will then actively observe the approaching thunderstorms and report the occurrence of severe weather to the local NWS office, who in turn can either verify an existing warning (which may have been based on radar evidence) or issue a new warning if the report is the first indication of severe weather. In addition, reports may come in from members of the public, including storm chasers. Responsible storm chasers will always make an effort to notify the local weather office of severe weather.

The normal method of disseminating a warning is via the National Oceanic and Atmospheric Administration (NOAA) Radio All Hazards, along with television and radio media. The NOAA radio is the preferred method as the user can set it up to receive warnings for their area only, which should then allow them to make informed decisions and take appropriate action. Many communities also have emergency civil defence sirens. Although there are several types of situation where these might be used, local emergency management personnel will usually sound them when a tornado warning has been issued or sometimes if they deem conditions severe enough even in the absence of a warning. It is a common misconception that sirens are the main indicator of an approaching tornado or a tornado warning; however, the reliance on the siren for this purpose can lead to a false sense of security. For example, most sirens are activated remotely and are powered from the local electricity grid. If the approaching storm causes a power outage, there will be no method of sounding them. Therefore, although there is a coordinated warning network, residents still have to take some responsibility for ensuring they keep up to date with the current situation when severe weather approaches.

12.4 Severe Storm Forecasting Elsewhere

Globally, many national weather services will issue forecasts of severe weather to their residents. However, in some countries, this does not extend to severe convective weather, including tornadoes. Some examples of countries where government agencies issue warnings of severe thunderstorms include Canada and Australia. Across Europe, an organisation called the European Storm Forecast Experiment, or ESTOFEX,[7] has been issuing daily convective outlooks for the European region for a number of years. The goals include enhancing the understanding of severe convective storms in Europe as well as verifying the forecasts.

In the United Kingdom and Ireland, TORRO has been producing forecasts of severe convective storms since around 1992. At first, these were for distribution among the organisation's staff only, as they were deemed too experimental for public use. The forecasts were used by the staff as a 'heads-up' for severe weather, as well as being a way of testing out forecasting methods. The forecasts, at the time, were produced by David Reynolds. There was a period where production of forecasts ceased, but from around 2004 onwards, they were reinstated, largely by Paul Knightley. Again, these were for internal use only for a couple of years, although they were posted onto TORRO's members-only forum. However, the TORRO staff decided in 2006 that the forecasts should be published more widely, and thus, they became publicly available. Initially, this occurred through posting the forecasts on other weather forums. However, forecasts also started to be published on TORRO's own web page.

The reasons for putting these forecasts out in the public domain included attempting to give the general populace some indication as to when and where severe thunderstorms might be expected as well as encouraging the reporting of severe weather back to TORRO.

12.5 Forecasting Techniques

Early attempts at forecasting severe thunderstorms and tornadoes were, as explained earlier, based on empirical techniques: that is, forecasters would study weather patterns and certain parameters and base their forecast on what had happened during previous severe weather events. In time, these techniques led to the development of certain severe weather indices.[8] If enough events have occurred, it is possible to develop an index that is based on the values of a combination of meteorological parameters. Indeed, indices have been developed for a number of hazardous weather types in an attempt to forecast them, for example, fog, thunderstorms and strong wind gusts. Although such indices can be useful, a disadvantage is that they can give the impression that the occurrence, or otherwise, of severe weather is dependent on whether the index value exceeds certain (rather arbitrary) thresholds. In reality, though, the likelihood of an event may vary as the index value varies. There are no such 'magic numbers' in meteorology to paraphrase Dr Chuck Doswell III (e.g. Moller *et al.*, 1994; Doswell *et al.*, 1996). In more recent times, an approach known as the 'ingredients-based' method has been employed by many meteorologists, when forecasting severe thunderstorms and tornadoes.

[7] www.estofex.org
[8] For example, the K-index and the Showalter index. For a few others, see, for example, http://www.teachingboxes.org/avc/content/Severe_Weather_Indices.htm.

12.6   The Ingredients-Based Approach

This methodology entails looking at predictions of the three basic ingredients required for deep, moist convection: moisture, instability and lift. Rather than taking the approach where one looks for values of these ingredients exceeding arbitrary thresholds, this method involves looking for areas in which a suitable combination of them is present.

12.6.1   Moisture

Moisture is key to the development of thunderstorms. For storms which develop from the boundary layer (including most, if not all, tornadic storms), the amount of low-level moisture is most important. In many cases, forecasters will assess the amount of moisture using the dew point temperature. In simple terms, the dew point temperature is the temperature the air needs to be cooled to in order for condensation to occur; it is an indication of the total amount of water vapour present in the air. When water changes state from gas to liquid or from liquid to solid, heat is released – this is called 'latent' heat. This heat is then effectively stored in the water vapour and is then released when condensation occurs. Deep moist convection derives most of its upward momentum (buoyancy) from the condensation of vast quantities of moisture in the developing cloud. Thus, it follows that, all else being equal, more abundant moisture (i.e. higher dew point temperatures) can lead to warmer, more buoyant updraughts and therefore more vigorous thunderstorm activity, although it should be noted that the relative amount of moisture compared to the ambient temperature and the vertical state of the atmosphere is very important. For example, a wintertime thunderstorm will have a relatively low amount of moisture in its inflow, but given that the temperature aloft is low, the storm can still be fairly powerful.

12.6.2   Instability

Typically, the temperature of the atmosphere decreases with height. This is because air obeys the ideal gas law, which states temperature and pressure are proportional, for example, if the pressure of a parcel of air decreases, it follows that the temperature will too. In a 'stable' atmosphere, the vertical temperature profile is such that if a parcel[9] is 'pushed' either up or down, it returns to its original position. In this case, the initial temperature of the parcel is the same as the air around it – when it is displaced up, it becomes cooler than the surrounding air and thus denser and sinks; if it is pushed down, it becomes warmer and more buoyant and rises. In an 'unstable' atmosphere, the temperature of the parcel, after displacement, is higher than the surrounding air which makes it lighter and more buoyant. It thus rises and continues to rise until the temperature around it is higher. In the real atmosphere, a rising parcel can acquire additional warmth by the release of latent heat when the moisture contained within it condenses into a cloud. More moisture means additional heat release and a stronger updraught, hence why moisture is important, as discussed previously. Finally, the actual buoyancy of a rising parcel of air is further determined by the *environmental lapse rate*; this is the actual temperature profile of the atmosphere and varies depending on the large-scale weather pattern. For example, in fine, settled conditions in an area of high atmospheric pressure, the air over a large area slowly descends and warms – thus, the vertical lapse rates are very low, and a rising bubble of air may find itself cooler than its surroundings and thus will rise no more.

12.6.3   Lift

Lift, in this sense, is ascent of the atmosphere on a wide variety of spatial scales, which may occur in association with several different processes. Lift in itself does not generate thunderstorms, that is, it is not forcing the cloud to develop per se. Rather, it conditions the atmosphere for convection to develop. One form of lift (or 'forcing for ascent' as it is sometimes known in the meteorological world) is the approach of a large-scale weather disturbance. In short, such disturbances result in the widespread synoptic-scale (i.e. on a horizontal scale of over 1000 km or so) ascent of the mid and upper[10] atmosphere. Lifting air upwards causes it to cool (remember the ideal gas law mentioned earlier). If the air above a warm, moist layer is forced to cool, the rate of temperature decreases with height increases (the 'lapse rate'), which tends to increase the instability; eventually the lower-level air may explode upwards into deep, moist convection.

Low-level lift may be present, perhaps where winds converge along a shallow boundary. Here, moist parcels of air may be forced upwards until their level of free convection is reached. Beyond this level, the parcels rise under their own buoyancy – they've just been helped to reach this level by the forcing. In practice, then, it is the relative combination of moisture, instability and lift which determines whether deep, moist convection will occur. The role of the forecaster and numerical weather models is to determine how the three parameters will combine and whether this will result in thunderstorms.

12.6.4   Wind Shear

For severe convection, we must consider a fourth ingredient: vertical wind shear. Wind shear is the change in the wind velocity vector with height. In a basic sense, this is the change in wind direction and speed as one moves up through the atmosphere. Briefly, increasing wind speed with height results in the updraught leaning over, allowing precipitation to fall away from the low-level inflow and preventing the updraught from 'choking' on cool, precipitation-filled air. Also, there is a dynamic effect which can cause a developing updraught to acquire rotation. A persistent, rotating updraught is called a *mesocyclone*, and a thunderstorm containing one is known as a *supercell thunderstorm*, or often just *supercell* (e.g. Browning and Donaldson, 1963; Weisman and Klemp, 1982). A typical lower threshold for surface-based supercell development is a vertical wind shear of 40 knots or more between the surface level and around 6 km above ground. However, the amount of instability also influences the type of convection. For example, a very unstable atmosphere

---

[9] When describing processes in the atmosphere from a scientific viewpoint, meteorologists refer to an imaginary 'parcel' of air. It is used more generally in the discussion of all fluids. See http://en.wikipedia.org/wiki/Fluid_parcel

[10] The part of the lower atmosphere (the *troposphere*) between approximately 3 and 12 km above sea level.

could support supercells with a vertical wind shear of less than 40 knots. Directional shear (i.e. changes of wind direction with height) is very important for supercells with strong low-level rotation.

For tornadoes, the low-level wind shear appears especially important. A strongly veering (turning clockwise in the Northern Hemisphere) wind vector with height in the lowest 1 km of the atmosphere can signal that tornadoes may develop, if surface-based thunderstorms are expected.

12.7 TORRO's Forecasts

Several types of forecast have been issued by TORRO over the years. Initially, there were two broad types: advisories and watches. Warnings also formed part of the system, but in practice, attempting to warn in real time, in the absence of minute-by-minute ground truth such as might be provided by the storm-spotting community in the United States was never really a realistic prospect. The two-tier system of advisory and watch was designed to give a distinction between marginal and/or isolated severe weather and well-organised severe weather, the latter bringing a higher risk to the public. However, there was some ambiguity with the terminology, and it was decided to remove 'advisory' and replace it with 'convective discussion'.

Through the late 2000s to the present day, TORRO has issued the following types of forecast:

- Convective discussions
- Severe weather watches, which are further divided into:
  - ° Tornado watches
  - ° Severe thunderstorm watches

A *convective discussion* means that conditions are either favourable for marginally severe weather or that severe weather is possible but the overall coverage is expected to be very low or that the storms will be poorly organised. Such storms tend to be short-lived – while they may produce instances of severe weather, any which occur will be brief and confined to a fairly small area.

A severe weather watch means that severe thunderstorms are possible across the watch area. In situations where the prime threat is expected to be either large hail or damaging 'straight-line' winds, a severe thunderstorm watch will be issued. Where one or more tornadoes are deemed possible and the parent storms are expected to be well organised, or widespread, a tornado watch will be issued. It should be noted that the other types of severe weather (large hail and damaging non-tornadic winds) may also be possible in this type of watch, as is the case for tornado watches issued by the SPC in the United States.

These *severe weather forecasts* consist of a map showing the outline of the discussion or watch area, along with a written forecast. As well as listing the affected areas, the text will give a general overview, or synopsis of the current weather situation, as well as a discussion of the threats, and how these are expected to evolve. It will also give a period of validity. Given the inherent uncertainties, the exact outline given by the map should be treated as a guide only. Being in the centre of the box does not mean you are more likely to have severe weather than someone on the edge!

In recent years, TORRO has embraced the world of social media, and its forecasts are posted on Facebook[11] and Twitter,[12] as well as the organisation's web page and forum. Use of social media has allowed a much larger outreach to the general populace. TORRO's staff also post explanatory notes on the Facebook page, so as to make sure the readers are clear about the overall risks. While the point of the forecasts is to make people aware that severe weather is possible, it has to be made clear that, for most, it is not *probable*. Later in this chapter, a case study of a rather remarkable severe weather event will show how social media helped to get the message out.

Since 2004, TORRO has verified its forecasts against severe weather reports and has presented the results at its annual Spring Conference, as well as publishing the reports in the *International Journal of Meteorology* (IJMet). Verification of forecasts against actual reports has, in general, been more successful for tornadoes, since tornadoes seem to be reported more readily than other types of severe weather. With this in mind, verification is not generally undertaken for hail and non-tornadic wind damage.

Table 12.1 shows the most recent verification statistics for TORRO's tornado forecasts. It should be noted that the number of tornadoes in each year is the provisional figure *at the time the verification was undertaken* and not a final figure. Indeed, annual tornado totals can never really be considered final, since reports will sometimes come in many years later! In this table, tornadoes are considered within *all* TORRO forecast types (i.e. including convective discussions and severe thunderstorm watches) and not just tornado watches.

In 2012, tornadoes were reported in six tornado watches, and no tornadoes were reported in two tornado watches. This gives a probability of detection (POD – the % number of watches containing at least one tornado) of 75% and a false alarm rate (FAR) of 25%. The final column in Table 12.1 summarises the annual POD for the verification period. Within the successful watches, 12 tornadoes occurred. This means 43% of 2012s tornadoes occurred within watches. The final column summarises the remaining years.

12.8 Forecasting Severe Weather: 28 June 2012

In order to illustrate the process by which TORRO compiles and disseminates forecasts, a case study of the severe thunderstorms of 28 June 2012 (see also Chapter 3) is given.

12.8.1 Background

During the days preceding the event, conditions were monitored closely as the overall synoptic condition looked increasingly favourable for hot, humid air to move from the south ahead of an eastward-moving upper trough. Deep instability, lift and moisture were expected

---

[11] https://www.facebook.com/pages/Torro-the-UKs-Tornado-and-Storm-Research-Organisation/249176491778442
[12] https://twitter.com/torroUK

Table 12.1 *Tornadoes within TORRO forecasts.*

Year	Tornadoes	No. of Tornadoes Within TORRO Forecast	No. of Tornadoes Not Forecast	POD (%)	(%) of Tornadoes in Tornado Watch
2004	51*	26 (51%)	25 (49%)	n/a	n/a
2005	63*	41 (65%)	22 (35%)	35	33
2006	70†	39 (56%)	31 (44%)	29	32
2007	51*	32 (63%)	19 (37%)	27.5	31
2008	13*	7 (54%)	6 (46%)	24	38
2009	39	21 (54%)	18 (46%)	19	13
2010	25	9 (36%)	16 (64%)	11	4
2011	26	15 (58%)	11 (42%)	37.5	35
2012	28	15 (56%)	13 (44%)	75	43
2004, '05, '06, '07, '08, '09, '10, '11, '12	366	205 (56%)	161 (44%)	32 (not incl. 2004)	28.6 (not incl. 2004)

*Figures based on provisional figures for those years, at the time the reviews were written.
†Five more occurred, but due to forecaster absence, are not included. Note figures for 2004–2006 are based on those used in these years' forecast reviews and may not match the actual, final tornado numbers.

to be present across parts of England. Over the same area, wind shear was expected to be favourable for organised and severe deep, moist convection. TORRO has increasingly been using social media to help disseminate its watches and discussions, as well as stimulating conversation on all aspects of severe weather. On 27th June, the following post was placed onto the TORRO Facebook page:

> Keep an eye out on this page and our forecast page over the next 24 hours, as we may be issuing watches or discussions – there is the potential for some severe thunderstorms although until it becomes fairly clear which areas are most at risk, and from what, we'll hold off.

TORRO TORNADO WATCH 2012/004

The numbering convention includes the year and weather watch number – they are ordered sequentially, with the count reset at the start of each new year

A TORNADO WATCH has been issued at 0740 GMT on Thursday 28 June 2012

Valid from/until: 0740–2000 GMT on Thursday 28 June 2012, for the following regions of the United Kingdom & Eire:

Parts of (see map – Figure 12.1)
    Wales
    Midlands
    E Anglia
    N England
    Southern and central Scotland

This section lists the areas covered by the watch as also the period of time for which the watch is valid.

THREATS

Tornadoes; hail to 40 mm diameter; wind gusts to 65 mph; Frequent CG lightning

This section explains which threats are possible. The wind gust values refer to non-tornadic 'straight line' winds which can

This was the first 'heads-up' for severe weather.

On the morning of 28th June, numerical weather model output was scrutinised, and it quickly became apparent that an outbreak of severe thunderstorms was likely. The ingredients across central and eastern England were of a fairly rare potency for this part of the world. A small cluster of elevated thunderstorms developed across Wales around 0700 GMT, coincident with a north-eastward-moving shortwave upper trough.[13] As such, the following tornado watch was issued at 0740 GMT. It has been annotated where appropriate (in italics) to further explain the layout and terminology:

occur with severe thunderstorms. CG lightning is cloud to ground lightning.

SYNOPSIS

This section explains, in somewhat more scientific terminology, the current weather situation, and how this is likely to further develop through the duration of the watch.

Plume of high theta-w air has advected across much of Britain overnight with a large upper trough approaching from the west. At the surface, a cold front will move north-eastwards through the day.

The text above sets the scene for the forecast – theta-w is shorthand for 'wet bulb potential temperature', a measure of atmosphere moisture. In this case, a large plume of very warm and moist air has moved, or 'advected' in meteorological parlance, across Britain.

Diffluent upper flow ahead of the trough is bringing forcing for ascent across portions of Wales already, and a number of thunderstorms have developed. At this stage, they are likely marginally elevated above a coolish boundary layer, but strong cloud-layer shear suggests large hail to 20–30 mm is possible with this early activity, with elevated supercells possible, with clusters of storms likely. Gusty winds are also possible.

---

[13] A 'trough' is an elongated region of relatively low air pressure. An upper trough is simply such a trough in the upper troposphere – the term 'shortwave' simply refers to the size of the trough. A shortwave trough may be on the order of a one to a few hundred kilometres, whereas a 'longwave' trough would typically be on the order of several thousand kilometres.

This section explains what is currently happening, referring to a cluster of existing thunderstorms across Wales. These storms derived their inflow from air parcels above the layer of air in contact with the ground, known as the 'boundary layer'. Such 'elevated' storms can be severe, but rarely if ever produce tornadoes. This is because the air beneath them is usually too cool or stable to support them.

This activity is expected to continue to move NNE and NE through this morning. Although there is a fair amount of mid-level cloud associated with the widespread large-scale ascent ahead of the trough, enough diurnal heating of the boundary layer is expected to allow storms to become surface based as the cluster of storms moves through the Midlands into N England and later Scotland. With 30–40 knots of deep layer shear, well-organised multicells and embedded supercell structures are likely. 20–30 mm diameter hail and gusts to 60 mph are possible. In addition, isolated tornadoes are possible.

This section explains what that cluster of storms is anticipated to do over the next few hours. With the boundary layer warming up due to daytime (diurnal) heating, it was anticipated that new thunderstorm updraughts would increasingly draw their inflow from lower levels of the atmosphere. The extra heat and moisture in the boundary layer would enhance the strength of the updraughts. In addition, the fact that the airflow in the lower layers was coming from the south-east (with flow aloft coming from the south-west) meant that these 'surface-based' storm updraughts would be subject to enhanced speed and directional wind shear, increasing the chances of them becoming supercell thunderstorms.

In the wake of this activity, and across the more south-eastern parts of the WATCH area, further thunderstorms are expected to develop and the move north-east, although these will be more isolated with south-eastern extent. Modified ascents for afternoon temperatures suggest around 1200–1800 J/kg of surface-based convective available potential energy (SBCAPE). With 0–6 km shear remaining around 40 knots, and perhaps increasing a little, well-organised multicells[14] and isolated supercells seem possible. With 0–1 km storm-relative environmental helicity (SREH) of approximately 150 J/kg and fairly low lifted condensation level (LCL), a few tornadoes may develop with this activity. Indeed, should an isolated supercell develop in an area with decent surface heating, a strong tornado cannot be ruled out. Hail to 40 mm or so and wind gusts to 65 mph are possible with supercells too; otherwise, 20–30 mm hail and 55 mph gusts. If this activity can develop into northern parts of England too, it may tend more towards a squall line, with an associated wind threat.

This section explains that once the initial activity moved away to the north-east, further surfaced-based thunderstorms could be expected. SBCAPE stands for Surface Based Convective Available Potential Energy – values between 1000 and 2000 J/kg are towards the upper end of what can typically be expected across the U.K., especially with strong wind shear. The vertical wind shear, calculated from the surface to 6 km above sea level was expected to be around 40 knots – that is, a vector difference of 40 knots between low level and 6 km winds. This value is sufficient for supercell development. SREH is Storm Relative Environmental Helicity – in short, this can give an indication of the potential for updraught rotation, due to the storm ingesting air parcels tracing a helical path in the lower atmosphere. This can be calculated over various depths – in this case, it has been calculated over the lowest 1 km of the atmosphere. Values of 150 J/kg indicate a fairly strong likelihood of low-level storm rotation. Low LCLs (lifted condensation level – effectively the cloud base) can enhance the chance of tornado development – downdraughts in such storms tend to be weaker and therefore less likely to undercut or reduce the instability of the updraught.

The activity will end from the south-west through the afternoon and evening as the cold front moves in.

Forecaster: RPK

Not all of TORRO's forecasts are so verbose and comprehensive. However, in this case, the fairly rare combination of strong vertical wind shear and deep instability was felt to necessitate a detailed forecast. This watch was issued on the TORRO forecast page, the TORRO forum, and the TORRO Facebook page. Several Facebook updates were provided to users during the course of the event. A couple of examples are presented here. At 1259 GMT, this was issued:

Supercell storm crossing Melton Mowbray now, heading towards Grantham – expect large hail, strong winds, and perhaps a tornado.

This was closely followed by another update. This second update also included an annotated radar image (Figure 12.2), in order to demonstrate where the most dangerous portion of the storm was located and where it was likely to move in the next 30–60 minutes. A resident of Sleaford, Lincolnshire, filmed a storm and unwittingly gained what is perhaps the best (to date) footage of a supercell thunderstorm taken in the United Kingdom. A tornado was also documented.[15] Several other people also filmed the same tornado from different vantage points. This event was, as far as is possible in severe storms forecasting, a fairly obvious one. The ingredients for severe thunderstorms were all present in ample quantities, and numerical weather guidance indicated that severe thunderstorms would develop. Larger-scale 'global' weather models cannot explicitly model the convection. Therefore, one tends to predict the risk of severe convection by looking at where the ingredients required for severe convection are present and by assessing whether the models develop convective precipitation in these areas.

Higher resolution models can explicitly develop convection, and this can then be acted upon by the wind shear within the model. In this way, it is possible for the model to, in effect, simulate rotating updraughts/supercell thunderstorms. The author is a meteorologist at MeteoGroup UK and thus has access to their high-resolution 'MG, incorporating WRF® model'.[16] Operationally, at the time of

---

[14] A 'multicell' thunderstorm is one where there are several updraughts and downdraughts in close proximity. Many thunderstorms exhibit multicellular characteristics, sometimes including those which are classed as supercells.

[15] The footage can be viewed at http://www.youtube.com/watch?v=zG8WoamSCys

[16] WRF® is a registered trademark of the University Corporation for Atmospheric Research (UCAR).

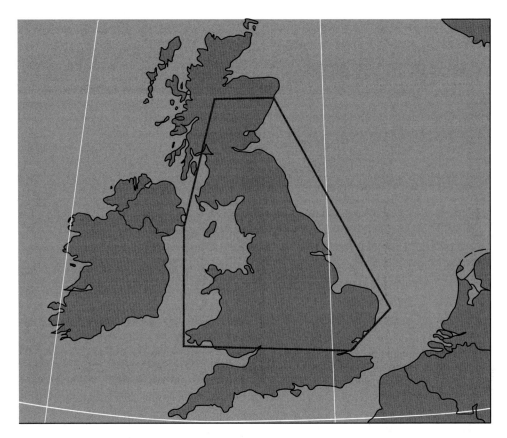

Figure 12.1 *The tornado watch area for 28 June 2012. © TORRO.*

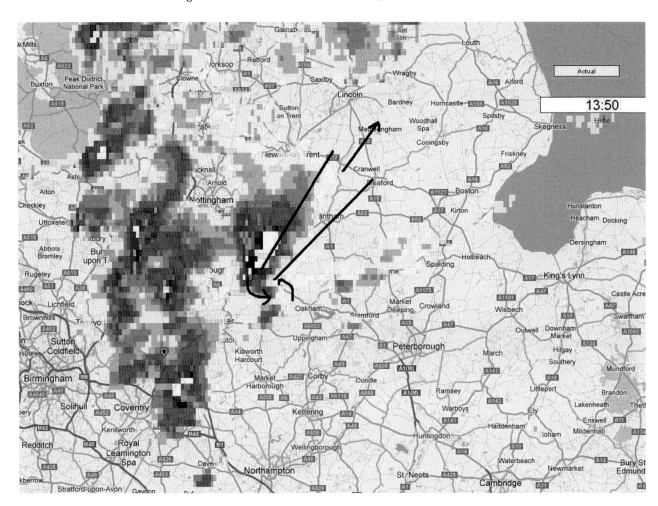

Figure 12.2 *The storm of 28 June 2012. The hook echo shows the approximate location of strong low-level rotation (arrows), and the black lines show the approximate track the storm will take. A tornado remains possible with this storm, along with large hail. Image courtesy of MeteoGroup, UK Ltd. (2012), www.raintoday.co.uk. Data from UKMO. (See insert for colour representation of the figure.)*

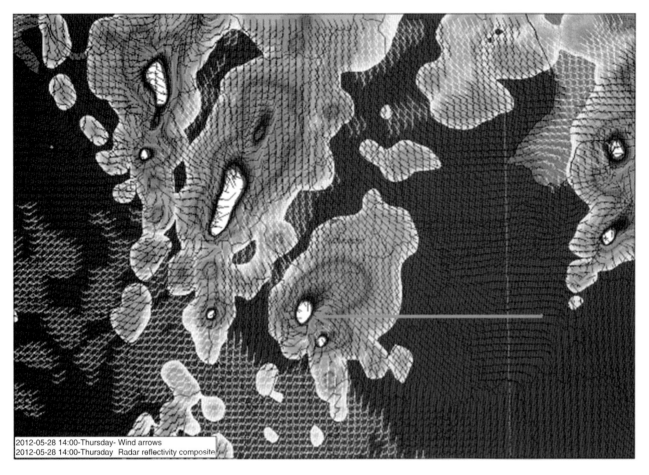

2012-05-28 14:00-Thursday- Wind arrows
2012-05-28 14:00-Thursday   Radar reflectivity composite

Figure 12.3   *A simulated supercell thunderstorm. This image shows the output of the MG, incorporating WRF® model for 1400 GMT on the 28 June 2012. Image courtesy of MeteoGroup, UK Ltd. (2012). WRF® is a registered trademark of the University Corporation for Atmospheric Research (UCAR). (See insert for colour representation of the figure.)*

this severe weather outbreak, it was run twice per day at a resolution of 3 km. The output in Figure 12.3, courtesy of MeteoGroup, shows what can be interpreted as a supercell thunderstorm across parts of the Midlands of England. The figure shows simulated radar reflectivity (effective precipitation intensity) along with surface wind barbs, the latter in standard meteorological notation. It can be seen that in the heaviest area of simulated precipitation (white), the winds show a distinct cyclonic (anticlockwise) 'swirl'. This indicates that the model has generated a rotating thunderstorm. The 'blob' of simulated radar activity, indicated by the blue arrow, depicts a thunderstorm. The yellow shading shows lighter precipitation areas, the blue and white areas show more intense precipitation. The shafts show the direction and speed of the 10 m above ground level wind in standard notation. A shaft with the 'barbs' on the left-hand end indicates a wind from the west. A full-length barb indicates a wind speed of 10 knots; a half barb indicates 5 knots. In the region of the white part of the thunderstorm, note the evidence of a storm-scale circulation, for example, the wind is from the east or south-east on the eastern side of the storm but from the west or south-west further west. This is further demonstrated by succeeding hours in the output, which continued to show rotation in the same simulated thunderstorm (not shown). As mentioned earlier, this output gave

the author increased confidence that supercell thunderstorms were possible across the Midlands of England on this day.

This model output, along with the knowledge that larger-scale conditions would be favourable for severe storms, as assessed using the ingredients-based approach, gave high confidence that an outbreak of severe thunderstorms could be expected across portions of central and eastern England during the afternoon of 28 June 2012. A tornado watch was issued, and numerous incidents of severe weather, including several tornadoes, occurred (e.g. Brown and Meaden, 2012; Clark and Webb, 2013). In the future, the continued growth of computing power and more frequent updates of high-resolution models should give further confidence to meteorologists when forecasting the occurrence of severe local storms.

12.9   Concluding Remarks

The forecasting and monitoring of severe local storms, including tornadoes, have come a long way since the early pioneering efforts of forecasters such as Finley, Fawbush and Miller. For example, high-resolution models can simulate supercell thunderstorms, giving forecasters added confidence that severe weather is possible. However, we are far from being able to forecast individual thunderstorms and tornadoes hours in advance. The best that can be done is that a fairly large area can be placed under a 'watch' for the next several hours, but individual towns and cities are only likely to receive warnings which can be measured on the scale of minutes rather than hours.

The future of severe storms forecasting is likely to see increased usage of short-term 'ensemble' models.[17] These may allow forecasters to highlight smaller areas of risk than can be achieved currently, perhaps being able to give useful probabilistic predictions of severe weather in the next few hours on the scale of several counties. The real-time monitoring of severe thunderstorms and tornadoes, including the issuance of warnings, requires further investment in high-resolution radar equipment. At the present time, fixed position weather radar cannot resolve the circulation of a tornado.[18] A large number of short-range radar sites across a city could allow forecasters to track, with reasonable precision, a tornado, perhaps being able to highlight neighbourhoods in the tornado's path. This can then allow emergency services to quickly attend locations which have been affected.

A perennial problem for weather forecasters is how the public respond to weather warnings (e.g. Golden and Adams, 2000). Forecasting, detecting and warning against a tornado are only useful if the end user is educated in what to do when confronted with the situation. The use of social media has certainly helped to get the message out when severe weather is expected. However, it is far from conclusive that it has changed the behaviour of individuals during severe weather. Indeed, it could be argued that the desire to report in real time, including taking photographs and video, actually *increases* the risk to individuals who engage with the medium.

In the United Kingdom, deaths caused by tornadoes are thankfully exceedingly rare. Other types of weather, including severe gales, flooding caused by heavy rainfall and lightning, are much more likely to cause fatalities than tornadoes. A question can then be posed: is the forecasting of tornadoes actually worthwhile in the United Kingdom? On the one hand, it can give warning of a potentially life-threatening situation in a country which is not used to especially damaging tornadoes but, by the same token, could lead to panic if the risk is not communicated correctly. Brooks and Doswell (2001) highlight the risk of major disasters caused by tornadoes in regions where routine tornado forecasts are not issued. Official severe weather forecasts in the United Kingdom routinely warn against the types of severe weather mentioned previously, but, as yet, not severe thunderstorms and tornadoes. It may ultimately be concluded that such warnings are just not viable. However, TORRO has demonstrated that tornadoes, at least more

significant ones, can be forecast to a reasonable level of accuracy, using the technique of issuing 'watches', as is currently done in the United States.

Acknowledgements

The author would like to acknowledge David Reynolds, Tony Gilbert and Nigel Bolton who have all been key contributors to the forecasting of severe convective storms in the United Kingdom and Ireland; the staff and members of TORRO for collating reports of severe weather and undertaking site investigations of damage to determine whether tornadoes have been involved, which ultimately allows verification of TORRO's forecasts to be achieved; MeteoGroup Ltd, for allowing the author time while on shift to compile the TORRO forecasts, using their dataset; Lindsey and Bob, the author's parents, for putting up with/nurturing weather interests in his younger years; and finally Helen Rossington, the author's partner, for accompanying him to the Great Plains to go storm chasing year after year!

References

Brooks, H. and Doswell III, C. A. (2001) Some Aspects of the International Climatology of Tornadoes by Damage Classification. *Atmos. Res.*, **56** (1), 191–201.

Brown, P. R. and Meaden, G. T. (2012) TORRO Tornado Division Report: June 2012. *Int. J. Meteorol.* **37**, 106–113.

Browning, K. A., Donaldson, R. J. (1963) Airflow and Structure of a Tornadic Storm. *J. Atmos. Sci.*, **20**, 533–545.

Clark, M. R. and Webb, J. D. C. (2013) A Severe Hailstorm Across the English Midlands on 28 June 2012. *Weather*, **68**, 284–291.

Corfidi, S. F. (1999) The Birth and Early Years of the Storm Prediction Center. *Weather Forecast*, **14**, 507–525.

Doswell, C. A., III, Weiss, S. J. and Johns, R. H. (1993). Tornado forecasting: a review. In *The Tornado: Its Structure, Dynamics, Prediction, and Hazards*. Washington, DC: American Geophysical Union. Geophysical Monograph, vol. **79**, 557–571.

Doswell, C. A. III., Brooks, H. E. and Maddox, R. A. (1996) Flash Flood Forecasting: An Ingredients-Based Methodology. *Weather Forecast*, **11**, 560–581.

Finley, J. P (1881) The tornadoes of May 29 and 30, 1879, in Kansas, Nebraska, Missouri, and Iowa. *Prof. Paper No. 4, U.S. Signal Service*, 116 pp. Available at: http://www.spc.ncep.noaa.gov/publications/galway/finley1.pdf (accessed on 21 May 2015).

Galway, J. G. (1985) J.P. Finley: The First Severe Storms Forecaster. *Bull. Am. Meteorol. Soc.*, **66**, 1389–1395.

Golden, J. and Adams, C. (2000) The Tornado Problem: Forecast, Warning, and Response. *Nat. Hazards Rev.*, **1** (2), 107–118.

Grazulis, T. (2001) *The Tornado: Nature's Ultimate Windstorm*. Norman: University of Oklahoma Press. 352pp.

Moller, A. R., Doswell, C. A., III., Foster, M. P. and Woodall, G. R. (1994) The Operational Recognition of Supercell Thunderstorm Environments and Storm Structures. *Weather Forecast*, **9**, 327–347.

Newton, C. W., Miller, R. C., Fosse, E. R., Booker, D. R. and McManamon, P. (1978) Severe Thunderstorms. Their Nature and Their Effects on Society. *Interdiscipl. Sci. Rev.*, **3**, 71–85.

Weisman, M. L. and Klemp, J. B. (1982) The Dependence of Numerically Simulated Convective Storms on Vertical Wind Shear and Buoyancy. *Mon. Weather Rev.*, **110**, 504–520.

---

[17] 'Ensemble' forecasting is the technique of varying the starting conditions of a numerical model in such a way as to account for the inherent uncertainty in ascertaining the initial conditions and to account for errors due to imperfections in the model formulation. This gives a 'spread' of possible outcomes.

[18] If the tornado is very close (e.g. within just a couple of kilometres) of the radar site and is large, it is possible that the circulation may be visible. However, this will still not be at ground level.

13

Extreme Flooding in the United Kingdom and Ireland: The Early Years, AD 1 to AD 1300

Robert K. Doe[1,2]

[1] *School of Environmental Sciences, Department of Geography and Planning, University of Liverpool, Liverpool, UK*
[2] *Tornado and Storm Research Organisation (TORRO)*, Dordrecht, the Netherlands*

13.1 Introduction

The United Kingdom and Ireland have one of the longest documented histories of flooding. We know this from the translations of numerous chronicles, diaries and annals, the Irish Annals (*Annales Quatuor Magistrorum*[1]) providing some of the earliest dated evidence (Britton, 1937). Although this somewhat legendary evidence was no doubt passed from generation to generation, the accounts make interesting reading. For example, in 2668 BC, there was '*an eruption of Lake Con and Lake Techet*'. Lake Techet is known today as Lake O'Gara (Lough Gara). There was a '*poisonous inundation of the sea upon the Kingdom … whence seven lakes overflowed … and therefore they were designated Lake Cuan*' in 2654 BC. Lake Cuan is known today as Strangford Lough, located on Ireland's north-east coast. It was in 1544 BC when the first river floods were mentioned, Subna, Torannia and Cullan having overflowed. This was later followed in 1449 BC by the overflowing of the Flaesc, Man and Labrand. Similar inundations are reported up to 506 BC. As the centuries progressed, the accounts of flooding became more numerous and indeed, more detailed in their descriptions of causes, locations and impacts. It is the aim of this chapter to highlight some of the more notable events, discuss reliability of sources and show that the United Kingdom has a long association with the impacts of water (Doe, 2006).

13.2 Sources of Evidence

Much detective work is needed to interpret the descriptions provided by the various diarists, annalists and chroniclers, as well as assess reliability. It is well known that the chroniclers generally

*http://www.torro.org.uk/

[1] For further details, see O'Conor (1826).

had little hesitation in transcribing and embodying in their own works the writings of their predecessors; it was indeed held among the monastic annalists to be a perfectly legitimate and, more often than not, a necessary practice. A large portion of what we know today and which is described in this text is reliant on their interpretation and transcribing. Many chroniclers wrote about floods and other meteorological phenomena such as thunderstorms, hailstorms, lightning and tornadoes. It is common to find such descriptions where fatalities or damage occurred, especially during important events like coronations, invasions, battles, funerals and weddings. It is very difficult to ascertain exactly where the chroniclers obtained their information about floods. It would seem unlikely from some of the very detailed descriptions that these were pure invention. However, as with all such historical sources that describe meteorological and hydrological events, the facts should be treated with due caution. Some of the events will undoubtedly be attached to legend, folklore and mythology, while others will have been passed down ancestral generations orally, and although we must consider that these may contain some truth, no confidence should be placed in any exact dates assigned to events. Sometimes the same event is reported in different years by different chroniclers. Sometimes the exact year, month or date is not reported at all. Some sources, like Lamb (1977), converted all dates to the new-style calendar for a consistent approach; some make no mention of whether the new or old style convention was being used at all, and some medieval writers were known to have constructed their own dating style. Sometimes the date of reporting is interpreted as the actual date of the event. Historical dating presents complexity, and therefore much corroboration is often needed and the sources should naturally be assessed with appropriate caution, especially with regard to the changes in calendar. The dates discussed herein are noted without conversion.

Extreme Weather: Forty Years of the Tornado and Storm Research Organisation (TORRO), First Edition. Edited by Robert K. Doe.
© 2016 John Wiley & Sons, Ltd. Published 2016 by John Wiley & Sons, Ltd.

In 1928 C.E.P. Brooks and J. Glasspoole published *British Floods and Droughts*, in which they rightly question the reliability of some of the early sources: '*It is very difficult to form an opinion as to the amount of credence which should be placed in the sensational reports of the early chroniclers*'. But more importantly they also admit that many of the historical floods are so striking that they are definitely worth discussion. The two authors did think positively about early flood sources and occasionally suggested some justification for events, for example with regard to the Thames flood of AD 9. They pondered how this could have been remembered, but speculate that it may have been possible from news that crossed the Channel to Gaul and then made it to Roman ears.

Very useful, well-documented, historical flood information is found in the compilation and corroboration provided by C.E. Britton (1937) in his report to the then Meteorological Office, Air Ministry, *A Meteorological Chronology to AD 1450*. This chronology mentioned not only floods but a whole multitude of environmental conditions, including aurora, cold winters, drought, frosts, heavy rainfall, hot summers, lightning, severe gales, thunderstorms, tornadoes and wet seasons, to name a few. It is an interesting source of information.

Sources that refer to the 1st century AD provide little detail of the causes and locations affected by flooding in the United Kingdom. This is because the majority of early sources were Roman in origin, and although the first official Roman presence in Britain was that of Julius Caesar in 55–54 BC, the Romans did not invade Britain until AD 43. Some early accounts refer to rainfall; for example Short (1749)[2] described '*a rain of blood in London lasting five hours*' which occurred in AD 4. It did not rain blood, but this description probably referred to rain mixed with volcanic dust or fine desert sand which had been transported in the atmosphere. Britton (1937) projected much doubt on pre-Roman sources, and his concerns are valid. He suggested that prior to the Roman invasion any reporting of events should be considered 'legendary'. Therefore, such evidence is based on oral tradition which inevitably leads to exaggeration or distortion. Examination of early sources is by no means an easy task!

13.3   The Early Years – 1st–10th Centuries

Early flood events do provide rather interesting reading, especially with regard to the locations noted. One of the first notable floods is thought to have occurred in AD 7 (AD 9 in some sources, or possibly there were two separate events) when extensive floods along the River Thames and Thames estuary were described as causing '*heavy casualties*'. This also indicated that there was some moderate settlement along the banks and on the floodplain at this time. This early event could have been the result of a storm surge, as reports of flooding were also noted further along the east coast and around the Humber estuary. More specific localities were soon named by Short (1749). In AD 14 the West Midlands suffered '*inundations of the Severn with great damage*' and there was a '*great flood*' in the Trent Valley in AD 29. A river flood

along the Dee in AD 33 caused much damage at Chester, and similar reports are noted in AD 37 concerning flooding in the Medway, Kent.

The *Annals of Tacitus* (1980) recorded how two Roman legions under the command of *Publius Vitellius* marched from the River Ems to the Rhine along the shores of the North Sea in AD 15 when:

> Before long … a northerly gale, aggravated by the equinox, during which the Ocean is always at its wildest, began to play havoc with the column. Then the whole land became a flood; sea, shore, and plain were a single aspect. … The companies became intermingled, the men standing one moment up to the breast, another up to the chin, in water; then the ground would fail beneath them, and they were scattered or submerged. … Words and mutual encouragement availed nothing against the deluge; there was no difference between bravery and cowardice, wisdom and folly, circumspection or chance; everything was involved in the same fury of the elements. Many men were drowned; the survivors reached rising ground, and day brought back the land.

Rivers played an important role during the Roman settlement period and their influence provided further detailed recording of floods. Rivers were regarded as defence barriers and often formed the boundaries between early kingdoms (later shires). In this context, rivers are mentioned in the conquering of parts of northern Britain by *Ostorius Scapula*, the second governor of Roman Britain (AD 47–52). *Ostorius Scapula* conquered the heartland between the rivers Severn and Trent. The River Severn is mentioned in AD 14 because it, too, was an important border – between England and Wales. Similarly, the upper Thames was considered the effective border between Mercia and Wessex. The Thames (and Thames Valley) was the subject of one of the first significant reports specifically detailing flood-related casualties in Britain in AD 48, estimated (although most probably exaggerated) at 10,000. Brooks and Glasspoole (1928) suggested that this flood may have been due to very high rainfall, but this seems dubious for such a high fatality figure. Clearly the figure is estimated and is likely to include both humans and livestock, but the flood also affected a large portion of the east coast and Thames estuary and it was more than likely the combined result of storm related rains and tides.

Short (1749) suggested that 20 years later an extreme flood changed the landscape of Britain for ever. In AD 68, inundations on the south coast apparently 'separated' the Isle of Wight from Hampshire, changing the coastline configuration dramatically! Very little is known about this event, which was speculated to be either flood- or earthquake-related. It is more likely a gross exaggeration and more probable that there was a storm surge where a spit or sand bar was breached leading to some localised separation. There is no indication as to where Short obtained such an incredible account, and therefore this should be noted with interest only. Later, in AD 80, floods along the River Severn cost the lives of many cattle and in AD 95 extensive floods in the Humber estuary resulted in 50 miles of land being inundated. A deluge of heavy rain that lasted for what was said to be an 'incredible 9 months' in AD 107 resulted in the corn being washed out of the ground. These long periods of heavy rain and associated crop damage inevitably led to widespread famine and further casualties. Short (1749) recorded that there were floods along the River

---

[2] It is important to note that Short (1749) often described events as if they were fact, but cited no supporting references. It is therefore recommended that caution is used in assessment.

Severn in AD 115; Luckombe (1800) noted that similar events occurred around the Humber in AD 125. Marine inundations along the south coast of Dorset in AD 131 were also catalogued in Short (1749), as was the overflowing of the River Trent for 20 miles on each side in AD 218. Large-scale loss of life was reported as a result of floods along the Tweed, also in AD 218.

Floods continued to be reported through the rest of the 3rd century along the Humber, Tweed, Severn and Dee, including inundations in Thanet, Lancashire, Cheshire and along the Northumberland coast, to name a few. Despite date and source ambiguities, the naming of flood-prone locations does give a good indication of early vulnerability in the United Kingdom. In Cheshire, a lowland county in the north-west, an estimated 5000 people and 'innumerable quantity of cattle' were reported drowned in AD 353. It is very difficult to speculate exactly where in the county such a large number of people would have been lost, but locations in and around Chester and the River Dee could be suggested, as Roman settlement by this time was well established and the Dee already had a long association with flooding.

The Anglo-Saxon period was soon under way by the middle of the 5th century and London entered the flood history books with a Thames flood 'exceeding 10 miles above and 10 miles below' the capital in AD 479. Cardiganshire, a west-coast county of Wales bordering the Irish Sea, was subjected to a storm surge in Cardigan Bay in AD 520. There were further inundations in the Humber (AD 529) and Tweed (AD 536), which resulted in heavy casualties. AD 564 was noted as a year of 'great rain floods'. East Anglia received a coastal deluge in AD 575, and similar inundations occurred at Anglesea (Anglesey), an island county of north-west Wales, in AD 580. There were floods around the North Sea during May AD 586, and then along the north-east coast in AD 589 affecting Hartlepool. There is some obvious disharmony between writers and chroniclers regarding the Hartlepool event. Seller (1696), in his text *The History of England*, mentioned flooding as follows: '*and the sea breaking in near Hortle Pool* [Hartlepool] *in the Bishoprick of Durham swept away divers villages, drowning many people and cattle*'. But Britton (1937) suggested that this event was probably quite mythical. This may be a presumptuous suggestion by Britton based on his opinion of sourcing by Seller. However, such coastal and fluvial flooding (the River Tees and its tributaries have a long history of overflowing) in and around Hartlepool during this year should not be considered implausible. By the 7th century London was well under flood attack from the Thames (in AD 630) and it was Scotland which featured prominently in reports in the 700s, when inundations at Edinburgh did great damage in AD 730. Short (1749) reported that some 400 families were drowned at Glasgow in AD 738.

The 9th and 10th centuries brought more detailed reporting and interpretation of floods but these were cold centuries, notable for frozen rivers rather than flooded ones, but this did not stop the assaults of south-westerly weather systems which generated a wind-storm and flood on 23–24 December 800. Britton (1937) quoted the translated works of Simeon of Durham (see also Stevenson, 1855), who described a great wind that, '… *by its indescribable violence destroyed and threw to the ground cities, many houses and numerous villas, innumerable trees also were torn up by the roots and thrown to the earth. An inundation of the sea burst beyond its bounds not fulfilling what the psalms say, "Thou has set them their bounds which they shall not pass"*'. This resulted in the deaths of many cattle in England, and across the Channel, much of Heligoland was lost, with a high number of fatalities. Similar inundations were to follow in 864 affecting the Humber and again down into the Netherlands.

During the 10th century, attention is drawn to Ireland, where in 920 (some sources suggest 918) flood waters reached the Abbey of Clonmacnoise and to the Road of the Three Crosses in Ulster. Clonmacnoise, in Offaly County, Ireland, suffered again in 942 when another great flood was said to have completely demolished the eastern part of the settlement, which flanked the side of the River Shannon, in a shallow valley close to the floodplain. St Ciaran, the son of an Ulsterman who had settled in Connaught, chose the site of Clonmacnoise in 545 because of its 'ideal location' at the junction of the river!

13.4 Extreme Flooding in the 11th Century

There were a number of very long, cold, severe winters this century. Early records suggested a severe frost in Ireland lasting some 3 months between January and March 1008. An English Channel gale wrecked ships and took many lives off Sandwich, Kent, in 1008. Britton (1937) discussed the marine flooding of September 1014 as reported in *The Anglo-Saxon Chronicle*: '*on St. Michael's mass eve, came the great sea flood widely through this country, and ran so far up as it never before had done, and drowned many vils [villages], and of mankind a countless number*'. These could be the same floods mentioned by the Scottish historian Hector Boece, whose Latin history of Scotland was translated at around 1531 into Scots prose by John Bellenden (and in 1535 into a verse chronicle, *The Buik of the Croniclis of Scotland*[3]): '*The sea, in the simer, rais further on the land, than evir it was sene afore in ony time*'. The sea not only flooded the coasts of England but also had devastating consequences on Walcheren (an island in Zeeland province, the Netherlands, in the estuary of the River Schelde) and in Flanders, including the low-lying borders of northern France. It is also interesting to note that this year was known for an 'atmospheric calamity' when a '*heap of cloud fell and smothered thousands*'. This description clearly contains some colourful distortion for what was conceivably a flash flood as a result of a heavy rainstorm and cloudburst.

Great inundations were also reported along the River Severn in 1046, which were responsible for cattle fatalities. A very wet year in Ireland in 1050 produced a flash flood in which there was damage to buildings and property, when 'milk, fruit and fish' were carried away, although no exact locations are mentioned in the sources. In 1076 a heavy rainfall and flash flood in Lincolnshire was described by Abbott Ingulphus in *the Chronicle of the Abbey of Croyland* (Riley, 1908) with a poignant reference to biblical times, '… *when, behold! A most dense cloud covered the sun in his course, and brought on, as it were, the shades of night, while the heavens poured forth such a deluge of rain, that, from the flowing of the waters, the days of Noah were thought to have come again*'.

Records in *the Anglo-Saxon Chronicle* (Thorpe, 1861) described how in 1086 a '*very heavy and toilsome and sorrowful year in England, through murrain of cattle, corn and fruits were*

---

[3] For further details, see Stewart (1858).

at a stand, and so great unpropitiousness in weather as no one can easily think'. The weather at the time was characterised by thunder and lightning which 'killed many men', presumably struck by lightning as opposed to falling victim to any flooding; however, the conditions did lead to localised heavy rains and flash floods and there were reports of great inundations which caused losses in many places when 'rocks were loosened and overwhelmed several towns in their fall'. This event echoes memories of 20th century floods at Lynton and Lynmouth in Devon, where the flood waters carried large rocks and damaging debris in their wake to seal the fate of the village and ultimately cost many lives. The loss of life from these thunderstorms was paralleled in 1087, when accounts referred to the post-storm famine and diseases that ensued. *The Anglo-Saxon Chronicle* (Thorpe, 1861) noted, *'Such a malady fell on men that almost every other man was in the worst evil, that is with fever, and that so strongly that many men died of the evil. Afterwards there came, through the great tempests which came as we have told before [1086], a very great famine all over England, so that many hundred men perished by death through famine'*.

Heavy rains and swollen rivers influenced the demise of Scottish King Malcolm III in 1093. King Malcolm III travelled south into Northumberland, seeking vengeance for his treatment at the court of King William Rufus (forcing him to pay homage in 1091 and then seizing the border city of Carlisle and Cumbria in 1092). His huge army encamped just north of Alnwick and it was here that Robert de Mowbray, Governor of Bamburgh Castle, rode out with a small contingent to repel them. Mowbray managed a surprise attack on the Scots that threw them into complete confusion, but a sudden deluge of rain and the floods that resulted were another surprise for Malcolm, whose *'army either fell by the sword, or those who escaped the sword were carried away by the inundations of the rivers, which were more than usually swollen by the winter rains'* (Britton, 1937). The floods in the valleys prevented their retreat and King Malcolm was slain in the battle and his army defeated.

The south-east coast of England was subjected to a notable storm surge in November 1099, in which a sea flood sprang up to such a degree and did so much harm as *'no man remembered that it ever before did'*. This storm surge affected not only the coast of Kent and the Thames estuary, but it pushed down the North Sea towards the Netherlands with tragic consequences. In total, some 100,000 people were estimated (possibly overestimated) to have drowned from this event, in what can only have been the most terrifying inundation of water to the inhabited coastal lowlands. The east coast of Kent took a fair share of the storm surge and legend has it that this 1099 coastal storm was responsible for the formation of the infamous Goodwin Sands. The sands are a stretch of shoals and sandbars about 10–12 miles (16–19 km) long and 5 miles (8 km) across at their widest, which now lie around 4–6 miles (6–10 km) off Deal on the east coast of Kent. The series of sandbanks was named after the original landowner, Earl Goodwin (or Godwin), the Earl of Wessex and father of King Harold II, who was said to have owned what was then the island of Lomea, encompassing 4000 acres of fertile but low-lying land with a sea wall all the way around. But the sea wall became neglected and did not last long. Seller (1696) suggested that many parts of Kent 'lay'd under water' in 1099, including the land that had been Earl Goodwin's. As a result, villages were washed away

and many people and livestock drowned. Although it was thought that the events of 1099 were responsible for the formation of the sands, which were to become known as the 'shippe-swallower' and 'the widower', events in 1092 and 1097 are also likely to have been highly influential in their creation. In fact, it would be highly plausible that their origins were influenced by a number of successive coastal storms. The great inundation of 1099 would have been a further, highly influential, event. The debate on their exact formation will continue, but what is certain is that the sands have been responsible for a large loss of life and shipping, particularly in 1703, when around one-third of the British navy's home fleet crashed on to the sands resulting in the loss of some great warships (it was estimated that as many as 13 men-o'-war, and over 2000 lives were lost). To this day the sands constantly change in configuration as a result of the strong winds and tidal currents, making them difficult to chart with accuracy.

13.5   Extreme Flooding in the 12th Century

According to *the Anglo-Saxon Chronicle*, 1116 and 1117 were very wet years, in which heavy rainfall and localised flooding proved detrimental to the corn and fruit crops. Lamb (1977) noted that in 1125, many villages were destroyed and suggested that this was the result of a great sea flood. This could be an event in its own right or based on events described in *the Anglo-Saxon Chronicle* (Thorpe, 1861), which recorded that the most notable flood was on 10 August, *'on so great a flood on St. Lawrence's mass day that many towns and men were drowned and bridges shattered, and corn and meadows totally destroyed'*. Such a devastating summer flood resulted in diseases to both men and cattle and the widespread failure of the fruit crop, leading to famine. It would seem that the floods in this year were flash-flood related, induced by fearsome thunderstorms. *The Chronicle of Peterborough* mentioned a severe thunderstorm with damaging hail during the summer of 1125, and this was repeated by Griffiths (1983) in *A Chronology of Thames Floods*, which noted that the summer of 1125 was very wet.

The year 1135 saw the death of King Henry I, and a number of writers commented on the weather around this time, particularly during the autumn, which produced a 'terrible tempest' between 27 and 28 October 1135, affecting the European continent, as it was known and shaped at this time. In France, wind-storms damaged Normandy, and there was extensive flooding along the coasts of north-west Germany and the Netherlands. The 12th century was prolific in sea floods, especially in 1145 and 1151; in Ireland the winter of the latter year was described as very tempestuous, boisterous and portentous, with 'great overflowings of the sea'. A summer flood on 11 August 1156 as a result of persistent heavy rains hindered the gathering of the harvest and the subsequent sowing of seed.

Even Henry II could not fight the heavy rains and flash-flooding in Wales during his war campaigns in 1164/5. Between October 1164 and September 1165, Henry II was in Wales three times: at Abergavenny, at Rhuddlan and in the Berwyn mountains. Caradoc of Llancarvan (the suggested writer of the early part of the *Brut y Tywysogion*) is referenced by Britton (1937): *'and there the king encamped, with his advanced troops, in the mountains of Berwyn. And after remaining there for a few days, he was overtaken by a dreadful tempest of the sky, and extraordinary torrents of rain'.* There is some difficulty over the exact date of the storm but it seems this is also the storm referred to by Giraldus Cambrensis, who suggested that King Henry II, as a result of a sudden and violent fall of rain, was forced to retreat with his army. Giraldus Cambrensis or Gerald de Barri (Gerald the Welshman or Gerald of Barry) was a writer and cleric. He was one of the most intriguing figures in the history of medieval Wales. During his education in Paris, he gained a superb command of Latin. However, his expectations of succeeding his uncle as bishop in 1176 were quashed. Despite this rejection, he became chaplain to King Henry II of England in 1184, and was chosen to accompany one of the King's sons, John, on an expedition to Ireland. This visit to Ireland was the catalyst for his literary career, his account being published as *Topographia Hibernica*. He followed it up, shortly afterwards, with an account of Henry's conquest of Ireland, the *Expugnatio Hibernica*. It was his earlier texts that described the Welsh flood that influenced the King's retreat. Indeed, it was actually Giraldus's editor who placed this event in 1165. It would seem that this date is plausible, as other sources suggest that it was in August 1165 that Henry's great army marched westwards up the Ceiriog valley towards Corwen on the other side of the Berwyn mountain range. Twentieth-century writers like Gregory (1993) still subscribe to this: '*As the invading army slowly trudged up the Berwyn Mountains, the incessant rain and the buffeting wind helped the darting raids of Welsh guerrilla bands, making life unbearable for Henry's army. Progress became impossible through the mud and moorland bogs, so the Norman army retreated back to England'.* Giraldus became a prolific writer and his work reflected experience gained on his travels as well as his knowledge of the authorities on learning.

August 1165 was unsettled, and thunderstorms and heavy rain falls affected Yorkshire, including, notably, Scarborough. *The Chronicle of Melrose* (Stevenson, 1856) described events with interesting symbolism; '*There was a great tempest in the province of York during that same month [August]. Many people saw the old enemy taking the lead in that tempest: he was in the form of a black horse of large size, and always kept hurrying towards the sea, while he was followed by thunder and lightning, and fearful noises and destructive hail. The footprints of this accursed horse were of a very enormous size, especially on the hill near the town of Scardeburch [Scarborough], from which he gave a leap into the sea; and here for a whole year afterwards, they were plainly visible, the impression of each foot being deeply graven in the earth'.*[4]

Marine inundations during March 1170 are discussed by Britton (1937), who referred to an account by Robert de Monte which detailed an unusual discovery during Lent when, '*... the sea passed over its usual limits, in consequence of which the lands on the coast which had been sown with corn were destroyed in many places, the sea having swept over them. This washing away of the land exposed to view in England the bones of a giant, whose body is reported to have been 30 feet [9 m] long'.*[5] An east-coast storm surge was recorded in January 1176 (although sourced as 1177/8

by some chroniclers and writers), affecting a large part of Lincolnshire where the highest walls, built of turfs (peat cut for fuel), were knocked down by the inundation of the sea. In the Netherlands, the devastation was profound as the sea broke through the dykes, '*which of old had been raised against the tempestuous force of the waves'*, and broke into the low-lying flat land, drowning cattle as well as 'a multitude of men'. The rest of the inhabitants with difficulty survived, by climbing trees and on to rooftops. It was reported that the flood waters took 2 days to recede! The 12th century ended with what was dubbed a very wet year. In 1199 there were such '*vast floods of water that bridges, mills and houses were carried away'*. The chronicler Roger de Hoveden described how the bridge at Berwick, Northumberland, was carried away by the River Tweed, and the subsequent disputes involved in its rebuilding.[6] As a chronicler, Roger was considered fairly impartial and accurate and his reporting of environmental events during the 12th and early 13th centuries makes fascinating reading. His chronicle ended rather abruptly in 1201 and it is therefore assumed he died suddenly.

13.6 Extreme Flooding in the 13th Century

The year 1200 produced its fair share of thunderstorms and heavy rain, one particular storm at Newmarket killing a monk en route to Bury St Edmunds, Suffolk. He was struck by lightning together with his horse, both died, apparently without any visible wounds. There was particularly inclement weather in November 1200, known from an account of the transition of the body of Bishop Hugh (bishop and founder of Lincoln cathedral) from London to Lincoln, where lighted tapers struggled to remain lit for the 4-day journey. It was the violent thunderstorms in June and July 1201, with their high rainfall, that caused flash-flooding. On 25 June, a '*violent tempest of thunder, lightning and hail arose, with deluging rain, by which men, animals and crops were destroyed'*. Fifteen days later, on 10 July, another storm arose, not unlike the first, so that meadows could not be cut and '*hay was carried away in the rapid flow of waters'*. It was reported that a great number of fish died as a result of the fouling of the flood water due to decomposing hay. Ralph of Coggeshall,[7] Abbot of Coggeshall Cistercian Abbey (abbot 1207–1218), wrote, '*Also so great an inundation was projected forth from the clouds, notably in divers neighbouring provinces, that bridges were broken down, crops and hay carried away, some even submerged, and several people were afraid that in this outpouring of waters the deluge of God was seen again'*. Other writers referred to these storms, and the *Waverley Annals*[8] noted that the abbey was deeply flooded. Amazingly, the *Margan Annals*[9] suggested that it rained continuously from 13 May to 8 September, which seems somewhat exaggerated, but all sources confirm flood damage to agricultural communities.

---

[4] The same storm system is reported to have destroyed a mill on the River Severn, with all inhabitants, except a single monk. This event also described one of the earliest known tornadoes in the United Kingdom.

[5] On 3 November 1170, further coastal inundations occurred along the Dutch coast, and as a result, much of Borkum was lost to the sea.

[6] Roger de Hoveden was probably a native of Hoveden or, as it is now called, Howden, in Yorkshire. He was King's clerk (*clericus regis*) during the time of Henry II.

[7] For further details, see Radulphus, Abbot of Coggeshall (1738) and Stevenson (1875).

[8] For further details, see Luard (1865).

[9] For further details, see Luard (1864a).

Locations badly affected included those in and around Exeter, Devon, and St Ives, Cornwall.

Frequent thunderstorms continued to be reported in August 1202, but it was the April floods of 1203 that 'suddenly swelled up way beyond normal levels around a number of places in England', but interestingly, 'only small rains preceded so great a flood'. A disastrous and extreme flood occurred in Perth, Scotland, on 9 September 1210, with many casualties, including members of the Scottish nobility. The Queen herself is reported to have had a narrow escape from drowning during this flood. 'So violent was the torrent that the whole town was undermined, the houses levelled, and many persons of both sexes lost their lives'. Coates (1916) rightly noted there may be some embellishment of 'picturesque exaggeration' in this account, but historians agree that the inundation was a terrible one.

Walter of Coventry was an English monk and chronicler, most likely connected with a religious house in York. He is known to us only through the historical compilation which bears his name, *Memoriale fratris Walteri de Cowntria* (later edited by William Stubbs, 1873). In this it is noted that on 8 September 1214 'the unusually high flow of the sea tide caused damage in England'. This looks like a North Sea storm surge, the body of water moving so fast and amplified to such an extent that at Brugge, in Belgium, the sea flowed over 4 miles (6.4 km) inland. November 1218 produced a devastating weather system between the 17th and 29th. Matthew of Westminster wrote, 'On the night of St. Andrew an incomparable and unprecedented storm with thunder, lightning, wind and rain burst fearfully upon the whole world in both eastern and western parts and the sea suddenly gathered and swelled like the Nile'. Matthew Paris added, 'This tempest was so general that in England and Scotland it engendered unprecedented thunder and lightning'.

The Annals of Loch Cé[10] declared 1219 a very wet year in Ireland, noting remarkable rain the whole year, except for a few days. The start of 1219 was exceptionally stormy, creating a catastrophic North Sea storm in January which claimed an estimated 36,000 lives along the Friesland coast (west Jutland and Schleswig-Holstein), and acres of land were permanently lost to the sea at Nordstrand (on the coast of Schleswig). Thunderstorm-induced flash-flooding continued to remain prevalent in the records. In 1222 there was a severe storm on 14 September, followed by what was reported as 'great rains' continuing until 2 February 1223: 'At the exaltation of the holy cross, there was much thunder throughout all England and this was followed by deluges of rain, with whirlwinds and violent gusts, and this tempestuous weather, together with an unseasonable atmosphere, continued until the Purification of St. Mary'. Time references made here are to key dates in the church calendar and provide a reasonable guide to the timing of events during the medieval period. The Feast of the Presentation of Christ in the Temple, also known as The Purification of Saint Mary the Virgin, fell on 2 February. These 'deluges of rain' produced widespread damage, especially to farmland and crops. Ravensdale (1974) noted that 13 arable acres could not be ploughed, 'which were in meadow and pasture because of too much rain'.

The power and intensity of thunderstorms are well reported in the 13th century and such descriptions aid our understanding of why short periods of heavy rain caused so much flood damage. On 30 November 1222, a severe thunderstorm in Warwickshire destroyed churches and church towers, houses and outbuildings, along with the walls and ramparts of castles. A number of sources mention that two small towers of Worcester cathedral were demolished by this storm, and in the town of Pilardeston, Warwick, the storm destroyed the house of a knight, burying his wife and eight other people. This stormy period continued until 12 December, when a 'sudden storm of wind arose, which raged more fiercely than the before mentioned tempest', where it 'threw down buildings as if they were shaken by the breath of the devil, levelled churches and their towers [including Dunstable church and Merton Church] to the ground, tore up the roots of trees of the forest and fruit trees, so that scarcely a single person escaped without suffering loss'. It was also mentioned by Short (1749) that 'inundations due to a very high tide' occurred this year. Given the nature of the driving wind-storm towards the end of the year, tidal floods are quite plausible. This was a prelude to what was a very wet 1223, with many river and flash floods as a result of prolonged rainfall. Matthew Paris wrote of the 'rain and overflowing of waters continuing in every month of the year that it greatly hindered the seasons and the fruits were very late in maturing, so much so that in the month of November there were hardly any crops to lay up'. Indeed, intense rainfall as a result of a cloudburst was said to 'choke many people'. This could indicate drowning or possibly be a reference to the sulphurous air produced by the thunderstorm. The clear, crisp smell of air after a thunderstorm is the smell of ozone, the charged odour of air.[11]

An 'awful and strange shower' in Connaught, Ireland, in 1224 led to 'terrible diseases and distempers among cattle that grazed on the land where the shower fell, and their milk produced, in the persons who drank it, extraordinary internal diseases'. In 1227 the Dee Bridge at Chester collapsed during a flood, and the *Annals of Worcester*[12] noted that the winter of 1227–1228 produced serious river floods in December, January and February, such that 'no one then living had ever seen the like in their time'.

Heavy summer rainfalls returned to many parts in July 1233 and great floods resulted. Described as 'sudden inundations', the level of flood water was extraordinary in many parts, notably in southern Britain (as mentioned in the *Annals of Winchester*[13]). As a result, people and animals drowned, buildings, mills and bridges were irreparably destroyed and irrecoverable damage was done to crops and fruit; famine and disease ensued. One location to report flood damage was the House of Waverley, a monastery near

[11] It is theorised that when ozone is charged, creating an unstable element, it becomes highly reactive and sulphur as an odour is smelt. Other theories are more specific and assume that sulphurous air can be produced by two atoms of oxygen being smashed into one atom of sulphur. The debate concerning the exact scientific cause of the smell of sulphur during storms continues, but historical descriptions of this distinct pungent smell have been noted by many writers, for example in *the Odysseys of Homer* (Homer, 1857), it was written, 'Zeus thundered and hurled his bolt upon the ship, and she quivered from stem to stern, smitten by the bolt of Zeus, and was filled with sulphurous smoke'.

[12] For further details, see Luard, H.R. (ed.) (1865).

[13] For further details, see Stevenson (1870).

[10] For further details, see Hennessy (1871).

Farnham, Surrey. Steeped in history, the Cistercian monastery was founded on 24 November 1128 by William Giffard, Bishop of Winchester, who brought over from the Abbey of Aumône, in Normandy, 12 brethren with their abbot to form the first colony. The location of the abbey, in the Wey Valley, was not ideal, as the monks suffered from regular flooding and crop failure. During times of severe hardship, the community at Waverley was forced to abandon its home and seek asylum in other monasteries. But it was on 11 and 12 July 1233 that the *Annals of Waverley* (Luard, 1865) recorded: ... *a terrible tempest, violent beyond precedent, raged. Stone bridges and walls were broken down and destroyed, rooms and all the offices were violently tumbled together, and, even at the new monastery, there was flooding in several places to the height of 8 feet. Damage and inconvenience to the same house was such that, in the building in which manifold things both interior and exterior were lost, no one is able or certain to value them.*

This is one of the first flash flood accounts that mention both inundation levels (8 ft, or 2.4 m) and an initial assessment of economic loss – the monastery was not the wealthiest in the land but clearly their belongings at the time were priceless or irreplaceable, or both. Repeat of such summer flooding occurred on 16 July 1234 and affected many areas between Bedford and Norfolk: '... *there suddenly arose a great storm of thunder, lightnings, and whirlwinds, attended by inundations of rain and hail*'. 'Whirlwinds' possibly refers to a tornado recorded at Abbotsley, Cambridgeshire, in July 1234. This storm system had major implications for crops as '*the corn in the fields was lifted up by a blast from hell, the cattle and birds, with everything growing in the fields, were destroyed as if trodden down by carts and horses*'. Roger of Wendover described the passage of, and area affected by, this storm in good detail and suggested that it commenced on the boundaries of Bedford, moving east through the Isle of Ely and then Norfolk, before moving out to sea.

London and the River Thames soon became a focus of interest during the early 13th century as the early timber London Bridge over the River Thames was replaced with a stone version, completed in 1209 after some 33 years of construction. The bridge included houses, shops and even a chapel built at the centre, and illustrations show it crowded with buildings of up to seven storeys high. But the medieval bridge had 19 small arches, which influenced the passage of water. The width between the pillars was fairly narrow and restricted the flow of water through the structure. Effectively, it was holding up the flow of water and the difference in water level each side of the structure was at times around 6 ft (1.8 m), especially during times of heavy rainfall. Brazell (1968) suggested that the bridge blocked a staggering 80% of the river flow, which was obstructed by the addition of water-wheels under the two north arches to drive water pumps, and under the two south arches to provide the power for grain mills. It was no surprise that ferocious rapids were produced between the bridge supports. Only 'the brave or foolhardy' attempted to navigate a boat under the bridge, and many were drowned trying to do so.[14] It was not long into 1236 that flooding

was being well reported, especially when Westminster Palace was entered by a deluge from the River Thames as 'monsoon-like rains poured from leaden skies' during January, February and early March 1236, indicating prolonged heavy rains. As the Thames overflowed it must have been an incredible sight to see small boats navigating the forecourt of Westminster Palace, where 'residents went to their chambers on horseback'. But what followed was one of the driest summers, being described as 'unbelievable' after such torrential spring downpours.

The summer of 1236 produced a lengthy drought with 'an almost intolerable heat which continued for 4 months or more'. The result of this was that rivers, deep pools and ponds dried up and water mills stood useless and ultimately profitless. As the water resource became exhausted, large cracks appeared in the ground and the corn crop in many places grew barely above 2 ft (0.6 m). The drought ended with heavy October rains, particularly in northern England, which led to flooding, so much so that streams and lakes overflowed, damaging a number of bridges and mills. But worse was to come on 12 November 1236 during a devastating storm surge which coincided with a high tide; there was flooding to low-lying coastal areas and harbours between The Wash and the Thames estuary (as far inland as Woolwich). In one village, north of Wisbech, Cambridgeshire, around 100 people, along with all the sheep and cattle, drowned. This was colourfully described by Matthew Paris (1872–1884, vol. 3): ... *a wonderful inundation of the sea broke suddenly in the night, and a great wind roared with the seas, and there were great and unusual floods in the rivers which, especially in places near the sea, forced ships from their harbours by dragging their anchors, drowned a multitude of men, destroyed flocks of sheep and herds of cattle, tore trees up by the roots, overturned houses, and ravaged the coast. And it ascended the shores increasingly, flowing for two days and the intervening night, which was unheard of, and it did not flow and ebb in the usual way but was prevented, as people said, by the very great violence of the opposing wind. The corpses of the drowned were seen unburied near the shore, thrown by the sea into caves, so that at Wisbech and the neighbouring villages, and along the shore and sea coast, a great number of men perished. Indeed, in one village, not a populous one, over 100 bodies were consigned to the tomb in one day.*

The chronicler Raphael Holinshed[15] also described how a great tidal surge arrived from the east, '*for several days with unabated fury, washing up the ocean in such tremendous waves that the banks gave way and the whole country lay completely exposed to its awful fury*'. The year ended with notable thunderstorms and flash floods which 'shook towers and buildings' on 23 December. This affected many parts of England, and in Wales it was reported that '*one night before Christmas Eve there arose a remarkable wind to break down an immense number of houses and churches, and to injure the trees and kill many men and animals*'.

Between December 1236 and March 1237, there were 'great rains and floods'. A climax of events occurred in February 1237 in which the River Thames (and also the River Seine in France) allegedly created a flood of such proportions that the swollen rivers 'destroyed cities, bridges and mills' and made lakes out of dry land. It may be a slight exaggeration to suggest that cities were

---

[14] In 1429 the Duke of Norfolk was thrown into the water, and in 1628 Queen Henrietta also had a close encounter with the rapids. Therefore, it soon became renowned as the bridge '*for wise men to pass over, and for fools to pass under*'.

[15] For further details, see Holinshed (1577).

destroyed, but this account is based on an inundation period of some 8–15 days. Similar floods were repeated a year later and recorded by a number of sources, including Matthew Paris and the *Waverley Annals* (Luard, 1865), during the autumn and winter of 1238, when rivers of England '*unaccustomedly and unnaturally burst forth violently upon numerous fields, level ground and dry and waterless places, and increased rapidly to swift torrents so that fish were taken in them*'. It is interesting to note that just 1 year later the floods were still described as 'unaccustomed' and 'unnatural', but without their locations it is difficult to say whether they refer to new places of inundation or indeed refer to flooding in what we know today as a floodplain, an area of land around a river or watercourse which does flood naturally from time to time. The River Severn continued to flood prolifically and created many problems at Gloucester in 1240, but this was surpassed by yet another 'huge deluge' in London, in which the River Thames overflowed after a thunderstorm on 19 November 1242 (and conditions continued very unsettled until the 27th); there was flooding in Westminster Hall. Conditions were described by Paris (1872–1884) as '… *continually unsettled weather lasting for many days, and sad disturbance of the air. Deluges of rain fell, so that the river Thames, overflowing its usual banks, caused floods for six miles around Lambeth, and unseasonably took possession of houses and fields there*'.

The coastal floods and accompanying windstorms of 1 October 1250 produced widespread flooding around the Humber, the Wash and the low-lying North Sea coasts in the south-east. The marine floods that year also made their way around the south coast to the town of Old Winchelsea, East Sussex (near Rye). Recorded as Winceleseia in 1130 and Old Wynchchelse in 1321, the town and port stood on the shingle barrier and formed a low, flat island exposed to the elements, particularly those of the wind and sea, which ultimately sealed its fate. At its peak, Old Winchelsea was a prosperous fishing village with some 700 houses and about 50 inns and taverns, implying a population of around 3000–5000 people. It had a well-established infrastructure known for the landing of kings as well as pirates, along with its own monetary mint. So when one storm and flood affected half of the town, the results were devastating: '… *beside the hovels of the salt makers and the storehouses of the fishermen, and the bridges and mills, more than 300 houses in the same village with a number of churches, by the violence of the rising water were submerged*'.

The chronicler Raphael Holinshed wrote about the accompanying windstorm, of which '*the like had not been lightlie knowne, and seldome, or rather never, heard of by men alive*'.[16] But it was his description of the mountainous sea, its far encroachment and power that were most colourful: '*The sea, forced contrary to its natural course flowed twice without ebbing, yielding such roaring that the same was heard a far distance from the shore*'. The chronicler noted how in neighbouring Hert-burne, tall ships perished. During the 13th century, the sea continued to breach the shingle barrier, which extended across the present area of Rye Bay from Fairlight to Dungeness, and the flood of 1 October was

the start of many incursions, leading to a complete demise around 1287, when the settlement subsequently moved inland.

In the spring of 1251, on 21 March, a coastal flood '*overflowed its usual bounds in parts of England, inflicting no little damage with the shores covered to a depth of 6 feet [1.8 m] more than has ever been seen*' (Paris, 1872–1884), which flooded areas in Lincolnshire and Kent. The summer of 1251 was deemed a very hot one, with a good and early harvest. The heat soon spawned many thunderstorms in May, June, August and September, and these were accompanied by intense rains and episodes of flash-flooding in a number of places throughout England and Ireland. The violent thunderstorms started on 19 May 1251 and Matthew Paris described the formation thus: '*A thick cloud arose in the morning over the whole world, as it seemed, darkening east as well as west, south as well as north, and thunder was heard as if at a great distance, with lightning preceding it. … The thunder and lightning approached, one clap being more terrible than the others, as if the sky bore down on the earth, the ears and hearts of all who heard being suddenly struck dumb*'. The storm had near-fatal consequences for Queen Eleanor of Provence when a single bolt of lightning struck a chimney attached to the Queen's bedroom at Windsor Castle (where she was staying with King Henry III and her sons), reducing the fireplace to rubble. In neighbouring Windsor Forest, 35 oaks were struck by lightning and many left splintered.

Heavy rains and flash-flooding followed on 29 June, particularly at Roscommon in Ireland, which killed men and cattle at Erinn. The *Annals of Loch Cé*[17] noted that '*a great shower fell on the festival day of Peter and Paul so that a boat sailed all round the town at Kilmore on the Shannon, and a mill could grind in the stream which flowed from the arch to Ath-nafaithcha*'. Similar reports of flooding are mentioned on 17 September, where '*the darkness was dreadful and could be felt, and there happened a great deluge of rain so that the channels of the air were uncovered to the cataracts and the clouds were seen to pour out upon the earth to destroy it*'.

Winchelsea experienced the violent throes of the sea once again in January 1252 during an easterly gale, at a time when the harbour was deemed a vital one to England and especially to London. The day before the flood, the storm was said to have raged day and night with a strong easterly wind (although a veering south-westerly wind occasioned much damage too), and '*with horrible roaring blasts and violent beating, drove back the waves of the sea from the shores, [and] unroofed or destroyed houses*'. A great wind-storm ensued, uprooting the largest oaks, stripping lead from church roofs, sinking large ships and inflicting '*great and irreparable damage*', which '*however great on dry land, was manifestly 10 times greater at sea*'. Back at Winchelsea, the sea, '*as if indignant and furious at being driven back on the day before, covered places adjacent to the shores, and washed away and drowned many men*'. The summer of 1252 was very hot and dry. A number of chroniclers, including Matthew Paris and Robert of Gloucester,[18] are in agreement with regard to the extremity and longevity of the heat, which started in April and lasted through to July. The persistent high temperatures led to widespread drought, fruit crops failed, rivers dried up and fish died in large quantities.

---

[16] The same storm was equally devastating along the Friesland coast, which sustained what was described as 'irreparable damage' as the rivers flowing into the sea were forced back by the surging sea, flooding meadows, mills, bridges and houses.

[17] For further details, see Hennessy (1871).

[18] For further details, see Gloucester (1887).

The monk and writer John de Taxater (sometimes Johannes de Taxter, de Taxster or Taxston) noted that many people died from the excessive heat that year. In Ireland, residents were even able to cross the River Shannon without getting their feet wet. During periods of such drought, the soil becomes highly susceptible to increased run-off, as high intensity rainfall cannot be quickly absorbed into the ground. Therefore, it was no surprise when the first 'great' autumn rains of 1252 arrived: '*the water floods covered the face of the earth, since the excess of the dryness of the earth could not absorb the waters, and rivers flooded so that the bridges and the mills and the houses adjoining the rivers were broken, and the woods and orchards were stripped*'.

During October 1253, coastal flooding affected the north-east coasts of England. Holderness and Lincolnshire suffered inundations far inland, many houses and their inhabitants being washed away. The Thames estuary suffered a similar fate, with the loss of homes, lives and a great deal of property. Robert Fabyan[19] commented that the River Thames rose so high that it flooded many waterfront houses, causing the loss of much valuable merchandise. In summing up 1253, Matthew Paris poignantly wrote, '*The past year was abundant in crops and fruits ... but the gain to the earth was lost to the sea, which overflowed its banks*'. This inundation had further implications in 1254 as, after the floods had receded and the months passed, farming returned to normal. However, the previously fertile land had become so saturated with salt that the crops never took and subsequently failed. Even where the sea had encroached into woodlands and orchards (some distance inland), these failed to become green, leaf, flower or bear fruit. The fruit that was produced was so dried up that it had to be destroyed. Similar conditions were experienced in Flanders, also in 1254.

The poor summer harvest of 1254 was not helped by two summer thunderstorms which produced heavy rainfall and flooding, the first in July, when it was noted that '*quite suddenly inundations of rain broke forth with violent hail of a kind not seen before, which lasted for an hour or more*'. The intensity of the driving rain and sizeable hail broke the tiles on houses and stripped branches off trees. Some 6 weeks later, in August, another, similar thunderstorm produced 'an unusual inundation of rain' particularly around St Albans. Any records of flooding were overshadowed by a lightning strike to the tower of the church of St Peter in St Albans, which '*penetrated into the upper part with a horrible crash, twisted the oaken material like a net, and what was marvellous, ground it into fine shreds*'. It was said that the lightning left the whole tower smoking with an intolerable stench. But worse was to come on 20 November with the flooding of a large part of Bedford.

The *Burton Annals* (Luard, 1864a) noted, '*a sudden inundation of waters, not on the account of rain, happened this same year before the feast of St. Edmund, king and martyr, by which the greater part of Bedford was carried away, and several villages in the fens, and many people of both sexes, old and young and also infants in cradles, were submerged and lost*'.

Summer thunderstorms produced flash-flooding in July 1256. Described by Matthew Paris (1884) as an extraordinary storm of wind, driving rain, hail, thunder and vivid lightning which filled the souls of men with fear: *The mill wheels were seen to be wrenched off their axles, and transported by the force of the waters great distances, destroying neighbouring houses. And what the waters did to the water mills, the wind did not spare to do to the windmills. The piles of bridges, stacks of hay, huts of fishermen with their nets and poles, and even babies in cradles, were carried away, so that it looked as if the floods of Deucalion were come again.*

Bedford, which was served by the Ouse, also suffered, damage being described as 'beyond estimation'. A block of six houses was lifted and carried away by the flash flood, the inhabitants only just able to evacuate them in time, indicating the sudden nature of the event. The Deucalion floods were again remembered on 28 December 1256, by Paris (1884), who wrote, '*There was such a deluge of rain that the surface of the earth was covered and it looked as if the time of Deucalion were returned. Whence the furrows assumed the appearance of caves, or caverns of water, and the rivers crossed the meadows, and the whole country making it look like seas*'. Paris mentioned that one river in northern England carried away seven large wooden and stone bridges, mills and neighbouring houses. Although he does not identify the river or give the location of the bridges, he does mention what appears to be a thunderstorm, which may have spawned a tornado at the time: '*on this memorable day [28 December 1256], a certain fierce whirlwind disturbed the air and darkened it like unto night, accompanied with violent hail; and the clouds became thickened, emitting lightning and darting forth coruscating and terrible flashes. The thunder obviously sounded a sad prophecy. For it was in the middle of winter and the cold was more like that of February*'.

The heavy rains and floods between August 1256 and February 1257, in the words of Paris (1884), were 'beyond measure'; '*From the time of the Assumption of the Blessed Virgin until the Purification rain did not cease to fall in deluges daily, the roads became impassable and the fields were made sterile, whence by the end of autumn the grain putrefied in the ear*'. This period of storminess is corroborated by *the Lanercost Chronicle*,[20] which noted, '*In this year there was throughout all England and Scotland much corruption of the air and inundation of rain*'.

Writing about the weather of August 1257, de Oxenedes (1859) described how a 'fast and procession', in desperate times, stopped the incessant deluge, '*When a great and numerous congregation of the faithful, and the convent, celebrating a fast in solemn procession, carried the martyr Alban to St. Mary in the meadows, as was usual in such times of peril. By this the tempest was made immediately to cease and so, I may add, the crops and fruits were saved and were in general abundant*'. The martyr Alban referred to here is St Alban, a protomartyr of Britain.[21] The summer downpours and associated floods did not last long (although

---

[19] For further details, see Fabyan (1533) and (1811).

[20] For further information, see Stevenson (1839).

[21] According to Bede's *Ecclesiastical History*, Alban was a pagan living at the Roman settlement of Verulamium (St Albans), who converted to Christianity and was executed by beheading on a hill above the town. St Albans Abbey, Hertfordshire, was later founded near this site. Bede (1883) tells several legends associated with the story of Alban's execution, one of which says that on his way to the execution, Alban had to cross a river and, finding the bridge full of people, he made the waters part and crossed over on dry land. The executioner was so impressed with Alban's faith that he converted to Christianity on the spot and refused to kill him. Another executioner was quickly found and legend has it that his eyes dropped out of his head when he finally executed Alban!

probably not through the divine intervention of the martyr Alban) and this led to a good late-August and September harvest, but what was not harvested by early autumn perished. The uncharacteristic warmth of the autumn and winter, along with persistent rainfall, led to what some perceived as a complete shift in the season, with the absence of the expected cold and frost. Paris (1884) wrote, '*The past year was sterile and meagre, whatever was sown in the winter, whatever was shewing promise by flowering in spring, whatever ripened in summer, the floods of autumn choked it. For there was neither a temperate nor serene day, nor was the surface of the lakes hardened up by the frost, as is usual, nor were icicles hanging, but the continued inundations of rain thickened the air until the Purification of the Blessed Virgin* [about 2 February]'.

Summer thunderstorms in 1258 brought unhappiness to a large part of western Britain between Shrewsbury and Bristol, including areas around Bridgnorth, Worcester, Tewkesbury, Gloucester and along the Bristol Channel. A thunderstorm appeared to have started in the west and moved towards the midlands, slowly building as it moved inland towards Shrewsbury to the west of Birmingham. During the passage of this thunderstorm, there was a high volume of localised rain along the high ground and tributaries of the River Severn, and by the time the storm reached Shrewsbury, the rivers were high, starting to flow swollen and fast towards Bristol and the Bristol Channel. The *Annals of Tewkesbury* (Luard, 1864b) recorded: *A great tempest of flooding rain, snow, ice, heavy thunder and horrible lightning happened suddenly before the feast of St. John the Baptist [24 June] covering the country beyond the banks of the River Severn from Shrewsbury to Bristol. And by this flood all grass likewise adjoining the Severn and the crops were lost and destroyed, displeasing both to men and animals. Submerged in the same were horses, and many men and children of both sexes.*

Continual and persistent rain in the autumn of 1258 led to misery and famine as the corn crop became rotted and ruined, resulting in desperate times for many: '*the poor devoured horse flesh, the bark of trees and things still worse, while multitudes died of starvation*'. The autumn flooding remained long in the memory of those who endured it. The Annals of both Tewkesbury and Dunstable could not help but compare the ferocity of the flooding to that of previous years: '*this past year was very dissimilar to all previous, that is, it was unhealthy and mortal, stormy and exceedingly rainy, so much so, that, although in summer time the harvest and crops and an abundance of fruit seemed promising, yet in the time of autumn continual heavy rains choked the corn, fruit and also vegetables*'.

The events of August 1265 changed history, but the influence of a sudden storm added to the climax of one of the most notable battle campaigns of the 13th century: the Battle of Evesham, Worcestershire.[22] However, this was no straightforward battle, the weather of the day clearly influenced morale and produced very difficult conditions. On 4 August 1265, a thunderstorm moved south-east towards Evesham from north Wales, accompanied by rain, thunder and lightning. The darkness was described as being so great that during dinner the inhabitants of Evesham could hardly see what they were eating. Conditions during the battle were described by Bartholomew de Cotton (Luard, 1859), who wrote, '*The day became dark, the sun withdrew his rays, there was an earthquake at the place, thunder muttered and frequent lightning shone forth*'. Britton (1937) quoted the verse of Robert of Gloucester, who drew biblical comparisons with the weather at the time of the crucifixion of Christ:

And there with Jesus Christ was very ill pleased
As he showed by tokens both terrible and true,
For as it to himself befell, when he died on the cross,
Also while the good men at Evesham were slain,
There arose, as in the north-west, a dark storm
So black and so sudden that many were terrified;
And it overcast all the land, so that we might hardly see;
A more fearful than it might not on earth be.
A few drops of rain exceeding large there fell.

It was perceived that such a storm was a displeasing sign from the heavens. Summer rainfall such as this at the height of the battle must have had some influence, particularly on the terrain and visibility, as well as on morale. Such high-rainfall events are often short-lived but it is clear that the coincidence of this with the climax of the campaign was deemed a fearful signal. In the end, Simon de Montfort and his army were heavily outnumbered and eventually defeated.

In April 1268, the *Oseney Annals* recorded 'inundations of rain and runnings of water, very terrible, and enduring for fifteen days'. This was followed in September by an account by William of Newburgh of a '*very great tempest of wind and rain, and a commotion of the air, by land and sea, through which much evil resulted in divers parts. For houses were felled, trees uprooted, seed shaken out of the ground and rotted, many cattle killed and ships sunk and broken to pieces. There was this year a wet and stormy winter and one full of diseases*'.

Forty days of heavy rain, inviting comparison with biblical times, caused flooding to areas surrounding the Thames during February and March 1270. The *Oseney Annals* referred to '*so much rain and pouring out of water, that before or after the deluge of Noah, such a terrible one had not been seen, and, it was said, not of such duration, for it lasted for forty days and more*'. The River Thames was said to have risen so high in London that waterside sellers drowned and all merchandise perished or was lost. Apart from the obvious human tragedy, the economic loss of such merchandise would also been very costly; by the late 13th century, London had become one of the trading and manufacturing capitals of Europe, producing everything from woollen cloth to weapons.

A thunderstorm of 'frightening proportions' affected Canterbury, Kent, in 1271. Some writers mentioned April, others September (11th or 18th); indeed these could be two separate events, but the descriptions of the events are very similar and certainly well merit discussion because of the extreme nature of the flooding and the loss of life that ensued. Both sides of the English Channel were affected, and in Burgundy, France, '*There were such rains and earthquakes that it was thought to be like a second Deluge, and*

[22] Following his battle victory at Lewes in 1264, Simon de Montfort, Earl of Leicester, had controlled the kingdom, while Henry III was effectively his prisoner along with his son Prince Edward (who later became Edward I). The prince soon escaped custody and joined forces with Gilbert de Clare (Gilbert the Red), eighth Earl of Gloucester, who, with the nobles of the Welsh Marches, assembled what was to be a strong and victorious army to defeat de Montfort. Simon de Montfort was eventually killed, decapitated and mutilated, and his head was displayed around the country as a warning of what happened to people who rebelled against the King!

houses were carried off by its inundations, and stone bridges broken; crops and vines laid to waste and submerged'. More graphic accounts are contained in writings of John Everisden, who noted that violent rain fell suddenly on Canterbury, that the greater part of the city was suddenly inundated, and 'there was such swelling of waters that the crypt of the church and the cloisters of the monastery were filled with water'. Walter de Hemingburgh (Hamilton, 1848–1849, vol. 1) noted:

> There was at Canterbury such a flood of rain, with thunder, lightning and tempest such that two very old men had never heard or seen anything like it for prolonged thunder, for it was as if one horrible clap sounded for the whole of the aforesaid day and night, and such floodwater followed that trees and hedges were overthrown, whereby to proceed was not possible either to men or horses, and many were imperilled by the force of the waters flowing in the streets and in the houses of citizens. A very great famine followed throughout the whole kingdom.

Cambridge and Norwich were flooded to such an extent in 1273 that it was said to have been *'almost equalled to those of 1258 ... while in some parts of England they appeared to have exceeded in violence'*. This is an interesting comparison between events some 15 years apart. The rains lasted 'a night and a day', and in Cambridge, the flood waters rose 5 ft (1.5 m) above the main bridge in the city.

An 'earthquake' signalled coastal floods in September 1275 as Gervase of Canterbury (Stubbs, 1879) mentioned: *'a great earthquake happened in many kingdoms, and chiefly in England, and floods of water also about maritime towns'*. Walter de Hemingburgh noted, *'In the year of the Lord 1275 on the ides of September, there was a general earthquake in London and in the kingdom of England, both in camps and towns, habitations and fields'*. It is difficult to establish whether the earthquake and the marine inundations are directly related, as the word 'earthquake' was often used to describe thunder. It could have been a tremor, and therefore this should not be discounted. There have been a number of similar tremors, especially in eastern Britain, and those in 1089, 1382, 1580, 1692 and 1884 are well recorded.

The Chronicle of Bury St Edmunds reported localised flooding in Bury St Edmunds, Suffolk (where the storm was thought to be at its strongest), as well as into Essex and Cambridgeshire around 10 October 1277,[23] which apparently lasted 2 days and a night *'Such a flood followed this cloud-burst that in some places men and cows, sheep and other domestic animals in the fields were overcome at night by the storm and drowned. Houses, walls, trees and some buildings which stood in the way of the flood were completely overthrown'*.

Cheshire experienced a marine and fluvial flood during February 1279, when there was flood damage at Stanlow. The reports of flooding at this location originate largely from the Cistercian abbey that was founded at Stanlow (or Stanlaw) in 1178. Built on the low-lying marsh, Stanlow Point, the abbey had an eventful history spanning nearly 400 years. The floods in 1279 were described by Wallis (1923): *'the inundations of the sea were of an extraordinary nature. The church and the buildings were flooded to the depth of 4 or 5 feet with salt water, part of the site*

---

[23] Later in December 1277, the sea inundated Reiderland, in the Netherlands, where 50 villages were washed away with great loss of life.

---

was washed away, and there was imminent danger of the whole becoming permanently water-logged and uninhabitable'. Later, the tower of the abbey was felled by a great gale in 1287 and further floods (and a fire) in 1289 caused much damage; by 1296 all the monks had left. The site of Stanlow Abbey was between the junction of the rivers Gowey and Mersey, but is largely inaccessible today as the site has become a major oil refinery. What was described in the Chester Annals as a very unusual and excessive current resulted in 'the bridge' at Chester being destroyed and carried away. This was probably a reference to the Dee bridge. It is difficult to establish the full sequence of events but it seems that either a coastal inundation from the Irish Sea funnelled up the rivers Mersey and Dee or perhaps a high fluvial outflow from excess rainfall could have amplified the river as described.

It was the historic city of Oxford that experienced a sudden and violent inundation of flood waters in April 1280, the like of which 'had not been seen in this time of year for 30 years or more'. Then followed an August flash flood; the exact locations affected are unclear but the sources hint towards a storm which moved between Cambridgeshire and Norfolk. It was described in the text of Bartholomew de Cotton (Luard, 1859), a Norwich monk and great historical writer for Norfolk: *'there was such an inundation of rain, and such violent floods of water followed, that men and women, old and young, and herds in the fields were drowned, mills and bridges, houses and trees, were submerged, and hay and corn in most places was carried away'*.

If there was one year that was notable for its succession of devastating floods, it was probably 1287, when notable events occurred in the United Kingdom during January, February and December of that year. The cusp of 1287 saw the first of the great inundations and these are mentioned by a number of sources, including *the Chronicle of Bury St Edmunds* and *the Chronicle of Florence of Worcester* (Forester, 1854): *On the night of the Circumcision [meaning the Feast of the Circumcision of Christ, 1 January] the wind was so violent, and the sea so stormy, at Yarmouth, Dunwich, Ipswich and other places in England, as well as on the coasts of other countries bordering on the sea, that many buildings were thrown down, especially in that part of England called the Fens: nearly the whole district was converted into a lake and unhappily a great number of men were overtaken by the floods and drowned.*

Bartholomew de Cotton suggested that the flood started at around midnight on 31 December: *'There was a great inundation of the waters and the sea, so that, by the floods, boats to the number of 180 at Kirkhale and one sailor perished'*. It is more than likely that Kirkhale (Kirkharle) is a reference to what was then known as Kirk Harle parish, Northumberland. The *Osney Annals* are connected to Osney Abbey, Oxford, and contain the history of the monastery between 1016 and 1347. These annals also mention events of 31 December and 1 January:

'On the feast of the Lord's Circumcision, a most strong wind began to blow from the eastern points (which is called Eurus) and ... the greater part of the town of St. Botolph's was submerged and men and a numberless multitude of cattle perished. There was a similar inundation in Essex and in several other eastern parts; nor was such a dreadful sea flood seen in those parts for thirty years past or more'. It is interesting to note the Greek mythological reference to Eurus, who was the god of the east wind and was known as the wind that brought warmth and rain.

The *Dunstable Annals* also mentioned this rain and flood: '*there was a great inundation of rain and the northerly sea from the Humber to Yarmouth overflowed its usual bounds, and in some parts to a breadth of three leagues, in other to four, everything being submerged. A great magnitude of men, sheep and cattle were destroyed*'. In theory, a league expresses the distance a person (or a horse) could walk in 1 hour, usually about 3 miles.[24] Therefore, the flooding covered a distance of some 9–12 miles (14–19 km) at Yarmouth in that year.

On 17 December 1287, a storm surge travelled down the east coast of England with the most tragic of consequences: '*the sea, as well by the violence of the winds as by their boisterousness and raging impetuosity, began to be disturbed by a dense cloud and, with huge rushes, broke through the flat shores, disturbing its customary bounds, and occupying fields, villages and other places on its confines, and also inundating parts which no one remembered seeing before covered by the waters of the sea in a cycle of a thousand ages past*'. Bartholomew de Cotton wrote, '*about midnight thunder was heard and on the Wednesday following there was a great inundation of waters of the sea*'. This inundation coincided with high tide and the flood affected the low-lying coast between the Humber and Kent, and particularly large areas of East Anglia around the Norfolk Broads. *The John of Oxnead Chronicle* (de Oxenedes, 1859) described how the night flood waters caught many by surprise and at midnight men and women rushed out of their houses in bed clothes, holding babies asleep in cradles and seeking a safe place of refuge. Some tried to climb trees, but the strength of the flood was too much to contend with, and they drowned. Large numbers of cattle and freshwater fish were destroyed through the vast inundations of sea water, and houses along with all of their contents 'were removed from their deep foundations with irreparable damage'. In the village of Hickling, Norfolk, 20 people drowned as the flood waters rose a reported 1 ft (0.3 m) above the high altar of the Priory Church (often referred to as the Priory of Canons). All the canons, except two, fled; the two remaining were said to have sheltered horses in their dormitories for safety. Bartholomew de Cotton noted that, as well as Hickling, the coastal villages of Horsey and Waxham and inland Martham (as well as a few other smaller neighbouring villages) suffered casualties, including around 200 people reported drowned. These locations circle Hickling Broad, and examination of these on a map shows that it is possible that the flood affected a much wider area around this low-lying stretch of county.

Further south-west along the Broads, the Abbey of St Benedict of Hulme in Horning was infiltrated by a river torrent which surrounded the building. de Oxenedes (1859) quoted what could have been poignantly said: '*Abissus vallavit me, et pelagus operuit caput meum*', meaning 'the sea has walled me in, the ocean has covered my head'. This can be interpreted as a distinct reference to the prophecy of Jonah: 'And thou hast cast me forth into the deep in the heart of the sea, and a flood hath compassed me: all thy billows, and thy waves have passed over me'. The abbey has an interesting history and in 1020 the manor of Horning

was given by King Canute (or Knut) to the Abbey of St Benedict of Hulme. It was the same King Canute who wisely said, after failing to command the sea to go back, '*Let all men know how empty and worthless is the power of kings. For there is none worthy of the name but God, whom heaven, earth and sea obey*'. Interestingly, the village sign today at Horning reads, 'The name [Horning] means, "The folk who live on the high ground between rivers"'.

Further south-east of Horning, on the coast at Great Yarmouth, around 100 people drowned and the stone wall of the main cemetery was destroyed by the sea for a length of 60 ft (18 m). In total, it was thought that some 500 lost their lives along the east coast. The gale-driven North Sea storm surge made its way down the coast and towards the Straits of Dover. During the 13th century, Old Winchelsea was attacked several times by French marauders and a by number of powerful coastal storms; it stood strong, but it was the incursions of the sea which proved its greatest enemy and by the end of 1287 the fate of the town was sealed as the storm surge made its way into the English Channel and removed all signs of what was left of town and beach.[25] It was said that this one storm created 'a tempest of all tempests' at Winchelsea, with waves of an unprecedented height totally removing all in their wake, including sea walls that offered little or no protection to what was left. Old Winchelsea was soon gone forever and a new town of the same name was later constructed some 2 miles away on higher ground.

The weather of the year 1288 was known for two things: the long hot summer that lasted at peak for some 5 weeks and led to abundant harvests (and also a drought that claimed many lives), and, more devastating, the sea floods of 4 February 1288. A number of writers and chronicles, including Everisden, Gervase of Canterbury and the *Oseney Annals*, mentioned the storms of the 3rd and floods of 4th February, and the consensus is that these occurred in February 1288 as stated. It should be noted that the *Worcester and Dunstable Annals* record this in 1287, but it is possible that these are being confused with the floods of the previous December. The storm system started on the night of 3 February and *the Chronicle of Florence of Worcester* (Forester, 1854) recorded:

> On the third of the nones of February, about nightfall, flashes of light were suddenly and unexpectedly seen at St. Edmunds [Bury St Edmunds], there having been no signs prognosticating it: and at the same instant, there was a tremendous crash, I will not say of thunder, followed by an insufferable stench. The storm was accompanied by visible sparks of fire, which fearfully dazzled the eyes of the beholders. The tower of the church of Barnwell was set on fire by the violence of the thunderstorm and further damage was done to the convent there, and one third part of the town was prey to the flames.

---

[24] The unit of measurement, the league, although no longer officially used in western measurements, was devised in Ancient Rome and came to the Romans via the Greeks and the Persian parasang (which was an ancient Persian unit of distance corresponding to ~3.5 miles).

[25] In Europe, particularly Friesland (northern coastal parts of the Netherlands), Flanders (Belgium and southern coastal Netherlands) and Denmark, the dykes that held back the North Sea failed. Hundreds of acres of low-lying land were affected and tragically it was estimated that 50,000–80,000 people lost their lives. A new bay, called Zuiderzee (Southern Sea), was created over former farmland and this started yet more years of building dykes and creating polders to push back the flood waters of the Zuiderzee.

On the 4th, the sea rose '*to such an extent in Thanet and round about, and in the marshes of Romney and all the adjacent parts, that all the dykes were demolished and nearly all the ground was covered from the great dyke at Appledore to Winchelsea both towards the east and the west*'. The Thames estuary suffered also, and the inundation from the seas was described as so violently powerful that 'for one day and night, contrary to its ordinary usage and the course of nature, it was seen to flow and reflow four times alternately'. This resulted in some substantial flooding in London along the banks of the Thames in areas which, up until this time, usually withstood the force of the sea. Hundreds of properties were flooded and the *Oseney Annals* described this occurring 'over a much greater distance than any mortal at any time had seen'. Further down the Thames, villages and fields were '*swallowed up and countless men and innumerable cattle ... were drowned*'. Along the east coast, at Spalding, a monastery was damaged by the '*impetuous deep waters which flowed and covered them*'. The building of the friars preachers at Yarmouth (which at the time was next to the sea) was submerged unexpectedly by the flood waters. The neighbouring town of Yarmouth also experienced high levels of coastal flooding, in which it was said that buildings were 'submerged and blotted out'. The North Sea coast between Spalding and the Thames estuary appeared to have suffered the most during the storm surge in 1288, but it also continued down the North Sea and towards the Netherlands, where similar levels of inundation were experienced.

Thunderstorm-induced flood waters on 19 July 1289 (and similarly during the summer of 1290) caused such crop destruction that the price of wheat increased sevenfold – from three pence to two shillings. William of Rishanger[26] described how a sudden inundation of rain, thunder and lightning drowned the crops especially those of the valued wheat, 'so for nearly 40 years, until the death of King Edward there existed a dearness of crops and especially wheat'. According to *the Chronicle of St Werburgh*, vast areas of cultivated land alongside the Mersey were abandoned to floods in 1294. This was most likely the result of the very wet August, September and October, when persistent rain meant that little or no corn could be harvested. In October the River Thames flooded a number of areas around Rotherhithe, Bermondsey, Tothill and Westminster. The *Bermondsey Annals* recorded that '*torrents of the waters of the Thames overpassed their usual limits ... and there happened a great breach at Rotherhithe and the plains at Bermondsey. ...similarly it reached the cottages of the merchants at the fair of Westminster and forced them to strengthen the dwellings with taller stakes*'. These Thames-side tradesmen were great sufferers, and as the 14th century approached, so did the severe, cold, hard winters and dangerous frozen waters of the River Thames.

13.7 Concluding Remarks

Historical floods in the United Kingdom and Ireland have resulted in large loss of life and social and economic impacts in multiple locations, covering vast areas. This text has detailed many flood accounts including fluvial, flash flooding and coastal inundations,

caused by a number of forcing factors, including prolonged heavy rains, short intense rainfall and storm surges. Without the influence of water, the United Kingdom would not be the shape it is today. It is sometimes hard to visualise the mechanisms which took place in separating the United Kingdom from the continent of Europe, a time when the River Thames was a tributary of the Rhine. The encroachment of water through rises in sea level, erosion by the waves, surges of sea, flash flood waters and the channelling of rivers has played an important part in our geomorphological history.

This selected chronology of early floods has shown that coastal and inland locations have suffered considerably. Much attention has been given to the assessment and validity of the sources and data discussed. As with any text that examines historical sources of evidence relating to physical processes, it would be excessively sanguine to expect absolute infallibility in reporting and recording. This said, much research has gone into providing a 'best known' interpretation. Reading through this chronology will prompt the reader to ask – what would happen if a similar event were to occur today? Events of similar magnitude already have, and maybe worse will come in the future. The fact remains, many communities continue to be vulnerable to the threat of flooding, with the potential for loss of life and high social, economic and environmental impacts.

Acknowledgements

This chapter is reprinted in part, and contains edited and verbatim extracts from: Doe, R. (2006) *Extreme Floods: A History in a Changing Climate*. Sutton Publishing, Stroud, U.K. and is reprinted with rights and permission. The author would like to express gratitude to researchers at the Tornado and Storm Research Organisation (TORRO) especially those who have submitted historical sources and reports; the reviewers of this text for valuable feedback; the National Meteorological Library and Archive; Met Office, U.K.; the Bodleian Library; University of Oxford; The British Library, London, for sources of information.

References

Bede (1883) *Interpolations in Bede's Ecclesiastical History and Other Ancient Annals Affecting the Early History of Scotland and Ireland*. J. Watson, Peebles.

Brazell, J.H. (1968) *London Weather*. HMSO, London.

Britton, C.E. (1937) *A Meteorological Chronology to A.D. 1450*. Geophysical Memoirs No. 70, Meteorological Office Air Ministry, HMSO, London.

Brooks, C.E.P. and Glasspoole, J. (1928) *British Floods and Droughts*. Ernest Benn Limited, London.

Coates, H. (1916) 'Floods and Droughts of the Tay Valley'. *Transcriptions of the Perthshire Society of Natural Science* **6**, 103–26.

de Oxenedes, J. (1859) *Chronica Johannis De Oxenedes*. Longman, London.

Doe, R. (2006) *Extreme Floods: A History in a Changing Climate*. Sutton Publishing, Stroud.

Fabyan, R. (1533) *The Chronicles of Fabyan*, **2** vols, London, W. Rastell.

Fabyan, R. (1811) *The New Chronicles of England and France*. J. Rivington, London.

---

[26] For further details, see Riley (1865).

Forester, T. (ed.) (1854) *The Chronicle of Florence of Worcester*. H.G. Bohn, London.

Gloucester, R. (1887) in W.A. Wright (ed.), *The Metrical Chronicle of Robert of Gloucester*, **2** vols, Rolls Series, London.

Gregory, D. (1993) *Wales Before 1536 – A Guide*. Gwasg Carreg Gwalch, Llanrwst.

Griffiths, P.P. (1983) *A Chronology of Thames Floods*. 2nd edn, Report No. 73. Thames Water, London.

Hamilton, H.C. (ed.) (1848–1849) *Chronicon Domini Walteri de Hemingburgh*, **2** vols, English Historical Society, London.

Hennessy, W.M. (tr. and ed.) (1871) *The Annals of Loch Cé*, **2** vols, Longman, London.

Holinshed, R. (1577) *The first volume of the chronicles of England, Scotland, and Ireland*. John Harrison, London.

Homer (1857) *The Odysseys of Homer*. tr. G. Chapman, **2** vols, J.R. Smith, London.

Lamb, H.H. (1977) *Climate: Present, Past and Future. Vol. 2: Climatic History and the Future*. Methuen, London.

Luard, H.R. (ed.) (1859) *Bartholomaei de Cotton, Monachi Norwicensis, Historia Anglicana: AD 449–1298*, pp. lxxviii, 493, Rolls Series, London.

Luard, H.R. (ed.) (1864a) 'Monasterium de Burtona super Trent. Annales Monasterii de Burton, 1004–1263', in H.R. Luard (ed.), *Annales Monastici*, **5** vols, Rolls Series, London.

Luard, H.R. (ed.) (1864b) 'Annales Monasterii de Theokesberia, 1066–1263', in H.R. Luard (ed.) *Annales Monastici*, **5** vols, Rolls Series, London.

Luard, H.R. (ed.) (1865) 'Annales Monasterii de Waverleia, 1–1291', in H.R. Luard (ed.) *Annales Monastici*, **5** vols, Rolls Series, London.

Luard, H.R. (ed.) (1869) *Annales Prioratus de Wigornia. A.D. 1–1377*, Longman, London.

Luckombe, P. (1800) *The Tablet of Memory, Shewing Every Memorable Event in History, from the Earliest Period to the Year 1792*. 10th edn, J. Johnson and J. Walker, London.

O'Conor, C. (ed.) (1826) 'Annales Quatuor Magistrorum', in C. O'Conor (ed.), *Rerum Hibernicarum Scriptores Veteres*. J. Seeley, Buckingham.

Paris, M. (1872–1884) in H.R. Luard (ed.), *Matthaei Parisiensis, Monachi Sancti Albani, Chronica Majora*, **7** vols, Rolls Series, London.

Radulphus, Abbot of Coggeshall (1738) 'Ex Radulphi Coggeshale Abbatis Chronico Anglicano', in M. Bouquet *et al.* (eds), *Recueil des Historiens des Gaules et de la France*, **24** vols, Palmé et Cie., Paris, tomes 13 and 18.

Ravensdale, J.R. (1974) *Liable to Floods: Village Landscape on the Edge of the Fens, AD 450–1850*. Cambridge University Press, Cambridge, UK.

Riley, H.T. (tr. and ed.) (1908) *The Chronicle of the Abbey of Croyland*. Bohn's Antiquarian Library, Bell & Sons, London.

Riley, H.T. (ed.) (1865) *Willelmi Rishanger, Quondam Monachi S. Albani … Chronica et Annales, Regnantibus Henrico Tertio et Edwardo Primo, AD 1259–1307*. Rolls Series, London.

Seller, J. (1696) *The History of England*. John Gwillim, London.

Short, T. (1749) *A General Chronological History of the Air, Weather, Seasons, Meteors, & c. In Sundry Places and Different Times*, **2** vols, T. Longman, London.

Stevenson, J. (ed.) (1839) *Chronicon de Lanercost: 1201–1346*. Maitland Club, Glasgow.

Stevenson, J. (tr. and ed.) (1855) 'The Historical Works of Simeon of Durham'. in *Church Historians of England*, Pre-Reformation Series, Seeleys, London.

Stevenson, J. (tr. and ed.) (1856) 'The Chronicle of Melrose', in *Church Historians of England*, Seeleys, London.

Stevenson, J. (tr. and ed.) (1870) *Annals of the Church of Winchester from the Year 633 to the Year 1277*. Rolls Series, London.

Stevenson, J. (ed.) (1875) *Radulphi de Coggeshall Chronicon Anglicanum*. Rolls Series, London.

Stewart, W. (1858) in W.B. Turnbull (ed.). *The Buik of the Croniclis of Scotland; or, a Metrical Version of the History of Hector Boece*, **3** vols. Rolls Series, London.

Stubbs, W. (ed.) (1873) *The Historical Collections of Walter of Coventry*. Longman, London.

Stubbs, W. (ed.) (1879) *The Historical Works of Gervase of Canterbury*, **2** vols, Longman Trübner & Co., London.

Tacitus (1980) *The Annals*. tr. C.H. Moore, Loeb Classical Library, Cambridge, MA.

Thorpe, B. (ed. and tr.) (1861) *Anglo-Saxon Chronicle According to the Several Original Authorities*, **2** vols, Rolls Series, London.

Wallis, J.E.W. (1923) *Whalley Abbey*. Society for Promoting Christian Knowledge, London.

14

Extreme Rainfall and Flash Floods in the United Kingdom and Ireland: Synoptic Patterns and Selected Case Studies

John Mason[1], Paul R. Brown[2], Jonathan D.C. Webb[3], and Robert K. Doe[4]

[1] *Department of Natural History, National Museum Wales, Cardiff, UK*
[2] *Tornado Division, Tornado and Storm Research Organisation (TORRO)*, Bristol, UK*
[3] *Thunderstorm Division, Tornado and Storm Research Organisation (TORRO)*, Oxford, UK*
[4] *Department of Geography and Planning, School of Environmental Sciences, University of Liverpool, Liverpool, UK*

14.1 Introduction

There is no fixed definition of 'extreme rainfall', but any very large rainfall that leads to severe and costly flooding is a reasonable starting point. Some places flood regularly, hence the term 'flood-plain': a good example is the Dyfi Valley in central Wales. Here, there are valley-scale floods two or three times every year, which lead to the closure of the A487 trunk road for a day or two and the abandonment of cars. Buildings are typically unaffected, as the regularity of the flooding means that awareness of the flood risk zone is known to the local population. Such floods result from rainfalls of 50–100 mm over the Dyfi's mountainous catchment in periods varying from 24 to 36 hours. These steady dynamic rainfalls are a feature of the United Kingdom's hilly western seaboard. However, many parts of the United Kingdom have been devastated by severe floods on varying scales. These events have been influenced by a number of meteorological and topographical factors. To understand the meteorological variables better, this chapter explores severe dynamic and convective rainfalls with reference to relevant case studies.

14.2 Severe Dynamic Rainfalls

Two factors are of critical importance in dynamic rainfall events in the western United Kingdom. Firstly, the west is directly exposed to incoming moist air on the prevailing south-westerly winds. Situations of particular interest involve the Azores High being south of its average position and high pressure extending north-eastwards across Iberia and central Europe. In such a set-up, zonal conditions establish themselves, with a succession of low-pressure systems moving up from the south-west – these essentially being waves in a long-fetch south-westerly airflow whose origins are far down in the tropics. The Clausius–Clapeyron[1] relation shows how, for every degree Celsius of added warmth, air can carry up to 7% more water vapour. In short, the warmer the air, the more moisture it can carry if it can make use of an evaporative source, such as an ocean surface. Thus, such long-fetch south-westerlies are particularly moist and bring very mild but wet weather during the winter months, when they are most frequently observed (an informal term for them is 'warm conveyors'). Secondly, when such airflows reach the United Kingdom's western coast they are forced up over the hills, resulting in ascent of 500–1000 m and leading to rapid forced cooling. The Clausius–Clapeyron relation then works in reverse as the cooled air sheds its moisture, which condenses as enhanced precipitation. In such set-ups the rainfall radar typically shows light rain over, for example, the Irish Sea and the coast, but moderate to heavy rain over the mountains inland. This effect, determined by the topography, is known as orographic enhancement.

Warm conveyors that lead to serious flooding are those that remain in position for prolonged periods of 24 hours or more, with a background of steady rains periodically augmented by frontal waves moving through. A good example of a particularly

---

[1] The Clausius–Clapeyron equation is used to estimate the vapour pressure (absolute humidity) at any temperature.

Extreme Weather: Forty Years of the Tornado and Storm Research Organisation (TORRO), First Edition. Edited by Robert K. Doe.
© 2016 John Wiley & Sons, Ltd. Published 2016 by John Wiley & Sons, Ltd.

prolonged (48-hour) warm conveyor rainfall event occurred on 3–4 February 2004 over Wales (Figure 14.1). During these 2 days, 261 mm of rain fell at Capel Curig and flooding was widespread and locally severe and life-threatening. Another disastrous warm conveyor occurred over 18–19 November 2009 (Figure 14.2), leading to severe flooding, with some loss of life and tremendous property damage, over large areas of Ireland, North Wales and, worst of all, Cumbria. The Lake District Fells in particular received phenomenal amounts of rainfall: Seathwaite, in Borrowdale, recorded 316.4 mm in 24 hours, setting a new UK record for any 24-hour period (Eden and Burt, 2010; Sibley, 2010). The corresponding 48-hour fall of 396 mm is also a UK record. The northwest

highlands of Scotland present an extensive range of slopes exposed to warm conveyor episodes, and the area has experienced the events of largest spatial extent. The event of 26–27 March 1968 deposited more than 100 mm of rain across a total area of 12,500 km², that of 16–17 December 1966 across 11,000 km² and the storm of 5–6 February 1989 across 10,000 km². The last produced a Scottish 2-day rainfall record of 306 mm at Kinloch Hourn, and torrents feeding the headwaters of the Ness and Spey resulted in serious flooding downstream around Inverness, including the destruction of the 127-year-old Ness railway bridge (Law, 1989).

Most major dynamic rainfalls are associated with the synoptic patterns of late autumn and winter; however, they can also occur

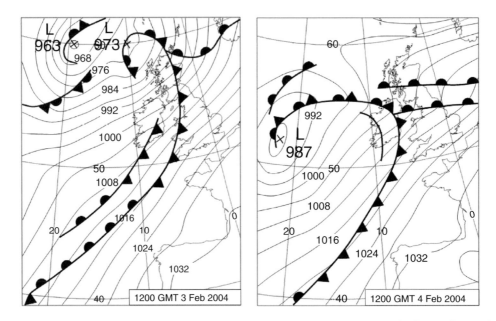

Figure 14.1   *Synoptic charts for 1200 GMT 3 and 4 February 2004. Redrawn from the Meteorological Office surface analyses for the same times. © Paul Brown.*

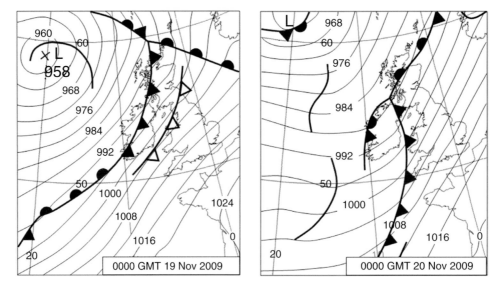

Figure 14.2   *Synoptic charts for 0000 GMT 19 and 20 November 2009. Redrawn from the Meteorological Office surface analyses for the same times. © Paul Brown.*

during summer and, with warmer air involved, they can have spectacular consequences. The events of June 2012 in mid-Wales were especially noteworthy. The weather had been unsettled for some time and the ground was already fairly saturated. At 0000 GMT on 7 June a deep low, for the time of year, with a central pressure of 986 hPa, was heading into the Southwest

Approaches (Figure 14.3). Twenty-four hours later it had deepened to 980 hPa and was centred over Cardigan Bay. By 0000 GMT on the 9th it had drifted northeast and filled a little, with a centre of 992 hPa over the North Sea some 80 miles off Tyneside. For much of 8–9 June its associated occlusion lay directly over mid-Wales. Rain set in on the night of 7–8 June accompanied in places

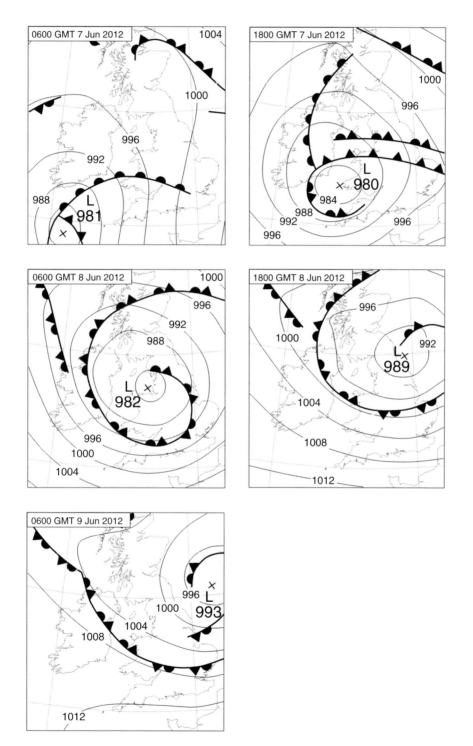

Figure 14.3   *Synoptic charts at 12-hour intervals from 0600 GMT 7 June to 0600 GMT 9 June 2012. Redrawn from the Meteorological Office surface analyses for the same times. © Paul Brown.*

by severe gales, especially along south-western coastal areas during the 8th. The rain finally dissipated by late morning on the 9th, almost 36 hours after it started.

In this example, it was the combined duration and intensity of the rainfall that led to severe problems. Rain fell at a rate of over 10 mm/hour for a considerable part of the event's duration. Whilst not especially intense compared with the rainfalls that accompany thunderstorms, the fact that it continued hour upon hour over the hills to the northeast of Aberystwyth was critical. Environment Agency rain gauges recorded some impressive totals, with 183 mm at Dinas Reservoir near Ponterwyd, and 198 mm at the nearby Nant-y-moch Reservoir. However, the upslopes further to the west received significantly more: on the 8th, cumulative radar plots indicated 92–144 mm, and up to 144–192 mm locally, while on the 9th another 48–56 mm fell before the rain dissipated during the morning (Figure 14.4). The result was the worst flooding in living memory along the catchments of some of the rivers draining these upslopes.

The worst affected location was the village of Talybont, approximately 7 miles north-northeast of Aberystwyth, where the main A487 trunk road crosses Afon Leri, a medium-sized river with a short (c. 12 km) catchment containing several small tributaries that flow out of the hills immediately to its east. The river rose steadily during the 8th and continued to rise overnight, but on the morning of the 9th it burst its banks and flooded the lower parts of the village. Some houses had well in excess of a metre of water flowing through their ground floors. Locals only recall one flood that came close to this, in the 1960s, when heavy rain combined with a great thaw of lying snow and a large tree jammed underneath the bridge caused the river to overflow. The June 2012 flood was due to the volume of rain alone. The neighbouring Afon Ceulan, which joins the Leri just west of the A487 bridge, downstream from the village, likewise surged over its banks and tore a path through the village public house's beer garden, its car park and the neighbouring village green, ripping up tarmac in the process.

Below Talybont, there are a number of caravan sites situated close to the Leri, and despite having flood defences these were severely inundated on the morning of the 9th, when people stranded in caravans had to be rescued by helicopter and the local inshore lifeboat. Further flooding affected parts of the coastal village of Borth, near where the Leri flows along a diverted canal down to the Dyfi Estuary. The worst of the rainfall affected only the southern side of the Dyfi catchment; nevertheless, the whole valley saw a flood comparable to those of winter, with blockage of main roads. The Rheidol catchment, to the south of the Dyfi and Leri, was less fortunate, although the morning flooding was localised in a relatively sparsely populated area. On the afternoon of the 9th there was a high tide at Aberystwyth, where the Rheidol flows out into Cardigan Bay. The tide had the effect of ponding the floodwaters, and damaging flooding affected the flood plain upon which Aberystwyth's Retail Park is situated. Several large retail premises were inundated as a consequence. The total cost of the event in terms of damage is thought to have been several millions of pounds. Ceredigion Council's costs alone for clean-up and repair to damaged roads and bridges was in excess of half-a-million pounds (*see also* Webb, 2013).

Severe dynamic rainstorms with orographic enhancement can also affect eastern regions of the United Kingdom, especially when an area is located on the north or northwest side of a depression, the sector of a low which Hand *et al.* (2004) identified as especially at risk from prolonged precipitation. The most noteworthy of such events have all occurred in late summer or early autumn, the period of the year when the North Sea is relatively warm. A recent example occurred on 5–6 September 2008 when a depression moved rather slowly east-northeast across central England (Met Office, 2013). There was severe flooding in Morpeth, Northumberland, where the 2-day rainfall at the long-established weather station at Cockle Park was a record 152 mm and around 1000 properties in the town were flooded. Orographic enhancement produced a fall of 226 mm at Goldscleugh in the Cheviots (305 m) over the same 48-hour period.[2] A depression following a similar track was associated with the devastating Tweed Valley floods of 12 August 1948 when 158 mm fell in 24 hours at Kelso (Met Office, 1950). Eastern Ireland is occasionally affected by such orographically accentuated dynamic rainstorms at this time of year – when the Irish Sea is also at its warmest. An outstanding event, undoubtedly made even more severe by the copious tropical air, was associated with the passage of ex-hurricane 'Charley' across the Celtic Sea and Wales on 25–26 August 1986. At Kilcoole (Co Wicklow), on the coast just south of Dublin, 200 mm of rain were recorded on the 25th, while more than 250 mm were recorded in the Wicklow Mountains (Meskill, 1986; Graham, 2006). The storm also caused considerable flooding in western Wales (Mayes, 1986), significantly in some areas which usually escape the worst impacts of the more typical warm conveyor rains. Aber, Bangor, recorded 135 mm on the 25th and there was severe local flooding in Pembrokeshire.

14.3 Hybrid Rainfalls – Dynamic Precipitation with Embedded Convective Cells

The prediction '*rain spreading up from the south, locally heavy and thundery*' is something heard in weather forecasts during the warmer and especially summer months in the United Kingdom. It refers to extremely moist, and in some cases unstable, air being advected up from the south or south-west, the result being widespread wet conditions, and in some cases widespread severe weather. 'Thundery rain', as it is often called, can range in nature from the odd embedded thunderstorm in an area of largely dynamic rainfall up to what is known as a mesoscale convective system (MCS), a giant thunderstorm cluster covering hundreds of square kilometres, often accompanied by prolific lightning. Such weather systems occur in most years across the United Kingdom, but the summer of 2007 achieved notoriety for its heavy downpours and severe flooding. By the end of the year insurance claims for flood-damaged properties and businesses approached £3 billion. The events of 19–21 July were particularly noteworthy for their great extent and severity. These involved a mixture of convective and dynamic, orographically enhanced, rain and the resultant floods had some interesting hydrological aspects.

Prior to this event, slack areas of low pressure circulated over northwest Europe, while the polar jetstream was some distance south of its usual summer position. An upper trough lay to the

[2] 226 mm was recorded from a tipping bucket rain gauge, the manual gauge recorded 249 mm.

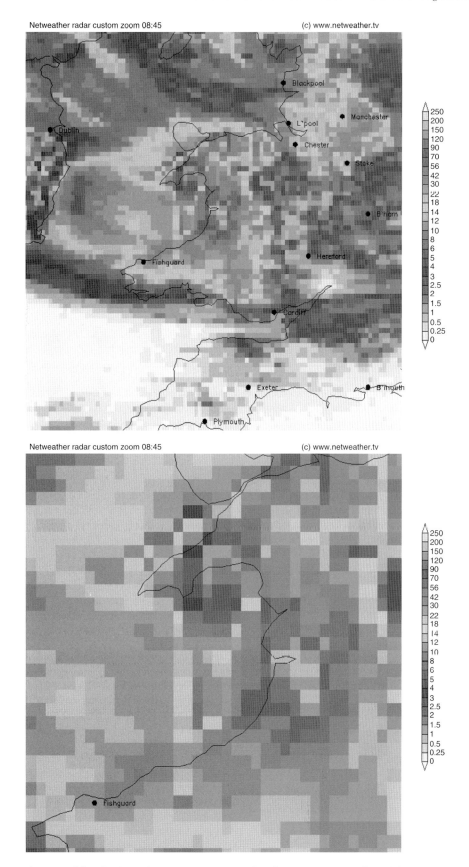

Figure 14.4   *Left: cumulative rainfall radar for Wales on 8 June 2012. Right: close-up on region affected showing up to 144–192 mm locally (left image) Image supplied. © NetWeather. (See insert for colour representation of the figure.)*

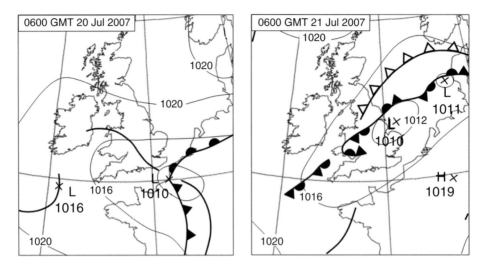

Figure 14.5 *Synoptic charts for 0600 GMT 20 and 21 July 2007. Redrawn from the Meteorological Office surface analyses for the same times.*
© Paul Brown.

west of the United Kingdom, over an abnormally warm eastern Atlantic, setting the scene for persistent low pressure development. There was plenty of atmospheric moisture and the precipitable water value for the midnight sounding at Herstmonceux on the 20th was as much as 29.93 mm, a good 10 mm more than a typical value – and convective instability.[3] On 19 July thunderstorms developed widely over England, Wales and Ireland, with local flash flooding. The saturated ground from the already very wet summer readily facilitated rapid run-off of rainwater. By the 20th low pressure was centred over south-east England and heavy, 'thundery rain' affected a wide area of England from the Midlands southwards (apart from Devon and Cornwall), moving north and west throughout the day as the parent low-pressure centre drifted over the Midlands (Figure 14.5) (Webb and Pike, 2010). By the afternoon the area of rainfall had ceased to produce lightning, but local very intense cores remained, indicating the continued presence of embedded convective activity (Figure 14.6). High hourly totals of 30–40 mm were widely recorded, while the highest daily totals included 147.0 mm at Sudeley Lodge near Winchcombe (Gloucestershire) and 135.2 mm at Pershore (Worcestershire) (Figure 14.7). Severe flash flooding occurred widely across the southern Midlands, the Home Counties (including London) and Central Southern England, and there were a number of fatalities. Helicopter evacuations were necessary in places and hundreds of road (including motorway) and rail closures were reported.

Overnight the rain moved into Wales where, in the reverse of the more typical pattern, the moist air approached the Welsh mountains from the east. As this occurred, orographic enhancement set in leading to a prolonged heavy spell of rain over the over-saturated Severn catchment. Already in flood following the tremendous totals that had fallen in its middle and lower catchments, it now received a second surge of run-off moving downstream from the Welsh mountains – a common occurrence in the winter months but the last thing the area needed. Environment Agency river gauges recorded the peak flow as it made its way downstream. In the Worcester district the peak occurred on 22 July, whilst closer to Gloucester it occurred on the 23rd. The peak flow at Haw Bridge, between Worcester and Gloucester, was estimated to be nearly 1400 m³/s, a phenomenal amount of water. The resultant flood was said to be as bad as (or possibly even worse than) that of March 1947, when after a bitter winter, a change to much milder conditions was accompanied by prolonged heavy rain over frozen ground covered in very deep (50–120 cm) snow. The result led to an abrupt thaw on a huge scale and near-total run-off – a highly contrasting set of conditions.

Flooding such as in July 2007, in a major low-lying river basin, takes time to clear, and it was well over a week before water levels began to subside significantly. As well as thousands of people being rendered temporarily homeless, those still able to remain in their homes had to rely on bowsers (mobile water tanks) to obtain safe drinking water, as purification stations had also been inundated. Prolonged power cuts added to what amounted to a civil emergency affecting this and other parts of southern England.

Another classic dynamic/convective hybrid 'thundery rain' event, on 20–21 September 1973, was described by Webb and Pike (2014). A prolonged downpour deposited 172 mm of rain in 19 hours at Manston, Kent, and this included severe thunderstorm activity which produced 23 mm in 15 minutes at the height of the event. This and other outstanding warm season dynamic/convective hybrid episodes over the past 100 years are presented in Table 14.1, which gives details of spatial extent, preceding conditions (a key factor for flood risk) and synoptic background. Figure 14.8 shows the most severe summer rainfall events (both those classified as dynamic and dynamic/convective hybrids) in terms of spatial extent; those where 100 mm of rain fell across areas exceeding 2500 km².

---

[3] Convective or Potential Instability occurs when the wet bulb potential temperature decreases with height through a layer of air. This instability is released by widespread forced ascent such as over high ground or (most commonly) ahead of a frontal zone or upper trough. When lifted, the moister air at the bottom of the layer reaches saturation (with latent heat release) quicker than the drier air above, this process resulting in an increased environmental lapse rate.

Figure 14.6 *Visible satellite image, 20 July 2007, 1200 GMT. The image shows a clear slot in the 'eye' of the depression, while the deep embedded convective clouds form a conspicuous wall on the back edge of the precipitation area. Courtesy of Robert Moore. © Eumetsat (2007).*

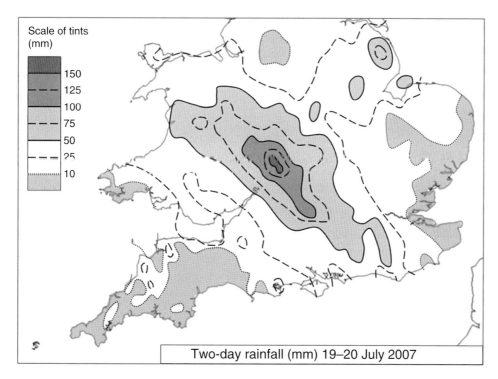

Figure 14.7 *Rainfall totals in millimetres for the 48-hour period beginning at 0900 GMT 19 July 2007. For further details, see Webb and Pike (2010). © Paul Brown.*

Table 14.1   *Extreme dynamic/convective 'hybrid' summer rainfall events (May to September) in England (other criteria >100 mm across area exceeding 1000 km².).*

Event Date and General Location	Area with Rainfall 100 mm or More, km²	Area with Rainfall 150 mm or More, km²	Max 48 hour Fall, mm (Max 24 hour Fall in Brackets)	References	Preceding Rainfall	Synoptic Development
28–29 June 1917 (North Wessex)	2100	240	245 (243) Bruton, Somerset	Mill and Salter (1918) and Clark and Pike (2007)	Dry 1–26 June (16 mm at Mere)	Complex depression moved ENE along Eng Channel. Extreme rainfall in narrow zone between fronts converging from both N and S
9–10 July 1968 (Somerset, Bristol, Midlands and Fenland)	2250 (2600)*	83 +	176 (173) Chewstoke-Nempnett Thrubwell, Somerset	Bleasdale (1974), Hanwell and Newson (1970)	E & W June rainfall 94 mm (average 66 mm), local downpours 1–7 July. SW Eng/S Wales rain well above average for June and 1–8 July	Depression moved slowly NE from Biscay, associated warm front became slow-moving across S Eng. Depression deepened significantly during evening and crossed SW Eng and S Midlands with associated triple point
14–15 September 1968 (SE England and E Anglia)	6250	660	217 (162)[†] Northchapel, W Sussex	Bleasdale (1974), Jackson (1977)	August wet in SE and E Anglia. Rainfall also above average for 1–13 September in southeast England	Depression over W France pushed occluding frontal system north, while a cold front moved S over N Sea reinforcing an undercutting NE wind. Small low formed in the E Eng Channel off Sussex by 0900, 15th.
28–29 July 1969 (SW England)	5250 (6975)*	–	146 Ellbridge, Cornwall & N Hessary Tor, Devon (145 Ellbridge, Cornwall)	Bridge, (1969), Bleasdale (1974)	Long dry spell 9–26 July	Depression in SW approaches at 0600 on 28th moved v slowly E into Eng Channel to be off Sussex at 0600, 29th. Warm front edged north converging on an old occlusion which had moved slowly SE overnight 27/28th. Sharp upper trough crossing England & Wales from west
20–21 September 1973 (Kent)	c. 1200	c. 150	209 (191) West Stourmouth, Kent	Webb and Pike (2014)	V wet 15–19 September (e.g. 40 mm Ashford, Kent), but dry 1–14 September	Depression moved NE from Normandy, 'splitting' with new triple point centre moving NNE, near Kent coast at 0300, 21st
19–20 July 2007 (Cent S Eng, SW Midlands)	c. 3500	c. 50	163 (147) Sudeley Lodge Glos	Prior and Beswick (2008), Eden (2008), Webb and Pike (2010)	E & W rainfall May–July 415 mm -highest recorded since 1766 for that specific 3 month period[‡]. Over 200 mm fell in parts of Worcs. over previous 5 weeks	Shallow depression formed over NE France, deepening as it moved NW to reach the Thames Valley; it then turned NNE and slowly filled

*For the events in July 1968 and July 1969, the first figure refers to one rainfall day (after Bleasdale, 1974) while the values in brackets are estimates for the two rainfall days based on the respective number of stations recording 100 mm or more over the 1- and 2-day periods.

[†]Both these 1- and 2-day readings in September 1968 were from Northchapel (West Sussex). The more extensive large rainfall totals were further east, where the highest 2-day readings were two of 201 mm near Tilbury (Essex); there was also a daily fall of 130 mm at Bromley (London) on 15th.

[‡]Rainfall in May–July 2007 was comfortably exceeded by some other 3-month periods, most recently December 2013–February 2014.

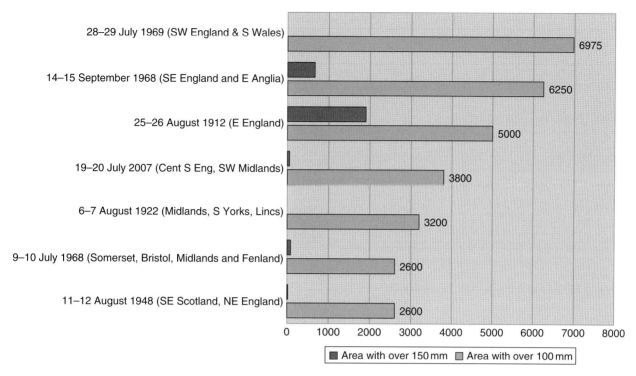

Figure 14.8   *Some extreme summer rainfall events by area (km²) affected, for the United Kingdom. For further information, see British Rainfall 1912, 1922, 1948 and 1968. (See insert for colour representation of the figure.)*

14.4   Severe Convective Rainfalls

Showery conditions are of frequent occurrence in the United Kingdom and can occur as a result of a variety of synoptic patterns. One of the most common occurs when low pressure is situated to the north-west of the mainland giving a brisk to strong airflow from a westerly point. In such set-ups, showers, both scattered and in more organised lines along surface troughs, with or without hail and thunder, tend to move through quickly, giving a few minutes of intense rainfall. Such conditions rarely cause problems beyond spray and temporary standing water on the roads – the rainfall moves through too quickly to allow the accumulation of much water in any one place.

Several factors may, however, complicate matters and instigate problems due to excessive rainfall. These may be brought about by slacker pressure gradients, leading to much slower-moving convective cells, by convective activity being concentrated along convergence zones and, in the warmer months, when there is more energy and moisture available, by the intensity of precipitation. All three factors have led to some notable flood disasters in recent years.

Flash flooding from slow-moving thunderstorms is difficult to predict accurately, but repeated observations based on radar plots show that even short-duration rainfalls in excess of 60 mm/hour can in minutes overwhelm local drainage systems. Such rates are relatively uncommon during the colder months; much winter convection involves polar maritime air, especially returning polar maritime air masses, in which the air may be appreciably unstable. However, the aforementioned Clausius–Clapeyron relation limits its moisture carrying capability, so that the heaviest winter

showers and thunderstorms tend to produce rates of 30–50 mm/hour locally, with precipitation rates that vary quickly. In the summer months the reverse is the case, and storms developing in warm, moist Spanish plumes can produce >60 mm/hour over a wide area. In such storms, slower movement increases the risk of severe flooding as any given point on the surface receives more rainfall.

Table 14.2a lists the most extreme recorded rainfalls in the United Kingdom for specific durations (see also Ross *et al.* (2009)), while Table 14.2b notes some additional estimated falls based on non-standard rainfall collectors, run-off evidence or radar accumulations. Cinderey (2004, 2005) presented a detailed first-hand account of the North Yorkshire storm of 10 August 2003, which produced record 8- and 10-minute rainfall. The Cheviot cloudburst of 2 July 1893, with an estimated localised fall of 229 mm (186 mm in an hour), fortunately occurred some distance from a populated area (Clark, 2005).

The extent to which drainage systems are overwhelmed by such storms depends on several factors. Land use determines how much of the rain will soak away, but soil type and underlying geology are also important. A clay soil is less permeable than a sandy soil, whilst limestone bedrock will take a great deal more water than shale. In populated and especially urban areas, much of the rain does not reach the soil where it is covered with concrete and tarmac. However, in rural areas none of these factors are that important if the ground is impermeable as a result of previous weather conditions – if it is highly saturated by previous rainfalls, or frozen solid, rainfall runoff will be very high. Topography plays an important role too, channelling floodwaters into and along natural landscape features. The north coast of Devon and

Table 14.2 *Extreme recorded rainfalls for specific short durations, United Kingdom.*

Date	Duration (mins)	Amount (mm)	Place	References
(a)				
10 August 1893	5	32	Preston (Lancashire)*	Symons (1894)
10 August 2003	5	30	Carlton-in-Cleveland (North Yorkshire)	Cinderey (2005)
10 August 2003	10	46	Carlton-in-Cleveland (North Yorkshire)	Cinderey (2005)
27 June 1970	12	51	Wisbech (Cambridgeshire)	Meteorological Office (1970)
18 July 1964	15	56	Bolton (Greater Manchester)	Met. Office (1972)
05 September 1958	20	63	Sidcup (London)	Rowsell (1963)
26 June 1953	30	80	Eskdalemuir (Dumfries & Galloway)	Met Office (1953) and (2011)
01 August 1980	45	97	Orra Begg, Co Antrim (Northern Ireland)	Woodley (1981)
09 June 1910	60	110	Wheatley (Oxfordshire)*	Mill (1911), Webb (2011)
08 August 1967	90	117	Middle Knoll, Dunsop Valley (Lancashire)	Jackson (1979) and Law (1968)
19 May 1989	120	193	Walshaw Dean Lodge (West Yorkshire)*	Acreman (1989), Acreman and Collinge (1991)
19 May 1989	150	193	Walshaw Dean Lodge (West Yorkshire)	
14 August 1975	150	169	Hampstead (London)	Keers and Westcott (1976)
19 May 1989	180	193	Walshaw Dean Lodge (West Yorkshire)	Acreman (1989)
07 October 1960	180	178	Horncastle (Lincolnshire)*	Met. Office (1963)
16 August 2004	240	197	Otterham (Cornwall)*	Doe (2004), Mason (2004), Burt (2005)
18 August 1924	300	203	Cannington (Somerset)	Glasspoole (1925)
(b) Estimates based on non-standard measurements, radar and run-off				
2 July 1893	60	186	Low Blakehope, Cheviot Hills (Northumberland)	Clark (2005)
16 August 2004	240	250	Hendraburnick Down (Cornwall)	Burt (2005)
30 October 2008	180	200	Ottery St Mary (Devon)	Clark (2011a), (2011b) and Grahame *et al.* (2009)

*'Visual estimates' or estimated by extrapolation with data from nearby recording rain gauge.

Cornwall is a good example of an area vulnerable to damaging flash floods, its steep-sided narrow river valleys channelling water seawards from the hilly coastal hinterland.

Large, and often slow-moving, severe thunderstorms occur somewhere in the United Kingdom most years. These storms are multicellular in nature, but less frequent supercells occur when shear profiles favour their development. A fine example of the latter occurred on 28 June 2012 (Clark and Webb, 2013). The synoptic situation at 0000 GMT on the 28th involved an Atlantic low pressure area to the west of the United Kingdom with a centre at 994 hPa some 600 miles northwest of the Iberian Peninsula (Figure 14.9). Twenty-four hours later the low had drifted north-east to be situated off the northern coast of Ireland and deepened to 988 hPa. An attendant cold front slowly pushed eastwards during the 28th. Convective storm development occurred in the warm plume of moist air ahead of the front. Beginning as elevated storms in the morning over South Wales, a cluster of rapidly growing cells tracked north-eastwards towards the Midlands where, taking root in the boundary layer (and hence no longer elevated), they took on supercellular characteristics, for example moving to the right with respect to the steering flow. The storms brought severe weather (including large hail and tornadoes) leading to damaging flooding to many parts of the East Midlands (see also chapters 4 and 9). Weather station data from Coalville in Leicestershire highlighted the ferocity of the precipitation:

3.6 mm in 1 minute (peak intensity) = 216 mm/hour
14.8 mm in 5 minutes = 177.6 mm/hour
26.2 mm in 10 minutes = 157.2 mm/hour

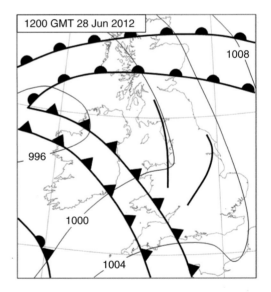

Figure 14.9 *Synoptic chart for 1200 GMT 28 June 2012. Redrawn from the Meteorological Office surface analysis for the same time. © Paul Brown.*

14.5 Gwynedd, North Wales, UK, 3 July 2001

Pulling onto the side of the forestry road the air was thick with a single smell – that of splintered conifer trees. I locked the Land Rover and scrambled down to the river, a familiar route that I have often used, weaving its way cunningly for several tens of metres along a small stream in dense undergrowth to end with a short traverse above a bluff of rock and the drop down onto the river bed where in times past I had often panned for gold. Except that this time something was different. The undergrowth ended abruptly some way before it should have done. Where there had been soil, grass, rhododendron and bramble, naked rock gleamed in the sunshine. Scrambling down, I saw that the river, whilst again at its normal level, had an unusual, almost milky, cloudiness to the water. Upstream there is a bluff of rock – tilted Cambrian siltstones on a thick sill of intrusive rock, on top of which a large Douglas Fir stands. It was still there, but the first three metres of the trunk were thickly and tightly wrapped with all manner of debris – branches, tree trunks, lengths of two-inch alkathene pipe bright blue against the browns and buffs of splintered, bark-stripped wood. Scrambling up the tight cone of flood-borne debris, I turned and looked down. It was a good ten metres from the top of the debris-wrap down to the water. This was no ordinary flood. (Figure 14.10)

This extract is an account from author John Mason's visit to Afon Mawddach in the Coed y Brenin forest, near Dolgellau in Gwynedd, North Wales, on 5 July 2001. Two days previously a severe thunderstorm had affected the region, and had the area been well populated, the resultant flash flood would have made the national headlines, but instead it largely went unnoticed outside of the United Kingdom meteorological community. It served, in Mason's case, acutely to raise awareness of what can happen when the required meteorological ingredients come together. There are insufficient data available with which to classify the events on the evening of 3 July 2001. Whether a supercell or a multicell, it was much slower-moving than the 2012 example described previously. John Mason watched it approach the Machynlleth district where there was moderate flooding and some damage from large hailstones. The storm featured an impressive shelf cloud ahead of the precipitation together with copious thunder and lightning; however, none of these features are diagnostic of storm type. The synoptic pattern at 1800 GMT on 3 July 2001 involved a slack area of low pressure developing over the western United Kingdom, between the Azores High and an eastward-drifting anticyclone centred over Germany (Figure 14.11). A trailing cold front to the west of Ireland and a pre-frontal trough were moving slowly eastwards. By the following midnight a coherent low-pressure system had developed and was centred over and south-west of the United Kingdom, with a main centre over the English Channel (1006 mb).

Storms developed in the afternoon and tracked north-northeast across Wales to bring devastation to the area between Ffestiniog, Bala and Dolgellau. At Machynlleth 89 mm of rain fell in 3 hours, and down-track, beyond the devastated area, at Betws-y-Coed the day total was 105 mm. Neither area suffered anything like the damage caused in the main impact zone, which is dominated by five rivers – from north to south the Prysor, the Eden, the Gain, the Mawddach and the Wen. The last four all converge in the Coed y Brenin forest north of Dolgellau, beyond which is the Mawddach flood plain and in a short distance its estuary. As with the historic Cheviot cloudburst of July 1893 (Clark, 2005), there were few habitations in the area most affected by the 3 July 2001 storm, which was probably why the event was hardly mentioned in the news. Eyewitnesses along the Mawddach spoke of a wall of water and debris thundering down the valley, and debris wraps around surviving trees along its banks indicate that the water level rose widely to more than 5 m above normal and in some narrower gorges with sharp bends it locally reached 10 m. Many large trees were ripped from the banks and carried away, some to end up strewn on the river's flood plain. Debris accumulated beneath the toll bridge on the estuary at Penmaenpool, whilst other debris was carried all the way out into Cardigan Bay, which created a hazard to shipping. A number of bridges, some dating back to the 19th

Figure 14.10 *Aftermath of 3 July 2001, above Ganllwyd the river Mawddach flows through the Coed Y Brenin Forest, along a number of narrow gorges interspersed with wider, gentler sections. In one such gorge the river turns a bend; here the bank had been stripped up to 10 m above normal water level. In this image a surviving conifer tree, high on the bank, has a debris-wrap of smaller trees and branches. © John Mason.*

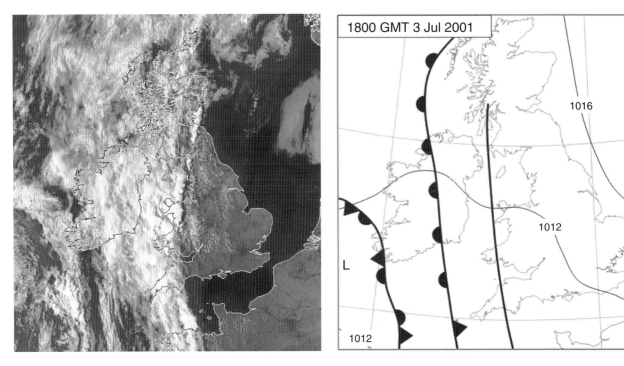

Figure 14.11 *Left: visible satellite image for 3 July 2001, 1630 GMT, indicating severe thunderstorm development across north Wales. © University of Dundee/Paul Brown. Right: synoptic chart for 1800 GMT 3 July 2001. Redrawn from the Meteorological Office surface analysis for the same time.*

century or earlier, were destroyed, and on the Afon Wen a sturdy footbridge completely disappeared and one of its 5 m steel girders was found, badly bent, 400 m downstream. It had successfully negotiated two gorges, a waterfall and a deep pool to arrive there. Given rainfall totals from relatively undamaged locations up- and down-track of the area worst affected by this storm, the rainfall required to inflict a flood of such magnitude must have been enormous. The nature of the damage was similar to that at Boscastle 3 years later, although that event was produced by another subtype in the thunderstorm family, the train echo.

A train-echo is a quasi-stationary line along which one convective storm forms after another. On a rainfall radar plot, the cells run along a line, like carriages moving along a railway line, hence the term. The focussing of convection along such a line is due to low-level convergence of airflows (though the subsequent release of convective instability may be at mid-levels). When the atmosphere is potentially unstable any slight difference in airmass properties, critically temperature, will let one airflow undercut and lift another, so that it reaches its level of free convection and the storm-forming process begins. Meanwhile, more air continues to feed into the convergence zone, thereby repeating the process over and again.

Such set-ups can deposit very significant rainfall totals in a matter of hours. Pike (1994) described a remarkable succession of thunderstorm cells which deposited 129 mm of rain in less than 5 hours in southwest Oxfordshire early on 26 May 1993. On this occasion the warm plume of air was subjected to simultaneous undercutting from both the southwest and northeast, along a quite narrow convergence zone of repeated cumulonimbus development. Dublin experienced a similar deluge from a slow-moving thunderstorm on 11 June 1963 (Morgan, 1971). The suburb of

Mount Merrion recorded 184 mm of rain that day, of which 83 mm fell in 65 minutes. The first thunder was heard rumbling over the Wicklow Mountains around 1130 GMT and the general storm movement subsequently appears to have been very slowly northwards, the storm being centred over the south of the city at 1400 GMT. Vivid lightning and 'ricocheting cannonades of thunder' accompanied the downpour (Anon, 1963). Many houses were flooded and vehicles were stranded in floodwaters 1 m deep. Charts at the time indicated a classic surface col across the United Kingdom and Ireland between lows to the east and west and anticyclones over the Norwegian Sea and the Azores while an upper trough was situated just west of Ireland. Forced ascent was also provided by a cold front which edged slowly east across Ireland during the day. Indeed, the broad-scale situation and slack surface gradients were similar to those associated with the 1975 Hampstead storm (which also developed in 'stagnant' air ahead of a front that was moving slowly eastward) and the cloudbursts in the upper Thames Valley on 9 June 1910 (see Chapter 7). In each case, the slack surface pressure would have been especially favourable to the development of local convergence zones promoting the development of heat lows and sea breeze fronts to supplement larger-scale ascent ahead of a surface front and upper trough.

14.6 Boscastle, Cornwall, UK, 16 August 2004

The convergence zone that hosted the train-echo that devastated Boscastle in 2004 resulted from a slack low pressure (centred off west Ireland, 1002 hPa, at 0000 GMT on the 16th and deepening to 994 hPa 24 hours later), which maintained an unstable, warm moist south-westerly airflow over the southern United Kingdom (Figure 14.12). Convective storms developed quite widely, but what influenced the Boscastle event was frictional backing of the wind over land, curving it round from southwesterly to south-southwesterly. Once it had backed so, it was converging with the southwesterly winds along the north Cornish coast, along a line that came ashore near Newquay and affected areas towards Barnstaple. Storms developed along the northern flanks of Bodmin Moor and, running along the convergence line, produced their most intense rainfall over the catchments that run down into Boscastle and neighbouring coastal villages before raining themselves out down-track.

Heavy rainfall began in the early afternoon and by just after 1500 GMT the Valency River was reported to be at bank height. By 1530, it was overtopping its banks. Half an hour later the first of a number of surges, some as much as 3 m in height, occurred (attributed by some locals to temporary dams of trees and other debris forming, ponding up water then giving way). The floods peaked around 1700 GMT, by which time a number of buildings had been badly damaged or destroyed and dozens of cars had been swept into the harbour or out to sea (Figure 14.13). At the peak, even the steeply inclined roads that climb out of the valley were like fast-flowing rivers. A major air-rescue effort was required to help people trapped in buildings to safety.

Rainfall totals for the day reached 200.4 mm to the east of Boscastle, at Otterham. Much of this fell over a 4-hour period which included some very high (but not record-breaking) rates. It was the cumulative total that did the most damage as cell after cell developed and moved along the convergence line over several hours (Doe, 2004; Burt, 2005). By contrast, locations barely 10 miles (16 km) south recorded less than 2 mm. In fact, much of Cornwall had a dry and fine sunny day, with a maximum temperature of 22.6°C (72.7°F). It is important to note that this flash flood was not the first time there had been such an occurrence in the region. Notable events had been recorded in 1957 (when 203 mm fell at nearby Camelford on 8 June), 1958, 1963, 1968 and 1981, to name but a few (HR Wallingford, 2005). However, this flash flood event was certainly one of the most severe the village has experienced.

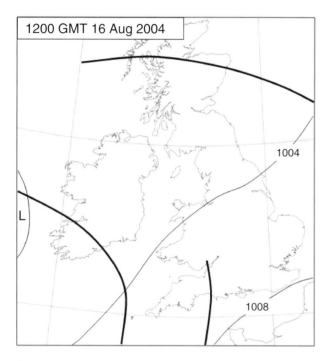

Figure 14.12 *Synoptic chart for 1200 GMT 16 August 2004. Redrawn from the Meteorological Office surface analysis for the same time.* © Paul Brown.

Figure 14.13 *Boscastle 16 August 2004, moment of impact, buildings flooded and cars being swept out to sea. From Doe (2004). Reproduced with permission from the International Journal of Meteorology.* © Art Mason.

While intense rainfall is essential to the onset of flash flooding, the drainage and topography of the surrounding area influence the scale and impact of an event. The proximity of high ground (Bodmin Moor) and morphology of Trevose Head and Pentire Point to the south-west of Boscastle played an important role in the prolonged *in situ* downpour. In environments like Boscastle, steep-sided valleys accentuate flooding by acting as huge funnels for run-off; they channel the water very quickly down to the sea. The high rainfall falling in such a short time could not be absorbed into the ground and instead surged through the village of Boscastle with great force and speed (Doe, 2004). The steeply incised narrow wooded river valleys that characterise this section of the Cornish coast are classic examples of what are often described as 'flashy' catchments, with very rapid run-off following high volumes of rainfall. The Valency and Jordan rivers that reach the sea at Boscastle, and the Ludon with its mouth at Crackington Haven, can become raging torrents (Mason, 2004).

In comparison with the Lynmouth flood (15 August 1952), the hydrological nature of the events was fairly similar, although the large loss of life was not repeated. The synoptic backround to the former was very different with a much larger area of heavy rain affecting southwest England (McGinnigle, 2002). Moreover, although the two events unfolded over broadly similar timescales (Boscastle, 5 hours: Lynmouth, 7 hours), a major factor in terms of mortality was that the Lynmouth flood occurred at night, whereas the misfortune that befell Boscastle happened in mid-afternoon. Floods at night automatically make rescue operations much more difficult and residents would have been much less aware of the developing situation, given that communications were much more limited in 1952 than in 2004, when by the time the flood was peaking, millions of people were watching it live on television (Doe, 2004, 2006).

14.7   Holmfirth, Yorkshire, UK, 29 May 1944

Communications were also restricted during the Second World War, which meant this severe event at Holmfirth was little commented on at the time. A detailed report of this event can be found in Doe and Brown (2005). The small town of Holmfirth, to the east of the southern Pennines, lies in the steep-sided valley of the River Holme, which trends from southwest to northeast to the town before bending northward to Huddersfield. Holmfirth itself is 400–500 feet (c. 130 m) above sea level, and the moors surrounding it rise above 1000 feet (300 m), peaking to the southwest at 1908 feet (582 m) on Black Hill. Its position would therefore make it liable to sudden floods when heavy rain falls on the hills. Indeed, such floods are recorded from the 18th century onwards, for example in 1738, 1777, 1852, 1867, 1933. That of 5 February 1852 was by far the most destructive for human life, but was in part man-made, when the accumulation of repeated frontal rains resulted in the bursting of the Bilberry Reservoir, which had been allowed to fall into disrepair: 81 people died in the resulting flood.

The last days of May 1944 (the week before D-Day) were very hot. Warm moist air in a southwesterly warm sector had dried out as the airflow backed to southeasterly, to produce some of the highest May temperatures ever recorded in this country, 91°F (32.8°C) being reached in parts of eastern England. By Monday

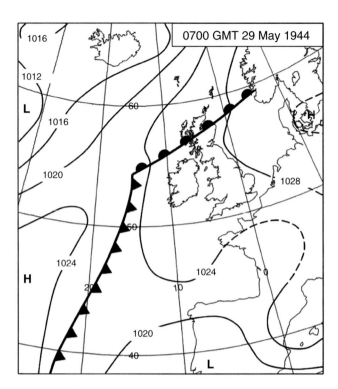

Figure 14.14   *Synoptic chart for 0700 GMT 29 May 1944. Redrawn from the Daily Weather Report of the Meteorological Office, including dashed lines where isobars were interpolated over occupied Europe. © Paul Brown.*

29 May a light anticyclonic easterly flow covered England associated with a high centre over Denmark. At the same time a shallow trough moved northwest from France across England (Figure 14.14), and despite barometric pressure being everywhere above 1020 hPa, this was sufficient to induce thunderstorms when accompanied by afternoon temperatures of 30°C or more (Figure 14.15).

Storms first appeared in central southern England early in the afternoon (Bower, 1944), then travelled north through the Midlands to reach their peak over west Yorkshire in the early evening, before dissipating. *The Times* of 9 June 1944 printed a report of the storms (wartime secrecy meant it could not be published earlier), and a collection of eyewitness accounts was later compiled by Beveridge (1982), who carried out a site investigation. After a brief early shower, the cloudburst broke in the late afternoon over the high moors west and southwest of Holmfirth, then descended into the town, where it reached its peak between 1530 and 1730 GMT (1730–1930 BDST[4]), before easing off by 1800 GMT. Intense thunder and lightning, hail, and strong squalls (possibly tornadic) accompanied the violent rainstorm. 'Three large black twisted columns' were reported 'coming to ground at Digley, Hinchliffe Mill and Hillhouse'. These are most likely descriptions of the rain core affected by turbulence on the edge of the updraft. The third column was described as 'a solid thunderspout 100 feet wide and composed of intense drops driven at high

---

[4] British Double Summer Time.

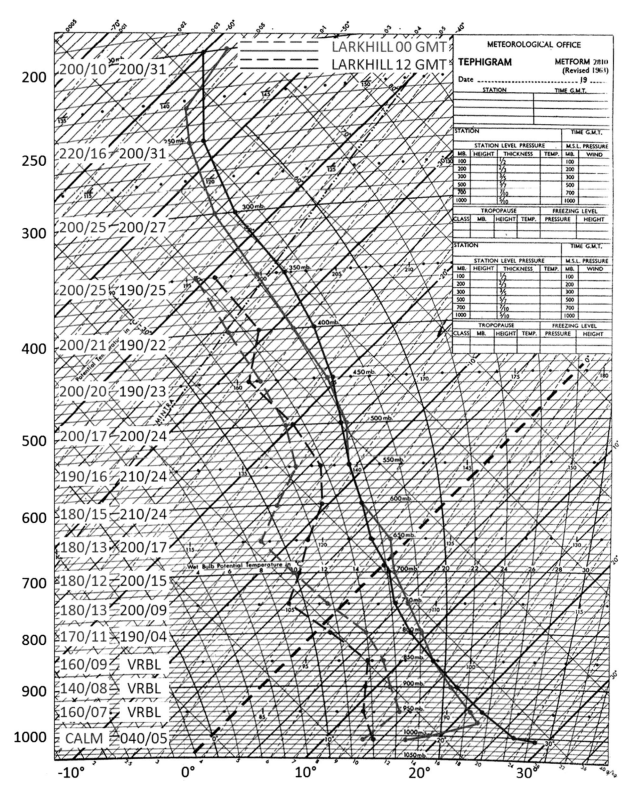

Figure 14.15 *Larkhill tephigrams (dry bulb and dew point temperatures) and upper winds for 0000 GMT (blue) and 1200 GMT (red) 29 May 1944. Drawn from data published in the Upper Air Section of the Daily Weather Report. © Paul Brown. (See insert for colour representation of the figure.)*

velocity by the wind'. Some significant lightning damage to properties is suggested, and hailstones 'the size of mothballs' fell at Meltham at 1615 GMT.

It was not long before reports of flooding were received. Water was described as, 'coming down opposite hillsides in a continuous sheet though stronger in some places than others, like the waves of the sea'. The Postmaster at Dewsbury described the moors as 'one sheet of water'. Some of this resulting flood went west down the Longdendale Valley, but the greater part ran east through the Holme Valley, where the river rose rapidly by 4–5 feet in a few minutes, eventually becoming a torrent 80 feet (24m) wide and over 15 feet (4.5 m) deep. As it rushed towards Holmfirth it burst its banks, and at one point changed its course to pass through a mill-yard, where it washed away sheds containing £10,000 worth of rabbit skins. It eroded the foundations of an adjacent main road (A6024), causing masses of stone and earth to fall into the river; and water pipes and a main sewer were also damaged. There was also much damage to roads farther up in the hills in the Marsden Clough area.

In Holmfirth itself, where the torrent arrived at about 1600 GMT (1800 BDST) (Figures 14.16 and 14.17), bales of wool were washed out of warehouses, and coping stones dislodged from walls, to be carried up to 4 miles downstream; several bridges, two banks, a grocery shop and a milliner's shop, together with mills and workshops, were either wholly or partly demolished by the water (Figures 14.18 and 14.19). About 200 houses were said to have been seriously damaged, and more than 100 people had to be evacuated from homes and shops which were flooded to a depth of up to 6 feet. At Bridge Road, Mr Norman Marsh saw the river rise behind Bottoms Mill and cover the fields in a matter of seconds. This was a result of the water having been temporarily

dammed up against the wooden buildings of Riverside Mills, which then collapsed. A powerful stream swept through the yard, and in a few minutes the stone walls on the southeast bank were 'collapsing like a pile of dominoes'. As the flash flood surged through the town it carried waterborne debris in its wake. Bales of cotton from the mills were thrown against houses.

Among those who were flooded out of their homes was a 76-year-old woman, who climbed onto her garden wall for safety. When the wall began to crumble, Geoffrey Riley, a 14-year-old boy, who had been watching the events with his father, went into the water to rescue the woman, but both of them became overwhelmed by its force. The father then waded in, in an attempt to rescue the other two, but while the three of them were edging their way to safety the wall suddenly collapsed, and they were swept along in the torrent. Geoffrey eventually managed to pull himself out, but his father and the woman were carried away to their deaths. Geoffrey Riley was later awarded the Albert Medal (subsequently replaced by the George Cross) in recognition of his gallantry. A third person in Holmfirth (a woman) was also reported to have drowned, but the circumstances are unclear.

Because intense convective rainfalls such as this are so localised there is often no official measurement of the maximum amount of precipitation produced by the storm. In the case of the Holmfirth storm, daily and short-duration values at rain-gauges in the general vicinity are presented in Table 14.3.

In connection with the flood at Glossop (on the opposite side of the moors from Holmfirth), *The Times* reported that between 5 and 6 inches (125–150 mm) had fallen in just over 2 hours. The authority for this statement is not known, nor are the exact whereabouts of the measurement or its reliability – but it sounds credible. The true maximum fall, however, cannot be known.

Figure 14.16 *Floodwaters surge through Hollowgate, Holmfirth, UK, on 29 May 1944. The image shows the church after the shops have collapsed. Image: © Bray & Son/Bamforth & Co. Ltd.*

Figure 14.17   *Floodwaters surge down through Victoria Square and Towngate, Holmfirth, UK, 29 May 1944. Image: © Bray & Son/Bamforth & Co. Ltd.*

Figure 14.18   *Collapsed buildings as a result of floodwaters at Victoria Bridge, Holmfirth, UK, 29 May 1944. © Bray & Son/Bamforth & Co. Ltd.*

As the flash flood surged through Holmfirth a number of records were made of the depth of the floodwaters at particular locations. Flooding started at 1530 GMT with water streaming down the surrounding hillside and creating two separate streams of floodwater. Between 1730 and 1745 GMT flooding of between 1 and 2 feet (0.3–0.6 m) was being experienced, and by 1755 GMT this had increased to 4 feet (1.2 m) in places. By 1830 GMT 6 feet (1.8 m) of floodwater was reported, and this soon rose to between

Figure 14.19 *Flash flood devastation outside the Valley Theatre, Holmfirth, UK, 29 May 1944. The picture shows debris and property damage at Market Walk. Image: © Bray & Son/Bamforth & Co. Ltd.*

Table 14.3 *Selected rainfall reports for 29 May 1944 in the south Pennines.*

Amount	Location	Source
4.50 inches (114.3 mm)	Glossop W.W. (Swineshaw)	British Rainfall 1944
3.10 inches (78.7 mm) (3.05 inches in 2 hours*)	Rhodes Wood Reservoir	British Rainfall 1944
2.62 inches (66.5 mm) (2.55 inches in 73 minutes*)	Hayfield (Kinder Filters)	British Rainfall 1944
2.33 inches (59.2 mm)	Ramsden Clough	TCO[†] report
1.94 inches (49.3 mm) (all in 2 hours)	Woodhead Reservoir	British Rainfall 1944
1.71 inches (43.4 mm)	Ramsden Clough	TCO report

*Classified as 'very rare'.
[†]Thunderstorm Census Organisation.

7 and 9 feet (2–3 m) around 1930 (2130 BDST). The maximum reported rise of the River Holme was between 12 and 14 feet (3.6–4.2 m); but following the storm it was claimed by locals that the river had risen to a height of more than 18 feet (5.4 m) in some places (Beveridge, 1982). It is interesting to note that the quoted rise of 10 feet (3.0 m) was '2 feet lower than in 1931'. The river was at its height between 1800 and 1850 GMT, and floodwaters started to abate (quite quickly) soon afterwards, following at least 4 hours of terrifying conditions. When the waters had subsided the fluvial landscape had changed considerably, some significant scouring having altered the bed and channel configuration.

14.8 Ottery St. Mary, Devon, UK, 30 October 2008

To conclude the accounts of extreme rainfall and their causes, we examine the back-building multicell storm of 30 October 2008 that caused severe flooding in and around Ottery St. Mary, a town in eastern Devon situated approximately halfway between Exeter and the Devon-Dorset border (Clark, 2011a and 2011b). The synoptic situation at midnight on 30th October involved a system of low-pressure centred just off southwest Wales at 989 hPa and an occlusion trailing down through southwest England to the Brest Peninsula with an associated moist southerly airflow followed by a cold northwesterly airstream cutting in behind it (Figure 14.20). Aloft was a deep pool of particularly cold air for the time of year, accompanied by a sharpening upper shortwave trough, setting the scene for strong convective instability and mass ascent of warm air along the frontal zone as it destabilised. By 0600 GMT on the 30th the low was centred off the south Devon coast, and the resultant rather slack airflow would ensure that any convective storms that did develop would be slow-moving.

The evening of the 29th October was wet, and according to an eyewitness report steady moderate rain fell until about 2345 GMT,

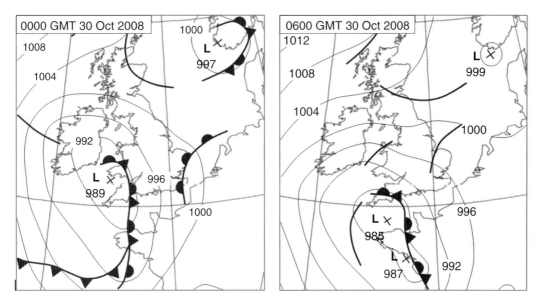

Figure 14.20   *Synoptic charts for 0000 and 0600 GMT 30 October 2008. Redrawn from the Meteorological Office surface analyses for the same times. © Paul Brown.*

Figure 14.21   *Accumulation of hailstones at Ottery St. Mary, Devon, UK, 30 October 2008. © Matt Clark.*

when the first flash of lightning was observed. By 0030 GMT rainfall was torrential, lightning was frequent and hail started to fall. By 0100 GMT precipitation was estimated to be about 50% hail. The storm continued until about 0300 GMT on the 30th.

The nearest Met Office rain gauge to Ottery, at Dunkeswell, provided data but the event total of 71 mm is considered to be unrepresentative of the storm as Dunkeswell was clearly situated outside the area of heaviest precipitation. However, two other sets of observations show how unusual this storm was. Firstly, analysis of 1 km × 1 km raw cumulative radar totals suggests that more than 200 mm fell over the upslope to the west of Ottery. Secondly, the Kings School, in the western part of Ottery, kept a rain gauge. This recorded 187 mm of rain between

0900 GMT on the 29th and 1200 GMT on the 30th of October. The data were verified as acceptable by the Environment Agency and are consistent with the radar estimates. Still more significantly, approximately 160 mm of this fell in just 3 hours, between 0000 and 0300 GMT on the 30th.

A significant part of this precipitation was hailstones, which, although small, fell in copious quantities over a prolonged period. Smaller hailstones are more buoyant and easily transported. There were reports of 'rafts' of hailstones being carried along by the floodwaters and piling up in low-lying areas to form banks up to 120 cm deep (Figure 14.21). There can be little doubt that the quantity of hailstones exacerbated what was already a serious situation. As they accumulated, local drainage systems were

Figure 14.22   *Accumulations of flood driven hailstones trap a car in Ottery St. Mary, Devon, UK, 30 October 2008. © Matt Clark.*

overwhelmed and in many cases blocked. Following the storm, as the vast accumulations of hailstones melted away, the floodwaters rose to as much as 1.5 m depth in places, the town was physically cut off from the outside world, and towards 0500 GMT many residents had to be evacuated, in some cases by air.

The severity of this incident was partly due to the synoptic situation, favouring strong, slow-moving convective storm development, but local topography played a role too. The southerly to south-southeast low-level airflow, bringing warm moist air from the English Channel, followed the north-south Otter valley (which enters the sea at Budleigh Salterton) and topographical gaps in the country to the south-east of the river. On reaching Ottery this low-level air would have been forced up over the higher ground to the west. However, the amount of forced ascent would not have been great, Ottery being about 40 m above sea level, whilst the higher ground to its west rises to little more than 100 m. It seems more likely that the slack pressure regime, with mesoscale circulations promoting a continued feed of warm, moist air into a zone of convective instability, allowed a back-building multicell storm to develop and last for several hours, until the circulation drifted out of the area and cut off its fuel supply in terms of warmth and moisture. Whatever the relative importance of these factors, the high rainfall total and exceptional hail fall (it has been estimated that 20–25 level centimetres of hailstones were produced) served to give the residents of this rural Devon town a memorable, and for some alarming night (Figure 14.22).

14.9   Concluding Remarks

How is our changing climate likely to affect extreme rainfall events across the United Kingdom and Ireland? That depends to a great extent on the regional response of synoptic weather-patterns to global warming. It is well-established that the climatic response to added greenhouse gases such as carbon dioxide and methane is an uneven one: for example, the Arctic is warming about twice as

fast as the rest of the world, a process that has been termed Arctic Amplification. Since the frequency and severity of floods are both perceived to have increased in recent years, many questions have been raised as to the reasons why, including changes in both climate and land-management (see also HR Wallingford, 2012). However, whilst the Clausius–Clapeyron relation clearly demonstrates that warmer air can carry more moisture, there are other factors to be taken into account.

Whilst the basic principles of climate change are well-understood, many of the details of regional responses to it are still being explored. For example, whilst in general terms the number of 'intense' rainfall events during summer months is thought likely to decrease, it is the intensity of the rain on the days that it does occur that is projected to increase when the requisite synoptic patterns are in place. While we cannot – at least currently – attribute any single weather event to climatic change, increasingly heavy summer rainfall events are in line with how climatic change could affect the weather of the United Kingdom.

Additionally, there are emerging signs that if Arctic sea ice declines further, allowing the Arctic Ocean to absorb more heat energy, the consequent long-term decline in the thermal gradient across the polar jetstream, slowing it down, will bring changes in its behaviour. Research has suggested that there is already evidence in some parts of the Northern Hemisphere for a shift away from zonality in favour of meridionality (Francis and Vavrus, 2012). In short, if the jetstream slows down, its longwave troughs and ridges will tend to push further south and north respectively. The overall result of such sluggish and increasingly meridional behaviour would be for weather-patterns to become 'stuck'. The consequences of such a synoptic pattern would depend critically on the phasing of the longwave troughs, which bring cold Polar air southwards, and the blocking ridges, which bring warm tropical air northwards. The bitter, prolonged cold weather of December 2010 across the United Kingdom showed what can happen if a trough is in the right place at the right time – right overhead in that case – with a blocking high occupying the

Eastern Atlantic and frigid Arctic air draining southwards down its eastern flank.

By contrast, the period December 2013–January 2014 saw a very active, zonal Atlantic with one major storm after another and a very strong polar jetstream overhead. During this period, a quasi-stationary blocking ridge was again involved, but in this case it was situated far away, over the western seaboard of North America, allowing very cold (well below 500 dam) Arctic air to spill deeply southwards down its eastern flank. Relatively unmodified, because it was mostly crossing high-latitude land, it met warm northbound air of tropical origin over the western Atlantic: the resulting steep thermal gradient across this airmass boundary spawned one major storm after another, accompanied by an intense jetstream containing winds reaching 200 knots at times. Clearly, then, the effects of any increase in meridionality will critically depend on where the blocking highs form and persist, with severe cold weather and prolonged zonality being two potential outcomes.

Prolonged zonality during the winter months can readily lead to flooding problems, not necessarily the result of any individual extreme rainfall totals but due to the frequency of the Atlantic storms, which in turn means widespread saturated ground and high run-off values. Added to this is the fact that in a warmer world air advected towards the United Kingdom from subtropical and tropical zones will potentially carry more moisture, setting the scene for much wetter conditions during periods of zonality. The worst situation in summer is a persistent upper trough just off our western seaboard, as occurred during parts of the very wet summer of 2007, because this not only leads to warm moist air advection from the Tropics but it facilitates convective activity, thereby encouraging thunderstorm development. Alternatively, prolonged ridging at any time of year brings with it the threat of drought. Clearly the response of the polar jetstream to Arctic Amplification will be of critical importance with respect to rainfall, and especially extreme rainfall patterns in the years and decades to come.

Acknowledgements

The authors express their appreciation to the following: Neil Lonie and Robert Moore for satellite imagery; Art Mason, Matt Clark, NetWeather, and Bray & Son – Bamforth & Co. Ltd. for pictures and illustrations; *The Journal of Meteorology* for text and image reproductions; and the many TORRO members who have contributed to reports of heavy rainfalls, flood events and site investigations over the years.

References

Acreman, M. (1989) Extreme rainfall in Calderdale, 19 May 1989. *Weather*, **44**, 438–446.

Acreman, M. C. and Collinge, V. K. (1991) The Calderdale storm revisited; an assessment of the evidence. British Hydrological Society, 3rd National Hydrological Symposium, Southampton, 16–18 September 1991, pp. 4.11–4.16.

Anon (1963) Dublin letter, 23 Pearce Street, Tuesday 11 June 1963. Cork Examiner, 12 June 1963.

Beveridge, J. P. (1982) *Like Waves of the Sea: Holmfirth Flood 1944.* [Typeset I.B., Halifax]. Superprint, Wakefield, 16pp.

Bleasdale, A. (1974) 1968: an outstanding year for multiple events and exceptionally heavy and widespread rainfall. *British Rainfall*, **1968**, 223–231.

Bower, S. M. (1944) Whit-Monday's Thunderstorm [Unpublished Report in the Archives of the Thunderstorm Census Organisation].

Bridge, G. C. (1969) The record breaking rainfall in southwest England on 28–29 July 1969. *Meteorol. Mag.*, **98**, 364–370.

Burt, S. D. (2005) Cloudburst upon Hendraburnick Down: the Boscastle storm of 16 August 2004. *Weather*, **60**, 219–227.

Cinderey, M. (2004) The storm with UK-record rainfall intensity at Carlton-in-Cleveland, 10 August 2003. *Convection* 4, TORRO, South Molton, Devon, 2004.

Cinderey, M (2005) The North Yorkshire – Teesside storm of 10 August 2003. *Weather*, **60**, 60–65.

Clark, C. (2005) The cloudburst of 2 July 1893 over the Cheviot Hills, England. *Weather*, **60**, 92–97.

Clark, M. R. (2011a) The 'Ottery St Mary' hailstorm of 30 October 2008: damage survey results and eyewitness accounts. *Int. J. Meteorol.*, **36**, 75–90.

Clark, M. R. (2011b) An observational study of the exceptional 'Ottery St Mary' thunderstorm of 30 October 2008. *Meteorol. Appl.*, **18**, 137–154.

Clark, C. and Pike, W. S. (2007) The Bruton storm and flood after 90 years. *Weather*, **62**, 300–305.

Clark, M. R. and Webb, J. D. C. (2013) A severe hailstorm across the English Midlands on 28 June 2012. *Weather*, **68**, 284–291.

Doe, R. K. (2004) Extreme precipitation and run-off induced flash flooding at Boscastle, Cornwall, UK – 16 August 2004. *J. Meteorol.*, **29** (293), 319–333.

Doe, R. (2006) *Extreme Floods. A History in a Changing Climate.* Sutton Publishing, Stroud.

Doe, R. K. and Brown, P. R. (2005) A sea on the moors. *J. Meteorol.*, **30** (299), 163–173.

Eden, P. (2008) *Great British Weather Disasters.* Continuum Books, London.

Eden, P. and Burt, S. (2010) Extreme rainfall in Cumbria, 18–20 November 2009. *Weather*, **65** (1), 14.

Francis, J. A. and S. J. Vavrus, (2012) Evidence linking arctic amplification to extreme weather in mid-latitudes. *Geophys. Res. Lett.*, **39**, L06801.

Glasspoole, J. (1925) The unprecedented rainfall at Cannington, August 18th, 1924, in: Meteorological Office (1925). British Rainfall 1924. HMSO, London, pp. 246–255.

Graham, E. (2006) 200 mm falls in Ireland. *Weather*, **60**, 151.

Grahame, N., Riddaway, R., Eadie, A., Hall, B., McCallum, E. (2009) Exceptional hailstorm hits Ottery St Mary on 30 October 2008. *Weather*, **64**, 255–263.

Hand, W. H., Fox, N. I., Collier, C. G. (2004) A study of twentieth-century extreme rainfall events in the United Kingdom with implications for forecasting. *Meteorol. Appl.*, **11**, 15–31.

Hanwell, J. D. and Newson, M. D. (1970) The great storms and floods of July 1968 on Mendip. *Occ. Pub. Wessex Cave Club* Ser. 1, (2), Wells, UK.

HR Wallingford (2005) Flooding in Boscastle and North Cornwall, August 2004, Phase 2 Studies Report, Report EX5160, Release 1. Wallingford, Oxford.

HR Wallingford (2012) The UK Climate Change Risk Assessment 2012. Report to Parliament, January 2012. Available at: http://randd.defra.gov.uk/Default.aspx?Module=More&Location=None&ProjectID=15747 (accessed 26 April 2015).

Jackson, M.C. (1977) Mesoscale and small-scale motions as revealed by hourly rainfall maps of an outstanding rainfall event, 14–16 September 1968. *Weather*, **32**, 2–16.

Jackson, M. C. (1979) The largest fall of rain possible in few hours in Great Britain. *Weather*, **34**, 168–175.

Keers, J. F. and Westcott, P. (1976) The Hampstead storm, 14 August 1975. *Weather*, **31**, 2–10.

Law, F. M. (1968) The flood of 8 August 1967. Report of the Fylde Water Board. Clitheroe, Lancashire.

Law, R. (1989) An Investigation of the Heavy Rainfall of 5–6 February 1989 in the Ness, Beauly and Conon catchments. A Report to the North of Scotland Hydro-Electric Board, Institute of Hydrology, Wallingford, Oxfordshire.

Mason, J. (2004) The influence of landscape and geology on the Boscastle flood, 16 August 2004. *J. Meteorol.*, **29**, 315–318.

Mayes, J. C. (1986) Charley comes to Wales: 25 August 1986. *J. Meteorol.*, **11**, 295–299.

McGinnigle, J. B. (2002) The 1952 Lynmouth floods revisited. *Weather*, **57**, 235–242.

Meskill, D. (1986) Ex-Hurricane Charley drenches eastern Ireland. *J. Meteorol.*, **11**, 318.

Meteorological Office (1950) British Rainfall 1943, 1944, 1945. HMSO, London, pp.62–63.

Meteorological Office (1950) British Rainfall 1948. HMSO, London, Part 1, pp. 42–45.

Meteorological Office (1963) British Rainfall 1959–1960. HMSO, London, pp. 61–62, 67.

Meteorological Office (1970) *Monthly Weather Report June 1970.* HMSO, London.

Meteorological Office (1972) British Rainfall 1964. HMSO, London, p. 74.

Meteorological Office (2011) Eskdalemuir flood. Available at: http://www.metoffice.gov.uk/media/pdf/b/1/Eskdalemuir_Flood_-_26_June_1953.pdf (accessed 17 March 2012).

Meteorological Office (2013) British Rainfall archive 1860–1968. Available at: http://www.metoffice.gov.uk/archive/british-rainfall (accessed 21 May 2015).

Meteorological Office, HMSO (1953) Thunderstorms of June 26, 1953 (Notes and News). *Meteorol. Mag.*, 82, 344–347. Available at: http://www.metoffice.gov.uk/climate/uk/interesting/sep2008 (accessed 7 April 2015).

Mill, H. R. (1911) *British Rainfall 1910.* Edward Stanford, London.

Mill, H. R. and Salter, C. (1918) British Rainfall 1917. Edward Stanford, London, Part 1, pp. 22–31.

Morgan, W. A (1971) Rainfall in the Dublin area on 11 June 1963. *Irish Meteorological Service Internal Memorandum.* Available at: http://www.met.ie/climate-ireland/weather-events/June_1963_Thunderstorms.pdf (accessed 7 May 2014).

Pike, W. S. (1994) The remarkable early morning thunderstorms and flash flooding in central southern England on 26 May 1993. *J. Meteorol.*, **19**, 43–64.

Prior, J. and Beswick, M. (2008) The exceptional rainfall of 20 July 2007. *Weather*, **63**, 261–267.

Ross, N. A., Webb, J. D. C. and Meaden, G. T. (2009) Daily rainfall extremes for Great Britain and Northern Ireland part two. *Int. J. Meteorol.*, **34**, 75–81.

Rowsell, E. H. (1963) Storms of 5 September 1958, in: Meteorological Office (1963) British Rainfall 1958. HMSO, London, Part 3, pp. 15–21.

Sibley, A. (2010) Analysis of extreme rainfall and flooding in Cumbria 18–20 November 2009. *Weather*, **65**, 287–292.

Symons, G. J. (1894) British Rainfall 1893, Rainfall and Meteorology of 1893. Edward Stanford, London, pp. 24–25.

Webb, J. D. C. (2011) Violent thunderstorms in the Thames Valley and south Midlands in early June 1910. *Weather*, **66**, 153–155.

Webb, S. (2013) Heavy rain and flooding in and around Aberystwyth on 8–9 June 2012. *Weather*, **68**, 162–168.

Webb, J. D. C. and Pike, W. S. (2010) Extreme thundery rainfall event in central and southern England, 19–20 July 2007. *Int. J. Meteorol.*, **35**, 183–195.

Webb, J. D. C. and Pike, W. S. (2014) Extreme thundery rainfall event over south east England 20–21 September 1973. *Int. J. Meteorol.*, **39**, 42–53.

Woodley, K. E. (1981) Exceptional rainfall of 1 August 1980 over the North Antrim Plateau. *Meteorol. Mag.*, **110** (1), 227–227; Meteorological Office, Bracknell.

15

Heavy Snowfalls Across Great Britain

Richard Wild

WeatherNet Ltd, Bournemouth, UK Tornado and Storm Research Organisation (TORRO)*, Bournemouth, UK

15.1 Introduction

Snow is solid precipitation, which occurs as minute ice crystals at air temperatures well below 0°C and as larger snowflakes at air temperatures near 0°C (Meteorological Office, 1991). Only 5% of the Earth's precipitation falls as snow (Hoinkes, 1968), and half of this falls into the ocean. On average 26% of the Earth's land surface is covered by snow or ice, ranging from about 39% in February to 11% in August. Snowfall, like any other weather element, fluctuates irregularly. Heavy snow impedes outdoor activity and is regarded as a danger (Chapman, 1947; Lester, 1948; Chandler and Gregory, 1976), adversely affecting transport and society in general; however this depends on snow depth, density, wetness and hardness, as well as exposure, topography and elevation (Perry, 1981).

The hazards created by heavy snow often make newspaper headlines. The depth of snow and length of time it remains on the ground have social, economic and environmental importance. Furthermore, the timing of heavy snowfalls (e.g. rush hour) can be critical in determining the effect on society (Ahrens, 1991). Much technology exists to counter these problems. All too frequently in Great Britain, however, the problems are caused by lack of preparation, out-of-season snowfall, the area it affects, or its unexpected intensity (Manley, 1974; Brown, 1987).

This chapter explores the main associations between Heavy Snowfall Events (HSEs) and frontal systems. It also examines spatial or geographical aspects of HSEs over Great Britain. For example, the origin and dissipation (i.e. the trajectory) of every depression associated with HSEs from 1870 to 1999 across the North Atlantic Ocean sector 40–65°N and 20°W–8°E are examined. These trajectories are scrutinised over monthly, seasonal and decadal periods. Finally, the spatial pattern of HSEs across Great Britain in relation to Lamb Weather Types (LWTs) is investigated.

*http://www.torro.org.uk/

15.2 Definitions

There is no general consensus as to what is meant by a 'snowstorm' or a 'blizzard'. For a fuller discussion of the evolution of the term 'blizzard', the reader is referred to selected publications by Wild (1995, 1996a, 1996b, 1997, 2005a). As the official and author's own definitions of a 'blizzard' have not been adopted in this chapter and no official definition of a heavy snowfall day (HSD) exists,[1] the following criteria have been adopted to identify heavy snowfall days (HSDs) between 1861 and 1999:

1. Thirteen centimetres (5 in.) or more of snow must have fallen somewhere in lowland Great Britain in 24 hours (not accumulated depths). The snow depth of 13 cm was chosen based on the literature cited for snowfall that interfered with normal life.
2. A particular snowfall has been described in the literature as a blizzard or a snowstorm (where no snow depth was known).
3. The snowfall has been described in the literature as heavy. For it to be included, the term 'heavy' is only used when the two criteria above are not known, especially in the early years of this study when the word 'blizzard' was not officially used or when snow depths were not recorded (Wild, 2005a).

15.3 Synoptic Systems and Heavy Snowfalls

While the formation of snow can be attributed to microscale and mesoscale processes, the origin and distribution of snowfalls are determined at the mesoscale and synoptic scale, combined with the effects of relief and altitude. The synoptic origins of snowfall in Great Britain are related to the same low-pressure systems

---

[1] Only an hourly heavy snowfall rate of ≥4 cm/hour is officially recognised (Meteorological Office, 1991).

Extreme Weather: Forty Years of the Tornado and Storm Research Organisation (TORRO), First Edition. Edited by Robert K. Doe.
© 2016 John Wiley & Sons, Ltd. Published 2016 by John Wiley & Sons, Ltd.

that bring rainfall, that is the North Atlantic depressions. Compared with rainfall, snowfall is much less dependent on convective storms, although convection can liberate and enhance snow showers on the edges of frontal systems such as cold fronts (Ahrens, 1991; Carlson, 1991; Hulme and Barrow, 1997).

HSEs within depressions are mostly associated with the occluded and warm fronts and surprisingly, least of all with the cold front (Wild *et al.*, 1996). The mid-latitude depressions and their frontal systems are in turn intensified by the presence of the circumpolar vortex and jet stream (Chandler and Gregory, 1976; Uccellini and Kocin, 1987; Ahrens, 1991; Carlson, 1991), which are at their most intense in the winter months, when most HSEs occur. One other synoptic system that is important for snowfall is the 'polar low'.[2]

15.4 Snowfall Climatology of Great Britain

The number of days with snow cover across Great Britain rises with altitude at the rate 1 day for every 15 m above 60 m, partly on account of lower temperatures and partly because of orographic effects, especially where there is exposure to northerly and easterly winds. The deeper the snow on the ground, the longer it will take to melt. Other factors include frequency, intensity, proximity to primary source of moisture and topography (Bonacina, 1915; Manley, 1940; Chapman, 1947; Ward and Robinson, 1990). Over most of Great Britain the snow season is from December to March; however, there are sporadic falls in November and April or May, especially across northeast Scotland (Parker *et al.*, 1992).

When snow falls across Great Britain it affects places differently, especially when there is a change of air mass. Snow can take the form of convective showers or squalls giving brief, localised accumulations or frontal cyclonic snow, which is longer lasting and usually more widespread. The sea temperatures around the British Isles exert significant influences on the amount and distribution of snowfall. The sea determines the rate of transfer of sensible and latent heat between it and eastward-moving depressions. The amount, intensity and geographical distribution of snowfall over coastal areas are determined by local relief, and by the stability and moisture content of the lower atmosphere as a result of the air mass trajectory. Wind direction and exposure are the principal controls on the mesoscale modification of snowfall. The sharpest transformation of prevailing weather over short ranges is that between the windward and leeward sides of the hills, for example the Highlands, the Pennines and the Snowdonia range (Belasco, 1952). Cloud amounts, the upper limit of clouds and the frequency and heaviness of snowfall from them are all increased on the windward side where the air is forced to ascend. In upland and lowland areas alike, it is the degree of exposure or shelter in a given air stream that governs much of the regional distribution of snowfall over Great Britain. The descending air over hills in a westerly airflow warms causing a rain-shadow effect, such as on the Moray Firth, resulting in less snowfall (Brooks, 1954). Total snowfall at any location will be determined by the air mass associated with it, as different air masses generally have different properties. Different places therefore receive differing proportions of a snowfall in any given synoptic type (Lamb, 1988).

The sheltered and exposed areas therefore shift according to the direction of the airflow. For example, in an easterly airflow the largest snowfalls might occur on east-facing slopes. In addition, whatever the airflow type, proximity to cyclonic systems will increase the snowfall if other conditions are suitable. Snow falling across southern areas of Great Britain is usually associated with an easterly or south-easterly airflow round a depression over northern France and a frontal system moving slowly from the south-west (Jenkinson and Collison, 1977). Other occurrences of snow are caused by a dry easterly air stream producing coastal instability in eastern districts as it crosses a warm sea. The warmth of the North Sea, like that of other waters surrounding the British Isles, is maintained by a continuous supply of warm water from tropical sources brought in by the Gulf Stream/North Atlantic Drift and the predominant southwest wind. Another significant cause of snow across Great Britain is where cold air in a depression moves southward across the country (Musk, 1989). This is usually associated with an Arctic Continental or Polar Continental air mass. In Scotland and other northern areas, snow can also fall in a northerly air stream, sometimes associated with a polar low (Chapman, 1947; Manley, 1969).

15.5 Sources of Data

The data sources and methodology contained in this chapter represent the outcome of data collection over a 10-year period (completed in 2005). The main aims were to make a detailed spatial and temporal analysis and to develop a comprehensive, original historical record, of all HSEs across Great Britain between 1861 and 1999. This was achieved by using observational/ documentary sources and instrumental records. Documentary sources include academic journals such as *Weather*, *The Journal of Meteorology* and the *Meteorological Magazine* and other publications and miscellaneous material including *British Rainfall*, the *Snow Survey of Great Britain*, personal conversations, newspaper cuttings and letters. Instrumental data have been gathered from meteorological station records or from publications such as the Daily, Monthly and Annual Weather Reports (Wild, 2005a). The year 1861 was chosen as the first year of this study as it coincided with the start of LWTs and the first complete year of the Daily Weather Report of the Meteorological Office.

15.6 Snow Depths and Days with Snowfall

A report of sleet or snow falling on a particular day refers to the 24-hour period beginning at 0000 GMT (Meteorological Office, 1991). A day of snowfall could be a day of a heavy snow lasting many hours or a day that produces merely a few flakes for a minute or so. It is difficult to determine the average number of days with sleet or snow falling, because it is difficult to keep a full 24-hour watch for the isolated snowflake (particularly during the night). Snow falling does not necessarily imply that snow was accumulating on the ground (Wild, 2005a).

A day of snow lying is defined as a day on which undrifted snow covers half or more of the ground to a depth of 0.5 cm or more at 0900 GMT in an open-level area easily visible from the meteorological station and as near as possible to the rain gauge (Meteorological Office, 1991). Ideally the mean of three such measurements at different locations within the meteorological site

---

[2] For a fuller discussion of a 'polar low', the reader is referred to a selection of publications (Eady, 1949; Stevenson, 1968; Suttie, 1970; George, 1972; Lyall, 1972; Bignell, 1981; Carlson, 1991; Pike, 1996; Wild, 2005a).

should be taken. Its measurement as rainfall (i.e. its water content) may be made in a suitable snow gauge (a device for the retention and measurement of snow) or by melting the snow caught in a normal rain gauge. Thirty centimetres of freshly fallen snow has about the same water content as 25 mm of rainfall. The amounts of snow that fall over Great Britain are measured as rain; separate statistics of snowfall amount are therefore not available. Moreover, it needs to be noted, that the amount of snow that is collected in the snow/rain gauge may not reflect the true amount if strong winds are blowing at any time during the snowfall (Wild, 2005a).

The snow on the ground could be a result of fresh snow that has fallen during the meteorological day or of snowfalls from earlier days or a combination of the two. True snow depth from one particular snowfall can therefore be somewhat subjective as found in the results collated of this research. Snow depth is measured by placing a graduated ruler vertically in level snow in a place free from drifting. The number of days of snow lying is governed by the amount of snow that falls and the rate it melts. It is also governed by the nature of the underlying surface on which the snow is falling and by the subsequent air and ground temperatures and radiation conditions. Snow may have fallen and lain on the ground during the HSE but completely melted before the following morning observation and therefore remain unrecorded. Statistics of snow lying in Great Britain refer only to occasions when it is present at morning observation (0900 GMT). The aforementioned limitations therefore result in a shortfall of recorded days with snow lying (Wild, 2005a).

Prior to the 1940s, snow lying and snow depth were only occasionally recorded; however, it became standard practice thereafter to record these parameters at the 0900 morning observation. Before 1920, we cannot say much about snow cover as no consistent records were kept. Of all the elements that make up the British climate, the frequency of snow falling and even more of snow lying show the widest variations (Wild, 2005a).

The trajectories of each depression associated with every HSE were manually recorded by viewing synoptic charts in the Daily Weather Report and Daily Weather Summary. The following technique was employed to map the trajectories. The timing of the active stage of each HSE was determined by using historical and instrumental records as mentioned above. From this information, the origin and dissipation of each depression associated with a HSE were identified by tracing its progress on each synoptic chart before and after the snowfall. This information was then transposed onto a 1° latitude by 1° longitude grid of the North Atlantic Ocean. The area of the grid was 40–65°N by 20°W/8°E, this being determined by the areas of the synoptic charts in the *Daily Weather Report* and *Daily Weather Summary*. Thus, the grid is only an approximation of the actual origin and dissipation of each depression, since a small number began and ended outside the chosen grid (Wild, 2005a).

15.7 Spatial Methodology for Heavy Snowfall Events

15.7.1 LWT Catalogue

One of the main techniques chosen in this chapter to define the heavy snowfall conditions of Great Britain is 'weather-type analysis'. This is used in preference to air masses (Belasco, 1952; Brooks,

1954; Chandler and Gregory, 1976), because of its more precise geographical definition. One useful but subjective weather type system is that devised by Professor Hubert Lamb. Lamb's Weather Type Classification and Catalogue for the British Isles is one of his most lasting memorials to climatology (Perry and Mayes, 1998). It is an unequalled technique for studying the atmospheric circulation patterns around the British Isles and is one of the most regularly quoted documents in the field of meteorology.[3]

First developed in 1950, it was revised in 1972 to include a classification of each day's weather since 1861 (Lamb, 1950, 1972). From 1972, it was updated by Lamb himself in various publications until it ended on 3 February 1997, shortly before his death, since when, the Hadley Centre has maintained the catalogue (Jones *et al.*, 1993). This classification is the longest daily listing of airflow types for any part of the world, in which each day is subjectively classified on the basis of surface atmospheric pressure patterns (O'Hare and Sweeney, 1993).

LWT recognises seven basic atmospheric circulation types; five of which are directional (easterly (E), northerly (N), northwesterly (NW), southerly (S) and westerly (W)) and two types of which are circulatory (anticyclonic (A) and cyclonic (C)). Hybrid types combining a directional and a circulatory type (19 in total, e.g. cyclonic south or CS) are used when appropriate (O'Hare and Wilby, 1995). If none of these types can be defined, then the day is termed 'unclassified' (U) (Jones and Kelly, 1982).

15.8 Heavy Snowfall Events over Great Britain

The research identified 609 HSEs covering 1137 HSDs over Great Britain between 1861 and 1999 based on the above criteria (Wild, 1996c, 1998, 1999, 2001a, 2004b, 2005a; Wild *et al.*, 2000). This may not be a complete list (the results are shown in Table B.1 (See Appendix, www.wiley.com/go/doe/extremeweather)). Some dates when the criteria above could have been met might have been excluded owing to insufficient information. In contrast, certain HSEs have been stated as 'heavy snowfalls' in the media when this was clearly incorrect. Out of the 609 HSEs, the largest 10 (Table 15.1) within the period 1861–1999 (by snow depth) are highlighted here as case studies.[4]

Table 15.1 *The largest 10 heavy snowfall events during the years 1861–1999 (by snow depth).*

Date of HSE	Snow Depth (cm)	Location of Highest Snow Depth Recorded
15–16 February 1929	200	Dartmoor
9–13 March 1891	150	Dartmoor
18–19 January 1881	120	Dartmoor
26–30 January 1940	120	Sheffield, South Yorkshire
18–20 February 1941	120	Consett, County Durham
20 February 1865	90	NE Scotland
11–12 January 1913	90	Perthshire
25 December 1923	90	Aberdeenshire
26–27 March 1941	90	Northern Scotland
8 February 1955	90	Northern Scotland

[3] Lamb Weather Type examples, benefits and advantages are given in Lamb (1972), Mayes (1991), Jones *et al.* (1993), Sweeney and O'Hare (1992), O'Hare and Sweeney (1993) and O'Hare and Wilby (1995).
[4] For detailed descriptions of the HSE history of Great Britain during this period, please see Wild (1995, 1996a, 1996b, 1997, 1998, 2005a).

15.9 Heavy Snowfalls 1861–1869

The decade 1861–1869 had 58 HSEs covering 102 HSDs. This decade experienced the highest number of HSEs of all decades from 1861 to 1999. It also had the highest number of HSEs (22) and HSDs (36) in January and (together with 1940–1949 and 1950–1959) the equal lowest number of HSDs (2) in November. In addition, it had the highest number of HSEs (32) lasting just a single day and was the only decade to have a HSE lasting seven consecutive days. Other interesting statistics from this decade are that 1864 (together with 1867 and 1985) had the equal highest number of HSDs (8) in January of all years from 1861 to 1999 and 1863 (together with 1947 and 1985) had the equal highest number of HSDs (14) in winter. The year 1862 was the only year within this decade not to have a HSD in March and spring, while 1868 was the only year without a HSD in winter. Of the largest 10 HSEs (by snow depth), one occurred in this decade: this was 20 February 1865 (Wild 1998, 2005a).

Very little has been written about the major snowfalls of this decade, mainly because few academic journals existed at the time. Most of the cases here were obtained from notes and summaries in the *Daily Weather Report*, *British Rainfall* and the *Meteorological Magazine*. A significant case study is outlined in the following (similar examples then follow each decade section where applicable).

15.9.1 20 February 1865

On the 20 February 1865, snow fell to a depth of approximately 90 cm in northeast Scotland, where drifts of 3 m blocked roads.

15.10 Heavy Snowfalls 1870–1879

This decade had the highest number of HSEs (33) and HSDs (14) in March, more than any other decade between 1861 and 1999. It also had the highest number of HSEs (13) and HSDs (27) in December. Other interesting statistics from this decade are that 1876 had the highest number of HSDs (25) of all years from 1861 to 1999. This was also the year with the highest number of HSDs in March (11) and spring (16) and (together with 1998) the equal highest in April (5) over the same period. The year 1875 was the only year within this decade not to have a HSE in spring, while 1877 was the only year not to have one in March. Of the largest 10 HSEs (by snow depth), none occurred in this decade (Wild, 1998, 2005a).

15.11 Heavy Snowfalls 1880–1889

The decade 1880–1889 had 34 HSEs covering 65 HSDs. This decade had the lowest number of HSDs (8) in February of all decades from 1861 to 1999. It also had the highest number of HSEs (2) and HSDs (3) in October. Other interesting statistics from this decade are that 1880 and 1883 had no HSE in winter, while 1880, 1882 and 1884 had none in spring. The year 1886 was the only year in this decade to have a HSE during April and May, while 1888 was the only year to have one during June (equally with the year 1986). Of the largest 10 HSEs (by snow depth), one occurred in this decade: this was on 18–19 January 1881 (Wild, 2005a).

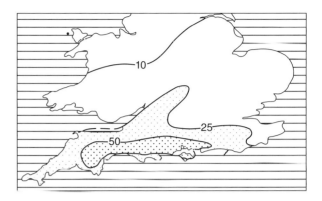

Figure 15.1 *Snow depths (cm) after the snowstorm of 18–19 January 1881. From Stirling (1997). Reproduced with permission of Giles de la Mare Publishers.*

15.11.1 18–19 January 1881

The year 1881 started with the second worst snowstorm of the 19th century during 18–19 January (Figure 15.1). This snowstorm affected the whole of England, with the exception of the four northernmost counties. It was most severe in southern England and the Midlands, depositing at least 30–40 cm of powdery snow from East Sussex to Cornwall.[5] Devon, Dorset and Wiltshire were among the counties worst hit and Hampshire and the Isle of Wight received over 60 cm (Eden, 1995). Conditions in Surrey, East and West Sussex, Kent, Essex, Dorset and Norfolk are discussed in detail by Davison and Currie (1991), Ogley *et al.* (1991a, 1991b), Currie *et al.* (1992), Ogley *et al.* (1993) and Ching and Currie (1997). The storm possessed defined warm and cold fronts, the warm sector being of tropical air, which had been drawn from low latitudes over the North Atlantic Ocean (Brazell, 1937). Many birds died as their food supplies became snow-covered and the storm also caused the deaths of 100 people, while in Plymouth, Devon, people were deprived of water for a week (Sowerby Wallis, 1881, 1882; Greely, 1888; Bowen, 1969). Snowdrifts 4½ m high completely buried trains and several people died in Berkshire (Currie *et al.*, 1994), while vehicles were buried by snow in Croydon, London (Davison and Currie, 1991). Snowdrifts 3–5 m high were found in the streets of Evesham, Worcestershire (Field, 1975; Jackson, 1977), Oxford Circus, London (Brazell, 1968; Mabey, 1983) and Portsmouth, Hampshire (Spink, 1947), and many ships in the English Channel were sunk (Eden 1995). Reported snow depths included 45 cm at Brighton, East Sussex, 35 cm at Exeter, Devon and 120 cm on Dartmoor (Stirling, 1997).[6]

---

[5] For further details please see: Bonacina (1928), Manley (1971), Burt (1978), Jackson (1978), Stirling (1997), Webb (1982), George (1990), Davison *et al.* (1993), Horton (1995), Hulme and Barrow (1997) and Wheeler and Mayes (1997).

[6] For accounts of how the snow affected the railway companies see: Anonymous (1881a, 1881b, 1882), Bell (1881), Birt (1881), Dykes (1881), Eddy (1881), Fenton (1881), Grierson (1881), Johnson (1881), Knight (1881), Mills (1881), Neele (1881), Noble (1881), Oakley (1881), Scott (1881) and Underdown (1881). Other references to this snowstorm include Sowerby Wallis (1882), Rand Capron (1883), Burder (1887), Bonacina (1917, 1928), Brooks (1954), Manley, (1969), Hardy *et al.* (1982), Wilson (1990) and Wild *et al.* (1996).

15.12   Heavy Snowfalls 1890–1899

The decade 1890–1899 had 27 HSEs (the lowest of any decade from 1861 to 1999) covering 68 HSDs. It had the lowest number of HSEs (4) in February, but also the highest number of HSDs in February and May (16 and 2, respectively); it also had the lowest number of HSEs (5) lasting just a single day. Other interesting statistics from this decade are that 1892 was the only year in the decade to have a HSE in April, while 1891 was the only year to have one in May. The years 1891 and 1896 had no HSE in winter. Of the largest 10 HSEs (by snow depth) in the period 1861–1999, one occurred in this decade: this was on 9–13 March 1891 (Wild, 2005a).

15.12.1   9–13 March 1891

The year 1891 had the worst snowstorm of the 19th century (by snow depth) and the second worst in the period 1861–1999 (Figures 15.2 and 15.3). Between 9 and 13 March, heavy fine

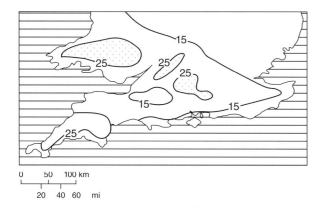

Figure 15.2 *Snow depths (cm) 9–13 March 1891. From Stirling (1997). Reproduced with permission of Giles de la Mare Publishers.*

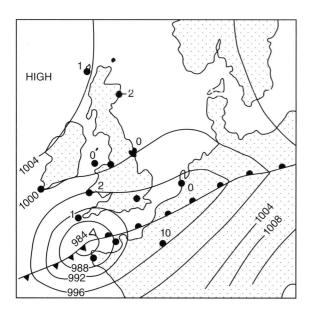

Figure 15.3 *Synoptic chart 10 March 1891 at 08:00. From Stirling (1997). Reproduced with permission of Giles de la Mare Publishers.*

powdery snow and strong easterly winds raged over SW England, southern England and Wales, where over half a million trees were blown down as well as many telegraph poles.[7] This storm caused much damage to transport, especially train services, many trains being buried in snowdrifts in SW England. Eden (1995) commented that the storm killed over 6000 sheep in Devon and Cornwall and over 200 lives and 63 ships were lost in the English Channel. A man was found dead at Dorking, Surrey, while drifts of 3½m were recorded at Dulwich, London and Dartmouth, Devon. At Torquay and Sidmouth, Devon over 30 cm of snow fell (Davison and Currie, 1991). On average, the snow in Cornwall and Devon reached 60 cm (Anonymous, 1891a; Stirling, 1997), while on Dartmoor, where drifts lasted until June (Manley, 1958, Mabey, 1983), it was up to 150 cm deep (Bonacina, 1928, 1948; Stirling, 1997).

15.13   Heavy Snowfalls 1900–1909

The decade 1900–1909 had 28 HSEs covering 55 HSDs (the lowest number of days in any decade from 1861 to 1999). It also had the lowest number of HSEs (3) and HSDs (5) in January and the lowest number of HSDs (9) in March, as well as the lowest number of HSEs (4) of just 2 days' duration. Other interesting statistics from this decade are that it did not have any HSEs in May and 1903 and 1907 did not have a HSE in winter. None of the largest 10 HSEs (by snow depth) in the period 1861–1999 occurred in the decade.

15.14   Heavy Snowfalls 1910–1919

The decade 1910–1919 had 36 HSEs covering 58 HSDs. It had the highest number of HSDs (10) in April of any decade from 1861 to 1999, but also (together with the decade 1920–1929) the equal lowest number of HSDs (6) in December. There were no occurrences of HSEs of 4 days' duration. Other interesting statistics from this decade are that 1915 was the only year in the decade to have a HSE in May, while 1911 and 1912 had none in winter or spring. Of the largest 10 HSEs (by snow depth), in the period 1861–1999, one occurred within this decade: this was on 11–12 January 1913 (Wild, 2005a).

15.14.1   11–12 January 1913

The year 1913 had a heavy fall of snow in southern Scotland and northern England on 11–12 January (Bonacina, 1913, 1928; Jackson, 1977; Stirling, 1997), especially in Perthshire, where drifts were up to 3 m in places. Railway and postal services were delayed and some areas had depths of up to 90 cm. Morpeth,

---

[7] For further details please see: Anonymous (1891a, 1891b), Backhouse (1891), Bonacina (1908, 1928), Breton (1928), Hawke (1937), Spink (1947), Brooks (1954), Brazell (1968), Bowen (1969), Manley (1971), Bonacina (1973), Holford (1976), Jackson (1977), Burt (1978), Whittow (1980), Hardy *et al.* (1982), Mabey (1983), Lobeck (1989), George (1990), Wilson (1990), Ogley *et al.* (1991a), Currie *et al.* (1992), Davison *et al.* (1993), Horton (1995), Ching and Currie (1997) and Hulme and Barrow (1997).

Northumberland and Crieff, Tayside sustained between 45 and 60 cm of snow (Wild, 2005a).

15.15 Heavy Snowfalls 1920–1929

The decade 1920–1929 had 34 HSEs covering 60 HSDs. It had the lowest number of HSEs (3) in both March and December of any other decade from 1861 to 1999. It also had (with 1990–1999) the equal highest number of HSEs (2) in May and (with 1920–1929) the equal lowest number of HSDs (6) in December. There were no occurrences of HSDs of 5 days' duration. Other interesting statistics from this decade are that 1928 was the only year in the decade not to have a HSE in winter, while 1925 and 1929 were the only years without a HSE in spring. Of the largest 10 HSEs (by snow depth), two occurred in the decade: these were 25 December 1923 and 15–16 February 1929. The last of these was the worst snowstorm of the 20th century (by snow depth) and the worst recorded as part of this study (Wild, 2005a).

15.15.1 25 December 1923

On 25 December 1923 Glasgow, Strathclyde had a snowfall of 20 cm. This was reported to be the heaviest in Glasgow for 33 years. In Aberdeenshire, snow fell to a depth of up to 90 cm (Hawke, 1937).

15.15.2 15–16 February 1929

A heavy snowfall over Scotland, Wales and SW England (Figure 15.4) occurred on the 15–16 February 1929 (Breton, 1930; Spink, 1947; Whittow, 1980), blocking many roads. At Dean Prior,

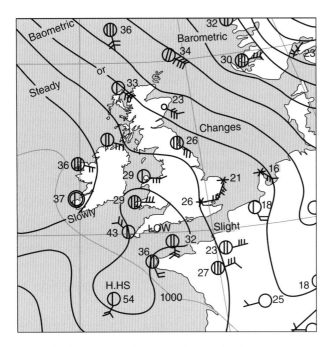

Figure 15.4 *Synoptic chart 15 February 1929 at 1800 GMT. Adapted from the Daily Weather Report of the Met Office, UK. © Crown copyright.*

Devon, up to 120 cm of snow fell in 15 hours. This is acknowledged as the heaviest known snowfall in 24 hours in Great Britain (excluding hills above 300 m). On the southern fringe of Dartmoor, 200 cm of snow was said to have fallen in 15 hours (Wilson, 1990). Eyewitnesses described the snow as falling as if it were being shovelled (Stirling, 1997). The weight of snow damaged many trees in the area. At Princetown, Devon, 20 cm of snow fell causing havoc for wildlife (Breton, 1930; Jackson, 1977).

15.16 Heavy Snowfalls 1930–1939

The decade 1930–1939 had 38 HSEs covering 77 HSDs. It had the highest number of HSEs (8) of 3 days' duration. Other interesting statistics from this decade are that 1935 was the only year in the decade to have a HSE in May, while 1932 was the only year without a HSE in winter. Of the largest 10 HSEs (by snow depth) in the period 1861–1999, none occurred in the decade.

15.17 Heavy Snowfalls 1940–1949

The decade 1940–1949 had 45 HSEs covering 81 HSDs. It had (with 1950–1959) the equal highest number of HSEs (14) in February of all decades from 1861 to 1999. It also had the lowest number of HSEs (1) in November. There were no HSEs in April during this decade. Other interesting statistics from this decade are that 1947 had the highest number of HSDs (9) in February of any year in the period 1861–1999 and (together with 1863 and 1985) the equal highest number of HSDs in winter (14) and in the year as a whole (20) during the 20th century. The years 1944, 1946 and 1949 had no HSEs in January, while 1943 and 1947 were the only years with a HSE in May, November and autumn. Of the largest 10 HSEs (by snow depth) in the period 1861–1999, three occurred in the decade: these were the 26–30 January 1940, 18–20 February 1941 and 26–27 March 1941 (Wild, 2004a, 2005a).

15.17.1 26–30 January 1940

The snowstorm of 26–30 January 1940 affected most of England and Scotland.[8] In Derbyshire and Cheshire 30–60 cm of snow fell, while in Sheffield, South Yorkshire, 120 cm was recorded. The West Highland railway line in Scotland was blocked by snow, some villages being isolated. By the 28th, 37 cm of snow lay at Pontefract, West Yorkshire (Brazell, 1968; Froggitt and Markham, 1993). On the 29th, a train became snowbound for 36 hours a few kilometres south of Preston, Lancashire (Bowen, 1969).

15.17.2 18–20 February 1941

The worst snowstorm of 1941 occurred in SE Scotland, NE England, the Midlands and East Anglia between the 18 and 20 February.[9] Wheeler (1991) described the snowstorm in detail, analysing the synoptic situation that caused it. The worst area of

---

[8] For further information see: Manley (1969, 1971), Jackson (1977), Whittow (1980), Stirling (1997), Mabey (1983), Clark (1986), Ogley *et al.* (1991a), Froggitt and Markham (1993) and Eden (1995).
[9] For further information see: Bowen (1969), Manley (1969), Jackson (1977, 1978), Whittow (1980), Pike (1991a, 1991b) and Wild *et al.* (1996).

snow was between the North Yorkshire coast and Berwick-upon-Tweed, Northumberland. The snow seemed to have penetrated 50 km inland. After 67 hours of continuous snow, Consett, County Durham had a snow depth of 120 cm, while at Durham Observatory, County Durham, the depth reached 107 cm. The city of Durham, County Durham was completely isolated for 3 days and six trains, with over 1000 people on board, were completely buried in drifting snow, north of Newcastle-upon-Tyne, Tyne and Wear. Approximately 1.5 million tons of snow fell in the town of Middlesbrough, Cleveland.

15.17.3   26–27 March 1941

On the 26–27 March 1941 in Sutherland and in parts of Ross and Caithness, snow fell up to 90 cm deep, with drifts up to 9 m (Jackson, 1977; Lobeck, 1989).

15.18   Heavy Snowfalls 1950–1959

The decade 1950–1959 had 51 HSEs covering 90 HSDs. It had (with 1940–1949) the equal highest number (14) in February of all decades from 1861 to 1999. It also had the highest number of HSDs (32) in February. Other interesting statistics from this decade are that 1952 was the only year in the decade to have a HSE in November and autumn, while 1955 was the only year with a HSE in May. Of the largest 10 HSEs (by snow depth) in the period 1861–1999, one occurred within this decade: this was the 8 February 1955 (Wild, 2004a, 2005a).

15.18.1   8 February 1955

The snowstorm of 8 February 1955 affected parts of northern Scotland, where many places received 90 cm of snow. The RAF dropped food supplies to isolated communities (Jackson, 1977; Wild, 2004a, 2005a).

15.19   Heavy Snowfalls 1960–1969

The decade 1960–1969 had 53 HSEs covering 93 HSDs. It had the highest number of HSEs (6) and HSDs (12) in November of any decade from 1861 to 1999. Other interesting statistics from this decade are that 1965 had the highest number of HSDs (5) in November and autumn of any year from 1861 to 1999. The years 1961 and 1969 were the only years in the decade not to have a HSD in January, while 1966 was the only year with HSEs in April (Wild, 2005a, 2005b). None of the largest 10 HSEs (by snow depth) in the period 1861–1999 occurred in this decade.

15.20   Heavy Snowfalls 1970–1979

The decade 1970–1979 had 52 HSEs covering 109 HSDs (the highest number of HSDs of any decade from 1861 to 1999). This decade had the highest number of HSEs (6) in April and the highest number of HSDs (24) in March. It also had (with 1990–1999) the equal highest number of HSEs lasting 2 days (22) and 4 days (4). Other interesting statistics from this decade are that 1974 and

1975 were the only years in the decade not to have a HSE in January, 1971 and 1975 were the only years without a HSE in February; 1973 was the only year with a HSE in May and 1975 and 1977 were the only years without a HSE in winter and spring. Of the largest 10 HSEs (by snow depth) in the period 1861–1999, none occurred in the decade (Wild, 2005a, 2005c, 2007a).

15.21   Heavy Snowfalls 1980–1989

The decade 1980–1989 had 45 HSEs covering 85 HSDs. It had the lowest number of HSEs (3) of 3 days' duration of all decades from 1861 to 1999. Other interesting statistics from this decade are that 1985 (with 1864 and 1867) had the equal highest number of HSDs (8) in January of any decade from 1861 to 1999 and 1986 (with 1888) was one of only 2 years to receive a HSE in June. The year 1985 (with 1863 and 1947) had the equal highest number of HSDs (14) in winter. The years 1983 and 1989 were the only years in the decade not to have a HSE in January; 1984 and 1985 were the only years without a HSE in spring. The years 1980 and 1988 were the only years with a HSE in November and autumn. Of the largest 10 HSEs (by snow depth) in the period 1861–1999, none occurred in the decade (Wild, 2005a, 2012).

15.22   Heavy Snowfalls 1990–1999

The decade 1990–1999 had 55 HSEs covering 93 HSDs. It had (with 1920–1929) the equal highest number of HSEs (2) and HSDs (4) in May of all decades from 1861 to 1999. It also had (with 1970–1979) the equal highest number of HSEs lasting 2 days (22) and (with 1930–1939) the equal highest number lasting 3 days (8). This decade had no HSEs lasting four or five days. Other interesting statistics from this decade are that 1998 (with 1876) had the equal highest number of HSDs in April (5) and 1993 had the highest number of HSDs (3) in May, of any year from 1861 to 1999. The years 1991 and 1992 were the only years in the decade not to have a HSE in January, while 1990 and 1991 were the only years without a HSE in spring. None of the largest 10 HSEs (by snow depth) in the period 1861–1999, occurred in the decade (Wild, 2005a).

15.23   Heavy Snowfall Frequencies in Great Britain 1861–1999

One of the clearest ways of expressing heavy snowfall frequencies (HSFs) is to look at the decadal numbers of HSEs and HSDs. Table B.2a and b (See Appendix, www.wiley.com/go/doe/extremeweather) shows the numbers of HSEs and HSDs (per decade) and their trends from 1861 to 1999. The highest number of HSEs was 58 in 1861–1869 and the lowest was 27 in 1890–1899 (Figure 15.5). From 1861 to 1999 the trend shows a slight rise in both HSEs and HSDs (linear regression trend line). Despite global warming there is no sign as yet of a significant drop in the number of HSE per decade.

A further examination of Table B.2a (See Appendix, www.wiley.com/go/doe/extremeweather) shows that the highest monthly totals of HSEs by decade were: 22 in January (1861–1869), 14 in February (1940–1949 and 1950–1959) and 14 in March (1870–1879). In February and December the lowest totals were four

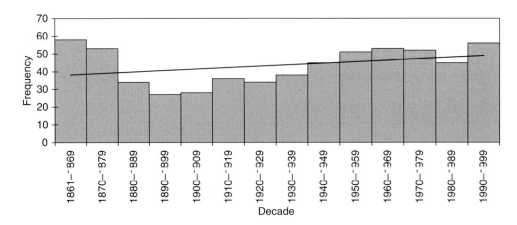

Figure 15.5   *The number of heavy snow events (HSEs) per decade across Great Britain 1861–1999.*

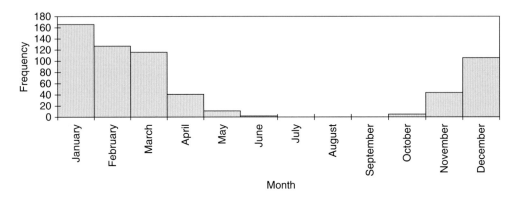

Figure 15.6   *The number of heavy snow events (HSEs) per month, 1861–1999 for Great Britain.*

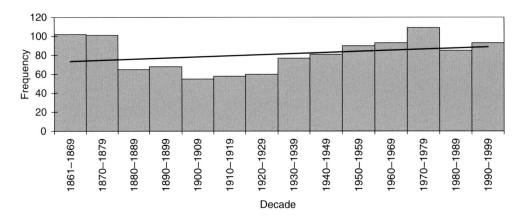

Figure 15.7   *The number of heavy snow days (HSDs) per decade, 1861–1999 for Great Britain.*

in February (1910–1919) and three in December (1920–1929). Table B.2a and b (See Appendix, www.wiley.com/go/doe/extremeweather) also highlights the fluctuating trend in the frequency of HSEs in January, although they have been quite steady since the 1940s, as have those of February. Other results from Table B.2a (See Appendix, www.wiley.com/go/doe/extremeweather) show that no HSE has occurred in October since the 1940s. November and December also show varying frequencies of HSEs, although they were quite steady in December between the 1920s and 1980s. Since the 1980s the HSEs have increased steadily in December.

Figure 15.6 shows the frequency of HSEs per month from the years 1861 to 1999. Of the 609 HSEs identified, January has the most at 166, followed by February with 127 and March with 116. At the other extreme, no HSEs occurred from July to September.

15.23.1   Numbers of Heavy Snowfall Days

Figure 15.7 (and Table B.2b, See Appendix, www.wiley.com/go/doe/extremeweather) shows the number of HSDs and trend (per decade) during the years 1861–1999. The highest number of

HSDs was 109 in 1970–1979 and the lowest was 55 in 1900–1909. Indeed, the period between 1880 and 1930 is notable for its generally low numbers of HSDs. The mean number of HSDs per decade over the period 1861–1999 was 81.2.

Table B.2b (See Appendix, www.wiley.com/go/doe/extremeweather) shows that HSDs in January had a fluctuating trend at first, followed by an increase from the 1930s, then a sudden drop in the 1990s. In February, HSDs also show a varying trend, although they have declined since the 1970s. The decrease in January and February could be due to the influence of global warming, although this is at variance with the general trend already noted in decadal frequencies. March also shows an undulating pattern, the highest numbers of HSDs occurring mainly in the later part of the 19th century. Other results from Table B.2b (See Appendix, www.wiley.com/go/doe/extremeweather) show that in April the highest numbers of HSDs generally occurred since the 1970s. November and December also show an undulating pattern of HSDs, although between the 1920s and 1980s they were quite steady, since when there has been an increase in December.

When annual numbers of HSDs are considered for the period 1861–1999, the same gently rising trend as in the HSEs is apparent. Figure 15.8 shows years with more than 20 HSDs which includes 1863, 1864, 1876 and 1947 (see also Table B.3 in Appendix, www.wiley.com/go/doe/extremeweather). Years with 2 days or less include 1862, 1880, 1884, 1903, 1905, 1907, 1910, 1911 and 1925. From 1925 to 1999, there was no year with fewer than three HSDs.

By month, the frequency of heavy snowfalls becomes somewhat complicated (Table B.3, See Appendix, www.wiley.com/go/doe/extremeweather). March perhaps unexpectedly, shows the highest number of HSDs in a single month, having eleven in 1876; the next highest monthly totals were nine in February 1947 and eight in January 1864, 1867, 1985, February 1973 and March 1864. Of the 1137 HSDs identified, January had the most (293), followed by February (238) and March (232) (see Tables B.2b and B.3, Appendix, www.wiley.com/go/doe/extremeweather).

Table B.3 (See Appendix, www.wiley.com/go/doe/extremeweather) indicates that 730 HSDs (64.2%) occurred in winter (December, January and February), 322 (28.3%) in spring (March, April and May), 2 (0.3%) in summer (June, July and August (both actually in June)) and 82 (7.2%) in autumn (September, October

and November). Also apparent from Table B.3 (See Appendix, www.wiley.com/go/doe/extremeweather) is that the average number of HSDs in a year is 8.2, of which 2.1 and 5.3 occur in January and winter respectively. From the 139 years researched, 39 years (28.0%) in the period 1861–1999 had no HSD in spring, whereas 1876 had 16 such days. Also, from the entire 139 years studied, 97 years (70.0%) had no HSD in autumn, whereas 1965 had five such days. Finally, 11 years (8.0%) had no HSD in winter (i.e. 1974–1975), however at the other extreme 1863, 1947 and 1985 had 14.

15.23.2 Mean Length (In Days) of Heavy Snowfall Events

Figure 15.9 shows the mean length and trend (in days' duration) of individual HSEs (per decade) from 1861 to 1999. The highest mean length, 2.52 days, was in 1890–1899 and the lowest, 1.61 days, in 1910–1919. The mean length over the whole period was 1.89 days. Since the 1970s, there has been a slight decline in the mean length of HSEs, which contrasts with a slight increase in their number, as shown in Figure 15.5.

A more detailed analysis of individual HSE durations shows a declining number of HSEs in relation to their durations (Figure 15.10). Of the 609 HSEs in total, 273 (44.8%) lasted just a single day, 219 (36.0%) lasted 2 days and just one HSE (0.2%) lasted 7 days.

Table B.4 (See Appendix, www.wiley.com/go/doe/extremeweather) shows the number of HSEs (in days' duration) per decade from 1861 to 1999. The highest number of HSEs lasting just a single day was 32 in 1861–1869 and the lowest was five in 1890–1899. For HSEs lasting 2 days the highest was 22 in 1970–1979 and 1990–1999 and the lowest was four in 1900–1909. For HSEs lasting 3 days the highest was eight in 1930–1939 and 1990–1999 and the lowest was two in 1910–1919. For HSEs lasting 4 days the highest was four in 1900–1909, whereas 1910–1919 and 1990–1999 had none. At 5 days' duration the highest was three in 1870–1879 and 1900–1909, whereas 1920–1929 and 1990–1999 had none. The decades 1870–1879, 1880–1899, 1950–1959, 1970–1979 all had one HSE lasting 6 days (the remaining decades had none). The only occurrence of a HSE lasting 7 days was in the decade 1861–1869 (25–31 January 1865).

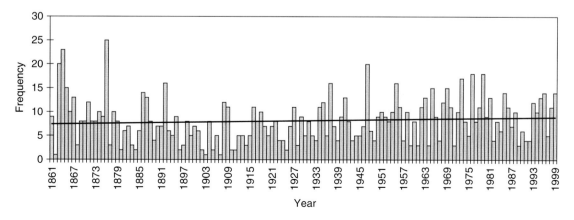

Figure 15.8 *The annual number of heavy snow days (HSDs) 1861–1999 for Great Britain.*

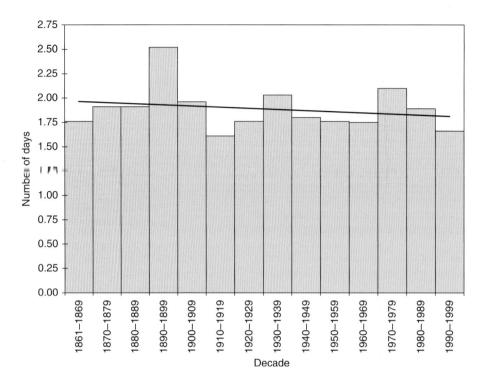

Figure 15.9 *Mean length and trend (in days' duration) of individual heavy snow events (HSEs) (per decade) 1861–1999 for Great Britain.*

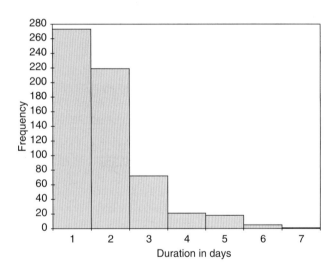

Figure 15.10 *Heavy snow events (HSEs) duration in days, 1861–1999 for Great Britain.*

Other statistics from Table B.4 (See Appendix, www.wiley.com/go/doe/extremeweather) show that there is an average of 43.5 HSEs in total per decade, of which 19.5 last for 1 day and 15.6 last for 2 days. HSEs lasting 1 day only show a rising trend since the 1970s, from 17 to 25 by 1990–1999, while those of 4 days' duration decline from four to zero over the same period. Note that there is a discrepancy of nine between the totals of HSEs in Tables B.4 and B.2a (See Appendix, www.wiley.com/go/doe/extremeweather) because of those that spanned 2 months and therefore appear twice in Table B.2a (See Appendix, www.wiley.com/go/doe/extremeweather).

The high numbers of HSEs during recent decades could be a reflection of not only greater frequency (despite global warming) but also improved data collection and monitoring.

15.24   LWTs and Heavy Snowfalls

This section examines the LWTs up to 5 days either side of a HSD. The period of 5 days before and after was chosen on the basis that outside this period circulations would be unrelated to that on the HSD and therefore unlikely to reveal any statistically significant relationship. This section examines only the seven primary LWTs (anticyclonic, cyclonic, easterly, northerly, north-westerly, southerly and westerly) and how they are related to the circulation patterns associated with HSDs. The hybrid and unclassified types were considered to be too variable and too few for any relevant statistical analyses, either on their own or when aggregated.

15.24.1   LWT Frequency by Decade

Table B.5 (See Appendix, www.wiley.com/go/doe/extremeweather) shows that, apart from the hybrid categories, the cyclonic type (264 occasions) is the most dominant circulation associated with HSDs in Great Britain. This is more than twice as frequent as the next highest, that is northerly (126) and easterly (117). The frequency of the remaining types (in descending order) is westerly (88), anticyclonic (65), north-westerly (40) and southerly (24).

Table B.5 (See Appendix, www.wiley.com/go/doe/extremeweather) also shows that HSDs have different LWT frequencies in different decades. For example, the anticyclonic type associated with HSDs was most prominent in 1870–1879 with 12 occurrences

(18.5%), whereas in 1920–1929 it only occurred once (1.5%). The cyclonic type was most prominent in 1980–1989 with 26 occurrences (9.8%) and lowest in 1890–1899 with seven (2.7%). The easterly type was at its highest with HSDs in 1970–1979 with 14 occurrences (12.0%), however only occurred once (0.9%) in 1900–1909. This marked decadal variation can also be seen for the other LWTs, including the hybrid type.

15.24.2 LWT Frequency by Month and Season

Table B.6 (See Appendix, www.wiley.com/go/doe/extremeweather) shows a monthly analysis of all 27 LWTs associated with HSDs from 1861 to 1999. The data demonstrate that the winter season is dominant in terms of HSDs and thus LWTs. Total frequencies show that January is the most significant month followed by February, March and then December. The analysis further shows that the cyclonic type has the highest frequency in all months and seasons except October, when it is replaced by the northerly type. The month and season with the highest occurrences of the cyclonic type are January with 64 (21.8%) and winter with 157 (21.5%). Other significant monthly LWTs are easterly with 36 occurrences (15.1%) in February and 29 (12.5%) in March; northerly with 26 occurrences (8.9%) in January; 24 (10.3%) in March and 20 (10.1%) in December; while westerly had 31 occurrences (10.6%) in January. At the other extremes, no HSD was associated with the anticyclonic south westerly type.

15.24.3 LWT Frequencies Before, During and After the Heavy Snowfall Day

From the analysis of primary LWTs before, during and after the HSD (Table B.7 in Appendix, www.wiley.com/go/doe/extremeweather), irrespective of month, the following points are made. Regardless of the LWT on the day of snowfall, the anticyclonic type tends to prevail 4–5 days earlier. This is then replaced first by the westerly and then the cyclonic type as the HSD approaches. On the day itself (as shown in Tables B.5–B.7, Appendix, www.wiley.com/go/doe/extremeweather), the types, in order of frequency, are cyclonic, northerly, easterly, westerly, anticyclonic, north-westerly and southerly. Regardless of LWT on the day, the cyclonic type tends to dominate for the first 3 days afterwards, followed by anticyclonic and to a lesser extent westerly, at 4–5 days. This LWT trend is the reverse of that for the circulation before the HSD.

Similar analyses at the monthly scale (Tables B.8–B.13, See Appendix, www.wiley.com/go/doe/extremeweather) shows a much more varied pattern. For the main months of snowfall (November to April) the results are as follows. In January (Table B.8, See Appendix, www.wiley.com/go/doe/extremeweather) the pattern of LWT in days preceding a HSD is similar to that described above. In February (Table B.9, See Appendix, www.wiley.com/go/doe/extremeweather) the anticyclonic type is replaced by the easterly type as the day approaches, while March (Table B.10, See Appendix, www.wiley.com/go/doe/extremeweather) and April (Table B.11, See Appendix, www.wiley.com/go/doe/extremeweather), the anticyclonic type is replaced by the northerly then the cyclonic type. In November (Table B.12, See Appendix, www.wiley.com/go/doe/extremeweather), the anticyclonic type is replaced first by the westerly then the northerly then finally the cyclonic type, while in December (Table B.13, See Appendix, www.wiley.com/go/doe/extremeweather), the sequence tends to be westerly then briefly

anticyclonic then cyclonic type as the day approaches. For the 5 days following a HSD (Tables B.8–B.13, See Appendix, www.wiley.com/go/doe/extremeweather), the individual monthly patterns are very similar to those for all months, as described in the previous paragraph.

15.25 Depressions and Heavy Snowfalls

Deep depressions are the main cause of heavy snowfalls. This section examines the link between depressions and HSEs in Great Britain during the years 1870–1999 (1870 was the first complete year for which North Atlantic synoptic charts were available). The first part explores the relationship between depression tracks and HSEs. The second part investigates the frequency of depressions associated with HSEs.

15.25.1 Depression Trajectories Associated with Heavy Snowfall Events

This analysis used the Meteorological Office's Daily Weather Report (1870–1980) and Daily Weather Summary (1981–1999). From these sources, the trajectory of every depression associated with a heavy snowfall was traced from 5 days before to 5 days after the HSE. The main geographical area chosen was 40–65°N/20°W–8°E. This was then subdivided into nine smaller areas: >60°N/>10°W, >60°N/10°W–2°E, >60°N/>2°E, 50–60°N/>10°W, 50–60°N/10°W–2°E, 50–60°N/>2°E, <50°N/>10°W, <50°N/10°W–2°E and <50°N/>2°E. On some occasions, more than one depression was associated with the HSE. If a depression was associated with a HSE, but its centre did not pass through the area (40–65°N/20°W–8°E), it was excluded from the survey.

Using the data in Table B.14 (See Appendix, www.wiley.com/go/doe/extremeweather) the following findings have emerged. Of the 680 depressions associated with HSEs, the majority (46 cases, 6.8%) originated in the region 50–60°N/>10°W (west of Ireland) and dissipated in the region 50–60°N/>2°E (North Sea, Netherlands and Norway region). Other significant regions where depressions originated then dissipated were 50–60°N/10°W–2°E (British Isles) then 50–60°N/>2°E (41 cases, 6.0%), and <50°N/>10°W (SW of the British Isles) then 50–60°N/>2°E (30 cases, 4.4%). Table B.14 (See Appendix, www.wiley.com/go/doe/extremeweather) shows that out of the 560 HSEs (1035 HSDs) in the period 1870–1999, 66 (9.7%) were not associated with depressions or were associated with depressions outside the area 40–65°N/20°W–8°E. Table B.15 (See Appendix, www.wiley.com/go/doe/extremeweather) shows the areas where no depressions associated with HSEs originated and dissipated. A decadal survey of areas where depressions originated and dissipated in association with HSEs from 1870 to 1999 shows that the highest number, 73 (10.7%), occurred in 1970–1979 and the lowest, 35 (5.1%), in 1900–1909.

15.25.2 Depression Traffic Associated with Heavy Snowfall Events

A more objective analysis of HSE depression tracks was conducted using 560 HSEs in the period 1870–1999. Instead of tracing the depression movements of this large sample, the passages of depressions were numerically recorded within a 1° latitude/longitude area. The total survey grid covers the region 40–65°N/20°W–8°E (Figure 15.11).

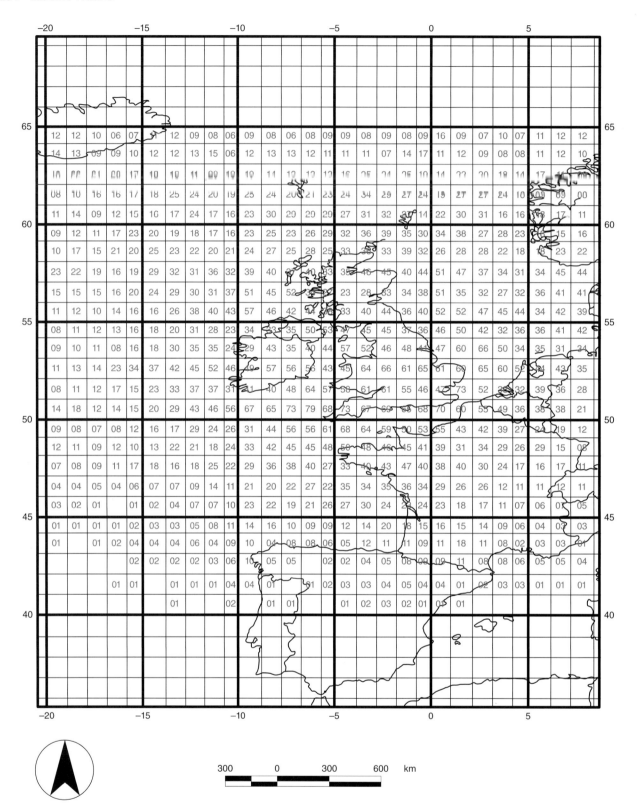

Figure 15.11 *Frequency of depressions that tracked through a 1° longitude/latitude geographical area (40–65°N/20°W–8°E) in association with heavy snowfall events in all months across Great Britain between the years 1870 and 1999.*

To better understand the general trends presented by the data, the survey grid defined earlier was subdivided into 5° frames, for example 50–55°N/5–10°W. The eastern limit was set at 3°E, for example Figure 15.12, to coincide with the eastern boundary of the synoptic charts used. Most HSE depression tracks occurred in the central region (horizontal zone) of the grid (50–55°N), where there were 5618 occurrences, while the southern zone (40–45°N) had the lowest at 615. The north-western sectors of the grid (55–60°N/20–15°W and 15–10°W) saw more HSE depressions (395 and 671, respectively) than the south-western sectors (45–50°N/20–15°W and 15–10°W), which had 180 and 378. The highest south to north depression tracks associated with HSEs was in the region 40–65°N/5°W–0° (3986) and the lowest was in the region 40–65°N/15–20°W (1250). The highest frequency within a 5° spatial frame occurred over England and Wales (1390), followed by the southern North Sea and Ireland (1256 and 1255, respectively). Depression numbers were greater over western France (1041) than over central and eastern Scotland (899).

Using the more detailed information from the 1° latitude/longitude grid, Figure 15.11 and Table B.16 (See Appendix, www.wiley.com/go/doe/extremeweather), it was found that the highest occurrence of depressions associated with HSEs passing through a specific latitude was 1318 (50–51°N). On the other hand, the highest number to pass through a given longitude was 835 (3–4°W). The lowest frequency of depressions associated with HSEs passing through a particular latitude was 16 (40–41°N) and the lowest number to pass through a particular longitude was 218 (19–20°W). At an even smaller geographical focus, the highest number of depressions passing through an individual 1° square of latitude and longitude was 79 (50–51°N/6–7°W), that is in the Celtic Sea just west of Cornwall.

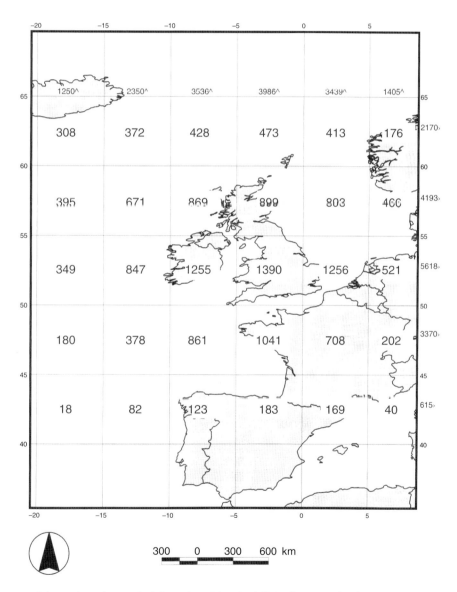

Figure 15.12   *Frequency of depressions that tracked through a 5° longitude/latitude geographical area (40–65°N/20°W–8°E) in association with heavy snowfall events in all months across Great Britain between the years 1870 and 1999.*

Table B.17 (See Appendix, www.wiley.com/go/doe/extremeweather) gives an outline of cyclogenesis in the eastern North Atlantic in relation to heavy snowfall across Great Britain from 1870 to 1999. It can be seen that HSEs are often the product of up to three depressions over the region on the same day. A single depression, however, is the most common situation; for example, of the 51 HSEs in the 1950s, 29 were associated with a single depression, 14 with two depressions and 3 with three depressions. On the other hand, in 1930–1939, 30 HSEs were associated with one depression and only 5 with two or three depressions. Table B.17 (See Appendix, www.wiley.com/go/doe/extremeweather) shows that cyclogenic activity associated with heavy snowfall might have increased since the 1950s, when a rise in the frequency of two-depression systems can be seen. It is quite possible that some of the apparent rise in depression activity in the latter half of the 20th century is a reflection of better recording and monitoring. A significant number of HSEs are not obviously related to depressions. This occurs most frequently when depression centres are outside the regional grid, but the region is affected by troughs of low pressure.

A monthly and seasonal analysis of the cyclogenic activity in relation to HSEs over Great Britain (Table B.18, See Appendix, www.wiley.com/go/doe/extremeweather) reveals the following points. As expected, winter is the most active season. In January, 92 of the 144 HSEs were related to a single depression, but as many as 29 (almost a third), were associated with two depressions. In February, 83 of the 116 HSEs were associated with a single depression, 16 were associated with two depressions and there were 6 associated with three depressions. Depressions that produce heavy snow are rare in other seasons. No heavy snowfalls in May, June and October were associated with three depressions (a reflection of the low numbers of HSEs in these months).

It is not surprising that the northern section of the grid (60–65°N) showed more depression activity associated with HSEs than the southern section (40–45°N). Depressions giving rise to heavy snowfalls are naturally more likely to originate in the colder north than the warmer south. This differs from rain-producing depressions, the majority of which come in through the south-west approaches. The maximum incidence of depressions producing heavy snowfall across England and Wales can be possibly explained by the convergence over this area of the main snowfall producing LWTs, for example cyclonic, northerly and easterly. Another reason for the high incidence of depressions associated with HSEs over England and Wales and indeed Ireland and Scotland, is orographic enhancement. Only those depressions that produce heavy snowfalls are included in this survey. Many of these snowstorms are enhanced by orographic uplift. Land areas are therefore a focus for depressions producing snowfall and have a high frequency of HSEs.

Figures 15.11 and 15.12 demonstrate the importance of depression tracks along the English Channel from the south-west approaches. Another track of importance is that from the Bay of Biscay northwards across Great Britain. These two depression tracks emphasise the importance of warm water surrounding the British Isles as a water vapour source for the production of snowfall. This category can also be extended to the polar low (Wild *et al.*, 1996), where the energy of southward-moving depressions is maintained by differences in temperature between progressively warmer seas and very cold air from the Polar Regions.

15.26 Fronts Associated with Heavy Snowfall Days (1937–1999)

This section investigates the links between meteorological fronts and the 578 HSDs (317 HSEs) during the years 1937–1999 (1937 being the first complete year when fronts were shown on synoptic charts). An objective analysis has been made of frontal association with the 317 HSEs (578 HSDs) using the same study area as above, that is Lamb's sector 50–60°N/10°W–2°E (Figure 15.13).

If one or more fronts occurred in association with a HSD the following factors were considered. If the front(s) did not enter the study area, or were only present between the times of the morning and evening synoptic charts, or the heavy snowfall was caused by troughs or showers, the snowfall was deemed (for this study) not to have been associated with fronts. On some occasions fronts occurred simultaneously in more than one of the Sectors 1, 2, 3 or 4 and this is clearly shown in the analysis. The different types of fronts used in this analysis were cold, warm and occluded; however it was not always possible to determine which type of front was responsible for the heavy snowfall, especially when more than one type was present.

Two aspects of this relationship are considered. Firstly, the spatial distribution of fronts within the study area associated with HSDs; finally fronts are related to LWTs to see which weather types recorded the lowest and highest frequencies (and combinations) of cold, warm and occluded fronts.

15.26.1 Geographical Variations in Meteorological Fronts

The fronts associated with HSDs were based on the synoptic charts of the Daily Weather Report (1941–1980) and Daily Weather Summary (1981–1999), together with individual North Atlantic synoptic charts (1937–1941), using both the morning and evening charts.

Table B.19 (See Appendix, www.wiley.com/go/doe/extremeweather) shows that the most frequent frontal positions associated with HSDs over Great Britain (morning situation only) were outside Lamb's sector, accounting for 176 (30.4%) of the 578 HSDs. This may be because the snowfall was a result of troughs or showers, or

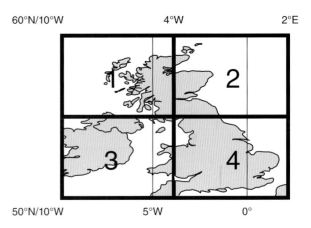

Figure 15.13 *The geographical area that sectors 1, 2, 3 and 4 represent.*

because of frontal systems that did not quite enter the area (as, for example, when precipitation falls ahead of a warm or occluded front while the frontal boundary remains further away). Another explanation may be that the snowfall was associated with fronts that passed through the area between successive synoptic charts and were thereby excluded from the study. The next most frequent situation was of occluded fronts alone, accounting for 128 occurrences (22.1%). This may be due to the fact that occluded fronts are normally closer than others to the central portions of depressions, where precipitation is often heavier than elsewhere. The third most frequent frontal arrangement for heavy snowfalls was of cold, warm and occluded fronts in combination, accounting for 91 occurrences (15.7%). For the evening period (Table B.20, See Appendix, www.wiley.com/go/doe/extremeweather), the same relative pattern prevailed, although frequency levels were different: 193 occurrences for frontal positions outside the study area, or for snowfall related to non-frontal activity within the area; 155 occurrences for occluded fronts and 69 occurrences for cold, warm and occluded fronts in combination.

When the Lamb's Sector is submitted into four (as defined earlier), Tables B.21 and B.22 (See Appendix, www.wiley.com/go/doe/extremeweather) shows that for the morning period (excluding occasions when the fronts were outside the area, or the heavy snow was associated with troughs or showers) the highest frontal frequency was 102 (17.6%) in sectors 3 and 4 combined, followed by sector 4 alone, which accounted for 42 (7.3%). For the evening period the highest frontal frequency was 93 (16.1%) in sectors 3 and 4, followed by sector 4 alone, which accounted for 44 (7.6%). Apart from one or two exceptions, similar trends were found in a monthly and seasonal survey (sector 3 and 4). This is because the heaviest precipitation (i.e. the frontal boundary) is normally close to the centre of the depression. These southern sectors are related to the southward movement of the jet stream to a position over southern parts of Great Britain in winter.

15.26.2 Relationship Between Fronts and LWT

Tables B.23 and B.24 (See Appendix, www.wiley.com/go/doe/extremeweather) show that cyclonic is by far the most frequent circulation type when frontal activity is associated with heavy snowfall. Accompanying this type on HSDs, the pattern of cold, warm and occluded fronts in combination occurs on 47 morning and 26 evening occasions; occluded fronts alone occur on 36 morning and 55 evening occasions and cold and warm fronts in combination on 17 morning and 6 evening occasions. When all possible frontal combinations are considered, the cyclonic type has total frequencies of 141 in both the morning and evening periods. The equivalent figures for other circulation types in morning/evening are easterly 71/71, northerly 63/63, westerly 43/44, anticyclonic 25/10, southerly 16/16 and north-westerly 3/20.

This section shows that dynamic LWTs, such as the cyclonic and northerly, are the ones most associated with frontal activity, showing strong links with cold, warm and occluded front combinations and with occluded fronts alone. There was also a high frequency of dynamic LWTs in situations where no front was present in Lamb's sector.

15.27 Concluding Remarks

This chapter has presented a discussion on heavy snowfalls or blizzards over Great Britain between 1861 and 1999. The terms 'snowfalls' and 'snowstorms' are difficult to define precisely. Definitions of the word 'blizzard' vary from one place and context to another. For this reason the term has not been used here. Instead, the terms heavy snowfall event (HSE) and heavy snowfall day (HSD) were employed. In relation to lowland areas (below about 300 m ASL) the former is mainly defined as 13 cm or more of fresh snow falling somewhere in Great Britain in 24 hours. This amount has proved useful when considering damage to property, effect on livelihoods and even loss of life. In upland areas (above 300 m), there is much more snow and a much smaller population. A HSD is simply a 24-hour period when heavy snow has fallen, whereas a HSE can produce heavy snow (≥13 cm/day) for several days. Recording of snow lying, however, is problematic as this can be the result of both accumulation and melting over different periods. Since snow depth is only measured once a day (at 0900 GMT), the reported amount is the aggregate of both recent and earlier accumulation and melting.

The number of days with snow cover in Great Britain rises swiftly with altitude (by 1 day for every 15 m above 60 m), partly because of lower temperatures on high ground and partly because of orographic enhancement of precipitation. The deeper the snow, the longer it will take to melt. Other influences include frequency and intensity and proximity to primary source of moisture and topography. Snowfall affects places differently in place and time, regardless of intensity, especially when one airflow changes to another. This snow can occur as convective showers giving short-lived, localised falls, or as frontal or cyclonic snow, which can be longer lasting and is usually more widespread. As this chapter has shown, sea temperatures around the British Isles affect the amount and distribution of snowfall across Great Britain, by determining the sensible and latent heat energy available in eastward moving depressions. Snowfall over coastal land areas is determined by local relief and by the stability and moisture content of the lower atmosphere. Wind direction and exposure are the principal controls on local variations in snowfall. The sharpest transformation of weather systems over short ranges is between the windward and leeward sides of the hills, such as the Highlands, the Pennines and the Snowdonia range. Total snowfall at any location will be determined by the moisture content of the air mass associated with it. Different places in Great Britain therefore receive different proportions of a snowfall depending on the synoptic situation.

Of the top 10 HSEs from 1861 to 1999, the three severest (by snow depth) were all on Dartmoor, the worst being that of 15–16 February 1929, when 200 cm of snow fell. The next seven severest snowstorms in order of snow depth were in either northern Scotland or northern England. Snow in southern areas of Great Britain is usually associated with an easterly or south-easterly wind round a depression centred over northern France and a frontal system moving slowly from the south-west. In eastern districts snowfall may be the result of a dry easterly airflow being warmed and established over the North Sea. Another situation in which snow occurs is when cold air in a depression moves on a westward track across the country. In Scotland and other northern areas snow can also fall in a northerly air stream, sometimes associated with a polar low.

It was not possible to determine precisely which fronts were responsible for the heaviest snowfalls, but it was found (surprisingly) that the most frequent situation during heavy snowfalls was when no fronts were present in the area. This may because of heavy snowfalls associated with troughs or showers; or because of snow associated with fronts that remain just outside the defined area of study; or because of fronts that passed through the area between successive (morning and evening) charts. When fronts are within the area on a HSD the most frequent type is the occluded front alone, followed by cold, warm and occluded fronts in combination. The LWTs most commonly associated with snow-producing fronts are cyclonic and northerly in association with cold, warm and occluded fronts in combinations.

The highest annual totals of HSDs occurred in 1863, 1864, 1876 and 1947 (all over 20). The highest decadal total of HSEs was in 1861–1869 (58) and the lowest in 1890–1899 (27). The highest decadal total of HSDs was in 1970–1979 (109) and the lowest in 1900–1909 (55). Over the whole period, however, numbers of HSEs and HSDs have not shown any significant upward or downward trends, although there has been a slight increase since the 1970s. In contrast, there has been a slight decadal decline in the durations of individual HSEs and of snow lying (features to be expected during global warming). Some of the changes identified could be the result of improved data collection and monitoring, which leads to uncertainty about some of the trends. The regions with the most occurrences of snowfall on HSDs were central and eastern England, NE England and northern and eastern Scotland; those with the fewest occurrences were Wales, NW England, SW and central southern England.

If the discovery of changes in the frequency and duration of snowfalls over Great Britain are a mark of this chapter, so too are the new data presented in relation to their geographical distribution. Snowfall distribution is strongly related to main primary LWTs; but whereas the LWTs responsible for the highest rainfalls are southerly, cyclonic and westerly, those associated with the deepest snowfalls are anticyclonic, easterly and southerly. While the highest frequency LWTs-associated with rainfall are westerly and anticyclonic, for snowfall they are cyclonic and easterly. New evidence is presented on the LWTs associated with HSDs before and after the event. Up to 5 days before a HSD the main types are anticyclonic, westerly and cyclonic (in that sequence); while up to 5 days after the event the sequence is reversed – cyclonic, then westerly, then anticyclonic.

When the westerly type occurs on the HSD it tends to dominate for the 5 days before and after, but other LWTs do not show any clear trends. In general, the LWTs associated with the more significant snowfalls tend to occur on a greater number of days before and after the event. The survey of the tracks of the main snow-producing depressions shows the importance of the track from the Bay of Biscay through the English Channel. Other significant tracks are across northern Britain from west to east and from the north as polar lows. The highest frequency of depression tracks associated with HSDs was found over England and Wales, followed by Scotland and western France (more in the colder north than in the warmer south). Part of the reason for higher densities of snow-producing depressions across the central region is land itself, where the mountains enhance orographic precipitation. A more detailed study revealed the importance of warm bodies of water leading to high depression densities over the Celtic Sea just west of Cornwall, Normandy and the western English Channel.

Acknowledgements

The events discussed in this chapter can be found on the website www.drrichardwild.co.uk and within the publication *The (International) Journal of Meteorology* (Wild, 2001b, 2002, 2003, 2004, 2005d, 2006, 2007, 2008). Future analysis for the period 2000 onwards will be available at www.drrichardwild.co.uk and in future publications proposed in *The (International) Journal of Meteorology*. The author would like to express his appreciation to Robin Stirling and Giles de la Marc Publishers for image reproductions from *The Weather of Britain* (ISBN 9781900357067), and to Claire Wootten for the support in the development and achievement of this chapter.

References

Ahrens, C. D. (1991) *Meteorology Today*. 4th Edition, West Publishing Company, St Paul, 576pp.

Anonymous (1881a) On the snow storm of January 1881. *Symons's Monthly Meteorological Magazine*, **16**, pp. 42–43.

Anonymous (1881b) On the snow storm of January 1881. *Symons's Monthly Meteorological Magazine*, **16**, pp. 43–45.

Anonymous (1882) The snow storms of January 17th–21st, 1881. *British Rainfall*, 1881, pp. 35–36.

Anonymous (1891a) The great snowstorm of March 1891. *Symons's Monthly Meteorological Magazine*, **26**, pp. 33–40.

Anonymous (1891b) The great snowstorm of March 1891. *Symons's Monthly Meteorological Magazine*, **26**, pp. 64.

Backhouse, T. W. (1891) The great snowstorm of March 1891. *Symons's Monthly Meteorological Magazine*, **26**, pp. 75.

Belasco, J. E. (1952) *Characteristics of Air Masses over the British Isles*. Meteorological Office, Geophysical Memoir, No. 87, HMSO, London, 34pp.

Bell, J. N. O. (1881) On the snow storm of January 1881. *Symons's Monthly Meteorological Magazine*, **15**, pp. 18.

Bignell, K. J. (1981) The April blizzard-was it a polar low? *Weather*, **36**, pp. 178–179.

Birt, W. M. (1881) On the snow storm of January 1881. *Symons's Monthly Meteorological Magazine*, **15**, pp. 18.

Bonacina, L. C. W. (1908) The Easter snowstorm of 1908. *Symons's Monthly Meteorological Magazine*, **43**, pp. 69.

Bonacina, L. C. W. (1913) The weather of January 11th 1913. *Symons's Monthly Meteorological Magazine*, **48**, pp. 10.

Bonacina, L. C. W. (1915) Remarks on snowfall study (geographical distribution). *Symons's Monthly Meteorological Magazine*, **50**, pp. 4–5.

Bonacina, L. C. W. (1917) The snowstorms of February and March 1916. *British Rainfall*, 1916, pp. 28–31.

Bonacina, L. C. W. (1928) *Snowfall in the British Isles during the half century, 1876–1925*. British Rainfall 1927, HMSO, London, pp. 260–287.

Bonacina, L. C. W. (1948) Summer retention of snow in the Cotswolds and elsewhere. *Weather*, **3**, pp. 253.

Bonacina, L. C. W. (1973) The West Country blizzard of March 1891. *Weather*, **28**, pp. 218.

Bowen, D. (1969) *Britain's Weather. Its Workings, Lore and Forecasting*. David and Charles, Newton Abbot, 310pp.

Brazell, J. H. (1937) The blizzard of February 27th–March 1st, 1937. *The Meteorological Magazine*, **73**, pp. 36–38.

Brazell, J. H. (1968) *London Weather*. HMSO, London, 249pp.

Breton, H. H. (1928) *The Great Blizzard of Christmas 1927, Its Causes and Incidents*. Hoyten and Cole, Plymouth, 55pp.

Breton, H. (1930) *The Great Winter of 1928–1929*. Hoyten and Cole, Plymouth, 30pp.

Brooks, C. E. P. (1954) *The English Climate*. English Universities Press Ltd, London, 214pp.

Brown, P. (1987) Attitudes to snow. *J. Meteorol.*, **12**, pp. 129–130.

Burder, G. F. (1887) The snow of March 15th. *Symons's Monthly Meteorological Magazine*, **22**, pp. 36–37.

Burt, S. D. (1978) The blizzards of February 1978 in south western Britain. *J. Meteorol.*, **3**, pp. 261–278.

Carlson, T. N. (1991) *Mid-Latitude Weather Systems*. Harper Collins, London, 507pp.

Chandler, T. J. and Gregory, S. (1976) *The Climate of the British Isles*. Longman Group, London, 390pp.

Chapman, D. L. (1947) The frequency and duration of snowfall in Great Britain. *Weather*, **2**, pp. 99–101.

Ching, M. and Currie, I. (1997) *The Dorset Weather Book*. Frosted Earth, Coulsdon, 100 pp.

Clark, J. B. (1986) Major snowfalls in Manchester since 1880. *Weather*, **41**, pp. 278–282.

Currie, I.; Davison, M. and Ogley, B. (1992) *The Essex Weather Book*. Froglets Publications Ltd, Kent, 160pp.

Currie, I.; Davison, M. and Ogley, B. (1994) *The Berkshire Weather Book*. Froglets Publications Ltd, Kent, 144pp.

Davison, M. and Currie, I. (1991) *The Surrey Weather Book*. Frosted Earth, Surrey, 104pp.

Davison, M.; Currie, I. and Ogley, B. (1993) *The Hampshire and Isle of Wight Weather Book*. Froglets Publications Ltd, Kent, 168pp.

Dykes, R. A. (1881) On the snow storm of January 1881. *Symons's Monthly Meteorological Magazine*, **16**, pp. 14–15.

Eady, E. T. (1949) Long waves and cyclone waves. *Tellus*, **1**, pp. 33–52.

Eddy, E. M. G. (1881) On the snow storm of January 1881. *Symons's Monthly Meteorological Magazine*, **16**, pp. 22–23.

Eden, P. (1995) *Weatherwise*. Macmillian Reference Books, London, 323pp.

Fenton, M. (1881) On the snow storm of January 1881. *Symons's Monthly Meteorological Magazine*, **16**, pp. 43.

Field, N. A. (1975) Severe winter weather. *Weather*, **30**, pp. 133–134.

Froggitt, B. and Markham, L. (1993) *The Yorkshire Weather Book*. Countryside Books, Newbury, 128pp.

George, D. J. (1972) The snowstorms of 4 March 1970. *Weather*, **27**, pp. 96–109.

George, D. J. (1990) *The Brecon Beacons National Park. Its Climate and Mountain Weather*. Cardiff Weather Centre, Information Booklet No. 1, Meteorological Office, Bracknell, 30pp.

Greely, A. W. (1888) *American Weather*. Dodd, Mead and Company, New York, 286pp.

Grierson, J. (1881) On the snow storm of January 1881. *Symons's Monthly Meteorological Magazine*, **16**, pp. 15–18.

Hardy, R.; Wright, P.; Gribben, J. and Kington, J. (1982) *The Weather Book*. Michael Joseph, London, 224pp.

Hawke, E. L. (1937) *Buchan's Days*. Lovat Dickson Limited, London, 231pp.

Hoinkes, H. C. (1968) Glacier variation and weather. *J Glaciol.*, **7**, pp. 3–19.

Holford, I. (1976) *British Weather Disasters*. David and Charles, Newton Abbot, 127pp.

Horton, B. (1995) *West Country Weather Book*. Barry Horton, Bristol, 159pp.

Hulme, M. and Barrow, E. (1997) *Climates of the British Isles, Present, Past and Future*. Routledge, London, 454pp.

Jackson, M. C. (1977) The occurrences of falling snow over the United Kingdom. *The Meteorological Magazine*, **106**, pp. 26–38.

Jackson, M. C. (1978) Snow cover in Great Britain. *Weather*, **33**, pp. 298–309.

Jenkinson, A. F. and Collison, F. P. (1977) *An Initial Climatology of Gales Over the North Sea*. Synoptic Climatology Branch Memorandum No. 62, Meteorological Office, Bracknell, 18pp.

Johnson, R. (1881) On the snow storm of January 1881. *Symons's Monthly Meteorological Magazine*, **16**, pp. 11.

Jones, P. D. and Kelly, P. M. (1982) Principal component analysis of the Lamb catalogue of daily weather types: Part 1, annual frequencies. *J. Climatol.*, **2**, pp. 147–157.

Jones, P. D.; Hulme, M. and Briffa, K. R. (1993) A comparison of Lamb circulation types with an objective classification scheme. *Int. J. Climatol.*, **13**, pp. 655–663.

Knight, J. P. (1881) On the snow storm of January 1881. *Symons's Monthly Meteorological Magazine*, **16**, pp. 11–12.

Lamb, H. H. (1950) Types and spells of weather around the year in the British Isles: Annual trends, seasonal structure of the year, singularities. *Q. J. R. Meteorol. Soc.*, **76**, pp. 393–429.

Lamb, H. H. (1972) *British Isles Weather Types and a Register of the Daily Sequence of Circulation Patterns, 1861–1971*. Meteorological Office, Geophysical Memoir No. 116, HMSO, London, 85pp.

Lamb, H. H. (1988) *Weather, Climate and Human Affairs. A Book of Essays and Other Papers*. Routledge, London and New York, 364pp.

Lester, R. M. (1948) *Everybody's Weather Book*. Marston and Co. Ltd, London, 235pp.

Lobeck, R. (1989) *Weather Wisdom, Fact or Fiction?* Geerings of Ashford Limited, Kent, 71 pp.

Lyall, I. T. (1972) The polar low over Britain. *Weather*, **27**, pp. 378–390.

Mabey, R. (1983) *Cold Comforts*. Hutchinson, London, 78pp.

Manley, G. (1940) Snowfall in the British Isles. *Meteorological Magazine*, **75**, pp. 41–48.

Manley, G. (1958) The Great Winter of 1740. *Weather*, **13**, pp. 11–17.

Manley, G. (1969) Snowfall in Britain over the past 300 years. *Weather*, **24**, pp. 428–437.

Manley, G. (1971) *Climate and the British Scene*. Collins, London, 314pp.

Manley, G. (1974) Central England temperatures: Monthly means 1659 to 1973. *Q. J. R. Meteorol. Soc.*, **100**, pp. 389–405.

Mayes, J. C. (1991) Regional airflow patterns in the British Isles. *Int. J. Climatol.*, **11**, pp. 171–176.

Meteorological Office (1991) *Meteorological Glossary*. 6th Edition, HMSO, London, pp. 43.

Mills, W. (1881) On the snow storm of January 1881. *Symons's Monthly Meteorological Magazine*, **16**, pp. 12–13.

Musk, L. F. (1989) *Weather Systems*. Cambridge University Press, Cambridge, UK, 160 pp.

Neele, G. P. (1881) On the snow storms of January 1881. *Symons's Monthly Meteorological Magazine*, **16**, pp. 22.

Noble, J. (1881) On the snow storms of January 1881. *Symons's Monthly Meteorological Magazine*, **16**, pp. 42–43.

O'Hare, G. and Sweeney, J. (1993) Lamb's circulation types and British weather: An evaluation. *Geography*, **78**, pp. 43–60.

O'Hare, G. P. and Wilby, R. (1995) A review of ozone pollution in the United Kingdom and Iceland with an analysis using Lamb weather type. *Geogr. J.*, **161**, pp. 1–20.

Oakley, H. (1881) On the snow storm of January 1881. *Symons's Monthly Meteorological Magazine*, **15**, pp. 16.

Ogley, B.; Currie, I. and Davison, M. (1991a) *The Kent Weather Book*. Froglets Publications Ltd, Kent, 160pp.

Ogley, B.; Currie, I. and Davison, M. (1991b) *The Sussex Weather Book*. Froglets Publications Ltd, Kent, 168pp.

Ogley, B.; Davison, M. and Currie, I. (1993) *The Norfolk and Suffolk Weather Book*. Froglets Publications Ltd, Kent, 168pp.

Parker, D. E.; Legg, T. P. and Folland, C. K. (1992) A new daily Central England temperature series 1772–1991. *Int. J. Climatol.*, **12**, pp. 317–342.

Perry, A. H. (1981) *Environmental Hazards in the British Isles*. George Allen and Unwin, London, 191 pp.

Perry, A. and Mayes, J. (1998) The Lamb weather type catalogue. *Weather*, **53**, pp. 222–229.

Pike, W. S. (1991a) The heavy early-season snowfall of 7–9 December 1990: Part I, historical precedents this century. *J. Meteorol.*, **16**, pp. 109–121.

Pike, W. S. (1991b) The heavy early-season snowfall of 7–9 December 1990: Part II, historical precedents this century. *J. Meteorol.*, **16**, pp. 149–152.

Pike, W. S. (1996) Abrupt visibility reductions in heavy wintry showers and a developing cold pool, 19 February 1996. *Weather*, **51**, pp. 391–392.

Rand Capron, J. (1883) Measurement of snow. *The Meteorological Magazine*, **18**, pp. 64.

Scott, A. (1881) On the snow storm of January 1881. *Symons's Monthly Meteorological Magazine*, **16**, pp. 13–14.

Sowerby Wallis, H. (1881) On the snow storm of January 1881. *Symons's Monthly Meteorological Magazine*, **16**, pp. 2–11.

Sowerby Wallis, H. (1882) On the snow storm of January 1881. *British Rainfall*, 1881, pp. 36–40.

Spink, P. C. (1947) Famous snowstorms 1878–1945. *Weather*, **2**, pp. 50–54.

Stevenson, C. M. (1968) The snowfalls of early December 1967. *Weather*, **23**, pp. 156–162.

Stirling, R. (1997) *The Weather of Britain*. Giles de la Mare Publishers Limited, London, 306 pp.

Suttie, T. K. (1970) Portrait of a polar low. *Weather*, **25**, pp. 504–506.

Sweeney, J. and O'Hare G. (1992) Geographical variations in precipitation yields and circulation types in Britain and Ireland. *Transactions of the Institute of British Geographers*, **17**, pp. 448–463.

Uccellini, L. W. and Kocin, P. J. (1987) The interaction of the jet streak circulations during heavy snow events along the East Coast of the United States. *Weather and Forecasting*, **2**, pp. 289–308.

Underdown, R. G. (1881) On the snow storm of January 1881. *Symons's Monthly Meteorological Magazine*, **16**, pp. 21–22.

Ward, R. C. and Robinson M. (1990) *Principles of Hydrology*. McGraw-Hill Book Company (UK) Limited, Maidenhead, 365pp.

Webb, J. D. C. (1982) The snowstorm of 8/9 January 1982 in and around Oxford. *J. Meteorol.*, **7**, pp. 120–121.

Wheeler, D. A. (1991) The great northeastern snowstorm of February 1941. *Weather*, **46**, pp. 311–320.

Wheeler, D. and Mayes, J. (1997) *Regional Climates of the British Isles*. Routledge, London, 343pp.

Whittow, J. (1980) *Disasters: The Anatomy of Environmental Hazards*. Penguin Books, London, 411pp.

Wild, R. (1995) Definition of the British blizzard. *Weather*, **50**, pp. 327–328.

Wild, R. (1996a) Frequency of blizzards and heavy snowfalls greater than 15 centimetres across Great Britain 1861–1995. *J. Meteorol.*, **21**, p. 217.

Wild, R. (1996b) Describing the origin of the word 'blizzard' and how a blizzard is defined in the United Kingdom. *Climate News*, **15**, pp. 1–2.

Wild, R. (1996c) Definition of the word blizzard within the United Kingdom. *Weather*, **51**, pp. 231–232.

Wild, R. (1997) Historical review on the origin and definition of the word blizzard. *J. Meteorol.*, **22**, pp. 331–340.

Wild, R. (1998) A review on the heavy snowfalls/blizzards/snowstorms/snowfalls greater than 13 cm in Great Britain between 1861–1996: Part 1: 1861–1899. *J. Meteorol.*, **23**, pp. 3–19.

Wild, R. (1999) A review on the heavy snowfalls/blizzards/snowstorms/snowfalls greater than 13 cm in Great Britain between 1861–1996: Part 2: 1900–1925. *J. Meteorol.*, **24**, pp. 19–32.

Wild, R. (2001a) Heavy snowfalls and blizzards division summary for the United Kingdom 2000. *J. Meteorol.*, **26**, pp. 181–183.

Wild, R. (2001b) Snowfalls/blizzards/snowstorms/snowfalls greater than 13 cm in Great Britain between 1861–1996: Part 3: 1926–1945. *J. Meteorol.*, **26**, pp. 92–100.

Wild, R. (2002) British Isles heavy snowfalls summary 2001. *J. Meteorol.*, **27**, pp. 177–180.

Wild, R. (2003) Heavy snowfalls division summary for the United Kingdom 2002. *J. Meteorol.*, **28**, pp. 199–200.

Wild, R. (2004a) Heavy snowfalls in the United Kingdom 2003. *J. Meteorol.*, **29**, pp. 221–224.

Wild, R. (2004b) Snowfalls/blizzards/snowstorms/snowfalls greater than 13 cm in Great Britain between 1861–1996: Part 4: 1946–1959. *J. Meteorol.*, **29**, pp. 17–26.

Wild, R. (2005a) Blizzards and heavy snowfalls in the United Kingdom 2004. *J. Meteorol.*, **30**, pp. 214–216.

Wild, R. (2005b) A spatial and temporal analysis of heavy snowfalls across Great Britain between the years 1861–1999. Unpublished PhD Thesis, University of Derby, 344pp.

Wild, R. (2005c) Snowfalls/blizzards/snowstorms/snowfalls greater than 13 cm in Great Britain between 1861–1996: Part 5: 1960–1969. *J. Meteorol.*, **30**, pp. 43–50.

Wild, R. (2005d) Snowfalls/blizzards/snowstorms/snowfalls greater than 13 cm in Great Britain between 1861–1996: Part 6: 1970–1975. *Int. J. Meteorol.*, **30**, pp. 363–367.

Wild, R. (2006) Heavy Snowfalls and Blizzards in the United Kingdom 2005. *Int. J. Meteorol.*, **31**, pp. 224–228.

Wild, R. (2007a) Heavy Snowfalls and Blizzards in the United Kingdom 2006. *Int. J. Meteorol.*, **32**, pp. 279–281.

Wild, R. (2007b) A review of heavy snowfalls/blizzards/snowstorms greater than 13 cm in Great Britain between 1861–1996: Part 7: 1976–1979. *Int. J. Meteorol.*, **32**, pp. 325–334.

Wild, R. (2008) Heavy snowfalls and blizzards in the United Kingdom 2007. *Int. J. Meteorol.*, **33**, pp. 193–194.

Wild, R. (2012) A review of heavy snowfalls/blizzards/snowstorms greater than 13 cm in Great Britain between 1861–1996: Part 8: 1980–1984. *Int. J. Meteorol.*, **37**, pp. 166–176.

Wild, R.; O'Hare, G. and Wilby, R. (1996) A historical record of blizzards/major snow events in the British Isles, 1880–1989. *Weather*, **51**, pp. 82–91.

Wild, R.; O'Hare, G. and Wilby, R. (2000) An analysis of heavy snowfalls/blizzards/snowstorms and snowfalls greater than 13 cm across Great Britain between 1861 and 1996. *J. Meteorol.*, **25**, pp. 41–49.

Wilson, F. (1990) *The Great British Obsession*. Jarrold Press, London, 144pp.

Appendix B
Selected Pictures from Conferences and Meetings

Plate B.1 *Bob Prichard speaking at the 1985 TORRO conference, Oxford, United Kingdom. © TORRO Archives.*

Extreme Weather: Forty Years of the Tornado and Storm Research Organisation (TORRO), First Edition. Edited by Robert K. Doe.
© 2016 John Wiley & Sons, Ltd. Published 2016 by John Wiley & Sons, Ltd.

Plate B.2   *Derek M. Elsom speaking at the 1985 TORRO conference, Oxford, United Kingdom. © TORRO Archives.*

Plate B.3   *(Left to right) Richard Barker and Keith Mortimore (with son and wife) examining a conference poster with TV's Michael Hunt at the 1985 TORRO conference, Oxford, United Kingdom. © TORRO Archives.*

Plate B.4   *(Left to right) Michael Hunt, Jacquie and Terence Meaden outside the 1985 TORRO conference hotel, Oxford, United Kingdom.*
© TORRO Archives.

Plate B.5   *Guest speaker David Brooks, Anglia TV weatherman, at the 1988 TORRO conference, Oxford, United Kingdom. © TORRO Archives.*

Plate B.6   *Poster presentations by Jonathan D.C. Webb at the 1988 TORRO conference, Oxford, United Kingdom. © TORRO Archives.*

Plate B.7   *Ball Lightning Division director and presenter Mark Stenhoff, at the 1988 TORRO conference, Oxford, United Kingdom. © TORRO Archives.*

Plate B.8   *Terence Meaden (left) with guest speaker David Brooks, Anglia TV weatherman (centre), and Derek M. Elsom (right) at the 1988 Oxford TORRO conference, United Kingdom. © TORRO Archives.*

Plate B.9   *John Tyrrell at the TORRO autumn conference, October 1999, Oxford, United Kingdom. © Derek M. Elsom.*

Plate B.10   *John Tyrrell at the 'First European Conference on Tornadoes and Severe Storms', Toulouse, France, February 2000. © Derek M. Elsom.*

Plate B.11   *Derek M. Elsom (seated) and Adrian James who is presenting on ball lightning at the TORRO conference, spring 2000, Oxford, United Kingdom. © TORRO Archives.*

Plate B.12   *(Left to right) 'In the field' TORRO staff meeting, June 2002, Devon, United Kingdom, and TORRO directors and executives: Tony Gilbert, Derek M. Elsom, Robert K. Doe, Steve Roberts, Nigel Bolton and Harry McPhillimy. © TORRO Archives.*

Plate B.13   *Tony Gilbert being interviewed for television news in 2002 following a tornado at Gosport, United Kingdom. Pictures courtesy of Tony Gilbert.*

Plate B.14   *TORRO staff meeting in South Molton, Devon, United Kingdom, April 2003. (Left to right) Harry McPhillimy, Chris Chatfield, Alan Rogers, Nigel Bolton and Wendy Rogers. © Adrian Mackey.*

Plate B.15   *Television filming at the October 2003 TORRO conference, Oxford, United Kingdom, with Derek M. Elsom (left) and TV weatherman Ian McCaskill (right). © TORRO Archives.*

Plate B.16 *The October 2003 TORRO conference, Oxford, United Kingdom. Paul Knightley (left), Terence Meaden (centre) and Wendy Rogers (right). © TORRO Archives.*

Plate B.17 *TORRO staff meeting, August 2004, Bradford on Avon, West Wiltshire, United Kingdom. (Left to right standing) Derek M. Elsom, Steve Roberts, Alan Rogers, Robert K. Doe, Nigel Bolton, Jonathan D.C. Webb and Tony Gilbert; (left to right sitting) Mike Rowe, Chris Chatfield, Terence Meaden, Ellie Gatrill-Smith, Mark Humpage and Wendy Rogers. © TORRO Archives.*

Plate B.18 *TORRO's Ian Brindley being interviewed by BBC News while on a tornado site investigation in Birmingham, United Kingdom, 28 July 2005. © Peter Kirk.*

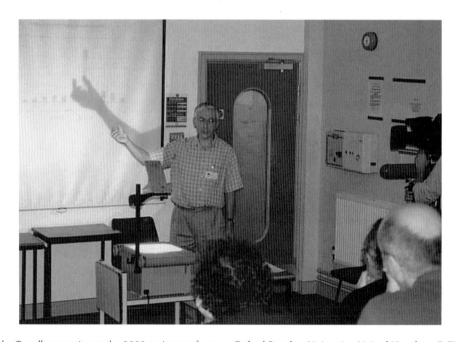

Plate B.19 *John Tyrrell presenting at the 2003 spring conference, Oxford Brookes University, United Kingdom. © TORRO Archives.*

Plate B.20 *Piers Corbyn – guest speaker on solar activity and forecasting at the 2003 spring conference, Oxford Brookes University, United Kingdom. © TORRO Archives.*

Plate B.21 *Paul Knightley presenting at the 2003 spring conference, Oxford Brookes University, United Kingdom. © TORRO Archives.*

Plate B.22   *Tony Gilbert presenting at the 2003 spring conference, Oxford Brookes University, United Kingdom. © TORRO Archives.*

Plate B.23   *Stuart Robinson presenting on the Birmingham tornado at the University of Birmingham, United Kingdom, 29 July 2006. © TORRO Archives.*

Plate B.24 *Chris Chatfield presenting on historical Birmingham area tornadoes at the Birmingham tornado conference, the University of Birmingham, United Kingdom, 29 July 2006. © TORRO Archives.*

Plate B.25 *Terence Meaden presenting on 'Tornadoes as a Weather Hazard in Britain' at the Birmingham tornado conference, the University of Birmingham, United Kingdom, 29 July 2006 (Paul Knightley, chairman, is seated on the left). © TORRO Archives.*

Plate B.26 *Steve Grogan, Birmingham City Council, Emergency Planning and Business Continuity Unit, presenting at the Birmingham tornado conference, the University of Birmingham, United Kingdom, 29 July 2006 (Paul Knightley, chairman, is seated on the left). © TORRO Archives.*

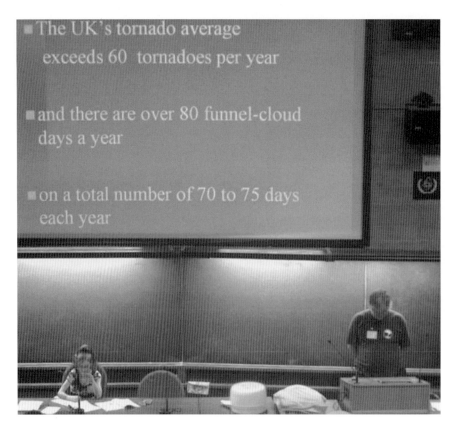

Plate B.27 *Presenting at the 'Fourth European Conference on Severe Storms', Trieste, Italy, September 2007, Sam Hall (seated) and Chris Chatfield (standing). © TORRO Archives.*

(a)

(b)

Plate B.28   *(a) Michalis Sioutas and Robert K. Doe presenting at the 'Fourth European Conference on Severe Storms', Trieste, Italy, September 2007, and (b) at the 'First International Summit on Tornadoes and Climate Change', Chania, Crete, May 2014. © TORRO Archives.*

Plate B.29   *TORRO staff meeting, May 2013, at the Royal Meteorological Society headquarters in Reading, United Kingdom. (Left to right) Paul Knightley, Terence Meaden, Robert K. Doe, Jonathan D.C. Webb, David Smart and Helen Rossington. © TORRO Archives.*

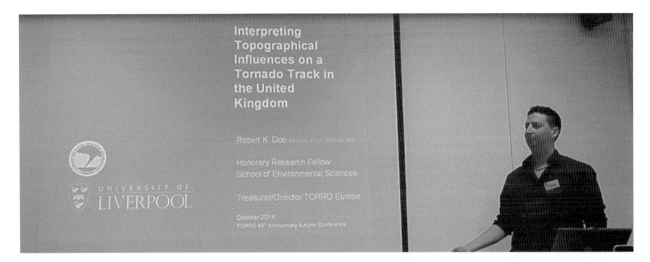

Plate B.30   Robert K. Doe presenting at the TORRO autumn '40th Anniversary' conference, October 2014. © TORRO Archives.

Plate B.31   Tim Prosser presenting at the TORRO autumn '40th Anniversary' conference, October 2014. © TORRO Archives.

Plate B.32   *The audience at the TORRO autumn '40th Anniversary' conference, October 2014. © TORRO Archives.*

Plate B.33   *Michalis Sioutas presenting at the TORRO autumn '40th Anniversary' conference, October 2014. © TORRO Archives.*

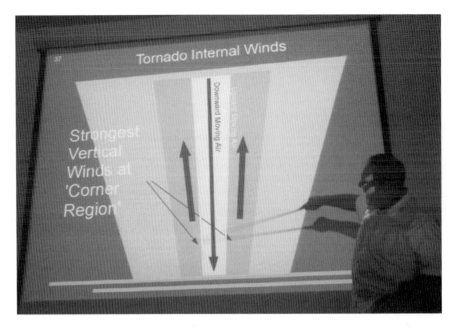

Plate B.34   *Tony Gilbert presenting at the TORRO autumn '40th Anniversary' conference, October 2014. © TORRO Archives.*

Plate B.35   *On the 40th anniversary of TORRO (Saturday, 18 October 2014), founder Dr Terence Meaden receives a commemorative plaque of appreciation from Dr Michalis Sioutas of ELGA–Meteorological Applications Centre, Greece, for inspiration and encouragement towards tornado research in Greece. © TORRO Archives.*

Appendix C

Tornadoes in the United Kingdom and Ireland 1054–2013

This map shows the distribution of tornado events from the complete TORRO dataset covering the years 1054–2013. The figure was composed by plotting the tracks of all known events over an image of the United Kingdom and Ireland derived from the NASA *Blue Marble* composite satellite image. Urban areas are coloured based on a land use atlas produced by the European Environment Agency (year 2000). Stronger tornadoes are shown in a lighter shade and are plotted on top of lesser events. Note that the track widths are exaggerated, so this does not represent actual land coverage. This map was produced by Tim Prosser.

Extreme Weather: Forty Years of the Tornado and Storm Research Organisation (TORRO), First Edition. Edited by Robert K. Doe.
© 2016 John Wiley & Sons, Ltd. Published 2016 by John Wiley & Sons, Ltd.

Tornado Intensity - TORRO Scale

Unknown
Waterspouts
0 1 2 3 4 5 6 7 8 9

Selected Name Index

Note: Page numbers in *italics* denote figures, when outside page ranges.

Subject Index

Note: Page numbers in *italics* denote figures, when outside page ranges.

Extreme Weather: Forty Years of the Tornado and Storm Research Organisation (TORRO), First Edition. Edited by Robert K. Doe.
© 2016 John Wiley & Sons, Ltd. Published 2016 by John Wiley & Sons, Ltd.